Chang Wen Chen, Zhu Li, and Shiguo Lian (Eds.)

Intelligent Multimedia Communication: Techniques and Applications

Studies in Computational Intelligence, Volume 280

Editor-in-Chief
Prof. Janusz Kacprzyk
Systems Research Institute
Polish Academy of Sciences
ul. Newelska 6
01-447 Warsaw
Poland
E-mail: kacprzyk@ibspan.waw.pl

Further volumes of this series can be found on our homepage: springer.com

Vol. 259. Kasthurirangan Gopalakrishnan, Halil Ceylan, and Nii O. Attoh-Okine (Eds.)
Intelligent and Soft Computing in Infrastructure Systems Engineering, 2009
ISBN 978-3-642-04585-1

Vol. 260. Edward Szczerbicki and Ngoc Thanh Nguyen (Eds.)
Smart Information and Knowledge Management, 2009
ISBN 978-3-642-04583-7

Vol. 261. Nadia Nedjah, Leandro dos Santos Coelho, and Luiza de Macedo de Mourelle (Eds.)
Multi-Objective Swarm Intelligent Systems, 2009
ISBN 978-3-642-05164-7

Vol. 262. Jacek Koronacki, Zbigniew W. Ras, Slawomir T. Wierzchon, and Janusz Kacprzyk (Eds.)
Advances in Machine Learning I, 2009
ISBN 978-3-642-05176-0

Vol. 263. Jacek Koronacki, Zbigniew W. Ras, Slawomir T. Wierzchon, and Janusz Kacprzyk (Eds.)
Advances in Machine Learning II, 2009
ISBN 978-3-642-05178-4

Vol. 264. Olivier Sigaud and Jan Peters (Eds.)
From Motor Learning to Interaction Learning in Robots, 2009
ISBN 978-3-642-05180-7

Vol. 265. Zbigniew W. Ras and Li-Shiang Tsay (Eds.)
Advances in Intelligent Information Systems, 2009
ISBN 978-3-642-05182-1

Vol. 266. Akitoshi Hanazawa, Tsutom Miki, and Keiichi Horio (Eds.)
Brain-Inspired Information Technology, 2009
ISBN 978-3-642-04024-5

Vol. 267. Ivan Zelinka, Sergej Celikovský, Hendrik Richter, and Guanrong Chen (Eds.)
Evolutionary Algorithms and Chaotic Systems, 2009
ISBN 978-3-642-10706-1

Vol. 268. Johann M.Ph. Schumann and Yan Liu (Eds.)
Applications of Neural Networks in High Assurance Systems, 2009
ISBN 978-3-642-10689-7

Vol. 269. Francisco Fernández de de Vega and Erick Cantú-Paz (Eds.)
Parallel and Distributed Computational Intelligence, 2009
ISBN 978-3-642-10674-3

Vol. 270. Zong Woo Geem
Recent Advances In Harmony Search Algorithm, 2009
ISBN 978-3-642-04316-1

Vol. 271. Janusz Kacprzyk, Frederick E. Petry, and Adnan Yazici (Eds.)
Uncertainty Approaches for Spatial Data Modeling and Processing, 2009
ISBN 978-3-642-10662-0

Vol. 272. Carlos A. Coello Coello, Clarisse Dhaenens, and Laetitia Jourdan (Eds.)
Advances in Multi-Objective Nature Inspired Computing, 2009
ISBN 978-3-642-11217-1

Vol. 273. Fatos Xhafa, Santi Caballé, Ajith Abraham, Thanasis Daradoumis, and Angel Alejandro Juan Perez (Eds.)
Computational Intelligence for Technology Enhanced Learning, 2010
ISBN 978-3-642-11223-2

Vol. 274. Zbigniew W. Raś and Alicja Wieczorkowska (Eds.)
Advances in Music Information Retrieval, 2010
ISBN 978-3-642-11673-5

Vol. 275. Dilip Kumar Pratihar and Lakhmi C. Jain (Eds.)
Intelligent Autonomous Systems, 2010
ISBN 978-3-642-11675-9

Vol. 276. Jacek Mańdziuk
Knowledge-Free and Learning-Based Methods in Intelligent Game Playing, 2010
ISBN 978-3-642-11677-3

Vol. 277. Filippo Spagnolo and Benedetto Di Paola (Eds.)
European and Chinese Cognitive Styles and their Impact on Teaching Mathematics, 2010
ISBN 978-3-642-11679-7

Vol. 278. Radomir S. Stankovic and Jaakko Astola
From Boolean Logic to Switching Circuits and Automata, 2010
ISBN 978-3-642-11681-0

Vol. 279. Manolis Wallace, Ioannis E. Anagnostopoulos, Phivos Mylonas, and Maria Bielikova (Eds.)
Semantics in Adaptive and Personalized Services, 2010
ISBN 978-3-642-11683-4

Vol. 280. Chang Wen Chen, Zhu Li, and Shiguo Lian (Eds.)
Intelligent Multimedia Communication: Techniques and Applications, 2010
ISBN 978-3-642-11685-8

Chang Wen Chen, Zhu Li, and Shiguo Lian (Eds.)

Intelligent Multimedia Communication: Techniques and Applications

Springer

Prof. Chang Wen Chen
Computer Science and Engineering
University at Buffalo
The State University of New York
201 Bell Hall Box 602000
Buffalo, NY, 14260-2000
USA
E-mail: chencw@buffalo.edu

Dr. Shiguo Lian
France Telecom R&D (Orange Labs) Beijing
2 Science Institute South Rd
Haidian District Beijing, 100080
China
E-mail: shiguo.lian@orange-ftgroup.com

Prof. Zhu Li
Department of Computing
Hong Kong Polytechnic University
China
E-mail: zhu.li@ieee.org

ISBN 978-3-642-11685-8 e-ISBN 978-3-642-11686-5

DOI 10.1007/978-3-642-11686-5

Studies in Computational Intelligence ISSN 1860-949X

Library of Congress Control Number: 2010920827

Typeset & Cover Design: Scientific Publishing Services Pvt. Ltd., Chennai, India.

Printed in acid-free paper

9 8 7 6 5 4 3 2 1

springer.com

Preface

Multimedia data are used more and more widely in human being's life, e.g., videoconferencing, visual telephone, IPTV, etc. Nearly most of the applications need multimedia transmission techniques that send multimedia data from one side to another side and keep the properties of efficiency, robustness and security. Here, the efficiency denotes the time cost of transmission operations, the robustness denotes the ability to survive transmission errors or noises, and the security denotes the protection of the transmitted media content. Recently, various intelligent or innovative techniques are invented, which bring vast performance improvements to practical applications. For example, such content transmission techniques as p2p, sensor network and ad hoc network are constructed, which adaptively use the peers' properties to improve the network's resources. Multimedia adaptation techniques can adjust the multimedia data rate in order to compliant with the network's bandwidth. Scalable encryption techniques can generate the data stream that can be correctly decrypted after bit rate conversion. Ubiquitous multimedia services make the user share any kind of content anywhere.

To the best of our knowledge, few books focus on intelligent or innovative multimedia transmission techniques. To access the latest research related to intelligent multimedia transmission, we launched the book project where researchers from all over the world provide the necessary coverage of the mentioned field. The primary objective of this project was to assemble as much research coverage as possible related to the field by defining the latest innovative technologies and providing the most comprehensive list of research references.

The book includes sixteen chapters highlighting current concepts, issues and emerging technologies. Distinguished scholars from many prominent research institutions around the world contribute to the book. The book covers various aspects, including not only some fundamental knowledge and the latest key techniques, but also typical applications and open issues. For example, the covered topics include the present and future video coding standards, stereo and multiview coding techniques, free-viewpoint TV techniques, wireless broadcasting techniques, media streaming techniques, wireless media transmission techniques and systems, and User-Generated Content sharing.

The diverse and comprehensive coverage of multiple disciplines in the field of intelligent multimedia transmission will contribute to a better understanding of all topics, research, and discoveries in this emerging and evolving field. Furthermore, the contributions included in this book will be instrumental in the expansion of the body of knowledge in this field. The coverage of this book provides strength to this reference resource for both researchers and also decision makers in obtaining a greater understanding of the concepts, issues, problems, trends, challenges and opportunities related to this field of study. It is our sincere hope that this publication and its great amount of information and research will assist our research colleagues, all faculties, their students, and our organizational decision makers in enhancing their understanding of this research field. Perhaps this publication will even inspire its readers to contribute to the current discoveries in this immense field.

Prof. Chang Wen Chen
University at Buffalo,
The State University of New York, USA

Prof. Zhu Li
Hong Kong Polytechnic University, China

Dr. Shiguo Lian
France Telecom R&D (Orange Labs) Beijing,
China

Acknowledgments

The editors would like to acknowledge the help of all involved in the collation process of the book, without whose support the project could not have been satisfactorily completed. Deep appreciation and gratitude is due to the authors of chapters, whose efforts make the high-quality project.

Special thanks go to the publishing team at Springer, whose contributions throughout the whole process from inception of the initial idea to final publication have been invaluable. In particular to Dr. Thomas Ditzinger, who continuously prodded via e-mail for keeping the project on schedule and to Prof. Janusz Kacprzyk, whose enthusiasm motivated us to initiate the project.

And last but not least, our families, for their unfailing support and encouragement during the months it took to give birth to this book.

November 2009 Editors

Contents

Rate Control and Error Resilience for Object-Based Video Coding

Paulo Nunes and Luis Ducla Soares

Abstract. The MPEG-4 audiovisual coding standard introduced the object-based video data representation model where video data is no longer seen as a sequence of frames or fields, but consists of independent (semantically) relevant video objects that together build the video scene. This representation approach allows new and improved functionalities, but it has also created new relevant problems in terms of typical non-normative parts of the standard, such as rate control and error resilience, which need to be solved in order to successfully transmit object-based video with an acceptable quality over networks that have critical bandwidth and channel error characteristics, such as mobile networks and the Internet. To deal with the specific problems of object-based video coding, rate control demands two levels of action: 1) the scene-level, which is responsible for dynamically allocating the available resources between the various objects in the scene (i.e., between the different encoding time instants and the different video objects to encode in each time instant), aiming at minimizing quality variations along time and between the various objects in the scene; and 2) the object-level, which is responsible for allocating the resources attributed to each object among the various types of data to code (for that object), notably texture and shape, and for computing the best encoding parameters to achieve the target bit allocations while maintaining smooth quality fluctuations. In terms of error resilience techniques, the object-based coding approach means that shape and composition information also have to be taken into account for error resilience purposes, in addition to motion and texture data. To do this, at the encoder side, the coding of video objects is typically supervised by a resilience configuration module, which is responsible for choosing the most adequate coding parameters in terms of resilience for each video object. This is important because the decoding performance will much depend on the protective actions the encoder has taken. At the decoder side, defensive actions have to be taken. This includes error detection and error localization for each decoded video object, followed by independent object-level error concealment. Afterwards, advanced scene-level error concealment is also performed, which has access to all the video objects in the scene and is used immediately before the final concealed video scene is presented to the user. In this chapter, the most recent basics, advances and trends in terms of rate control and error resilience for object-based video coding will be described.

1 Introduction

Some thirty years ago, digital video started to emerge. Back then, digital video was just a digital representation of the corresponding analog video, in the sense that the data model was the same. In fact, both analog and digital video consisted of a periodic sequence of (rectangular) frames or fields and the sole conceptual difference between the two models was the fact that in the analog representation each frame or field was

C.W. Chen et al. (Eds.): Intel. Multimedia Communication: Tech. and Appli., SCI 280, pp. 1–50.
springerlink.com

made of a number of (analog) lines, whereas in the digital representation the frames or fields corresponded to matrices of picture elements, also known as *pixels* (ITU-R BT.601 1986). Nowadays, this type of digital video is commonly referred to as "frame-based video".

More recently, a different video data representation model has been introduced: the object-based model. In this model, video data is no longer seen as a sequence of frames or fields, but consists of several independent (semantically) relevant video objects that together build the *video scene*. Object-based video coding schemes can also be called content-based video coding schemes because the representation entities in the model — *the objects* — are now very close to the video content since by having semantic value they can be subjected to semantically meaningful actions. This representation approach allows, in addition to the advantages already provided by the digital frame-based representation, new and improved functionalities in terms of interactivity, coding efficiency and universal access since, for the first time, the content is not only selectively processed but also independently represented, accessed, and consumed. This video representation model was first introduced in a large scale with the MPEG-4 standard (ISO/IEC 14496 1999).

With the advent of object-based video, new relevant problems have appeared in various fields, since new types of data, such as the shape and the scene composition information, have to be transmitted in addition to the motion and texture already used in previous frame-based coding systems. Two very important fields that have been affected are:

- **Video coding rate control** – Video coding rate control is understood here as the mechanism responsible for efficiently controlling the video encoder in order to guarantee that it meets the relevant constraints of the encoding framework, such as channel, delay/buffering, complexity, and quality constraints. The challenge in terms of video coding rate control is, therefore, to develop new methods that are able to allocate the available coding resources among the various objects in the scene using appropriate allocation criteria and constraining mechanisms. The major objectives of a rate control mechanism, whatever the coding architecture, can, therefore, be summarized as: i) *regulation of the video encoder data rate according to the application constraints*; and ii) *maximization of the subjective impact of the decoded video*. They involve long-term and short-term objectives. *Long-term objectives*, handled at time periods of several pictures, deal with prevention of buffer overflows and underflows (that may cause loss of data or non-optimal use of the available resources) and preservation of coded data rate limits. *Short-term objectives*, handled at time periods of one or few pictures (or even short time periods, such as a few macroblocks), deal with the minimization of picture quality variations (a key factor for maximizing the subjective impact of the decoded video) and the stability of the encoder control.
- **Video error resilience** – Video error resilience is understood here as the set of techniques that improve the capability of the video communication system to withstand channel errors and achieve an acceptable decoded video quality. To do this, the error resilience of both encoder and decoder has to be improved and, therefore, new techniques that can deal with object-based video are needed for both the encoder and decoder sides of the communication chain. This way, at the encoder, the

challenge is to develop new techniques that make the bitstreams of the various objects in the scene more resilient to errors, in order to allow the decoder to better recover in case errors occur; these techniques may be called *preventive error resilience techniques*. At the decoder, the challenge is to develop new techniques that make it possible for the decoder to take all the available received data (correct and corrupted) for the various video objects in the scene and decode it with the best possible video quality, thus minimizing the negative subjective impact of the errors in the video quality offered to the user; these techniques may be called *corrective error resilience techniques*.

This chapter presents the developments that have been achieved in terms of rate control and error resilience techniques for object-based video coding systems in the last decade or so. To do this, this chapter is organized as follows. In Section 2, the basic concepts behind the object-based video coding approach are briefly described. Section 3 describes some of the most relevant developments in the field of rate control for object-based video. This is followed by Section 4, where the most important developments in terms of error resilience are described. Finally, Section 5 concludes the chapter by summarizing the past developments and projecting the future of object-based video.

2 The Object-Based Coding Approach

Object-based video coding systems represent a paradigm change in video coding technology and for that a new data representation model is adopted. While in frame-based systems video data was represented as a sequence of rectangular matrices of picture elements (pixels), in object-based systems, video data is modeled as a composition of (semantically) relevant 2D arbitrarily shaped video objects. These objects can have one or more non-connected parts, which are typically referred to as regions. Therefore, in order to represent each video object, the traditional motion and texture data is no longer enough and a new video data type has to be added — *the shape*. This way, the complete data set required to represent a video object consists of shape, motion and texture data. In addition, some composition data is necessary to create a video scene where several objects may co-exist (Pereira and Ebrahimi 2002).

To better understand the concept of video object and object-based coding[1], imagine a news program, as the one shown in Fig. 1 (a), represented using an object-based approach instead of a frame-based approach (i.e., the scene is represented as a composition of several objects and not as a periodic sequence of rectangular matrices of pixels).

At first sight, this appears to be just a regular news program, consisting of a sequence of frames, with two anchorpersons in a studio reading the news, a large monitor in the back showing video footage related to the news being read and a textual logo on the bottom-left corner of the screen. However, it is not so. This scene is really the composition of four different video objects, also shown in Fig. 1. Going from left

[1] In reality, when the object-based representation approach is used, the objects that make up the scene can be video objects, audio objects or even audiovisual objects. However, in this chapter, only video objects are considered and, therefore, the other types of objects will not be taken into account.

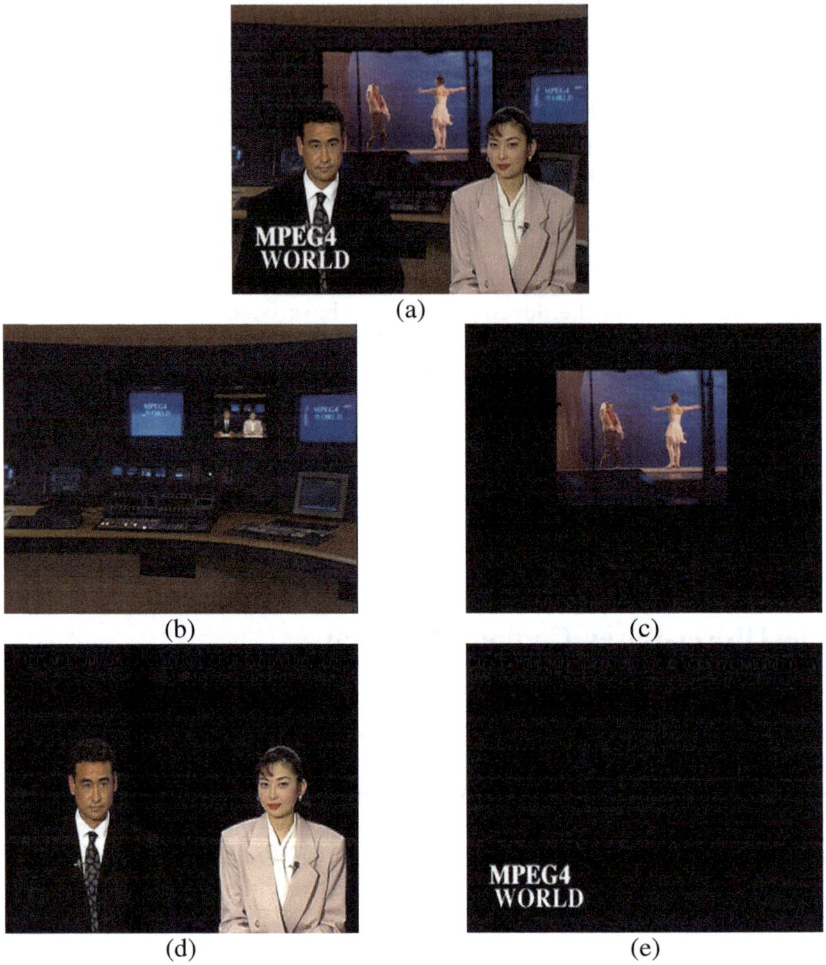

Fig. 1. News sequence and its video objects: (a) Composed scene; (b) Background; (c) Dancers object; (d) Speakers object; (e) Logo object.

to right, top to bottom, the first object is just a still image, corresponding to the background. The second object is the video footage of two ballet dancers, which are the subject of the news being read. The third object corresponds to two anchorpersons. And finally, the fourth object is a simple text overlay. With the object-based approach, these four objects are independently coded and transmitted to the receiver, along with the composition information, which will allow the receiver to take the four independent objects, decode them and compose the scene that can be displayed on a screen.

If the scene obtained at the receiver is apparently the same as the one that could be obtained with frame-based video, one might ask what is the advantage of using the object-based representation model. The answer is simple and lies in all the new and improved functionalities that can now be provided, which were not possible with

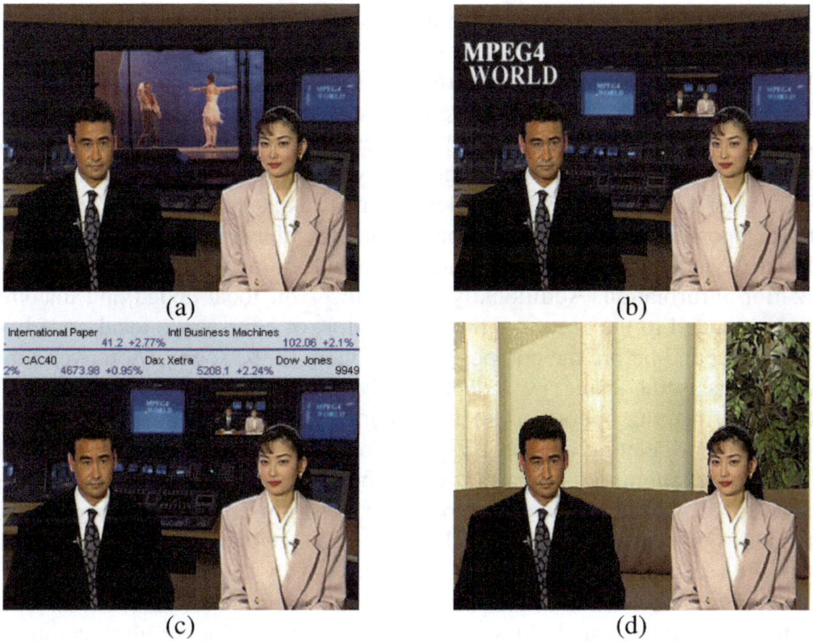

Fig. 2. Possible changes to the News sequence: (a) Logo is removed; (b) Logo changes position; (c) Logo is removed and ticker bar is added; (d) Speakers are placed on a different setting.

frame-based video. For instance, if the user finds that the logo object on the bottom-left corner is annoying, he/she can simply remove it, obtaining the scene in Fig. 2 (a). Alternatively, if the user does not want to eliminate the logo, he/she can simply drag it to a position that is considered less annoying, as well as remove the ballet dancers in the back, as shown in Fig. 2 (b). After these changes, if the user is still not satisfied, he/she can replace the logo with more useful information by including a new object such as a stock exchange ticker bar on top of the screen, as can be seen in Fig. 2 (c). If a more drastic change is desired, the two speakers can be moved to a completely different setting, which is shown in Fig. 2 (d). Additionally, hyperlinks can be provided allowing the user to access more information about the ballet dancers or the anchorpersons' biography, depending on where the user clicks on the screen. Moreover, the object-based approach allows the different objects to be encoded with techniques that are more adequate to their nature, which did not happen for frame-based systems. For instance, the text overlay can be coded with text coding techniques instead of using video coding techniques (which are very inefficient to code text), as was (and still is) the case in frame-based systems, such as MPEG-2 Video (ISO/IEC 13818-2 1996), where everything is coded together, using the same coding tools.

The object-based video representation model, which was adopted by the ISO/IEC MPEG-4 standard (ISO/IEC 14496 1999), is the cornerstone of all the new functionalities provided and represents the greatest conceptual difference with respect to previous frame-based standards, namely previous MPEG standards such as MPEG-1 (ISO/IEC 11172 1993) and MPEG-2 (ISO/IEC 13818 1996).

In Fig. 3, a simplified version of the architecture of an object-based coding system is presented. At the transmitter, the various objects in the scene are separately encoded; moreover the composition information is also created. The generated elementary streams (in addition to locally stored elementary streams) are then multiplexed to form a single (multiplexed) bitstream that is sent through the channel and includes all the information about the scene. At the receiver, the received bitstream is demultiplexed in order to obtain the various elementary streams corresponding to the objects and the composition information of the scene. The elementary streams are then decoded and passed on to the compositor, which will compose the scene based on the composition information. Additionally, at the receiver, local coded and uncoded objects can be added to the scene by the compositor; of course, the local coded objects have to be decoded before the compositor adds them to the scene. Moreover, since the various objects in the scene are independently coded, the end-user can also interact with them, performing operations such as changing the spatial position or the size of objects, adding or deleting objects, triggering pre-defined behaviors, following hyperlinks, or even changing color properties. Depending on the type of interaction, the necessary action will be taken either locally (i.e., at the decoder) or remotely (i.e., at the encoder). For instance, if the user chooses to change the spatial position of a given object in the scene, this can be simply taken care of by the compositor. On the other hand, if the user chooses to delete one object, this can be taken care of more efficiently by the encoder, which does not have to send it any longer, thus saving bit rate resources.

The fact that a given scene can be modeled as a composition of semantically relevant objects (i.e., relevant objects in the context of a given application) introduces a number of new technical and functional possibilities, namely:

- **Selective processing and coding of objects** – With the object-based coding model, it is possible to process and encode different objects with tools that are adequate to their different natures. This alone is enough to make the use of the object-based representation approach worth it because it can bring very significant coding gains or bit rate reductions. For instance, in older standards, subtitles were

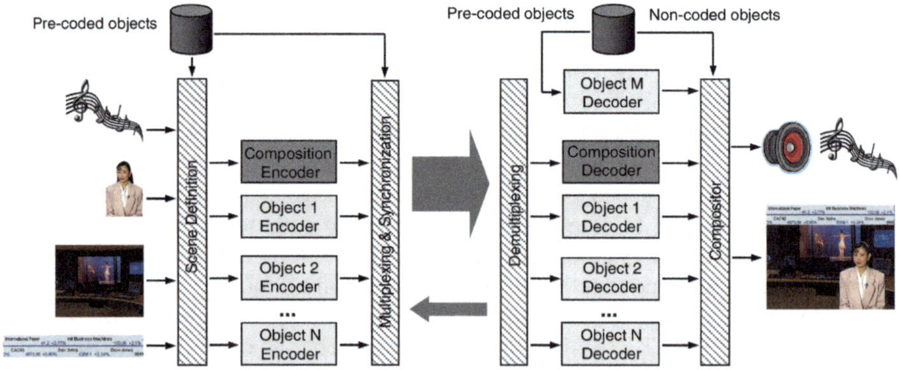

Fig. 3. Object-based system architecture.

often encoded using video coding tools. Since video coding tools are not optimized to code text, in order to achieve a reasonable quality, necessary to help the viewers read the subtitles, a lot of bits had to be spent on the subtitles alone. If a fixed bit rate was used, this inevitably resulted in a decrease of the global video quality. With the new object-based representation, subtitles can be independently encoded as text characters, such as Unicode (Unicode 2000), in addition to some formatting information. This will save an enormous amount of bits that can be spent elsewhere, providing at the same time a better quality. Of course, this does not apply only to subtitles and can also bring benefits when other types of objects are used, such as television station logos, graphics or, more generally, synthetic content.

- **Reusability of objects** – With object-based coding, since an independent bitstream is generated for each object, the objects in a given scene are independently accessible, which makes it very easy to reuse a given object from a scene in another scene. This offers many new possibilities in terms of content creation and makes it easier than ever. For instance, it becomes now possible to create new content based only on already existing objects, eliminating the need to film new similar ones, thus saving a lot of time and money (although maybe creating some new intellectual property management and protection issues). This way, by using objects stored in large databases and possibly adding new ones, if the means to create them are available (e.g., digital cameras), anyone may become a content creator, capable of creating rather complex and high quality content that can be easily distributed and published using the Internet.
- **Integration of synthetic and natural content** – With the object-based model, it is possible to efficiently integrate in the same scene natural (i.e., shot with a camera) and synthetic objects. For instance, it is possible to have scenes where cartoon characters coexist with live actors, each being encoded with adequate tools. The synthetic characters can be encoded as three-dimensional (3D) wire frame models, thus allowing new possibilities in terms of interaction, typically not available with natural objects. For instance, a synthetic 3D object can be easily rotated, thus uncovering regions of the object that were not previously visible. This is possible because the synthetic object is simply a fully specified 3D model, which is then rendered in the chosen position.
- **Interaction with and between objects** – Since the objects are independently coded, the user can interact with the scene in several ways. For instance, the user can change the spatial position of a given object, find more information about it by clicking on it, or even modify the perspective with which he/she is looking at the (3D) scene. In addition, objects can also interact with each other. For instance, it is possible for an object to have its trajectory changed due to a collision with another object. Therefore, with this type of interaction with and between objects, if some of the objects in the scene are buttons, it becomes possible to create many interesting applications where everything is encoded with a single standard (e.g., MPEG-4); to trigger a predefined behavior, all the user has to do is to click on the corresponding button.
- **Universal accessibility to the content** – With the object-based model, true universal accessibility may finally become a reality, meaning that the content is available through any type of network and accessed with any type of terminal. Since some of the targeted networks, such as mobile networks, have very critical channel error

characteristics, sufficient error robustness has to be provided. However, error resilience always comes at a price in terms of additional bit rate. Since in many of these networks the bit rate is also very limited, this usually meant for frame-based systems that in order to improve error resilience the uncorrupted quality had to be sacrificed, sometimes below acceptable levels. With object-based coding systems, other approaches can be used. For instance, error resilience can be selectively introduced, making more robust to errors the more important objects, without sacrificing their quality at all; the quality of less important objects can be decreased, instead. Still another possibility is to consider content scalability, which basically means that only the more important objects will be sent to the receiver, in order to guarantee that the received content has acceptable quality. This last approach may also be the solution when terminals with reduced complexity are used, which do not have enough resources to decode the whole scene. Therefore, an adequate allocation of the encoding resources between the various objects is necessary.

Due to these and other technical and functional possibilities, many new applications are expected to appear, which are based on the new object-based representation approach. The next sections will show the importance of rate control and error resilience tools for the deployment of object-based video applications due to the problems they solve.

3 Object-Based Rate Control

In a lossy video coding scenario where the compression ratio can be increased through an increase in the level of distortion of the coded video, the problem of rate control can be seen, in a first approach, as a typical rate-distortion problem that may be formulated as follows (Cover 1991):

> "Given a source distribution and a distortion measure, what is the minimum expected distortion achievable at a particular rate? Or, equivalently, what is the minimum rate description required to achieve a particular distortion?"

Whenever this rate-distortion formulation is valid, whatever the type of coding architecture, frame-based or object-based, a set of rate control constraints, dimensions, and strategies can be identified.

Generically, when designing a rate control mechanism for a given application, or set of applications, the following constraints have to be taken into account: delay, bit rate, and complexity.

Because a variable number of bits per frame can provide better subjective quality, a trade-off between the variability in the number of bits per frame and the necessary decoder buffering delay to support it is necessary. The larger the variability wanted, the larger the initial delay must be in order to ensure that the decoder receives all the bits necessary for decoding each frame before the scheduled decoding time. The rate control mechanism exploits the variability in the amount of bits per frame in order to maintain the spatio-temporal quality as constant as possible while fulfilling the buffering constraints.

The type of channel directly connecting the encoder and decoder typically characterizes the type of video encoding, i.e., constant bit rate (CBR), if the channel requires that a constant number of bits per time unit be sent to the channel, or variable bit rate (VBR), otherwise. Traditionally, video has been transmitted using CBR channels (Reibman and Haskell 1992). Thus, buffering is necessary both at the encoder and the decoder in order to translate the variable bit rate output of the video encoder into the constant channel rate, and to recover from the constant channel rate the variable number of bits per frame at the scheduled decoding times. Buffering is also necessary for practical VBR encoding, although in this case the channel capacity available to the end user is variable, allowing data to be transmitted as it is being produced (in the ideal case without buffering). However, it is not efficient from the point of view of network utilization and management that VBR networks be able to accommodate the bit rate variability of fully "uncontrolled" or "open-loop" video sources. Consequently, for both CBR and VBR encoding the rate control plays the fundamental role of adapting the encoding to the channel constraints.

Theoretical rate-distortion bounds depend on arbitrarily long data blocks and, eventually, unlimited computational power, which means that the actual rate-distortion characteristics of practical encoding systems are obtained under restricted conditions, particularly, small data block sizes, aiming at reducing the end-to-end delay and the computational power required at the encoder and, especially, at the decoder. Therefore, either implicitly or explicitly (Gormish and Gill 1993, Richardson and Zhao 2001, Zhao and Richardson 2002), the rate control mechanism is responsible for obtaining the best rate-distortion-complexity trade-off. This can be generically described as: minimize D, subject to $R \le R_T$ and $C \le C_T$, where D, R, and C are, respectively, the distortion, bit rate, and computational complexity, and R_T and C_T are, respectively, the target bit rate and target computational complexity.

Given a certain set of constraints, the basic dimensions (i.e., degrees of freedom) of rate control (frame-based and object-based) are intimately related to the traditional characteristic dimensions of the input data model, i.e., the coded spatial resolution, the coded temporal resolution, and the introduced texture data distortion (the coding errors). However, in an object-based coding framework, such as the MPEG-4 Visual architecture (ISO/IEC 14496-2 2004), the input data model for video is characterized by some additional dimensions, in comparison with the frame-based video data model, i.e., the shape data that defines the shape of each object and the scene description data that specifies the way that the scene is organized. Since, typically, the objects in the scene have semantic relevance, the major novelty here is the *semantic dimension of the data model* and, consequently, of the rate control since it becomes possible to perform actions such as not transmitting a less relevant object to save bits for the most semantically relevant objects. Therefore, the semantic dimension of the object-based representation paradigm opens new degrees of freedom to rate control such as the semantic resolution control and the amount of content control. The semantic resolution of a certain scene is related to the semantic detail provided; this means to the number of objects in which a certain amount of video data is organized (for the same amount of pixels, more objects mean higher semantic resolution and vice-versa). The amount of content in a scene is related to the number of objects and to the corresponding amount of data; decreasing the amount of content means reducing the number of objects in the sense that the pixels corresponding to a certain object are removed (not in the sense that objects are merged).

Frame-based and object-based rate control strategies are, therefore, designed within the boundaries provided by the corresponding rate control dimensions: while in frame-based coding, the rate control mechanism has the task to choose the best trade-off in terms of spatial and temporal resolution as well as to introduce distortion in the texture data, maximizing the global subjective quality, in object-based coding architectures, the rate control mechanism has the task to choose the best trade-off in terms of the amount of content to be coded, the corresponding semantic resolution, the spatial and temporal resolution for each object, and the amount of distortion in the texture and shape data for each object, in order that the global subjective impact in terms of the relevant requirements is maximized.

In object-based coding, since the various objects in a scene are now independent entities in terms of coding, although building together a scene, the rate control dimensions are dealt with by using two levels of action (Fig. 4):

- **Scene-level rate control** – Responsible for allocating the available resources between the objects in the scene, i.e., between the different encoding time instants and video objects to encode in each encoding time instant.
- **Object-level rate control** – Responsible for allocating the resources attributed to each object (in a rigid or dynamic way) among the various types of data to code (for that object), notably texture and shape, and for computing the best encoding parameters to achieve the target bit allocations while maintaining smooth quality fluctuations.

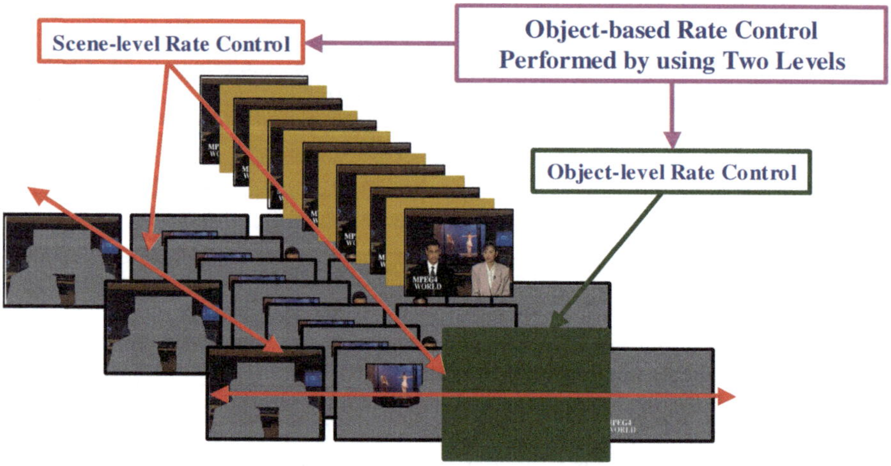

Fig. 4. Object-based rate control levels.

3.1 Object-Based Rate Control Architecture

To better understand the problems and solutions dealt with in object-based rate control, Fig. 5 presents an object-based coding architecture, showing the interfaces with the rate control module responsible for jointly controlling the encoding of multiple video objects composing a given scene. This architecture is composed of three major building blocks (Nunes 2007):

- **Scene encoder** – This block is responsible for encoding the original video content, i.e., the video objects (VOs) composing the scene, into a set of bitstreams (one for each VO[2]). This block is composed by a scene buffer where the original video object planes (VOPs) are stored; a symbol generator that reduces the redundancy and irrelevancy of the original video data generating adequate coded symbols; an entropy coder that reduces the statistical redundancy of the coded symbols converting them into bit codes; and, finally, a video multiplexer responsible for organizing the coded symbols according to the adopted video coding syntax.
- **Video buffering verifier** – This block is composed by a series of normative models[3], each one defining rules and limits to verify if the amount required for a specific type of decoding resource is within the bounds allowed by the corresponding profile and level specification (ISO/IEC 14496-2 2004). The rate control algorithm must use these rules and limits to define the control actions that will drive the scene encoder without violating this mechanism.
- **Rate controller** – This block corresponds to the mechanism responsible for controlling the scene encoder aiming at efficiently encoding the original video data while producing a set of bitstreams that does not violate the video buffering verifier mechanism. Essentially, this mechanism is composed by six modules (analyzed with more detail in Section 3.2) that, based on statistics computed from the input data stored in the scene buffer (feedforward information) and statistics computed through the different video buffering verifier models (feedback information), compute a multidimensional control signal (e.g., encoding time instants, macroblock coding modes and quantization parameters) that will command the encoding process. One fundamental task of this block is the *bit allocation* along time and between the various objects taking into account: i) the scene characteristics, obtained through the *scene analysis for resource allocation* module, and ii) the status of the various video buffering verifier models, provided by the *video buffering verifier control* module. Another important task of the rate controller is the *spatio-temporal resolution control* in order to achieve the best possible subjective quality given: i) the scene characteristics, ii) the available bit resources; and iii) the status of the various video buffering verifier models. The actual scene encoder control is performed through the *coding mode control* module, which is responsible for computing the adequate coding parameters given: i) a certain model of the scene encoder provided by the *rate-distortion modeling* module, ii) the characteristics of various objects, and iii) the status of the various video buffering verifier models (Section 3.3 provides a more detailed walkthrough of these modules).

---

[2] In case scalability is used, one bitstream for each video object layer of each video object is generated.

[3] The Video Reference Memory Verifier (VMV) is used to check the amount of decoding memory, the Video Complexity Verifier (VCV) to check the amount of decoding computational power, and the Video Rate Buffer Verifier (VBV) to check the bistream memory. This chapter will only consider VBV control; VMV and VCV control are addressed, for example, as in (Nunes and Pereira 2000, Valentim et al. 2002).

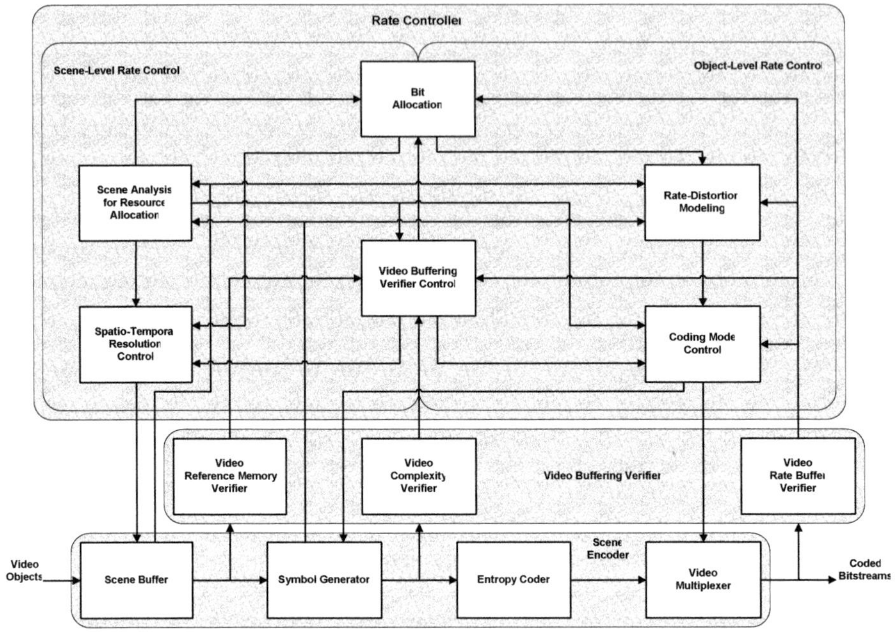

Fig. 5. Example of an object-based rate control architecture (Nunes and Pereira 2009) (© 2009 IEEE).

This chapter will consider the control of scene encoders operating under low-delay constraints, i.e., under low processing delay and low buffering delay. In terms of processing delay, it means that the rate controller can only process the VOPs corresponding to each encoding time instant, i.e., no future VOPs are analyzed, and a single pass encoding is performed, i.e., each coding unit is encoded only once. In terms of buffering delay, it means that the rate control algorithm should be able to cope with small buffer sizes for which the number of bits per frame needs to be kept nearly constant and, consequently, a very tight control of the VBV buffer[4] occupancy and fine coding mode control (e.g., quantization parameter selection) is required.

In this context, it is important to adopt adequate control techniques that relate the scene encoder output information with the required reference command signal in order to produce a multidimensional control signal that adequately drives the scene encoder, compensating its deviations regarding the ideal behavior.

Rate-distortion modeling is an important rate control technique that targets the design of adequate models for describing the rate-distortion behavior associated to the encoding system. These models must capture the statistical characteristics of the source and describe the encoding process as a function of some encoder control

[4] The VBV is a decoder buffer model, which means that for CBR encoding it is filled at constant bit rate and emptied impulsively and each VOP decoding time. Therefore, VBV underflows correspond to encoder buffer overflows and vice-versa. For rate control purposes the rate controller assumes that an equivalent buffer exists ate the encoder side — encoder bitstream buffer.

parameters (e.g., the quantization step), reflecting the lossy encoding rate-distortion trade-off.

In the ideal case, where the models are extremely accurate, the rate control mechanism could simply use these models to compute the scene encoder parameters to achieve a desired encoding objective (e.g., average bit rate, target video quality). However, usually, these models tend to deviate from the actual scene encoder behavior. Therefore, to deal with these deviations between theoretical models and actual coding results, it is necessary to develop: i) adequate compensation mechanisms (e.g., rate control decisions and actions) that are able to track these deviations and compensate them in order to allow a stable an efficient operation of the scene encoder, and ii) adequate adaptation mechanisms (e.g., estimation of model parameters) that are able to instantaneously represent the actual behavior of the scene encoder and its rate controller.

These two problems, i.e., compensation of the undesired behavior of the scene encoder and adaptation of the rate-distortion and rate controller parameters, aiming at building a robust and efficient rate control algorithm, are another set of important techniques that require proper attention in devising effective rate control mechanisms.

3.2 Object-Based Rate Control Techniques

Being a very complex mechanism, the design of a multiple video object (MVO) rate controller requires that the following aspects be addressed together:

- **Feedback compensation** – The rate control problem can be seen as a generic control problem where a given process (the video scene encoder) is controlled by a controller mechanism (the rate controller) in order to produce an output that follows closely a desired reference input command. Therefore, whenever process output deviations occur, adequate compensation actions are required. A simple solution is to relate the VBV buffer fullness with the MB quantization parameter as, for example, in the old ITU-T H.261 Reference Model 8 (RM8) (CCITT SGXV 1989)[5]. It is important to notice, however, that buffer feedback methods such as the RM8 compensation mechanism have a severe drawback, resulting from the direct relation of the quantization parameter with the buffer occupancy, which can lead to an equilibrium operational point at very high or very low buffer occupancies. In this case, a sudden scene change or a very low scene activity can lead to imminent buffer violations (respectively, an overflow or an underflow), which may require extreme actions to avoid and lead typically to large quality fluctuations. This drawback can be circumvented by avoiding a direct relation between the VBV buffer occupancy and the quantization parameter as in the MPEG-4 Video Verification Model 4.0 (VM4)[6] (MPEG-4 Video VM4 1996) rate control algorithm. The general idea of the VM4 compensation mechanism is to increase the quantization parameter by a given amount, whenever the number of bits spent in the previous VO VOP is higher than the average number of bits available to encode

[5] In this case, the decoder model buffer is called Hypothetical Reference Decoder (HRD) buffer.

[6] This algorithm has been superseded by (MPEG-4 Video VM5 1996, MPEG-4 Video VM8 1997) rate control algorithms.

the remaining VO VOPs. Conversely, the quantization parameter is decreased by a certain quantity, whenever the number of bits spent in the previous VOP is lower than the average number of bits available to encode the remaining VOPs. A major drawback of the VM4 algorithm results from the uniform bit allocation assumption, which neglects the VOP coding type, i.e., Intra (I), Inter (P), or Bidirectional (B), and coding complexity, which can lead to large quality fluctuations and to unstable control due to the frequent and large changes of the quantization parameter. This problem can be avoided by allocating to each VOP a number of bits proportional to its estimated complexity (depending on the VOP coding type and coding complexity). When the generated bits deviate from the nominal target, the target number of bits for the next encoding time instant(s) is updated. The feedback compensation loop is closed by relating the bit allocation with the quantization parameter value (low bit allocation corresponds to a high quantization parameter value and vice versa). This approach has been adopted, partially, in the context of the MPEG-2 Video Test Model 5 (TM5) (MPEG-2 TM5 1993) rate control algorithm, developed for frame-based video coding, and in most state-of-the-art object-based rate control algorithms, e.g., (MPEG-4 Video VM5 1996, MPEG-4 Video VM8 1997, Chiang and Zhang 1997, Vetro et al. 1999), since it is the most effective and robust method to compensate the undesired behaviors of video encoders. The bit allocation feedback compensation methods can still be improved with the additional use of buffer and quality (e.g., pixel distortion) feedback compensation mechanisms (Sun and Ahmad 2004, Nunes and Pereira 2007).

- **Use of feedforward information** – The basic idea of feedforward rate control is to plan in advance what will be the encoding results of a certain set of encoding parameters and act before deviations occur, i.e., based on all or a subset of the input data, the rate controller selects the set of encoding parameters that should produce the desired result. Feedforward methods can achieve very tight rate control by processing simultaneously large amounts of data and performing multiple encoding passes. However, besides being computationally very expensive, these methods are not usually suitable for real-time encoding since they generally involve high end-to-end delay. Some rate control algorithms rely exclusively on the feedback information from the buffer occupancy to adjust the quantization parameter; consequently, no feedforward information is used, i.e., the incoming data complexity is not taken into account in the control actions. However, being able to plan in advance what will be the encoding results of a certain set of encoding parameters and acting before deviations occur can bring significant gains in terms of rate control performance, e.g., based on all or a subset of the input data, the scene encoder rate controller can select the set of encoding parameters that should produce the desired result (e.g., target VBV occupancy and target scene quality). Therefore, it is convenient to extract relevant information from the input data to be encoded and, based on these data and adequate models of the scene encoder behavior, to compute the estimated behavior of the encoder. In the TM5 rate control algorithm (MPEG-2 Video TM5 1993), in order to take into account the lower sensitivity of the human visual system to the quantization error in highly texture zones and the higher sensitivity to the quantization error in uniform zones, the quantization parameter is further feedforward adjusted to the local characteristics of the picture using the MB texture activity computed from the luminance information of the MB being encoded. The MPEG-4 Visual Annex L rate control algorithms (ISO/IEC 14496-2 2004) use feedforward information into three different situations: i) *Frame Rate Control* — to compute the frame quantization parameter; ii) *Macroblock Rate Control* — to allocate the VOP target number of bits among the several MBs in the VOP; and iii) *Multiple Video Object Rate Control* — to distribute the allocated bits among the several VOs. In the first case, the VOP mean absolute difference (MAD), computed during the motion estimation step, is used in the VOP rate-quantization (RQ) model to compute the quantization parameter to encode the VOP. In the second case, the MB MAD is used in the MB-level feedback law to perform the MB bit allocation, and in the MB RQ model to compute the MB quantization parameter. Finally, in the third case, the MVO rate control algorithm uses feedforward information about the size, motion vectors amplitude, and VOP MAD to distribute the allocated number of bits for a given encoding time instant among the several VOs to encode for that time instant. Since the various VOPs to encode at each encoding time instant may have different sizes, and different complexities, it is important to take that into account by assigning a bit allocation weight to each VOP based on this feedforward information.

- **Sampling period of compensation and adaptation actions** – The sampling period of the rate controller is closely connected with the capabilities of this mechanism to achieve its goals. Lower sampling periods allow the rate controller to react faster to undesired deviations from the target behavior. However, frequent changes of the coding parameters may lead to instabilities or high quality variations. When the MB rate control is not used, the MPEG-4 Visual Annex L rate control algorithm (ISO/IEC 14496-2 2004) uses only a feedback loop at the VOP-level. If the MB rate control is activated, the algorithm uses two embedded feedback loops, corresponding to two different sampling periods: VOP-level and MB-level. Notice, however, that, when a very tight rate control is needed, e.g., in real-time interpersonal communications, the sampling period can be as lower as the DCT-level, involving skipping/stuffing actions.
- **Buffering control** – An important goal of a rate control algorithm is to keep the buffer occupancy within permitted bounds. There are, usually, two types of buffer control: i) soft buffer control and ii) hard buffer control. While the soft buffer control is usually achieved through the feedback compensation mechanisms by smoothly changing the bit allocations or the quantization parameter, hard buffer control requires drastic skipping or stuffing measures. Soft buffer control should usually be favored as it does not penalize so dramatically the spatio-temporal picture quality. This does not mean that hard buffer control should not be used. In some situations it can be the only method available to ensure that the encoding restrictions are not violated. The MPEG-4 Visual Annex L rate control algorithms (ISO/IEC 14496-2 2004), as other state-of-the-art object based rate control methods use these two types of buffer control.
- **Quality control** – Since the ultimate goal of a rate control algorithm is to maximize the subjective quality of the decoded video, it is important to incorporate this goal, as much as possible, in a direct way into the rate control actions. Although most rate control algorithms do not explicitly consider picture quality control actions, since no feedback information about the decoded picture quality is used to compute the control signal, there are two simple rate control mechanisms that aim

to provide some indirect quality control: i) quantization parameter variation range limitation between consecutive encoding time instants to avoid large picture quality variations between consecutive VOPs of the same VO; and ii) skipping VOPs in order to achieve a better trade-off between motion smoothness and spatial quality — when the bit rate resources are scarce, notably in low bit rate encoding, the target number of bits for the current time instant that emerges from the buffer control may not be sufficient to even encode the auxiliary data (i.e., header, motion vector, and shape data); thus, no bits would be left for encoding the texture data. In these cases, rate control algorithm should consider skipping these VOPs.

- **Parameter estimation** – Typically, adopting a linear feedback control model involves selecting a given operation point and designing the controller with constant parameters, e.g., in (Keesman et al. 1995) a simple PI-controller is used to perform VBV control in the context of a rate control mechanism. This approach is adequate for many applications since feedback systems are inherently insensitive to modeling errors and disturbances (Åström and Wittenmark 1995). This is typically a tuning problem, where the system is viewed as having constant but unknown parameters; thus, controller design consists in the off-line computation of the optimal controller parameters, e.g., PID parameters. However, sometimes the process dynamics is affected by variations of the operational conditions (e.g., scene changes, channel rate variation, etc.). In this case, constant-gain feedback controllers are usually insufficient due to stability problems, e.g., large picture quality fluctuations or imminent VBV violations. In this case, it is important to adapt the rate control mechanism to the unknowns of the process. In this context, the adaptation process consists typically in the estimation of the changing parameters during the control process itself.

The adequate balancing between feedback and feedforward rate control through different hierarchical levels of action and modeling is the core aspect of an efficient object-based rate controller and, therefore, of the six rate controller modules presented in Fig. 5. The following sections present the major features and major contributions available in the literature regarding each of these modules.

3.2.1 Scene Analysis for Resource Allocation

This module aims to extract relevant information for rate control as, for example, the size of the video data units (frames or VOPs), the object activity derived from the amplitude of the motion vectors, and the texture coding complexity derived from the prediction error variance (Nunes and Pereira 1997, Lee et al. 1997, Vetro et al. 1998, Lee et al. 2000).

The relevant data (e.g., MB organized) can then be feedforwarded to the other rate controller modules in order to guide their actions, notably the spatio-temporal resolution control, the bit allocation, and the coding mode control modules. These data can be used to compute the VOP coding complexity as a weighted sum of the various VOP features (Nunes and Pereira 1997) as follows

$$X_{VOP} = \alpha_T \times \left(\alpha_S \times \overline{S}_{VOP} + \alpha_A \times \overline{A}_{VOP} + \alpha_C \times \overline{C}_{VOP} \right) \quad (1)$$

where α_T is the VOP coding type weight $T \in \{I, P, B\}$, $\overline{S}_{VOP}$ is the normalized VOP size, $\overline{A}_{VOP}$ is the normalized VOP activity, $\overline{C}_{VOP}$ is the normalized VOP texture complexity, and α_S, α_A, and α_C are weights such that $\alpha_S + \alpha_A + \alpha_C = 1$.

The relative VOP coding complexities can be used by the bit allocation module to assign an adequate fraction of the bit rate to each VO for each encoding time instant. Further details of this approach can be found in (Nunes 2007, Nunes and Pereira 2009).

This module can also to be used to detect scene-changes during encoding in order that the rate controller can take adequate actions, such as, changing the bit allocations to handle a sudden high number of Intra MBs (Sun and Ahmad 2004) or to update the amount of previous data used for parameter estimation (Lee et al. 2000).

Since the various VOs composing the scene may have different visual importance, this module can also be used to assign priorities to the various VO through the computation of a bit allocation weight for each VO in the scene based on a visual attention model. In (Chen et al. 2006) the authors propose a visual attention model where the VO visual attention value (VAV) is a weighted sum of a static attention value (that depends on the visual saliency of the VOP pixels) and a motion attention value (that depends on the number of moving pixels in the VOP). The relative VAVs for the various VOs in the scene are then used for dynamic bit allocation.

3.2.2 Spatio-temporal Resolution Control

This module is conceptually responsible for deciding the appropriate spatial and temporal resolutions of the input video data, trying to achieve the best trade-off between spatial quality and motion smoothness based on information received from the other rate controller modules (see Fig. 5). However, the spatial resolution of the input data can only be changed during the scene encoding if this is supported by the encoder, e.g., MPEG-4 Advanced Real-Time Simple Profile (ISO/IEC 14496-2 2004). For temporal resolution control, for each possible encoding time instant, this module performs the following actions: i) receives input from the bit allocation module regarding the available number of bits for the next encoding time instant; ii) receives input from the video buffering verifier control regarding the estimated status of the different video buffering verifier buffers after the current set of VOPs had been encoded; iii) receives input from the scene analysis for resource allocation regarding the input data coding complexity; and iv) based on this information, this module decides to skip or to proceed with the encoding process for the current time instant (Nunes and Pereira 2001). For example, if the bit allocation is too small for a given encoding time instant it can be a better decision to skip the encoding of all or only some VOs for that time instant than encoding the VOs with very low texture quality. Skipping encoding (i.e., changing momentarily the temporal resolution) can also be the only solution when, based on the information coming from the scene analysis for resource allocation and the video buffering verifier modules, an imminent violation of any of the video buffering verifier models is foreseen.

Commonly, temporal resolution control is mainly based on the buffer occupancy criteria (Vetro et al. 1999, Ronda et al. 1999, Lee et al. 2000, Sun and Ahmad 2004).

3.2.3 Rate-Distortion Modeling

This module provides mathematical descriptions of the encoding process relating the video data rate and distortion characteristics with the coding parameters. In order to be able to predict in advance the behavior of the scene encoder, notably the joint behavior of the symbol generator and the entropy coder, it is advisable to have some mathematical description of this behavior that can be adapted to the actual encoding results during the encoding process, typically through parameter estimation. Rate-distortion (RD) models are used essentially to map bit allocations into coding parameters, notably the quantization parameter that will be used to encode each coding unit, i.e., a given VOP or MB inside a VOP.

The problem of rate-distortion modeling for accurate bit allocation, in the context of MPEG-4 video encoding, was first addressed by (Chiang and Zhang 1996, Chiang and Zhang 1997). In this work, the authors propose the quadratic rate-quantization (RQ) model that has been adopted in (MPEG-4 Video VM5 1996) for the frame and multiple video-object rate control algorithms:

$$R(Q) = \left(X_1 \frac{1}{Q} + X_2 \frac{1}{Q^2} \right) \cdot E_C \qquad (2)$$

where X_1, and X_2 are the model parameters, and E_C is the encoding complexity expressed by the mean absolute difference (MAD) between the VOP being encoded and its reference.

However, Intra and Inter RD models exhibit different characteristics since, while Intra RD models depend only on the current VOP characteristics and coding parameters (Nunes and Pereira 2004a), the RQ functions for Inter depend also of the reference VOPs and, consequently, become bi-dimensional, e.g., $R(Q, Q_{ref})$. Since these bi-dimensional rate and distortion functions are difficult to obtain, at least for a wide range of Q and Q_{ref} values, some assumptions need to be made. Therefore, (Nunes and Pereira 2004b) proposed a new type of Inter RD models for small Q variations between successive VOPs where the RQ function is approximated by a stationary component $R(Q)$, plus a delta component $\Delta R(Q, \Delta Q)$ that depends also on the difference between the stationary quantization parameter Q_0 and the quantization parameter of the reference VOP Q_{ref}, i.e., $\Delta Q = Q_0 - Q_{ref}$.

In order to achieve an adequate spatio-temporal encoding tradeoff it is also recommended to consider explicitly the rate-distortion characteristics of coded and skipped VOPs. This problem has been addressed by (Lee et al. 2003) where the distortion of the coded VOPs is modeled as

$$D_C(Q_i) = a \cdot 2^{-2R(t_i)} \cdot \sigma_{z_i}^2 \qquad (3)$$

where a is a model parameter, $R(t_i)$ is the average bit rate per sample at time instant t_i, and $\sigma_{z_i}^2$ is the variance of the input signal. The distortion of skipped VOPs is modeled as

$$D_S(Q_i, k) = D_C(Q_i) + E\left[\Delta^2_{z_{i,k}}\right] \tag{4}$$

where $D_C(Q_i)$ is the distortion of the last coded VOP and $E\left[\Delta^2_{z_{i,k}}\right]$ is an estimate of the interpolation mean square error for the skipped VOP at time instant t_k.

In (He et al. 2001a, He and Mitra 2001, He and Mitra 2002a) the authors proposed a new framework for modeling the RD characteristics of DCT video encoders where the coding rate and the pixel distortion are modeled as functions of the percentage of zeros in the quantized DCT coefficients, ρ, where the number of bits, R, and the pixel distortion, D, for a given VOP is modeled as follows

$$R(\rho) = \theta \cdot (1-\rho) \tag{5}$$

$$D(\rho) = \sigma^2 e^{-\alpha(1-\rho)} \tag{6}$$

where θ and α are model parameters, σ^2 is the picture variance, and ρ is the percentage of zero DCT coefficients on the current VOP as a function of the VOP quantization parameter.

For single video object (SVO) encoding, the VOP quantization parameter, Q, is computed after computing the $\rho(Q)$ mapping function for the given VOP and solving (5) for the target number of bits for the current VOP.

For MVO encoding, each VO has its own set of RD models. Therefore, after computing the target number of bits for each VO in the scene through the minimization of a RD cost function, (5) is solved independently for each VO (He et al. 2001b, He and Mitra 2002b).

Using a fixed quantization parameter for all the MBs in a VOP, computed based on a given target number of bits and the corresponding VO RQ model can lead in certain situations, notably, in low delay video encoding, to imminent buffer violations due to large deviations between the actual and the estimated number of encoded bits. In (Ribas-Corberas and Lei 1999) the authors proposed a RQ model for MB-level to circumvent this problem, allowing the quantization step to change from MB to MB inside a VOP. The MB rate control assumes that the encoder RQ function can be modeled as:

$$R_{MB}(Q) = \begin{cases} A_1 \dfrac{1}{Q^2_{MB}} MAD^2_{MB} & \Leftarrow R_{bpp} > \varepsilon_T \\ A_2 \dfrac{1}{Q^2_{MB}} MAD_{MB} + A_3 \dfrac{1}{Q_{MB}} MAD_{MB} & \Leftarrow R_{bpp} \le \varepsilon_T \end{cases} \tag{7}$$

where $R_{MB}(Q)$ is the number of bits to encode the texture data of MB i, A_1, A_2, and A_3 are the model parameters, Q_{MB} is the MB quantization parameter, MAD_{MB} is the mean absolute difference for the MB, R_{bpp} is the target number of bits per pixel for encoding the MB texture data, and ε_T is a control parameter (by default $\varepsilon_T = 0.085$).

In (Nunes and Pereira 2007) the authors also used a MB-level RQ model to improve the accuracy of the rate control mechanism defined as follows:

$$R_{MB}(Q) = \left(a/Q_{MB}\right)\cdot X_{MB} \tag{8}$$

where a is the model parameter and X_{MB} is the MB texture complexity. The MB RQ model (8), although very simple, supports a fine level of rate control, i.e., with a lower sampling period — MB period — and, consequently, a faster reaction to deviations relatively to the nominal operation, improving this way the performance of the rate control mechanism.

3.2.4 Bit Allocation

This module allocates the available bit rate resources taking into account the varying characteristics of the input video data compensating the deviations between the idealized behavior and the actual encoding results. Following a strategy of dividing to conquer, the bit allocation can be partitioned into several hierarchical levels, similarly to the syntactic organization of the encoded video data:

- **Group of Scene Planes (GOS)** – The set of all encoding time instants between two random access points, typically encoded with a constant number of bits.
- **Scene Plane (SP)** – The set of all VOPs of all VOs to be encoded at a particular encoding time instant. Not all VOs of a given scene may have VOPs to be encoded in every scene plane (see Fig. 6)
- **Video Object Plane (VOP)** – The sample of each video object at a particular encoding time instant.
- **Macroblock (MB)** – The smallest coding unit for which the quantization parameter can be changed.

At each level and for the corresponding time instant, this module allocates an adequate number of bits taking into account the channel bit rate, each VO target encoded temporal resolution, the buffer size, current buffer occupancy, target buffer occupancy, and number of bits for the previous VOPs of each VO.

For example, after allocating a certain number of bits for a given GOS, the bit allocation proceeds to the SP-level allocating an appropriate number of bits for each SP according to number and complexity of the VOs to encode in that SP. Afterwards, the bit allocation module distributes the available bits among the several VOPs in the SP, and, inside each VOP, among the several MBs, in a way the overall (perceptual) quality is maximized.

For low complexity and low encoding delay, the bit allocation for each level, N, should be composed by a nominal value at each level, $\overline{T}_N$, and a compensation mechanism that takes into account deviations between the generated bits and the nominal bit allocations. The following formulation is proposed by (Nunes and Pereira 2007):

$$T_N[n] = \overline{T}_N[n] + K_N[n]\sum_{k=1}^{n-1}\left(\overline{T}_N[k] - S_N[k]\right) \tag{9}$$

where K_N is an integral compensation factor, which may be constant or dependent on the encoding time instant n, and $S_N[k]$ is the number of bits spent for time

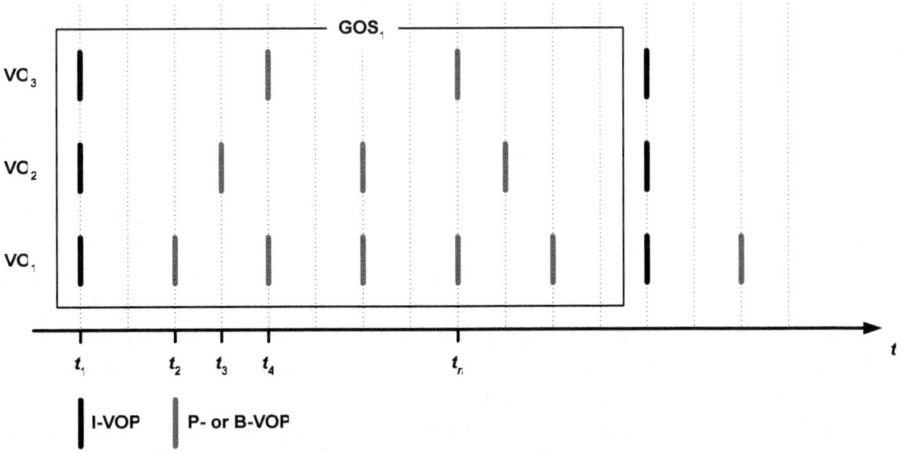

Fig. 6. Multiple video objects encoding with different VOP rates.

instant k . Therefore, for each source of uncertainty (i.e., for each bit allocation level), an adequate feedback compensation mechanism is provided, in a way that, together, the different levels contribute to the overall rate control goals.

For the case of MVO encoding, another important aspect to be considered is the fact that quality between the various VOs for each encoding time instant and along time should be as constant as possible.

For a given VO to reach an average distortion per pixel for the different VOP coding types approximately constant, i.e., $D_I \approx D_P \approx D_B$, the following relation should approximately hold

$$\frac{b_I}{\alpha_I} \approx \frac{b_P}{\alpha_P} \approx \frac{b_B}{\alpha_B} \tag{10}$$

where α_I, α_P, and α_B, are the VOP coding type weights, and b_I, b_P, and b_B, are the average number of bits per pixel for each corresponding VOP coding type.

Therefore, in (Nunes and Pereira 2009), it is proposed that before encoding each I-VOP, i.e., at the beginning of each GOV, of a given VO, α_I be estimated through the following equation

$$\alpha_I = \frac{\overline{b}_I}{\overline{b}_P}\left(\frac{\overline{D}_I}{\overline{D}_P}\right)^{\gamma_T} \tag{11}$$

where $\overline{b}_I$, $\overline{D}_I$, $\overline{b}_P$, and $\overline{D}_P$ are, respectively, the average number of bits per pixel and the average pixel distortion for I- and P-VOPs, computed over window sizes W_I and W_P of past I- and P-VOPs encoding results, and γ_T is a parameter that controls the impact of the average distortion ratios on the estimation of the VO coding weight; in (Nunes and Pereira 2009), $W_I = 3$ and $W_P = N_{SP} - 1$, and $\gamma_T = 0.5$.

Similarly, for B-VOPs, at the beginning of each B-VOP encoding, α_B should be updated according to

$$\alpha_B = \frac{\bar{b}_B}{\bar{b}_P}\left(\frac{\bar{D}_B}{\bar{D}_P}\right)^{\gamma_T} \tag{12}$$

A similar approach is also used in (Sun and Ahmad 2004, Sun and Ahmad 2005).

Encoding all the objects in the scene with approximately constant quality is an important goal that can hardly be achieved when only a pure feedforward approach is used to compute the VO weights used to distribute the SP target among the various VOPs in the given SP. This is the approach followed in (MPEG-4 Video VM5 1996), where there is no compensation for deviations on the bit rate distribution among the various VOPs for a given time instant.

Consequently, it is important to update the VO coding complexity weights along time and to compensate the bit allocation deviations through the feedback adjustment of these parameters to meet the requirement of spatial quality smoothness (Nunes and Pereira 2007). For this purpose, the coding complexity (reflecting the texture coding complexity and the coding mode) of a given VOP n in SP p of GOS m is, given by

$$X_{VOP}[n] = \alpha_T[n] \cdot \omega[n]\,,\ n = 1,\ldots,N_{VO} \tag{13}$$

where $\alpha_T[n]$ and $\omega[n]$ are, respectively, the coding type weight ($T \in \{I,P,B\}$) and coding complexity weight – reflecting the texture, shape, and motion (if applicable) coding complexity – of VO n in SP p of GOS m.

The coding complexity weight, $\omega[n]$, reflects its relative coding difficulty in the current SP, computed as follows

$$\omega[n] = \alpha_S \times \bar{S}_{VO}[n] + \alpha_A \times \bar{A}_{VO}[n] + \alpha_C \times \bar{C}_{VO}[n] \tag{14}$$

where $\bar{S}_{VO}[n]$, $\bar{A}_{VO}[n]$, and $\bar{C}_{VO}[n]$ are, respectively, the normalized values of size, activity, and texture complexity of VO n in the current SP, and α_S, α_A, and α_C are weights such that $\alpha_S + \alpha_A + \alpha_C = 1$ — e.g., $\alpha_S = 0.2$, $\alpha_A = 0.5$, $\alpha_C = 0.3$ as proposed in (Nunes and Pereira 1997).

Additionally, the SP average luminance pixel distortion is defined as the weighted sum of the various VOs distortions, i.e.

$$D_{SP}[p] = \sum_{k=1}^{N_{VO}} \left(N_{PIX}[k] \cdot D_{VOP}[k]\right) \Big/ \sum_{k=1}^{N_{VO}} N_{PIX}[k] \tag{15}$$

where $D_{VOP}[n]$ and $N_{PIX}[n]$ are, respectively, the average pixel distortion and the VOP size in pixels of VO n in SP p.

Using (15) as the reference SP distortion, a complexity weight adjustment can be computed for each VO as follows

$$\phi_D[p][n] = \left(\frac{b_{VOP}[p-1][n] / \alpha_T[p-1][n]}{\sum_{k=1}^{N_{VO}} b_{VOP}[p-1][k]} \times \frac{D_{VOP}[p-1][n]}{D_{SP}[p-1]}\right)^{\gamma_D} \tag{16}$$

where $b_{VOP}[p][n]$ is the VO n VOP number of bits in SP p, and parameter γ_D controls the impact of ϕ_D in the VOP bit allocation feedback compensation; typically, $0.1 \le \gamma_D \le 0.5$; in (Nunes and Pereira 2009), $\gamma_D = 0.2$ revealed to be an adequate choice.

From (13) and (16), the VO complexity and VOP target bits are, respectively, feedback-adjusted as

$$\eta_D[n] = \phi_D[n] \cdot X_{VOP}[n] \tag{17}$$

and

$$T_{VOP}[n] = T_{SP} \frac{\eta_D[n]}{\sum_{k=1}^{N_{VO}} \eta_D[k]} + \frac{\eta_D[n]}{\sum_{k=n}^{N_{VO}} \eta_D[k]} \sum_{k=1}^{n-1} \left(T_{SP} \frac{\eta_D[n]}{\sum_{k=1}^{N_{VO}} \eta_D[k]} - b_{VOP}[k] \right) \tag{18}$$

The main goal of this compensation mechanism is to smooth the spatial quality differences among the various VOs. Therefore, it can be seen as an SP spatial quality control also.

3.2.5 Video Buffering Verifier Control

This module guides the other modules decisions in order to avoid critical situations (e.g., encoder buffer overflows or underflows). For the video rate buffer verifier (VBV) mechanism (ISO/IEC 14496-2 2004), this module specifies adequate dynamic target VBV occupancies, B_T, for each encoding time instant taking into account the VBV size, the relative encoding time instant inside each GOS, the complexity of each SP, and the target number of bits for the current GOS. The deviation between B_T and the actual VBV occupancy, B, is used to softly compensate the bit allocation at SP-level — soft SP-level VBV control. If soft control is not enough to lead B to the nominal operation area, then hard SP-level VBV control is applied limiting the bit allocations to prevent VBV violations.

A common approach in MPEG-4 object-based rate control is to set the target VBV to half the buffer size, B_S, (MPEG-4 VM5 1996, MPEG-4 VM8 1997, Vetro et al. 1999, Sun and Ahmad 2004, Sun and Ahmad 2005). However, this is a rather sub-optimal approach since, when using different VOP coding types (e.g., I, P and B) or the various VOs in the scene are encoded with different temporal resolutions, a different target number of bits shall be allocated for each coding time and, consequently, a different target buffer occupancy shall be specified. This approach is followed in (Nunes and Pereira 2007, Nunes and Pereira 2009) where the authors define different target VBV buffer occupancies for each encoding time instant as follows

$$B_T[p] = B_S - \left[\frac{T_{GOS} \sum_{k=1}^{p-1} X_{SP}[k]}{X_{GOS}} - R\left(t_{SP}[p] - t_{SP}[1]\right) \right] - B_{udflw} \tag{19}$$

where B_S is the VBV buffer size, T_{GOS} is the target number of bits for the whole GOS, $X_{SP}[p]$ is the SP p complexity, X_{GOS} is the GOS complexity, $t_{SP}[p]$ is the time instant of SP p, R is the GOS average target bit rate, and B_{udflw} is a VBV underflow margin.

3.2.6 Coding Mode Control

This module decides the appropriate coding parameters for each coding unit e.g., MB texture and shape coding modes, motion vectors and quantization parameter, aiming at achieving accurate bit allocations while maximizing the spatio-temporal quality. Essentially, in terms of rate control, this module is responsible for determining the quantization parameter for each MB of each VOP. The first step of this module is to determine the target quantization parameter for the VOP, Q_T, using the VOP-level RQ models. Afterwards, if MB-level quantization adaptation is used, this module adjusts the MB quantization parameter, Q_{MB}, setting a trade-off between spatial quality smoothness (i.e., approximately constant Q_{MB}) and efficient control of bit allocation deviations. The two levels of RD modeling can be used in conjunction in the architecture of Fig. 5 (Nunes and Pereira 2009) leading to more efficient rate control in comparison with (MPEG-4 Video VM5, MPEG-4 Video VM8) where the suggested rate control methods use, alternatively, the two levels of modeling.

In (Ronda et al. 1999) the authors specify a set of different goals for the rate control. In this case the coding mode control is defined as a minimization of a cost function given a certain goal. In this context, the authors proposed three rate control variants with different goals and associated different cost criteria, namely:

1. **Weighted distortion rate control** – Aims at minimizing the weighted distortion of the various VOs in the scene, embedding VO priorities in the optimization process, i.e., the rate control goal becomes the minimization of the weighted sum of the individual VO distortions without imposing any relation between the different VOs in the scene.
2. **Priority-based rate control** – Aims at satisfying a prioritized list of VO target distortions, specifying minimum qualities for the various VOs, ordered according to their priority, i.e., the rate control goal is to achieve the minimum target quality specified by the user for each VO (for each encoding time instant), respecting the prioritized order.
3. **Constant distortion-ratio rate control** – Aims at achieving a constant distortion ratio for each VO, relatively to a reference VO, i.e., given a reference VO and a set of distortion ratios specified by the user, the rate control goal is to minimize the average VOs distortion, respecting these distortion ratios.

For each of the above rate control goals, the authors define a cost function that should be minimized under the constraint that a given bit allocation is met. Nevertheless, the coding mode control module aims essentially at controlling the quantization parameter variation for each VOP to be encoded.

3.3 Object-Based Rate Control Walkthrough

The main purpose of this section is to summarize the integration of the various rate control techniques described above in the form of rate controller modules in a way that they can be implemented with relative low degree of uncertainty for controlling a

MPEG-4 Visual scene encoder aiming at producing compliant bitstreams with a given MPEG-4 Visual Profile and Level (ISO/IEC 14496-2 2004). In this description, it is assumed that the VOs are available to the scene encoder through the scene buffer (Fig. 5)

It is worthwhile mentioning here, that some rate controller modules interact iteratively with each other (e.g., the spatio-temporal resolution control and the scene analysis for resource allocation modules or the coding mode control and the video buffering verifier modules). Therefore, a step-by-step description of such rate control mechanism can hardly associate each step to a single rate controller module without compromising the simplicity and legibility of this description.

For a given scene composed by a set of VOs, the rate controller requires from the user the following parameters:

- Profile and level for encoding the scene.
- Target temporal resolution for each VO.
- Common random access point period (I-VOP period).
- Target average channel bit rate, R .
- Video rate buffer size, B_S .

The proposed rate controller can be implemented through the following steps:

STEP 1 — Rate Controller Initialization
This step consists in: i) initializing the video buffering verifier model parameters; ii) checking the profile and level limits; iii) computing the scene base temporal resolution, SR ; and iv) setting the rate controller parameters.

STEP 2 — Spatio-Temporal Resolution Control
This step computes successively the next encoding time instant (next SP) – a multiple of $1/SR$. After reading each VOP from the Scene Buffer, the rate controller performs the following tasks: i) invokes the Scene Analysis for Resource Allocation module to estimate the size, motion activity, and texture coding complexity of each VO; ii) invokes the Video Buffering Verifier Control to check the estimated status for the next encoding time instant of the various Video Buffering Verifier model buffers — the next SP is skipped if any of these model buffers is foreseen to overflow, the VBV buffer is foreseen to underflow, or if $B < B_T - B$ for the considered SP, where B_T is the target VBV buffer occupancy for the current encoding time instant.

STEP 3 — Rate Controller Parameters Update
Before encoding, this step updates the video buffering verifier models for the current SP and, for each VO in the SP, updates the corresponding VOP-level RD Models.

STEP 4 — Bit Allocation (Scene-Level)
This step computes the bit allocation for each SP, T_{SP} , and for each VOP, T_{VOP} , inside each SP — for the first SP of each GOS, it also computes the target number of bits for the current GOS. Invokes the soft SP-level VBV control and hard SP-level VBV control if necessary — if T_{SP} is lower than the estimated header, motion, and shape bits, the current SP is skipped and returns to STEP 2. For each VO in current SP, computes

the VO feedback adjusted complexity, η_D (17), and the VOP target number of bits, T_{VOP} (18).

STEP 5 — Coding Mode Control (VOP Encoding)
This step is essentially responsible for computing the QP for each MB. If a MB-level coding mode control is used it should also invoke the MB-level model updates.

STEP 6 — GOTO STEP 2
Steps 1–6 describe a possible execution flow of the rate control modules described in the previous section. For low-delay and constant bit rate application scenarios the rate controller should be able to efficiently deal with deviations between the ideal and the actual behavior of the scene encoder. To deal with these deviations between the theoretical models and the actual coding results, it necessary to have adequate adaptation and compensation mechanisms that are able to track these deviations and compensate them in order to allow a stable an efficient operation of the encoder. These two problems (adaptation and compensation) are tackled along the different modules composing the rate control architecture presented in Fig. 5. In this context, the following two rate control approaches should be combined

- **Feedback rate control** – This rate control approach compares the actual coding results (e.g., VBV buffer occupancy, bit rate, quality, etc.) with the target results and take an action based on the difference, e.g., whenever the past encoding decisions resulted in a successive decrease of the VBV buffer occupancy, the rate control mechanism compensates this behavior by coarsely encoding the next incoming VOPs.
- **Feedforward rate control** – This rate control approach plans in advance what will be the encoding result of a certain set of encoding parameters and acts before deviations occur, e.g., based on all or a subset of the input data, the encoder selects the set of encoding parameters that should produce the desired result, e.g., target VBV occupancy and target scene quality.

Being a complex system, the rate control mechanism is composed, typically, of a large number of parameters. Consequently, setting the values for these parameters is a critical task when developing an efficient rate control algorithm, since some of these parameters can have a high impact on the algorithm performance, e.g., the bit allocation weights between I-, P-, and B-VOPs, and the bit allocation weights for the several VOs in the scene. This problem is circumvented by an adaptive approach based on the adaptation of the parameters describing the encoding process and the parameters of the rate controller during the encoding process.

4 Object-Based Error Resilience

The change in video representation from the frame-based model to the object-based model has a significant impact in terms of error resilience, since new solutions have to be found to deal with transmission or storage errors in the bitstreams. For instance, in frame-based systems, the problem of error concealment always corresponded to the concealment of rectangular objects, since the shape of the object is *a priori* known to be a rectangle with fixed dimensions. In object-based systems, however, arbitrary shape is added, as well as scene composition information, which brings an additional dimension to the problem of video error concealment. This means that, in these new systems, when errors occur, the shape data may have to be concealed as well as the data that describes the scene composition. Therefore, error resilience in frame-based systems can be seen as a subset of the general problem of object-based video error resilience in the sense that most techniques available may still be useful (even if some adaptation may be required) but new techniques, notably for shape and scene composition concealment, have to be developed. Frame-based resilience techniques have been extensively studied and much work has been published; a good review of this work can be found in (Wang and Zhu 1998, Wang et al. 2000, Kumar et al. 2006). However, not many papers are known dealing with object-based video error resilience and, therefore, much work still remains to be done to completely solve the general problem of object-based video coding error resilience. The known work on object-based video error resilience is briefly described here.

4.1 Object-Based Error Resilient Architecture

To better understand the problem of object-based error resilience, the general architecture of an error resilient object-based video coding system should be first considered. This system, which includes many different modules responsible for very specific tasks, is illustrated in Fig. 7.

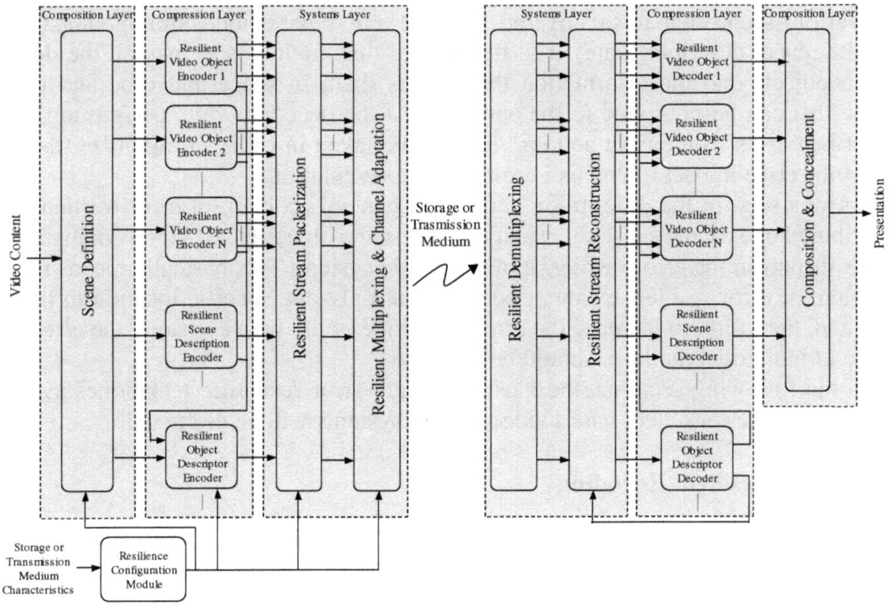

Fig. 7. Example of an error resilient object-based video coding architecture.

On the encoder side, the first step is to take the input video content and define a video scene with it. This can be done by segmenting an existing video sequence, by composing it with already existing pre-segmented video objects or a mixture of both. After the video scene to be coded has been defined, the elements that compose the scene are sent to the corresponding encoders. This means that each video object is sent to a video object encoder and the information that positions them in space and time is sent to the scene description encoder. In addition to this, the object descriptor data, which creates the link between the various elementary streams and the objects in the scene description structure, is sent to the object descriptor encoder. The output of all these encoders corresponds to various elementary streams containing encoded data, which will then be packetized, multiplexed and, finally, adapted to the storage or transmission channel(s) in question. All the previously described operations are supervised by the resilience configuration module, which is basically responsible for choosing the most adequate (in terms of resilience) coding parameters. This module is necessary because the encoding side of the communication chain plays an important part in error resilience; in fact, the decoding performance will depend a lot on the kind of "protective action" the encoder has taken. As for the scene definition module, it typically plays no part in error resilience since its role is to define the scene, possibly based only on semantic criteria. However, it can also help. For instance, by slightly smoothing the video objects, concealment may become easier at the decoder.

At the decoding side, the procedure is basically the opposite but, instead of having a scene definition module, a scene composition module is used. This way, the decoder starts by demultiplexing the received bitstream(s) from the storage or transmission channel(s) and then reconstructing the various elementary streams, which are sent to the corresponding decoders. After all the elementary streams have been decoded and the various scene elements recovered, they are sent to the composition module (except for the object descriptor data). The function of this module is to take all the decoded video objects and the information that relates them in space and time and build a scene that can be presented to the end-user. At the decoder, while performing all the described tasks, "defensive actions" have to be taken in order to minimize the negative subjective impact of channel errors in the presentation.

As was seen in the description above, the general problem of error resilience is a distributed one, in the sense that each module in the object-based video coding system plays its part in the error resilience of the whole system. This basically means that the problem of error resilience cannot be solved at only one specific location in the system and, therefore, to increase the error resilience of the whole system, the error resilience of individual modules should be improved.

In the following sections, the most important error resilience techniques available in the literature for object-based video coding systems will be discussed.

4.2 Error Resilient Encoding

There are several types of techniques that encoders may use to improve error resilience. However, not many of those techniques are specific to object-based video. Most have been proposed for frame-based video, but can also be extended for object-based video.

In particular, it is largely recognized that Intra coding refresh can play a major role in improving error resilience in video coding systems that rely on predictive (Inter) coding to remove temporal redundancy because, in these conditions, the decoded quality can decay very rapidly due to error propagation if errors occur in the transmission or storage of the coded streams. Therefore, in order to avoid error propagation for too long a time, the encoder can use an Intra coding refresh scheme to refresh the decoding process and stop (spatial and temporal) error propagation. The refresh process will certainly decrease the coding efficiency, but it will significantly improve error resilience at the decoder side, increasing the overall subjective impact (Soares and Pereira 1999).

4.2.1 Refresh Need Metrics

In order to design an efficient Intra coding refresh scheme for an object-based coding scheme, it would be helpful for the encoder to have a method to determine which components of the video data (shape and texture) of which objects should be refreshed and when, especially if based on appropriate refresh need measures. The refresh need for a given visual data entity, such as a shape or a texture MB for MB-based coding schemes, is a metric that measures the necessity of refreshing this visual data entity according to some error resilience criteria and can be used by the encoder to decide if a given visual entity should be refreshed or not at a certain point in time. In (Soares and Pereira 2003), shape refresh need (*SRN*) and texture refresh need (*TRN*) metrics have been proposed.

Since MPEG-4 is the only available object-based video coding standard, the metrics proposed in (Soares and Pereira 2003) were developed in the context of an MPEG-4 video coding system. However, any other object-based coding system could have been used; only small adjustments to the proposed refresh need metrics would be necessary, since they are directly related to the used coding syntax.

In MPEG-4 shape coding (ISO/IEC 14496-2 1999), the only way to completely eliminate the shape dependency on the past is by refreshing the entire shape of a VOP. Therefore, although shape coding is block-based, the shape refresh need metric has to be VOP-based since VOP Intra refresh is the only way to guarantee full shape refresh. This way, by considering the most important factors impacting the shape refresh need, the SRN_i parameter measuring the shape refresh need for VOP i in a video object sequence, was defined as

$$SRN_i = SEV_i \times SCD_i = SEV_i \times f_{SCD}\left(SSCD_i, TSCD_i\right) \quad (20)$$

where SEV_i is the shape error vulnerability (*SEV*), SCD_i is the shape concealment difficulty (*SCD*), $SSCD_i$ is the spatial shape concealment difficulty (*SSCD*), $TSCD_i$ is the temporal shape concealment difficulty (*TSCD*), all for VOP i. Additionally, f_{SCD} is a function that combines the values of $SSCD_i$ and $TSCD_i$ (both with values between 0 and 1) into a single *SCD* value.

SRN_i was defined as a product where the first factor (i.e., the shape error vulnerability) measures the statistical exposure of the shape data to errors, already considering the used bitstream structure, and the second factor (i.e., the shape concealment difficulty) expresses how hard the shape is to recover using concealment techniques. A product is used because the dependency of the shape refresh need on the shape error vulnerability and the shape concealment difficulty should be such that when either one of them approaches zero, so should the shape refresh need. For instance, if the considered shape is extremely easy to conceal (shape concealment difficulty approaching zero), the shape refresh need should also approach zero independently of the shape error vulnerability. Or, on the other hand, if the shape is not vulnerable to errors (because almost no bits were spent or the amount of channel errors is very small), the shape refresh need should also approach zero independently of the shape concealment difficulty.

To eliminate the texture dependency on the past it is not necessary to refresh the entire VOP as happens for MPEG-4 shape coding (ISO/IEC 14496-2 1999). Therefore, the texture refresh need metric can be defined at the MB-level. This way, by considering the most important factors impacting the texture refresh need, the TRN_i^k parameter measuring the texture refresh need of MB k in VOP i, was defined as:

$$TRN_i^k = TEV_i^k \times TCD_i^k = TEV_i^k \times f_{TCD}\left(STCD_i^k, TTCD_i^k\right) \tag{21}$$

where TEV_i^k is the texture error vulnerability (TEV), TCD_i^k is the texture concealment difficulty (TCD), $STCD_i^k$ is the spatial texture concealment difficulty ($STCD$), $TTCD_i^k$ is the temporal texture concealment difficulty ($TTCD$), all for MB k in VOP i. Additionally, f_{TCD} is a function that combines the values of $STCD_i^k$ and $TTCD_i^k$ (both with values between 0 and 1) into a single TCD value.

TRN_i^k was defined as a product where the first factor (i.e., the texture error vulnerability) measures the statistical exposure of the texture data to errors already considering the used bitstream structure; and the second factor (i.e., the texture concealment difficulty) measures how hard the texture is to recover using error concealment techniques. A product is used because the dependency of texture refresh need on the texture error vulnerability and the texture concealment difficulty is similar to the dependency of shape refresh need on the shape error vulnerability and the shape concealment difficulty.

The results presented in (Soares and Pereira 2003) have shown that both the shape refresh need and the texture refresh need metrics are able to correctly express the necessity of refreshing and, therefore, can be extremely useful in allowing the encoder to efficiently decide which video objects should have their shapes and texture MBs Intra coded to improve the subjective impact at the decoder side.

4.2.2 Shape and Texture Intra Refresh Techniques

Based on the refresh need metrics described above, an adaptive shape and texture Intra coding refresh scheme has been proposed in (Soares and Pereira 2004a). This scheme considers a multi-object video scene in which it has the target to efficiently distribute the available video error resilience resources (i.e., the resources that have been allocated to improve error resilience and not coding efficiency) among the various video objects. This basically corresponds to controlling the refresh rate for the various video objects depending on their refresh needs, as defined in the previous section. Due to the way the refresh needs were defined, the shape refresh is VOP-based, while the texture refresh is block-based. Therefore, only encoders that allow for this type of shape and texture refreshment are able to use the proposed scheme; this includes MPEG-4 encoders.

The efficient distribution of the available error resilience resources is expected to improve the overall video quality for a certain total bit rate when compared to the cases where less sophisticated refresh schemes are used (e.g., a fixed Intra refresh period is used for all the objects). This expectation is based on the principle that objects with low refresh need can be refreshed less often without a significant reduction in their quality, thus saving refresh resources. The saved resources can then be used to greatly improve the quality of objects with high refresh need by refreshing them more often.

The amount of error resilience resources used for refresh, also referred to as refresh resources in the following, is defined by means of two parameters (one for shape and one for texture). These parameters are specified by the user, at the beginning of encoding and changed later on, if an update is necessary and possible:

- **Target average shape refresh period** ($ASRP_{target}$) – This parameter corresponds to the target average of the individual shape refresh periods (*SRP*) of the various video objects in the scene, at a given time instant. The chosen value will very likely depend on the specific channel and content characteristics.
- **Number of texture MBs to be refreshed (*NTMR*) at each time instant** – This parameter indicates the total number of texture MBs (considering all the video objects) to be refreshed at a given time instant. As above the chosen value will very likely depend on the specific channel and content characteristics.

These two parameters establish the amount of shape and texture Intra refresh (or error resilience) resources to be used, implying that a certain amount of the total available bit rate is to be used for these refreshes and not to increase the error-free video quality. After these two refresh parameters have been specified, the encoder has to distribute the corresponding refresh resources, as efficiently as possible, among the various video objects to be refreshed targeting the maximum subjective impact.

The proposed algorithm performs an efficient allocation of the available refreshment resources as follows:

- **Shape** – At each time instant, the encoder has to rank the existing video objects according to their average shape refresh need (following the metric defined in the previous section). Those with higher refresh needs will have their shape refresh periods decreased, and vice-versa. The average of the shape refresh periods is maintained constant, for all time instants, and equal to the $ASRP_{target}$. Otherwise, the amount of shape refresh resources used would be higher than established, decreasing the error-free quality (in this case, the error-free quality of the texture because lossless shape coding is used).
- **Texture** – At each time instant, the encoder has to rank the existing video objects according to (the average of) the accumulated texture refresh need of their MBs (following the metric described in the previous section). Those with higher refresh needs will have higher Intra refresh rates assigned, and vice-versa. The number of texture MBs refreshed per time instant is maintained constant, for all time instants, and equal to *NTMR*. Otherwise, as above, the amount of texture refresh resources used would be higher than established, decreasing the error-free texture quality.

The encoder architecture adopted to implement the proposed Intra refreshment scheme targeting the best allocation of the initially defined refreshment resources is

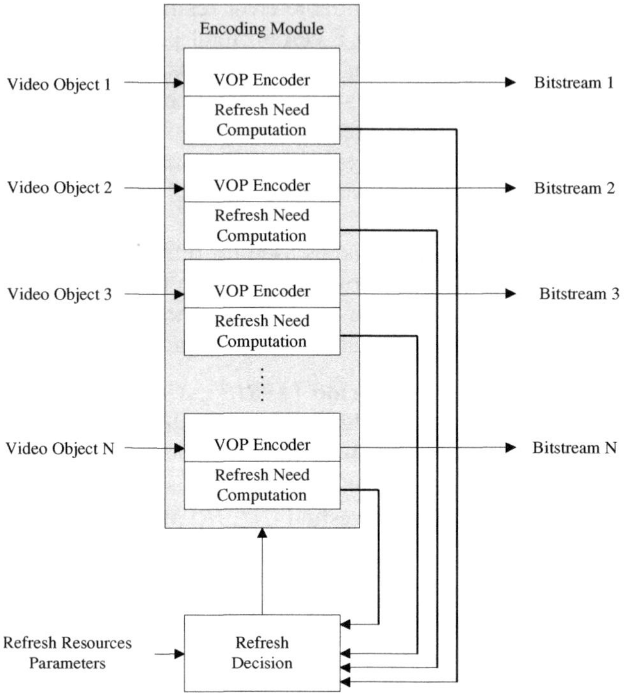

Fig. 8. Multi-object Intra refresh encoder architecture (Soares and Pereira 2004a) (© 2004 IEEE).

illustrated in Fig. 8. The encoding module is responsible for the encoding of the VOPs making up the scene, with or without Intra refresh, as well as for computing all the necessary refresh need parameters. As for the refresh decision module, it is responsible for deciding which VOPs will be refreshed in terms of shape and which MBs, within each VOP, will be refreshed in terms of texture. These decisions are taken based on the refresh need parameters provided by the VOP encoder module and the refresh resources parameters initially determined by the user.

Based on the results presented in (Soares and Pereira 2004a), which have shown that performance of the adaptive shape and texture refresh scheme is always equal or better than the fixed refresh scheme used as a reference, it was concluded that the use of the proposed adaptive shape and texture Intra refreshment scheme is worth considering in any object-based video encoder.

4.3 Error Resilient Decoding and Concealment

By using the encoder techniques described in Section 4.2 on their own, considerable improvements can be obtained in terms of the decoded video quality. However, further improvements are still possible by also using other sophisticated error resilience techniques at the decoder, in terms of both shape and texture. The error resilience techniques that can be used at the decoder basically fall under one of the following 3 types: error detection, error localization and error concealment (Soares and Pereira 1999). However, there are currently no known error detection and error localization techniques specific for object-based video; at the time of writing, only object-based error concealment techniques are known, which assume that channel errors have been adequately detected and localized.

As can be seen in Fig. 7, error concealment techniques can either be applied to individual objects right after the object is decoded by the video object decoder or they can be applied to the whole scene in the composition module. Therefore, existing error concealment techniques can be classified according to the two following levels:

- **Object-level concealment** – This level deals with error concealment for each single object. This means that each object is concealed as a stand-alone object (independently) and no information other than the one decoded for it is used.
- **Scene-level concealment** – This level deals with error concealment for the whole scene, notably for the composition of the individual objects. This type of concealment has to deal with the relationship between objects as well as with the composition information where their spatial and temporal positioning is encoded.

In addition to the object/scene dimension, error concealment techniques can also be divided into one of the following three categories, depending on the type of information that is used to perform it: spatial, temporal or spatio-temporal. These three categories of error concealment have been extensively investigated and numerous techniques have been proposed for frame-based video. However, in terms of object-based video, the interest of researchers working in the field is just awakening and it is expected that many new techniques will soon appear in the literature. In fact, a few object-based error concealment techniques are already available. These techniques shall be summarized in the following three sections.

These techniques assume that some kind of block-based technique has been used to encode the VOPs. This situation corresponds to the most relevant case in terms of object-based video coding, which is represented by the MPEG-4 Visual standard (ISO/IEC 14496-2 1999). This way, although video objects can have an arbitrary shape, they are transmitted in the form of a rectangular bounding box which is divided in 16×16 blocks. For each 16×16 block, shape data is transmitted, followed by texture data if needed. The total transmitted shape corresponds to a binary alpha plane, which dictates the parts of the transmitted texture that will be visible. In terms of bitstream errors, it is considered that they manifest themselves as bursts of consecutive lost 16×16 blocks, which have to be detected by an error detection stage before applying the concealment itself. This also corresponds to the situation in the MPEG-4 Visual standard (ISO/IEC 14496-2 1999), where several consecutive blocks (called macroblocks) are carried in a Video Packet (VP). Additionally, in the case of the texture concealment technique, it is also considered that the shape data of corrupted blocks has already been concealed (e.g., using one of the shape concealment solutions proposed here) and only the texture data needs to be recovered. Thus, these techniques can be directly applied to any object-based video coding system (including MPEG-4) that verifies these conditions, leading to an improved decoded video quality.

4.3.1 Object-Level Spatial Error Concealment

The error concealment techniques described in this section are spatial (or Intra) techniques, in the sense that they do not rely on information from other temporal instants: they only use the available information for the video object being concealed at the time instant in question. This approach guarantees that they can be applied both to still images and video object-based coded data. In fact, they can be applied to any corrupted still image or video object whose shape can be represented by means of a binary alpha plane, such as for MPEG-4 still images and video with binary shapes (ISO/IEC 14496-2 1999). Although the proposed techniques are very adequate for usage in MPEG-4 Visual decoders (for the profiles accepting arbitrarily shaped objects), they are by no means limited or anyhow tuned to this type of decoders.

In terms of error concealment techniques for object-based video coding systems, it is important to remind that they have to deal with both shape and texture data concealment. However, considering that the available (frame-based) texture concealment techniques can most of the times be adapted to work for object-based video, researchers have mainly invested in shape concealment techniques. This explains why mainly shape concealment techniques have appeared in the literature and only one texture concealment technique, specifically designed for object-based video coding systems, is known.

In terms of spatial shape concealment techniques, the first technique to appear in the literature was (Shirani et al. 2000). This technique is based on the use of a binary Markov Random Field (MRF) of shape, in conjunction with a maximum a posteriori (MAP) estimation. According to this algorithm, each missing shapel[7] in a missing 16×16 shape block is estimated as a weighted median of the 16 neighboring shapels (correctly decoded or concealed) shown in Fig. 9 (a). The weight assigned to each one of the neighboring shapels is adaptively selected based on the likelihood of an edge in that direction. To determine the likelihood of an edge in a given direction, the blocks surrounding the corrupted block are inspected and the number of edges in each of eight possible directions, shown in Fig. 9 (b), is counted. The rationale behind this selection is to weight more the neighboring shapels in the directions along which shapels have a higher tendency to be the same. For each missing shape block, this procedure is iteratively repeated until the algorithm converges. Additionally, if several consecutive missing blocks have to be concealed, the procedure is recursively applied to all the missing shape blocks.

In (Shirani et al. 2000), this technique was tested for relatively simple alpha planes (i.e., CIF versions of Akiyo, Bream and Weather), which have been MPEG-4 encoded with 5 to 10 video packets per VOP. In the shown results, some of these video packets were corrupted, corresponding to a percentage of lost shape blocks between 25 and 30%. Although this technique is able to conceal most of the fully opaque or transparent lost blocks with hardly any visible differences with respect to the original blocks, in most other cases (i.e., border blocks) it can lead to rather poor (subjective and objective) concealment results. This happens because the shapels of a given missing shape block are concealed based on the local shape data statistics of the surrounding

[7] A shapel is an element of the alpha plane; its name results from the contraction of the words *shape* and *element*. In a binary alpha plane, the shapels can be either opaque, if they are inside the object, or transparent, if they are outside.

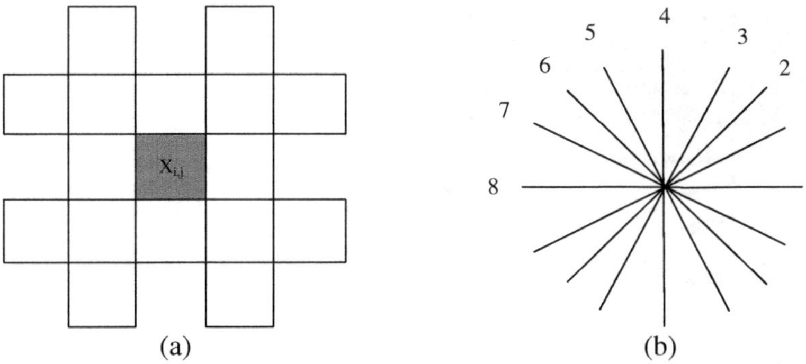

Fig. 9. (a) The 16 neighboring shapels used to estimate the value of the lost shapel (in gray) and (b) The 8 possible edge directions used to determine the weight associated with each of the 16 neighboring shapels.

shape blocks, which is an adequate solution for isolated lost blocks. However, when several consecutive lost blocks have to be concealed, which actually happens much more often than isolated blocks, many of the lost blocks are going to be concealed based on the statistics of surrounding blocks that have already been concealed themselves. Therefore, when many consecutive lost blocks exist, this technique tends to include severe error propagation. As later suggested in (Soares and Pereira 2004c), this can be solved by using the correctly decoded object contour to perform the shape concealment, instead of using such a block-based approach. This way, when many consecutive shape blocks have been lost, they can be concealed as a whole by considering the available correctly decoded contour, thus avoiding the error propagation and leading to better results. Another problem of the technique proposed in (Shirani et al. 2000) is that it can be computationally very intensive since it is iterative.

As explained in (Soares and Pereira 2004c), in alpha planes, such as the one shown in Fig. 10, the loss of shape information corresponds to broken contours that have to be interpolated. By interpolating the broken contours, it is possible to recover the complete shape with a good accuracy since most contours are fairly well behaved in natural video scenes, showing a rather slow direction variation. Naturally the quality of the concealment will mainly depend on the validity of this assumption. After the broken contours have been recovered, it should be fairly easy to recover the values of the missing shapels from the neighboring ones by using an adequate continuity criterion, and then filling in the shape. After all, to recover the complete alpha plane it is sufficient to have the contours and the value of a single shapel (either transparent or opaque).

The block diagram for the shape concealment technique proposed in (Soares and Pereira 2004c) is presented in Fig. 11; the input is a corrupted alpha plane with several lost blocks, typically arranged in bursts (see Fig. 12). The correctly decoded blocks in the same alpha plane will be used to extract useful information for the concealment process, whose output is a fully concealed alpha plane.

To help understand this technique, an illustrative example will be given. As explained above, the input of the concealment chain depicted in Fig. 11 is an alpha plane

Fig. 10. Cyclamen alpha plane – (a) Original; (b) Corresponding extracted contour (Soares and Pereira 2004c) (© 2004 IEEE).

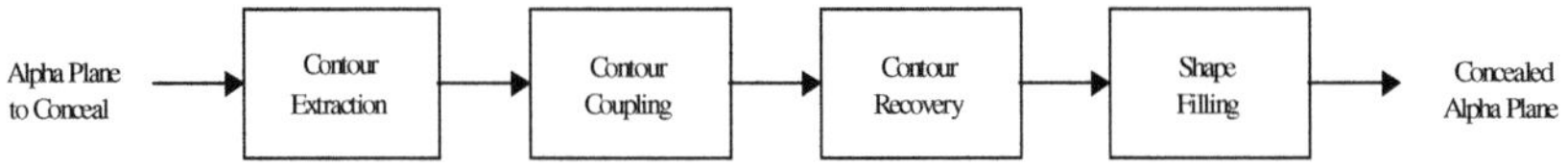

Fig. 11. Shape concealment process (Soares and Pereira 2004c) (© 2004 IEEE).

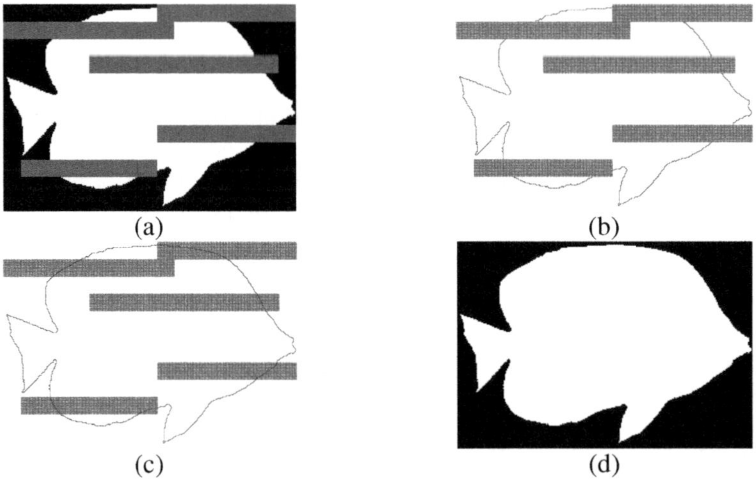

Fig. 12. Exemplifying the steps of the shape concealment process: (a) Lost blocks surrounded by the available alpha plane blocks; (b) Lost blocks surrounded by the available contours; (c) Interpolated contours inside the lost blocks; (d) Recovered alpha plane (Soares and Pereira 2004c) (© 2004 IEEE).

where some blocks have been lost (i.e., the lost area), as illustrated in Fig. 12 (a). The lost area is shown in gray throughout the example. The first step is to extract the contour of the corrupted alpha plane, which is illustrated in Fig. 12 (b), where the broken contours can be clearly seen. After that, the broken contour endings have to be coupled and then interpolated (with Bézier curves) in order to recover the contour, as shown in Fig. 12 (c). Finally, after the contour has been fully recovered, the shape can be filled in, as shown in Fig. 12 (d). Since the contour endings around a given lost area cannot be coupled with the contour endings of another lost area, the shape concealment processing of each lost area can and will be done independently.

Simultaneously, in (Schuster et al. 2004), the authors have proposed a very similar approach, the main difference being the fact that Hermite splines were used instead of Bézier curves. The achieved results were basically very similar to the ones of (Soares and Pereira 2004c).

In terms of texture data concealment, only one specific technique for object-based video is known. The existing frame-based techniques, such as those in (Wang et al. 1993, Hemami and Meng 1995, Aign and Fazel 1995), cannot be directly used because they are not able to deal with the new problems created by the fact that the visible texture data no longer corresponds to a simple rectangle of pixels and is now dictated by the shape data. However, the technique proposed in (Soares and Pereira 2005) is usable in object-based image and video coding systems. This technique, which is the only one available in the literature for object-based systems, consists of two steps. In the first step, padding is applied to the available texture. This is followed by the second step, where the lost pixel values are estimated; in order to use existing frame-based techniques for this, they would have to be modified, namely in terms of the surrounding blocks that are used. For the pixel value estimation step, two different approaches were proposed, for which results have been presented, showing their ability to recover lost texture data in a quite acceptable way. The first approach is based on the frame-based technique proposed in (Hemami and Meng 1995), where the missing transform coefficients in a block are recovered by interpolating the corresponding transform coefficients from the spatially adjacent blocks. In (Soares and Pereira 2005), however, the concealment is based on the closest available blocks to each side, which are not necessarily spatially adjacent to the block being concealed. The second approach uses the weighted median instead of the linear interpolation of the missing transform coefficients. This is done because, if one of the blocks used for the interpolation is significantly different from the rest, the results can be seriously negatively affected. The weighted median, however, has the ability to remove outliers and, therefore, will provide better results is many situations.

4.3.2 Object-Level Temporal Error Concealment

The shape error concealment techniques described in this section are temporal (or Inter) techniques, in the sense that they rely on information from temporal instants other than the current one. Since, in most video object sequences, the shape and texture data does not change that much in consecutive time instants, these techniques are typically able to achieve better concealment results than techniques that rely solely on the information from the current time instant (i.e., spatial techniques).

In terms of temporal concealment techniques, the first technique to appear for object-based video was a simple extension of the motion compensated error concealment used in frame-based systems, which simply corresponds to replacing a corrupted (texture) MB with the MB specified by the motion vector of the damaged block. However, since in object-based video both the shape and texture data can have motion vectors associated to it, this technique can be used for the texture data, as well as for the shape data. To employ this approach in object-based video, it is simply a matter of extrapolating the technique. This way, when errors occur in the shape data, the

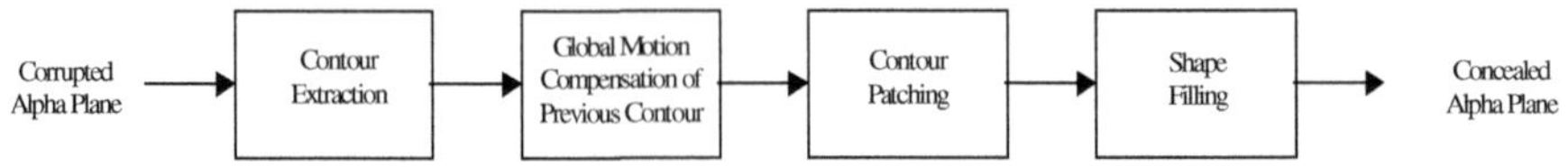

Fig. 13. Different modules of the global motion compensation shape concealment method.

decoder has to use the shape motion vectors (correctly decoded or estimated based on the correctly decoded information) and copy the indicated areas from the alpha plane in the previous time instant. For the texture data, the same procedure is followed but the texture motion vectors are used to copy the indicated texture area from the previous time instant. The first time this approach was used in object-based video was during the MPEG error resilience core experiments (Brady and Soares 1997) while developing the MPEG-4 standard.

Later on, more advanced techniques started to appear in the literature, such as the one proposed in (Salama and Huang 2002) for shape error concealment, which is based on a global motion compensation concealment approach. The technique described in (Salama and Huang 2002) assumes that the shape of a given video object, although it can move in time, is rather rigid, meaning that it can only suffer very little deformation in consecutive time instants. This way, the motion of the object shape can be fairly well described by one of the more common global motion models; this model is also used to conceal the current corrupted shape based on the previous shape.

In (Salama and Huang 2002), since block-based shape coding is assumed, bitstream errors manifest themselves as lost blocks (or blocks that have been declared lost since the decoded information was not considered reliable enough). Therefore, when corrupted, the decoded image will have several missing shape blocks, which have to be concealed. To conceal these lost blocks, the technique proposed in (Salama and Huang 2002) can be divided into four sequential tasks or steps that have to be performed by the decoder, as shown in Fig. 13.

The four steps that have to be followed in order to conceal the corrupted alpha plane are described next:

- **Contour extraction** – The first step to conceal the lost blocks in the alpha plane is to extract the contour from the corrupted shape data. An example of a broken contour, extracted from the shape of the Bream video object, is shown in Fig. 14 (a), where two separate broken contour segments have been identified.
- **Global motion compensation of previous contour** – The second step is to take the shape from the previous time instant (correctly decoded or already concealed), extract its closed contour and motion compensate it with the global motion parameters corresponding to the current time instant, which have been computed at the encoder and sent to the decoder. The obtained motion compensated contour will be referred to as the *reference contour* from now on.
- **Contour patching** – The third step corresponds to patching the broken contour with the available information from the reference contour. To do this, the decoder simply copies from the reference contour (i.e., the previous motion compensated contour) to the corrupted contour the missing segments, as illustrated in Fig. 14 (b), thus recovering the whole contour.

- **Shape filling** – Finally, the fourth step, which corresponds to recovering the shape from the contour, it is simply a matter of filling in the shapels while taking into account the recovered contour.

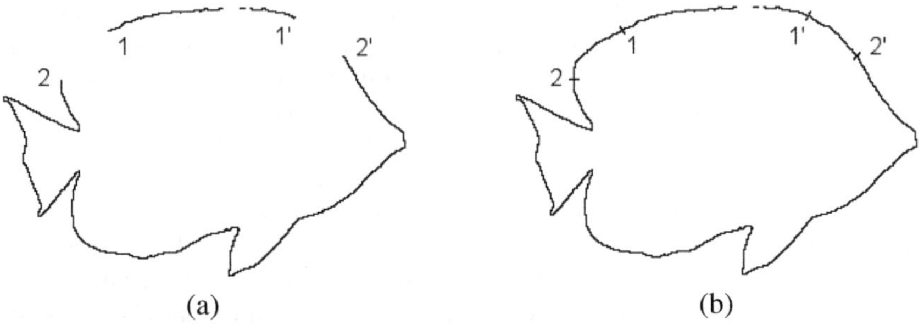

Fig. 14. Illustration of the global motion compensation method to recover the object contour: (a) Broken contour extracted from a corrupted shape of the Bream object; (b) Reference contour with overlapped end points from the corrupted contour.

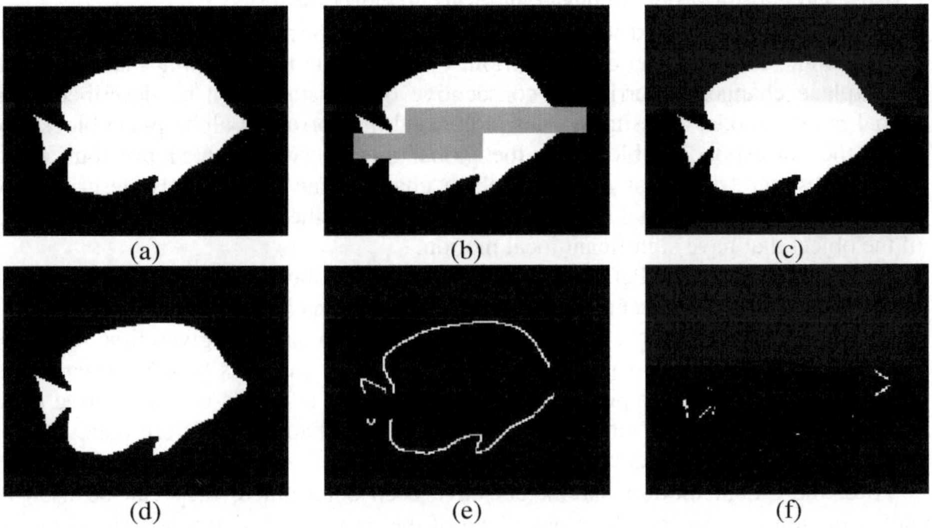

Fig. 15. Shape error concealment by global motion compensation as in (Salama and Huang 2002) - (a) Original shape; (b) Corrupted shape; (c) Reference shape; (d) Recovered shape; (e) Contour of the corrupted shape; (f) Differences between original and recovered shapes.

With this technique, the authors are able to achieve rather good results for well-behaved video object sequences in terms of global motion, such as the Bream sequence, for which results are shown in Fig. 15. These results correspond to the QCIF version of the referred video object sequence, MPEG-4 encoded (with lossless shape coding) at 122.4 kbps, using video packet sizes of 384 bytes. These video packets were then packed into ATM cells that were transmitted and subjected to a random packet loss of 3%. Under these conditions, the differences between the original shape shown in Fig. 15 (b) and the concealed shape shown in Fig. 15 (d) are very hard to spot by the naked eye, which can be explained by noticing the very little differences between the two shapes shown in Fig. 15 (f).

The main drawback of this technique is that the global motion parameters are estimated at the encoder, when the sequence is being encoded. Since these parameters are needed at the decoder for the concealment itself, they must be transmitted somehow. To do so, a separate stream, called USER_DATA stream in the MPEG-4 nomenclature (ISO/IEC 14496-2 1999), was used in (Salama and Huang 2002); this stream represents a 5% increase in terms of bit rate. This can be a serious limitation because this concealment technique can then only be used if both the encoder and decoder support it, which is very unlikely in such a competition-driven market with many manufacturers (and remind that concealment is non-normative). In addition, other issues would have to be considered, such as how to synchronize this new USER_DATA stream with the visual data stream itself and what should be done if this stream is corrupted. In this context, a concealment technique able to provide similar concealment capabilities without having to rely on specific processing and information received from the encoder would be highly desirable since it would be more generic and efficient. This is precisely what is done in the techniques proposed in (Soares and Pereira 2006b, Schuster and Katsaggelos 2006).

The technique proposed in (Soares and Pereira 2006b) is based on a combination of global and local motion compensation. First, it starts by assuming that the shape alpha plane changes occurring in consecutive time instants can be described by a global motion model and simply tries to conceal the corrupted alpha plane blocks by using the corresponding blocks in the global motion compensated previous alpha plane. However, since not all alpha plane changes can be perfectly described by global motion, an additional local motion refinement is then applied to deal with areas of the object that have significant local motion.

Since, in (Soares and Pereira 2006b), the global motion parameters are not computed at the encoder and sent to the decoder side, they have to be locally computed at the decoder. This is possible because the decoded video data at a given time instant is usually not completely corrupted, only some parts of it are. This happens because, as mentioned above, errors typically manifest themselves as bursts of consecutive erroneous blocks. With the remaining correctly decoded shape and texture data, the decoder can extract the global motion parameters.

After the global motion parameters have been determined, they can be used to global motion compensate the alpha plane of the previous time instant. The obtained global motion compensated alpha plane is considered as the reference alpha plane and it will be used to perform the concealment. This way, the erroneous shape blocks in the corrupted alpha plane being concealed can be simply replaced by the co-located shape blocks in the reference alpha plane. Assuming that the global motion model can accurately describe the shape motion, this alone should be able to produce very good results. However, in many cases, the shape motion cannot be accurately described only by global motion, due to the existence of local motion in some areas of the (non-rigid) shape. Therefore, to avoid significant differences when concealing erroneous blocks in areas with local motion, an additional local motion refinement scheme has been introduced. In this scheme, the available blocks surrounding an erroneous shape

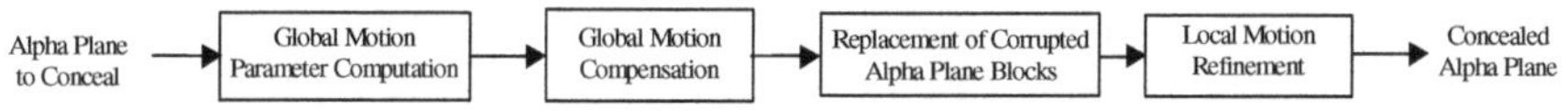

Fig. 16. Temporal shape concealment process.

block are used to determine if any local motion exists; if so, a local motion vector to be used to find a better replacement block in the previous alpha plane for the erroneous shape block is estimated (in this case, the global motion compensation will not be used, which means that the local motion supersedes the global motion).

The block diagram for the proposed temporal shape concealment technique is presented in Fig. 16, where each block corresponds to one of the four consecutive steps in the concealment process.

In order to better understand the shape concealment process, an illustrative example will be given. In this example, the alpha plane in Fig. 17 (a) has been corrupted, as illustrated in Fig. 17 (b). Based on the estimated global motion parameters (using the information in Fig. 17 (b)), the previous VOP is motion compensated and Fig. 17 (c) is obtained. After that, the corrupted blocks are replaced with the corresponding ones in the global motion compensated previous VOP, which gives the concealed alpha plane shown in Fig. 17 (d). In this case, since the shape and its motion are very simple, the global motion model is able to describe the shape motion accurately enough, and thus no refinement would be needed for local motion. However this is not always the case, notably for humans.

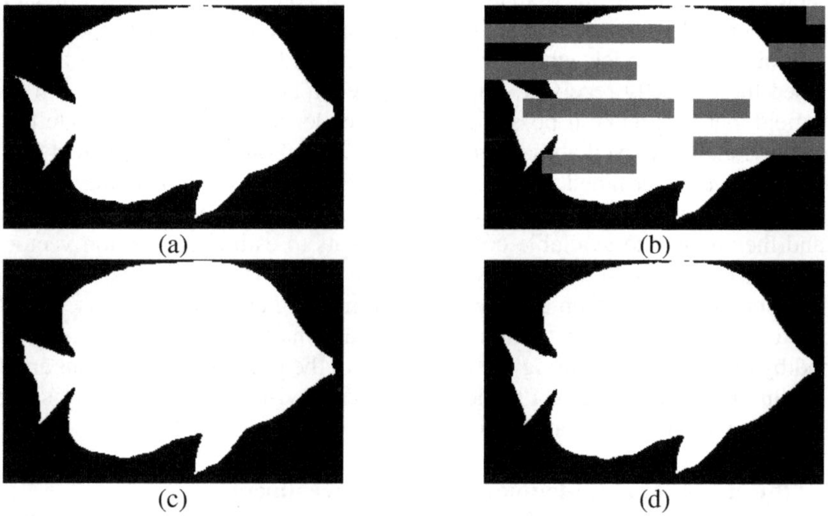

Fig. 17. Exemplifying the steps of the temporal shape concealment process for the Bream video object (CIF, 10 fps): (a) Original uncorrupted alpha plane; (b) Corrupted alpha plane; (c) Motion compensated previous alpha plane; (d) Concealed alpha plane without local motion refinement (Soares and Pereira 2006b) (© 2006 IEEE).

However, in some sequences, the objects may have a lot of local motion in addition to the global motion, as is the case in Fig. 18. As can be seen from Fig. 18 (d), by simply replacing the corrupted shape blocks with the co-located blocks from the global motion compensated previous alpha plane, this would lead to very annoying artifacts. Therefore, to avoid this, an additional refinement step is needed to deal with the local motion. The results with this refinement are shown in Fig. 18 (e), which no longer exhibits the most annoying artifacts.

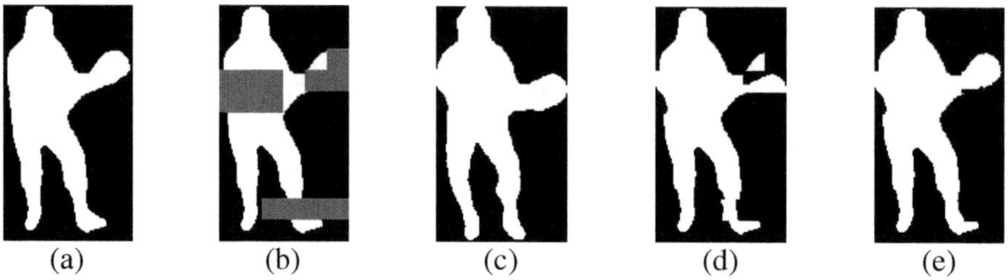

Fig. 18. Exemplifying the steps of the temporal shape concealment process for the Stefan video object (CIF, 15 fps): (a) Original uncorrupted alpha plane; (b) Corrupted alpha plane; (c) Motion compensated previous alpha plane; (d) Concealed alpha plane without local motion refinement; (e) Concealed alpha plane with local motion refinement (Soares and Pereira 2006b) (© 2006 IEEE).

In (Schuster and Katsaggelos 2006), the authors have followed a slightly different approach to conceal corrupted shape data from the one used in (Soares and Pereira 2006b). Instead of using the correctly received shape data to estimate the parameters of a global motion model, which will then be refined for local motion, the authors have used the correctly received shape data to estimate a maximally smooth motion vector field that will make it possible for the decoder to motion compensate the contour of previously received alpha plane into the current one being concealed (i.e., one motion vector is determined for each point of the object contour being concealed). This is done by first extracting the broken contour from the received corrupted shape data and then using the available contour segments to estimate a motion vector field with respect to the previously received object contour. Based on this motion vector field, motion vectors are then estimated for the missing contour segments, in order to achieve a maximally smooth motion vector field. Finally, the broken contour can be patched by motion compensating the contour from the previous time instant and filled in to obtain the concealed alpha plane. The obtained visual results are very similar to the ones achieved in (Soares and Pereira 2006b).

4.3.3 Object-Level Spatio-temporal Error Concealment

The main problem with the error concealment techniques described in the two previous sections is that they cannot be used with acceptable results for all possible situations. On one hand, spatial error concealment techniques, which only use spatially adjacent shape and texture information from the same time instant to perform the concealment, are especially useful when the shape and texture change greatly in consecutive time instants, such as when new objects appear or are uncovered. On the other hand, temporal error concealment techniques, which rely only on shape and texture information from other time instants to perform the concealment, are typically able to achieve better concealment results than spatial techniques when the shape and texture data does not change that much in consecutive time instants. Therefore, by considering information from both the current time instant and other time instants, it should be possible to improve the error concealment results even further.

The first technique to consider this approach appeared in (Frater et al. 1998) and corresponds to an extension of the simple motion compensated concealment presented in (Brady and Soares 1997) for object-based video. In (Frater et al. 1998), the authors consider the concealment of corrupted shape and texture data in object-based video coding systems, when the associated motion data has also been corrupted. Since the motion data is corrupted, the authors propose to locally estimate it based on the surrounding available data. This way, for each corrupted MB, the decoder starts by using the correctly decoded shape data in surrounding MBs to determine a motion vector for the shape. Then, the correctly decoded texture data in surrounding MBs is used to determine a motion vector for the texture. After these two motion vectors have been estimated, the concealment itself can be performed as in (Brady and Soares 1997). The shape motion vector is used to copy the indicated area from the alpha plane in the previous time instant, while the texture motion vector is used to copy the indicated texture area from the previous time instant.

Later, in (Soares and Pereira 2004b), the authors proposed a scheme that adaptively selects one of two shape concealment techniques depending on the shape characteristics: one technique is a spatial technique, while the other is a temporal one. By doing so, the authors managed to obtain the advantages of both solutions while compensating for their disadvantages and, thus, improve the final concealed shape quality. The block diagram for the adaptive spatio-temporal shape concealment technique proposed in (Soares and Pereira 2004b) is presented in Fig. 19. The first module is responsible for determining which corrupted areas in the corrupted alpha plane are to be concealed with each type of concealment technique, i.e., spatial or temporal. Afterwards, the concealment itself is performed according to the selected approach, as described in the previous two sections.

Experiments have shown that the decision of which concealment technique should be applied to a given corrupted area should be based on the length of the longest contour segment that has been lost inside the corrupted area in question and has to be

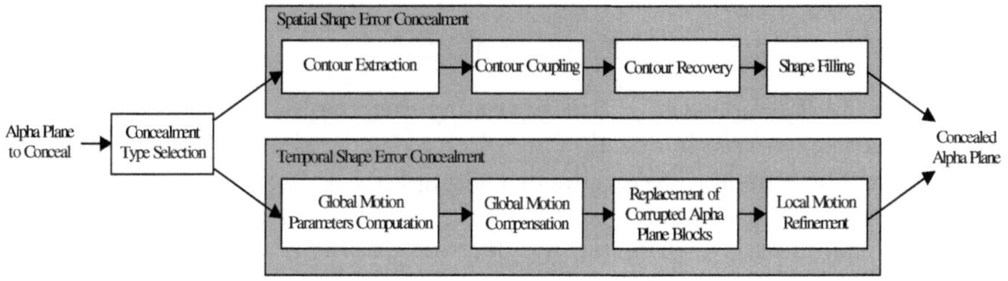

Fig. 19. Adaptive spatio-temporal shape concealment process.

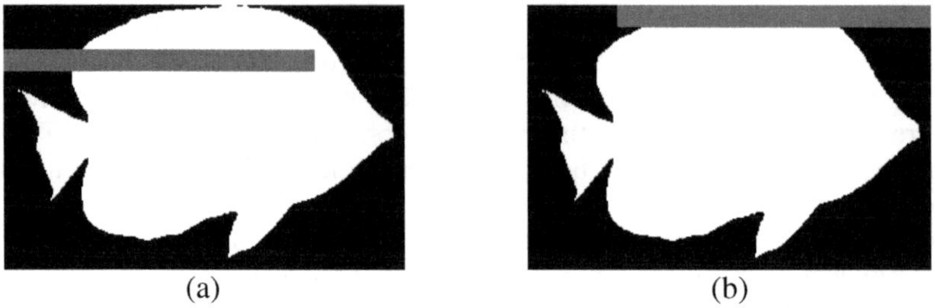

Fig. 20. Examples of corrupted alpha planes where the corrupted areas have the same size, but the length of the corrupted contour segment is very different – (a) Short corrupted contour segment; (b) Long corrupted contour segment (Soares and Pereira 2004b).

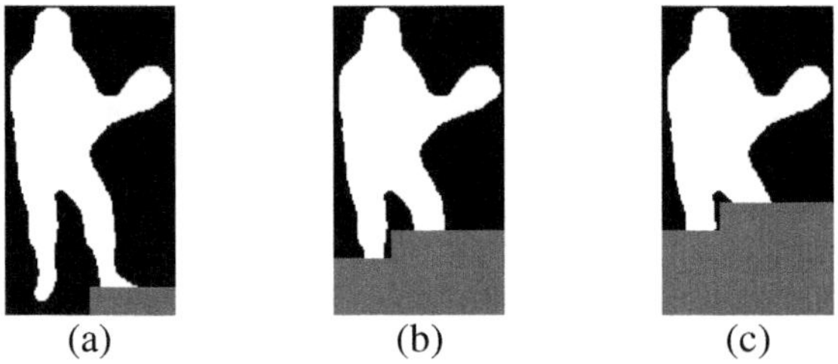

Fig. 21. Examples of corrupted alpha planes where the distance separating the coupled contour endings is similar, but the size of the corrupted areas is very different – (a) Small corrupted area; (b) Large corrupted area; (c) Larger corrupted area (Soares and Pereira 2004b).

concealed. The problem with this approach is that this length cannot be determined since the contour has been lost in the corrupted area. However, it can be estimated by considering the factors on which it depends. The first factor on which the length of the corrupted contour segment depends is clearly the distance that separates the two corresponding contour endings, as it is clearly illustrated in Fig. 20. As for the second factor, it is the size of the corrupted area itself. This happens because large corrupted areas may hide additional important information, which cannot be simply inferred from the contour endings alone and the distance that separates them, as is shown in Fig. 21.

Based on these two factors, a selection metric S was defined as:

$$S = d_{\max} \times A_{corrupted} \tag{22}$$

where d_{max} is the (Euclidean) distance separating the coupled contour endings farthest apart in the corrupted area in question and $A_{corrupted}$ is the size of the corrupted area in shape blocks. Since spatial concealment techniques can typically achieve better results when the corrupted contour is short while for long corrupted contours temporal concealment techniques work better, the decision of which shape concealment technique should be used for the best concealment of a given corrupted area is done by comparing the S metric with an experimentally determined threshold S_{th}, according to the following rule:

$$\begin{cases} \text{if } S \le S_{th}, & \text{use the proposed spatial shape concealment} \\ \text{if } S > S_{th}, & \text{use the proposed temporal shape concealment} \end{cases} \tag{23}$$

By using this adaptive spatio-temporal concealment technique, the shape concealment results can be significantly improved, as can be easily seen Fig. 22. In this particular example, when the adaptive concealment technique is used, the racket area is concealed with the spatial concealment technique, whereas Stefan's feet are concealed with the temporal concealment technique.

(a) (b) (c) (d) (e)

Fig. 22. Examples of concealed alpha planes for the Stefan video object – (a) Uncorrupted original alpha plane; (b) Corrupted alpha plane; (c) Alpha plane concealed with the spatial concealment technique; (d) Alpha plane concealed with the temporal concealment technique; (e) Alpha plane concealed with the adaptive spatio-temporal concealment (Soares and Pereira 2004b).

4.3.4 Scene-Level Error Concealment Techniques

In the previous sections, several object-based error concealment techniques, dealing both with shape and texture data, have been described. In these techniques, each video object was independently considered and how the objects fit in the video scene was never taken into account. This represents a serious limitation because the fact that a concealed video object has a pleasing subjective impact on the user, when it is considered on its own, does not necessarily mean that the subjective impact of the whole scene, when the objects are put together, will be acceptable. An example of this situation is given in Fig. 23, where a hole has appeared as a result of blindly composing the scene by using two independently concealed video objects.

(a) (b)

Fig. 23. Illustration of a typical scene concealment problem: (a) Original video scene; (b) Composition of two independently error concealed video objects (Soares and Pereira 2006c) (© 2006 IEEE).

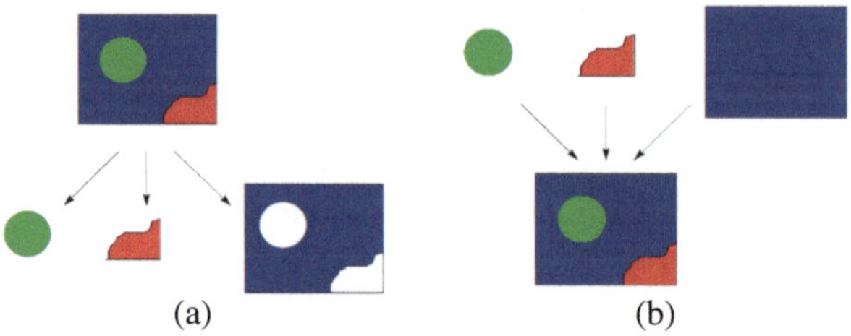

Fig. 24. Two different scene types: (a) Segmented scene; (b) Composed scene (Soares and Pereira 2006c) (© 2006 IEEE).

When concealing a complete video scene, the way the scene was created has to be considered since this will imply different problems and solutions in terms of error concealment. As shown in Fig. 24, the video objects in a scene can be defined either by segmentation of an existing video sequence (segmented scene), in which case all shapes have to fit perfectly together, or by composition of pre-existing video objects whose shapes do not necessarily have to perfectly fit together (composed scene). Additionally, it is also possible to use a combination of both approaches.

For the work described here, segmented video scenes (or the segmented parts of hybrid scenes) are considered since the concealment of composed scenes can typically be limited to object-level concealment.

Scene-level concealment techniques are intrinsically spatial techniques since, to conceal a given video object in the scene, they consider the adjacent objects in that same scene. This explains why only spatial techniques and spatio-temporal techniques have been found in the literature.

The first known scene-level error concealment technique to appear in the literature is the spatial shape error concealment technique presented in (Soares and Pereira 2006a). In this technique, where only the information from the same time instant is used, the fact that the existing video objects have to perfectly fit in together like the pieces of a jigsaw puzzle is exploited to conceal errors in the binary shape data. In many cases, it is possible to conceal the corrupted shape in a given corrupted video object by considering the uncorrupted complementary shape data from surrounding objects. For those parts of the corrupted shape for which complementary data is not available because it is corrupted, concealment will be much harder. In those cases, the various objects involved will be individually concealed with one shape concealment technique, such as the one from (Soares and Pereira 2004c), and a refinement is then applied to guarantee that the objects do not overlap and no holes exist between them.

Later, in (Soares and Pereira 2006c), texture concealment was also added to the technique described in (Soares and Pereira 2006a). This way, in (Soares and Pereira 2006a), the shape is concealed in basically the same way as before, but now the concealed shape data is also used to estimate the motion associated with the corrupted video object being concealed. The determined motion is then used to conceal the corrupted texture of the video object, as in (Soares and Pereira 2006b). Since this technique relies on information from the same time instant, as well as from other time instants, it can be considered a spatio-temporal error concealment technique.

5 Final Remarks

This chapter has reviewed the state-of-the-art on rate control and error resilience for object-based video coding systems. A brief description of the basic concepts behind the object-based video coding approach was also included at the beginning to help readers better understand the new problems that have to be dealt with when proposing rate control and error resilience techniques for object-based video coding systems.

In terms of rate control, after presenting its basic objectives, the new rate-control dimensions and strategies were analyzed and a generic object-based rate-control architecture has been presented. Afterwards, the state-of-the-art object-based rate control techniques for low-delay constant bit rate object-based video coding were reviewed. For these encoding scenarios, it is very important to deal with the deviations between the ideal and actual encoding results. Therefore, compensation and adaptation mechanisms play a very important role in object-based rate control to efficiently encode multiple video object content.

In terms of error resilience, some of the most important techniques currently available in the literature for both the encoder and decoder sides of the communication chain have been described. Although these techniques can be independently used, it is important that several of the complementary error resilience techniques be combined at both sides of the communication chain. This will globally increase the error resilience of the whole system and, thus, make it possible to deploy object-based video services over error-prone environments, such as mobile networks or the Internet, with an acceptable video quality.

Since real world networks, where object-based video services are expected to be used in the future, can have very critical bandwidth and error characteristics, it is clear that both rate control and error resilience techniques will be necessary to achieve the best possible decoded video quality. The techniques that have been described in this chapter correspond to some of the most important ones and, therefore, are good candidates to be used in future object-based video communication systems.

References

Aign, S., Fazel, K.: Temporal & spatial error concealment techniques for hierarchical MPEG 2 video codec. In: Proc of the IEEE International Conference on Communications (ICC), Seattle, WA, USA, vol. 3, pp. 1778–1783 (1995)

Åström, K., Wittenmark, B.: Adaptive control, 2nd edn. Addison-Wesley, Reading (1995)

Brady, N., Soares, L.D.: Error resilience of arbitrarily shaped VOs (CE E14). ISO/IEC JTC1/SC29/WG11 M2370. Stockholm MPEG meeting (1997)

CCITT SGXV, Description of reference model 8 (RM8). Doc. 525 (1989)

Chen, Z., Han, J., Ngan, K.N.: Dynamic bit allocation for multiple video object coding. IEEE Transactions on Multimedia 8, 1117–1124 (2006)

Chiang, T., Zhang, Y.-Q.: A new rate control scheme using quadratic rate distortion model. IEEE Transactions on Circuits and Systems for Video Technology 7, 246–250 (1997)

Cover, T., Thomas, J.: Elements of Information Theory. John Wiley & Sons, New York (1991)

Frater, M.R., Lee, W.S., Pickering, M., Arnold, J.F.: Error concealment of arbitrarily shaped video objects. In: Proc. of the IEEE International Conference on Image Processing (ICIP), Chicago, IL, USA, vol. 3, pp. 507–511 (1998)

Girod, B., Aaron, A., Rane, S., Rebollo-Monedero, D.: Distributed video coding. Proceedings of the IEEE 93, 71–83 (2005)
Gormish, M., Gill, J.: Computation-rate-distortion in transform coders for image compression. In: Proc. of the SPIE: Image and Video Processing, San Jose, CA, USA, vol. 1903, pp. 146–152 (1993)
He, Z., Kim, Y.K., Mitra, S.: ρ-domain source modeling and rate control for video coding and transmission. In: Proc. of the IEEE International Conference on Acoustics, Speech, and Signal Processing (ICASSP), Salt Lake City, UT, USA, vol. 3, pp. 1773–1776 (2001)
He, Z., Mitra, S.: A unified rate-distortion analysis framework for transform coding. IEEE Transactions on Circuits and Systems for Video Technology 11, 1221–1236 (2001)
He, Z., Mitra, S.: A linear source model and a unified rate control algorithm for DCT video coding. IEEE Transactions on Circuits and Systems for Video Technology 12, 970–982 (2002)
He, Z., Kim, Y.K., Mitra, S.: Object-level bit allocation and scalable rate control for MPEG-4 video coding. In: Proc. of the Workshop and Exhibition on MPEG-4 (WEMP), San Jose, CA, USA, pp. 63–66 (2001)
He, Z., Mitra, S.: Optimum bit allocation and accurate rate control for video coding via-domain source modeling. IEEE Transactions on Circuits and Systems for Video Technology 12, 840–849 (2002)
Hemami, S., Meng, T.: Transform coded image reconstruction exploiting interblock correlation. IEEE Transactions on Image Processing 4, 1023–1027 (1995)
ISO/IEC 11172-2, Information technology coding of moving pictures and associated audio for digital storage media at up to about 1,5 Mbit/s, part 2: video (1993)
ISO/IEC 13818-2, Information technology generic coding of moving pictures and associated audio information – part 2: video (1996)
ISO/IEC 14496, Information technology – coding of audio-visual objects (1999)
ISO/IEC 14496-2, Information technology – coding of audio-visual objects – part 2: visual, 3rd edn. (2004)
ITU-R BT.601-1, Encoding parameters of digital television for studios (1986)
ITU-T H.261, Video codec for audiovisual services at p 64 kbit/s (1993)
Keesman, G., Shah, I., Klein-Gunnewiek, R.: Bit-rate-control for MPEG encoders. Signal Processing: Image Communication 6, 545–560 (1995)
Kumar, S., Xu, L., Mandal, M.K., Panchanathan, S.: Error resiliency schemes in H.264/AVC standard. Journal of Visual Communication and Image Representation 17, 425–450 (2006)
Lee, H.-J., Chiang, T., Zhang, Y.-Q.: Scalable rate control for very low bit rate (VLBR) video. In: Proc. of the International Conference on Image Processing (ICIP), Santa Barbara, CA, USA, vol. 2, pp. 768–771 (1997)
Lee, H.-J., Chiang, T., Zhang, Y.-Q.: Scalable rate control for MPEG-4 video. IEEE Transactions on Circuits and Systems for Video Technology 10, 878–894 (2000)
Lee, J.-W., Vetro, A., Wang, Y., Ho, Y.-S.: Bit allocation for MPEG-4 video coding with spatio-temporal tradeoffs. IEEE Transactions on Circuits and Systems for Video Technology 13, 488–502 (2003)
MPEG Test Model Editing Committee, MPEG-2 test model 5 (TM5). ISO/IEC JTC1/SC29/WG11 N400, Sydney MPEG meeting (1993)
MPEG Video Group, MPEG-4 video verification model 4.0 (VM4), ISO/IEC JTC1/SC29/WG11 N1380, Chicago MPEG meeting (1996)
MPEG Video Group, MPEG-4 video verification model 5.0 (VM5), ISO/IEC JTC1/SC29/WG11 N1469, Maceió MPEG meeting (1996)

MPEG Video Group, MPEG-4 video verification model 8.0 (VM8). ISO/IEC JTC1/SC29/WG11 N1796, Stockholm MPEG meeting (1997)
Nunes, P.: Rate control for object-based video coding. Ph.D. Thesis. Instituto Superior Técnico, Lisboa, Portugal (2007)
Nunes, P., Pereira, F.: Rate control for scenes with multiple arbitrarily shaped video objects. In: Proc. of the Picture Coding Symposium (PCS), Berlin, Germany, pp. 303–308 (1997)
Nunes, P., Pereira, F.: MPEG-4 compliant video encoding: analysis and rate control strategies. In: Proc. of the 34th ASILOMAR Conference, Pacific Grove, CA, USA (2000)
Nunes, P., Pereira, F.: Scene level rate control algorithm for MPEG-4 video encoding. In: Proc. of the SPIE Visual Communications and Image Processing (VCIP), San Jose, CA, USA, vol. 4310, pp. 194–205 (2001)
Nunes, P., Pereira, F.: Rate and distortion modeling analysis for MPEG-4 video intra coding. In: Proc. of the Workshop on Image Analysis for Multimedia Interactive Services (WIAMIS), Lisboa, Portugal (2004)
Nunes, P., Pereira, F.: Rate and distortion models for MPEG-4 video encoding. In: Proc. of the SPIE 49th Annual Meeting: Applications and Digital Image Processing XXVII, Denver, CO, USA, vol. 5558, pp. 382–394 (2004)
Nunes, P., Pereira, F.: Improved feedback compensation mechanisms for multiple video object encoding rate control. In: Proc. of IEEE International Conference on Image Processing (ICIP), San Antonio, TX, USA (2007)
Nunes, P., Pereira, F.: Joint rate control algorithm for low-delay MPEG-4 object-based video encoding. IEEE Transactions on Circuits and Systems for Video Technology 19, 1274–1288 (2009)
Pereira, F., Ebrahimi, T. (eds.): The MPEG-4 book. Prentice-Hall, Upper Saddle River (2002)
Reibman, A., Haskell, B.: Constraints on variable bit-rate video for ATM networks. IEEE Transactions on Circuits and Systems for Video Technology 2, 361–372 (1992)
Ribas-Corbera, J., Lei, S.: Rate control in DCT video coding for low-delay communications. IEEE Transactions on Circuits and Systems for Video Technology 9, 172–185 (1999)
Richardson, I., Zhao, Y.: Adaptive algorithms for variable-complexity video coding. In: Proc. of the International Conference on Image Processing (ICIP), Thessaloniki, Greece, vol. 1, pp. 457–460 (2001)
Ronda, J., Eckert, M., Jaureguizar, F., García, N.: Rate control and bit allocation for MPEG-4. IEEE Transactions on Circuits and Systems for Video Technology 9, 1243–1258 (1999)
Salama, P., Huang, C.: Error concealment for shape coding. In: Proc. of the IEEE International Conference on Image Processing (ICIP), Rochester, NY, USA, vol. 2, pp. 701–704 (2002)
Schuster, G.M., Li, X., Katsaggelos, A.K.: Shape error concealment using Hermite splines. IEEE Transactions on Image Processing 13, 808–820 (2004)
Schuster, G.M., Katsaggelos, A.K.: Motion compensated shape error concealment. IEEE Transactions on Image Processing 15, 501–510 (2006)
Shirani, S., Erol, B., Kossentini, F.: A concealment method for shape information in MPEG-4 coded video sequences. IEEE Transactions on Multimedia 2, 185–190 (2000)
Soares, L.D., Pereira, F.: Error resilience and concealment performance for MPEG-4 frame-based video coding. Signal Processing: Image Communication 14, 447–472 (1999)
Soares, L.D., Pereira, F.: Refreshment need metrics for improved shape and texture object-based resilient video coding. IEEE Transactions on Image Processing 12, 328–340 (2003)
Soares, L.D., Pereira, F.: Adaptive shape and texture intra refreshment schemes for improved error resilience in object-based video. IEEE Transactions on Image Processing 13, 662–676 (2004)

Soares, L.D., Pereira, F.: Combining space and time processing for shape error concealment. In: Proc. of the Picture Coding Symposium (PCS), San Francisco, CA, USA (2004)
Soares, L.D., Pereira, F.: Spatial shape error concealment for object-based image and video coding. IEEE Transactions on Image Processing 13, 586–599 (2004)
Soares, L.D., Pereira, F.: Spatial texture error concealment for object-based image and video coding. In: Proc. of the EURASIP Conference focused on Signal and Image Processing, Multimedia Communications and Services (ECSIPM), Smolenice, Slovak Republic (2005)
Soares, L.D., Pereira, F.: Spatial scene level shape error concealment for segmented video. In: Proc. of the Picture Coding Symposium (PCS), Beijing, China (2006)
Soares, L.D., Pereira, F.: Temporal shape error concealment by global motion compensation with local refinement. IEEE Transactions on Image Processing 15, 1331–1348 (2006)
Soares, L.D., Pereira, F.: Spatio-temporal scene level error concealment for shape and texture data in segmented video content. In: Proc. of the IEEE International Conference on Image Processing (ICIP), Atlanta, GA, USA, pp. 2242–2252 (2006)
Sun, Y., Ahmad, I.: A robust and adaptive rate control algorithm for object-based video coding. IEEE Transactions on Circuits and Systems for Video Technology 14, 1167–1182 (2004)
Sun, Y., Ahmad, I.: Asynchronous rate control for multi-object videos. IEEE Transactions on Circuits and Systems for Video Technology 15, 1007–1018 (2005)
Unicode Consortium, The Unicode standard, Version 3.0. Addison-Wesley (2000)
Valentim, J., Nunes, P., Pereira, F.: An alternative complexity model for the MPEG-4 video verifier mechanism. In: Proc. of the International Conference on Image Processing (ICIP), Thessaloniki, Greece, vol. 1, pp. 461–464 (2001)
Valentim, J., Nunes, P., Pereira, F.: Evaluating MPEG-4 video decoding complexity for an alternative video complexity verifier model. IEEE Transactions on Circuits and Systems for Video Technology 12, 1034–1044 (2002)
Vetro, A., Sun, H., Wang, Y.: Joint shape and texture rate control for MPEG-4 encoders. In: Proc. of the IEEE International Symposium on Circuits and Systems (ISCAS), Monterey, CA, USA, vol. 5, pp. 285–288 (1998)
Vetro, A., Sun, H., Wang, Y.: MPEG-4 rate control for multiple video objects. IEEE Transactions on Circuits and Systems for Video Technology 9, 186–199 (1999)
Wang, Y., Zhu, Q.-F., Shaw, L.: Maximally smooth image recovery in transform coding. IEEE Transactions on Communications 41, 1544–1551 (1993)
Wang, Y., Zhu, Q.-F.: Error control and concealment for video communications: a review. Proceedings of the IEEE 86, 974–997 (1998)
Wang, Y., Wenger, S., Wen, J., Katsaggelos, A.K.: Error resilient video coding techniques. IEEE Signal Processing Magazine 17, 61–82 (2000)
Zhao, Y., Richardson, I.: Complexity management of video encoders. In: Proc. of the 10th ACM International Conference on Multimedia, Juan-les-Pins, France, pp. 647–649 (2002)

Representation and Coding Formats for Stereo and Multiview Video

Anthony Vetro

Mitsubishi Electric Research Laboratories, USA
avetro@merl.com

Summary. This chapter discusses the various representation and coding formats for stereo and multiview video in the context of next-generation 3D video services. Several application scenarios are discussed including packaged media such as Blu-ray Disc, as well as the delivery over cable, terrestrial and Internet channels. The various types of 3D displays are also described and the data requirements for each examined. A review of different representation formats for stereo and multiview video is given including multiplexed formats, full-channel formats and depth-based formats. The corresponding compression technology for these formats is then described. The chapter concludes with a discussion of future outlooks and research challenges.

1 Introduction

Delivering higher resolution video and providing an immersive multimedia experience to the home has been a primary target for industry and researchers in recent years. Due to advances in display technology, signal processing and circuit design, the ability to offer a compelling 3D video experience on consumer electronics platforms has become feasible. In addition to advances on the display side, there has also been a notable increase in the production of 3D contents. The number of title releases has been steadily growing each year; a number of major studios have announced all future releases in 3D. There are also substantial investments being made in digital cinema theaters with 3D capability. The push from both production and display side have played a significant role in fuelling a renewed interest in 3D video.

There are a number of challenges to make 3D video a practical and sustainable service that consumers will enjoy. For instance, it will be essential to determine an appropriate data format for delivery under different constraints. Interoperability among various devices will be essential. It will also be critical to ensure that the consumer has a high quality experience and is not turned off by viewing discomfort or fatigue, or the need to wear special glasses.

This chapter discusses the various representation and coding formats for stereo and multiview video that are available or being studied, which could be used

C.W. Chen et al. (Eds.): Intel. Multimedia Communication: Tech. and Appli., SCI 280, pp. 51–73.
springerlink.com

to drive next-generation 3D video services. Several application domains and distribution environments are discussed including packaged media such as Blu-ray Disc, as well as the delivery over cable, terrestrial and Internet channels. The various types of 3D displays are also described and the data requirements for each examined. A review of different representation formats for stereo and multiview video is given including multiplexed formats, full channel formats and depth-based formats. The corresponding compression technology for these formats is then described with particular focus on the recently finalized extension of the H.264/MPEG-4 Advanced Video Coding (AVC) standard [1] on multiview video coding. Future outlooks and research challenges are also discussed.

2 Application Domains

A diagram illustrating a system level view of the end-to-end delivery chain for 3D video, including production, distribution and consumption, is shown in Figure 1. Each of these domains will be discussed further in the following subsections.

2.1 Production

The main approaches to creating 3D content include camera capture, computer generated, and conversion from 2D video. Most 3D video that is captured use stereo cameras. While some multi-camera setups exist, large camera arrays coupled with the increased amount of data currently prohibit widespread use of multi-camera capture. Computer generated content is much easier to produce since scene information such as depth is an inherent part of production process, and the depth information enables great flexibility in editing and repurposing of the content. Yet another way to create 3D video is by taking conventional 2D video and adding depth information. The typical process is to identify a series of objects from the 2D image, assigning relative depth to each object, then fill in occluded areas. The creation of a depth map from 2D allows for the creation of multiple views through a rendering process.

An important element of the production domain is a master format. Whether the content is a 3D cinema production or a live event, the master format specifies a common image format along with high level metadata that are required to make sense of the data and prepare the data for distribution. The format is generally independent of any specific delivery channel.

In August 2008, the Society of Motion Picture and Television Engineers (SMPTE) formed a task force to investigate the production requirements to realize 3D video to the home. After a 6-month study, the final report of the task force recommended standardization of a *3D Home Master* which would essentially be an uncompressed and high-definition stereo image format, i.e., 1920×1080 pixel resolution at 60Hz per eye [2]. The mastering format will also specify metadata, e.g., signaling of left and right image frames, as well as scene information such as the maximum and minimum depth of a scene. The master format is also expected to include provisions to associate supplementary data such as pixel-level

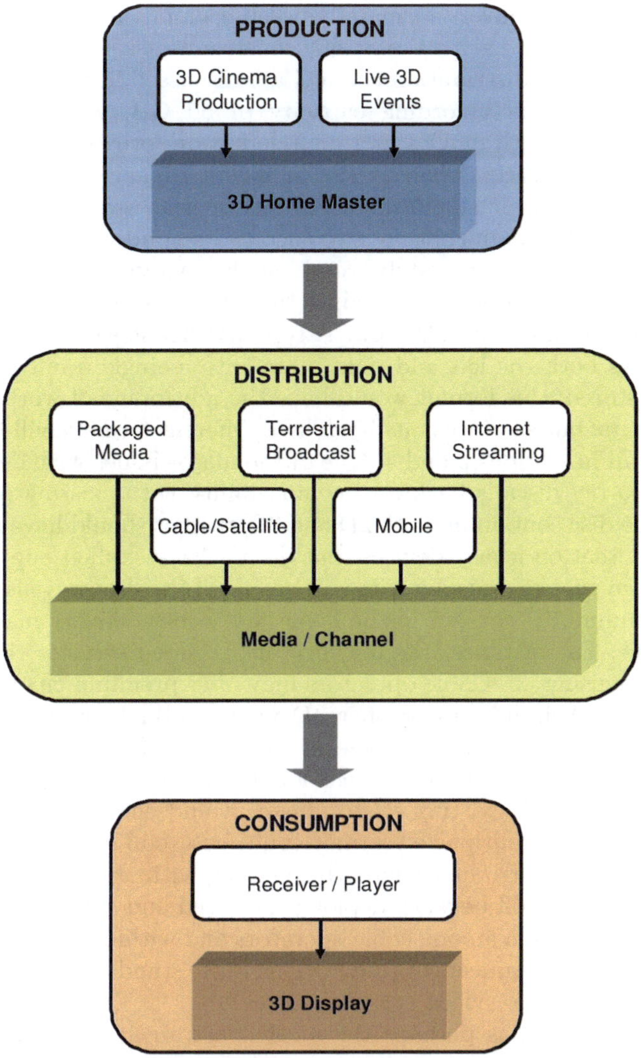

Fig. 1. Illustration of major areas and components in an end-to-end 3D video delivery chain, including production, distribution and consumption.

depth maps, occlusion and transparency data. SMPTE expects to finalize the mastering standard for 3D to the home by the end of 2010.

2.2 Distribution

It is expected that 3D content will reach the home through a variety of different channels, including packaged media such as Blu-ray Disc, through cable or terrestrial broadcast, as well as Internet streaming or download. It is an open question whether the delivery formats between each of these distribution channels could be harmonized given the unique constraints associated with each. The key challenges of each distribution channel are discussed in the following.

Blu-ray discs have a recording capacity of 50 GB, which is more than enough for feature-length movies in a high-definition format, which is typically 1920×1080 pixels at 24Hz. There is also sufficient capacity to store additional streams, which allows for significant flexibility in the way that 3D content is stored on Blu-ray discs. One option is to encode a second view independently of the 2D view and store it separately; combing the two views would then offer a stereoscopic viewing experience. To reduce the bit rate, the second view may also be compressed based on the 2D view. A separate 3D stream that sub-samples and multiplexes both the left and right views into a single frame could also be encoded and stored; this format would have the advantage of working with existing 2D players, but sacrifices quality. These different formats will be discussed in greater detail in sections 3 and 4. One of the major issues with delivery of 3D content on Blu-ray discs is backward compatibility with existing players. This means that the disc containing both 2D and 3D content should have no problems playing 2D content on legacy devices that do not have explicit support for 3D.

It is expected that packaged media will be one of the first ways for consumers to receive premium 3D content in the home, but other delivery means are very likely to follow. For instance, the delivery of 3D video services through cable seems very promising [3]. Cable operators may offer premium channels with 3D video as part of their line up or offer 3D video through on-demand services. While bandwidth is not a major issue in the cable infrastructure, the set-top boxes to decode and format the content for display is a concern. A 3D format that is compatible with existing set-top boxes would enable faster deployment of new 3D services; a multiplexed format could be useful for this purpose. New boxes could also be introduced into the market with full resolution and 3D capabilities. This tradeoff between deployment speed and 3D capabilities is part of the ongoing discussion among cable operators and within the Society of Cable Telecommunications Engineers (SCTE), which is the standards organization that is responsible for cable services.

Terrestrial broadcast is perhaps the most constrained distribution method. Among the organizations responsible for digital televisions broadcast are the Advanced Television Systems Committee (ATSC) in the US, the Digital Video Broadcast (DVB) Project in Europe and the Association of Radio Industries and Businesses (ARIB) in Japan. Many analog systems around the world have converted or are in the process of converting to all digital broadcast, where the allocated bandwidth for each channel is fixed and somewhat limited. Furthermore, most countries around the world defined their digital broadcast services based on the MPEG-2 standard, which is often a mandatory format in each broadcast channel. Between the limited channel bandwidth, the legacy format issues, costs associated with upgrading broadcast infrastructure and the lack of a clear business model on the part of the broadcasters to introduce

3D services, over-the-air broadcast of 3D video is likely to lag behind other distribution channels.

With increased broadband connectivity in the home, access to 3D content from web servers is likely to be a dominant source of content. There are already media servers on the market that ensure sufficient processing capability and interfaces to support 3D video, including interactive gaming media, and such capabilities could integrated into the television or other home appliances in the near future. Existing online services that offer content as part of a subscription or charge a fee for download or streaming of video are likely to add 3D services to their offerings in the near future. Mobile phone services are also expected to introduce 3D capable devices and offer 3D content as well.

In summary, while there are many obstacles in delivering 3D content to the home, there are also several viable options. With an increased demand for 3D in the home combined with the push for an additional revenue stream by services providers, we should be able to expect the availability of some 3D services in the very near future.

2.3 Consumption

There are a wide variety of different 3D display systems designed for the home user applications, starting from classical two-view stereo systems that require glasses to more sophisticated auto-stereoscopic displays that do not require glasses [4]. Auto-stereoscopic displays emit a large number of views, but the technology ensures that users only see a stereo pair from any particular viewpoint. In the following, some of the more prominent display technologies and their corresponding data format requirements are outlined.

There are two major categories of stereo displays: passive and active. The main difference between these devices is the way in which the light for each eye is emitted and viewed.

With passive devices, the images for left and right eyes are polarized in an orthogonal direction and superimposed on a display screen. To view the 3D scene, a pair of polarized glasses are used that match the orientation of the views being displayed for left and right eyes. In this way, the glasses are used to separate or filter the light with a specific orientation, i.e., only light with an orientation that matches that of the lens will be able to pass through. As a result, images that correspond to the left and right views can be properly viewed by each eye to achieve a 3D viewing effect.

There are a number of passive displays based on LCD technology available today. These displays are implemented using alternating lines for left and right views with opposite polarization. The native display format in this case is referred to as row interleaving.

In contrast to passive displays, active displays operate by rapidly alternating between the left-eye and the right-eye image on the screen. To maintain motion continuity, a frame rate of 120Hz is typically used. Active shutter glasses have lenses which turn from opaque to transparent, e.g., based on liquid crystals, in perfect sync with the image being displayed. To maintain synchronization, an active signal must be sent from the display to the glasses. Infrared emitters are typically used for this purpose.

Currently, active displays are implemented based on DLP or plasma technology, where each left and right frame are displayed at alternating time instants. The native display format for some active displays based on plasma technology is frame sequential, which outputs a full spatial resolution for each eye. The native display format for all DLP-based displays and some plasma displays is checkerboard, which is essentially applies a quincunx sub-sampling to each view. Next-generation LCD panels with higher frame rates may also enter the active display category.

Auto-stereoscopic displays do not require glasses and achieve 3D by essentially emitting view-dependent pixels. Such displays can be implemented, for example, using conventional high-resolution displays and parallax barriers; other technologies include lenticular sheets and holographic screens [4]. Each view-dependent pixel can be thought of as emitting a small number of light rays in a set of discrete viewing directions, typically between eight and a few dozen. Often these directions are distributed in a horizontal plane, such that parallax effects are limited to horizontal motion of the observer. Obviously, these multiview displays have much higher data requirements relative to conventional stereo displays that only require two views.

Since there are a number of displays already on the market that use different formats, the interface from distribution formats to native displays formats is a major issue. There is currently a strong need to standardize the signaling and data format to be transmitted from the various source devices in the home such as TV receivers and players, to the many types of sink devices, i.e., displays. HDMI v1.4 has recently been announced and includes support for a number of uncompressed 3D formats [5]. Efforts are underway to also update other digital interface specifications including those specified by the Consumer Electronics Associations (CEA). There are also new initiatives within CEA to standardize the specification of 3D glasses, as well as the interface between display devices and active glasses [6].

3 Representation Formats

3.1 Full-Resolution Stereo and Multiview

The format that most people first think of for stereo and multiview video are full-resolution formats. In the case of stereo, this representation basically doubles the data rate of conventional single view video. For multiview, there is an N-fold increase in the data rate for N-view video. Efficient compression of such data is a key issue and will be discussed further in Section 4.

3.2 Stereo Interleaving

Stereo interleaving is a class of formats in which the stereo signal is essentially a multiplex of the two views into a single frame or sequence of frames. Some

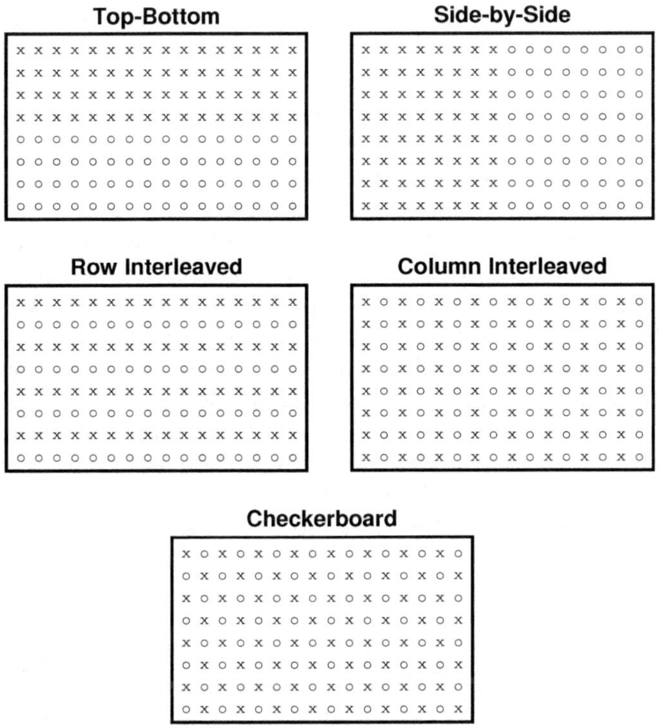

Fig. 2. Common spatial interleaving formats, where x represents the samples from one view and o represents samples from the another view.

common spatial interleaving formats are shown in Figure 2. Another common name for such representation formats are *frame-compatible* formats.

With a spatially multiplexed format, the left and right views are sub-sampled and interleaved into a single frame. There are a variety of options for both the sub-sampling and interleaving. For instance, a quincunx sampling may be applied to each view and the two views interleaved with alternating samples in both horizontal and vertical dimensions. Alternatively, the two views may be decimated horizontally or vertically and stored in a side-by side or top-bottom format, respectively.

With a time multiplexed format, the left and right views would be interleaved as alternating frames or fields. These formats are often referred to as frame sequential and field sequential. The frame rate of each view may be reduced so that the amount of data is equivalent to a that of a single view.

A major advantage of such formats is that the stereo video is represented in such a way that is compatible with existing codecs and delivery infrastructure. In this way, the video can be compressed with existing encoders, transmitted through existing channels and decoded by existing receivers and players. This format essentially tunnels the stereo video through existing hardware and delivery channels. Due to these minimal changes, stereo services can can be quickly deployed to capable display, which are already in the market.

The drawback of representing the stereo signal in this way is that spatial or temporal resolution would be lost. However, the impact on the 3D perception may be limited. An additional issue with interleaving formats is distinguishing the left and right views. To perform the de-interleaving, some additional out-of-band signaling is necessary. Since this signalling may not be understood by legacy receivers, it is not possible for such devices to extract, decode and display a 2D version of the 3D program. While this might not be so problematic for packaged media since special 3D players could be upgraded and new discs might be able to store both 2D and 3D formats, it is certainly a major issue for broadcast services where the transmission bandwidth is limited and devices cannot be upgraded.

The signalling for a complete set of interleaving formats has been standardized within the H.264/MPEG-4 AVC standard as Supplementary Enhancement Information (SEI). In general, SEI messages provide useful information to a decoder, but are not a normative part of the decoding process. An earlier edition of the standard already specified a Stereo SEI message that identifies the left view and right view; it also has the capability of indicating whether the encoding of a particular view is self-contained, i.e., frame or field corresponding to the left view are only predicted from other frames or fields in the left view. Inter-view prediction for stereo is possible when the self-contained flag is disabled. This functionality has been combined with additional signaling for the various spatially multiplexed formats described above as a new SEI message referred to as the Frame Packing Arrangement SEI message. This new SEI message has recently been specified as an amendment of the AVC standard [7].

3.3 Depth-Based Formats

ISO/IEC 23002-3 (also referred to as MPEG-C Part 3) specifies the representation of auxiliary video and supplemental information. In particular, it enables signaling for depth map streams to support 3D video applications. Specifically, the 2D plus depth format is specified by this standard and is illustrated in Figure 3. The inclusion of depth enables a display-independent solution for 3D that supports generation of an increased number of views as need by any stereoscopic display.

A key advantage of this representation format is that the main 2D video provides backward compatibility with legacy devices. Also, this representation is agnostic of the actual coding format, i.e., the approach works with both MPEG-2 and H.264/MPEG-4 AVC video coding standards. In principle, this format is able to support both stereo and multiview displays, and also allows adjustment of depth perception in stereo displays according to viewing characteristics such as display size and viewing distance.

The main drawback of this format is that it is only capable of rendering a limited depth range and was not specifically designed to handle occlusions. Also, stereo signals are not easily accessible by this format, i.e., receivers would be required to generate the second view to drive a stereo displays, which is not

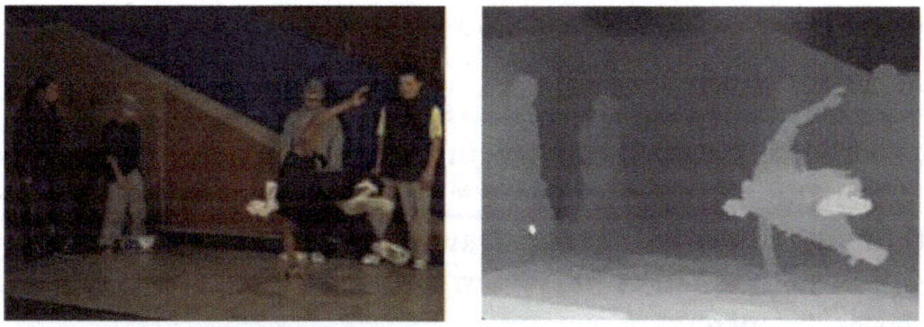

Fig. 3. Illustration of the 2D plus depth format.

the convention in existing displays. Finally, while automatic depth estimation techniques have been a heavily explored topic in the literature, their accuracy is still not sufficient to support the synthesis and rendering requirements of this format. Therefore, some semi-automatic means are needed to extract depth maps with sufficient accuracy, which could add substantially to production costs and may not be practical for live events.

To overcome the drawbacks of the 2D plus depth format, while still maintaining some of its key merits, MPEG is now in the process of exploring alternative representation formats and considering a new phase of standardization. The targets of this new initiative were discussed in [8] and are also illustrated in Figure 4 [9]. The objectives are:

1. Enable stereo devices to cope with varying display types and sizes, and different viewing preferences. This includes the ability to vary the baseline

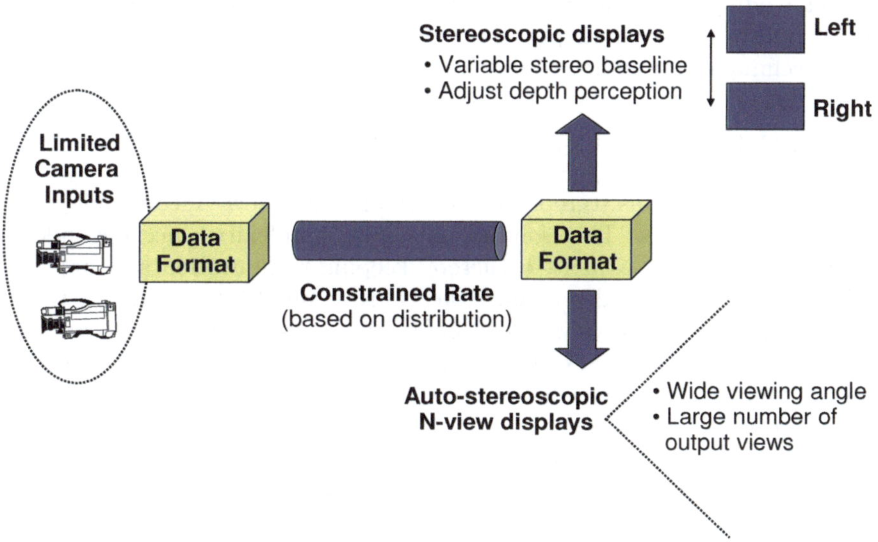

Fig. 4. Target of 3D video framework including limited camera input, fixed rate transmission and capability to support auto-stereoscopic displays and advanced stereoscopic processing.

distance for stereo video so that the depth perception experienced by the viewer is within a comfortable range. Such a feature could help to avoid fatigue and other viewing discomforts.
2. Facilitate support for high-quality auto-stereoscopic displays. Since directly providing all the necessary views for these displays is not practical due to production and transmission constraints, the new format aims to enable the generation of many high-quality views from a limited amount of input data, e.g. stereo and depth.

A key feature of this new 3D video (3DV) data format is to decouple the content creation from the display requirements, while still working within the constraints imposed by production and transmission. Furthermore, compared to the existing coding formats, the 3DV format aims to enhance 3D rendering capabilities beyond 2D plus depth, while not incurring a substantial rate increase. Simultaneously, at an equivalent or improved rendering capability, this new format should substantially reduce the rate requirements relative to sending multiple views directly. These requirements are outlined in [10].

4 Compression Technology

To realize an efficient coded realization of the 3D representation formats discussed in the previous section, compression of the scene data is required. This section describes related compression techniques and standards. In particular, the multiview video coding extension of the H.264/MPEG-4 AVC standard [1] is reviewed. Coding tools that have been proposed for efficient coding of multiview video, but have not been adopted to the standard, are covered as well. Finally, specific techniques for coding of depth map information are described.

4.1 Multiview Video Coding Standard

A straightforward means to represent stereo or multi-view video is independent encoding of each view. This solution has low complexity since dependencies between views are not exploited, thereby keeping computation and processing delay to a minimum. It also enables backward compatibility with existing 2D solutions since one of the views could be easily accessed and decoded for legacy displays. The main drawback of the simulcast method is that coding efficiency is not maximized since redundancy between views is not considered.

To improve coding efficiency of multiview video, the redundancy over time and across views could be exploited. In this way, pictures are not only predicted from temporal neighbors, but also from spatial neighbors in adjacent views. This capability has been enabled most recently as the Multiview Video Coding (MVC) extension of the H.264/MPEG-4 AVC standard [1]. Several key features of the standard are reviewed below.

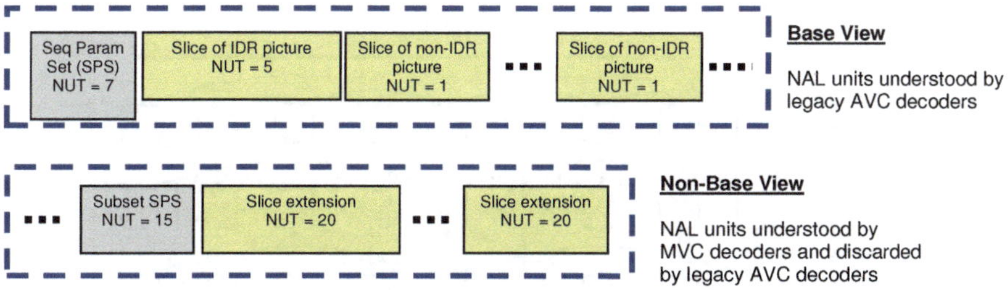

Fig. 5. Base structure of an MVC bitstream including NAL units that are associated with a base view and NAL units that are associated with a non-base view.

Bitstream structure

A key aspect of the MVC design is that it is mandatory for the compressed multiview stream to include a base view bitstream, which is independently coded from other non-base (or enhancement) views. Such a requirement enables a variety of uses cases that need a 2D version of the content to be easily extracted and decoded. For instance, in television broadcast, the base view would be extracted and decoded by legacy receivers, while newer 3D receivers could decode the complete 3D bitstream including non-base views.

As defined in the AVC standard, there exists a Network Abstraction Layer (NAL) and coded data is organized into NAL units. There exist many types of NAL units, some which are designated for video coding data, while others for non-video data such as high level syntax information. MVC extends the NAL unit types used for single video to provide backward compatibility for multiview video.

To achieve this compatibility, the video data associated with a base view is encapsulated in NAL units defined for single view video, and the video data associated with additional views are encapsulated in a new NAL unit type for multiview video. The base view bitstream conforms to existing AVC profiles for single view video, e.g., High profile, and decoders conforming to an existing single view profile will ignore and discard NAL units corresponding to the multiview data since it would not recognize those NAL unit types. Decoding the additional views with these new NAL unit types would require a decoder that conforms to one of the MVC profiles and recognizes the new NAL unit types. The basic structure of the MVC bitstream including NAL units associated with a base view and NAL units associated with a non-base view is shown in Figure 5. Further discussion of the high-level syntax is given below, and MVC profiles and levels are discussed further below.

Inter-view prediction

The basic idea of inter-view prediction, which is employed in all related works on efficient multiview video coding, is to exploit both spatial and temporal redundancy for compression. Since all cameras capture the same scene from different

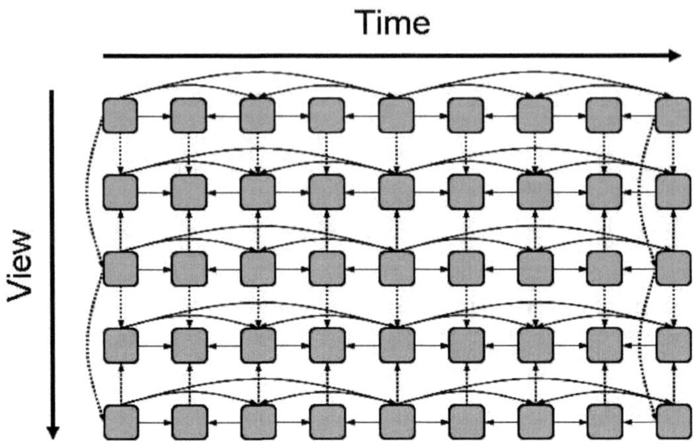

Fig. 6. Illustration of inter-view prediction in MVC.

viewpoints, inter-view redundancy is present. A sample prediction structure is shown in Fig. 6. Pictures are not only predicted from temporal references, but also from inter-view references. The prediction is adaptive, so the best predictor among temporal and inter-view references is selected on a block basis.

Inter-view prediction is a key feature of the MVC design and is enabled through flexible reference picture management of AVC, where decoded pictures from other views are essentially made available in the reference picture list. Specifically, a reference picture list is maintained for each picture to be decoded in a given view. This list is initialized as usual for single view video and would include any temporal references that may be used to predict the current picture. Additionally, inter-view reference pictures may be appended to the list and thereby available for prediction of the current picture.

According to the MVC specification, inter-view reference pictures must be contained within the same access unit as the current picture, where an access unit contains all the NAL units pertaining to a certain time instance. In other words, it is not possible to predict a picture in one view at a given time from a picture in another view at a different time. This would involve inter-view prediction across different access units, which would incur additional complexity for limited coding benefits.

To keep the management of reference pictures inline with single view video, all the memory management control operation commands that may be signalled through an AVC bitstream apply to a particular view. The same is true for the sliding window that is used to mark pictures as not being used for reference; this process of AVC is also applied independently for each view. Also, just as it is possible to re-order the reference pictures in a reference picture list including temporal references, the same can be done with reference picture lists including inter-view references. An extended set of re-ordering commands have been adopted to the MVC specification for this purpose.

It is important to emphasize that the core block-level decoding modules do not need to be aware of whether a reference picture is a temporal reference or an inter-view reference. This distinction is managed at a higher level of the decoding process.

In terms of syntax, the standard only requires small changes to high-level syntax, e.g., view dependency as discussed below. A major consequence of not requiring changes to lower block-level syntax is that MVC is compatible with existing hardware for decoding single view video with AVC. In other words, supporting MVC as part of an existing AVC decoder should not require substantial design changes.

Since MVC introduces dependencies between views, random access must also be considered in the view dimension. Specifically, in addition to the views to be accessed (target views), any dependent views also need to be accessed and decoded, which typically requires decoding time or delay. For applications in which random access or view switching is important, the prediction structure could be designed to minimize access delay.

To achieve access to a particular picture in a given view, the decoder should first determine an appropriate access point. In AVC, Instantaneous Decoding Refresh (IDR) pictures provide a clean access point since these pictures can be independently decoded and all the coded pictures that follow in decoding order can be decoded without temporal prediction from any picture decoded prior to the IDR picture. In the context of MVC, an IDR picture prohibits the use of temporal prediction for any of the views at that particular instant of time; however, inter-view prediction may be used to reduce the rate overhead. MVC also introduces a new picture type, referred to as an anchor picture. Anchor pictures are similar to IDR pictures in that they do not utilize temporal prediction, but do allow inter-view prediction from views within the same access unit. The difference between anchor pictures and IDR pictures is similar to the difference between the open GOP and closed GOP concept in MPEG-2, respectively, where closed GOPs do not allow pictures from one GOP to be used as a reference for pictures in a different GOP. In contrast, open GOPs effectively allow the I-frame of one GOP to be used as a backward reference for a B-frame that is earlier in display order. In MVC, both IDR and anchor pictures are efficiently coded and provide random access in both time and view dimensions.

High-level Syntax

The decoding process of MVC requires several additions to the high-level syntax, which are primarily signalled through a multiview extension of the Sequence Parameter Set (SPS) defined by AVC. Three important pieces of information are carried in the SPS extension:

- View identification
- View dependency information
- Level index for operation points

The level index is an indicator of resource constraints imposed on a decoder that conforms to a particular level; it is mainly used to bound the complexity of a decoder and discussed further below. In the context of MVC, an operating point corresponds to a specific temporal level and a set of views including those intended for output and the views that they depend on.

The view identification part includes an indication of the total number of views, as well as a listing of view identifiers. The view identifiers are important for associating a particular view to a specific index, while the order of the views identifiers signals the view order index. The view order index is critical to the decoding process as it defines the order in which views are decoded.

The view dependency information is comprised of a set of signals that indicate the number of inter-view reference for each of the two reference picture lists that are used in the prediction process, as well as the views that may be used for predicting a particular view. Separate information is maintained for anchor and non-anchor pictures to provide some flexibility in the prediction, while not over-burdening decoders with dependency information that could change for each unit of time.

For non-anchor pictures, the view dependency only indicates that a given view may be used for inter-view prediction. There is additional signaling in the NAL unit header indicating whether a particular view at a given time is used as an inter-view reference for any other picture in the same access unit. The view dependency in the SPS and this syntax element in the NAL unit header are used to append the reference picture list to include inter-view references as described earlier.

The final portion of the SPS extension is the signalling of level information and associated information about the operating points to which they correspond. An MVC stream with 8 views may include several operating points, e.g., one corresponding to all 8 views, another corresponding to a stereo pair, and another corresponding to a set of three particular views. According to the MVC standard, multiple level values could be signalled as part of the SPS extension, with each level being associated to a particular operating point (i.e., temporal level and target set of views). The syntax indicates the number of views that are targeted for output as well as the number of views that would be required for decoding particular operating points.

Profiles and levels

Consistent with prior MPEG standards, profiles determine the subset of coding tools that must be supported by decoder. There are two profiles currently defined by MVC with support for more than one view: Multiview High profile and Stereo High profile. Both are based on the High profile of AVC with a few differences.

- The Multiview High profile supports multiple views and does not support interlaced coding tools.
- The Stereo High profile is limited to two views, but does support interlaced coding tools.

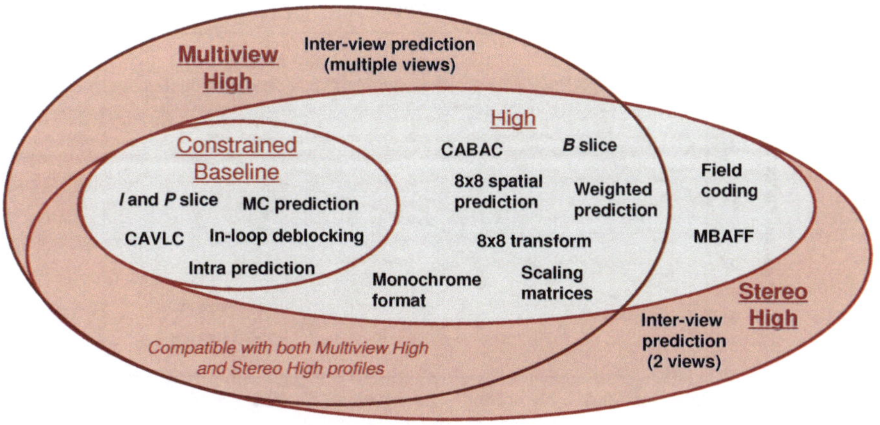

Fig. 7. Illustration of MVC profiles including Multiview High and Stereo High profiles, as well as region that is compatible with both profiles.

An illustration of these profile specifications relative to High profile of AVC is provided in Figure 7. It is possible to have a bitstream that conforms to both the Stereo High profile and Multiview High profile when there are only two views are coded and interlaced coding tools are not used. In this case, a flag signaling their compatibility is set.

Levels impose constraints on decoder resources and complexity. A similar set of limits as imposed on AVC decoders are imposed on MVC decoders including limits on the amount of frame memory required for the decoding of a bitstream, the maximum throughput in terms of macroblocks per second, maximum picture size, overall bit rate, etc.

The general approach to defining level limits in MVC was to enable repurposing the decoding resources of single-view decoders for multi-view decoders. In this way, some level limits are unchanged such as the overall bit rate; in this way, an input bitstream can be processed by a decoder regardless of whether it encodes a single view or multiple views. However, other level limits are increased such as the maximum decoded picture buffer and throughput; a fixed scale factor of 2 was applied to these decoder parameters. Assuming a fixed resolution, this scale factor enables decoding of stereo video using the same level as single view video at that resolution. For instance, the same Level 4.0 index could be used to decode single view video and stereo video at 1920×1080p at 24Hz. To decode a higher number of views, one would need to use either a higher level and/or reduce the spatial or temporal resolution of the multiview video.

Coding Performance

It has been shown that coding multiview video with inter-view prediction does give significantly better results compared to independent coding [11, 12]. For some cases, gains as high as 3 dB, which correspond to 50% savings in bit rate,

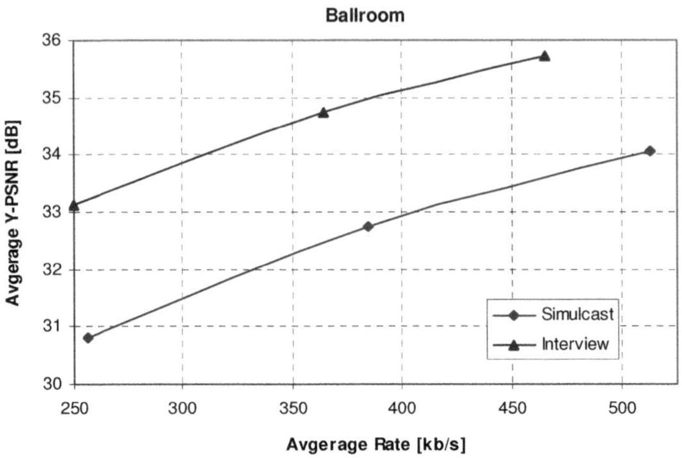

Fig. 8. Illustration of inter-view prediction in MVC.

have been reported. A comprehensive set of results for multiview video coding over a broad range of test material was also presented in [13]. For multiview video with up to 8 views, an average of 20% improvement relative to the total simulcast bit rate has been reported. In other studies [14], an average reduction of 20-30% of the bit rate for the second (dependent) view of typical stereo movie content was reported, with peak reduction up to 43% of the bit rate of the dependent view.

Fig. 8 shows sample RD curves comparing the performance of simulcast coding (without the use of hierarchical B-pictures) with the performance of the MVC reference software that employs hierarchical predictions in both spatial and view dimensions. Of course, the use of hierarchical B-pictures in the simulcast solution will also provide some gains, but they are not shown in this plot.

There are many possible variations on the prediction structure considering both temporal and spatial dependencies. The structure not only affects coding performance, but has notable impact on delay, memory requirements and random access. It has been confirmed that the majority of gains are obtained using inter-view prediction at anchor positions in Fig. 8. Rate penalties of approximately 5-15% could be expected if the spatial predictions at non-anchor positions are removed [15]. The upside is that delay and required memory would also be reduced.

Prior studies on asymmetrical coding of stereo, whereby one of the views is encoded with less quality, suggest that substantial savings in bitrate for the second view could be achieved. In this way, one of the views is significantly blurred or more coarsely quantized than the other [16] or coded with a reduced spatial resolution [17, 18], yielding an imperceptible impact on the stereo quality. With mixed resolution coding, it has bene reported that the an additional view could be supported with minimal rate overhead, e.g., on the order of 25-30%

additional rate for coding the right view at quarter resolution. Further study is needed to understand how this phenomenon extends to multiview video.

Additional considerations

MVC was designed mainly to support auto-stereoscopic displays that require a large number of views. However, large camera arrays are not common in the current acquisition and production environments. Furthermore, although MVC is more efficient than simulcast, the rate of MVC encoded video is still proportional to the number of views. Of course, this varies with factors such as scene complexity, resolution and camera arrangement, but when considering a high number of views, the achievable rate reduction might not be significant enough to overcome constraints on channel bandwidth.

Despite the challenges associated multiview video, MVC is still an effective format for the delivery of stereo contents. It has been shown that a good level of rate reduction could be achieved relative to simulcast and that backward compatibility with 2D systems could also be provided.

4.2 Block-Based Coding Tools for Multiview Video

Several macroblock level coding tools have also being explored during the MVC standardization process. It has been shown that additional coding gains could be achieved beyond the inter-prediction coding supported by MVC. However, these tools were not adopted to the standard since they would require design changes at the macroblock level. It was believed that this implementation concern outweighed the coding gain benefits at the time. The benefits of block-level coding tools may be revisited in the specification of future 3D video formats; the main ideas of the proposed block-level coding tools are reviewed in the remainder of this section.

Illumination Compensation

Illumination compensation is a block-based coding tool for multiview video that compensates for the illumination differences between views [19, 20]. This tool has shown to be very useful when the illumination or color characteristics vary in different views. This is a likely case since even cameras from the same manufacturer could acquire video with very different color properties. While conventional color normalization procedures could be applied prior to encoding, not all applications and content creation settings would allow for this step.

The proposed method determines an offset value that corresponds to the difference in illumination between a current block and its reference. This offset value is calculated as part of the motion estimation process. Rather than compute the typical sum of absolute differences (SAD) between blocks of pixels in the current and reference frame, a mean-removed SAD is computed instead, where there is a mean associated with the current block and a mean associated with a reference block. The difference between these two mean values is the offset that is used for illumination compensation in the decoder. The decoder simply adds this offset value as part of the motion-compensated prediction.

The illumination differences between views have been found to be spatially correlated. Therefore, rather than coding the offset value directly, a prediction from neighboring illumination offset values is used keep rate overhead to a minimum. Coding gains up to 0.6 dB have been reported in comparison to the existing weighted prediction tool of H.264/AVC.

It was also observed by Lai, et al. [21, 22] that there are other types of mismatches present in multiview video. In particular, an adaptive reference filtering scheme was proposed to compensate for focus mismatches between different views.

Motion Skip Mode

An effective method to reduce bit rate in video coding is to infer side information used in the decoding process, e.g., motion vectors for a particular block, based on other available data, e.g., motion vectors from other blocks. This is the basic principle of direct mode prediction in AVC.

In [23, 24], Koo, et al. proposed extensions to the conventional skip and direct coding modes for multiview video coding. Specifically, this method infers side information from inter-view references rather than temporal references. A global disparity vector is determined for each neighboring reference view. The motion vector of a corresponding block in the neighboring view may then be used for prediction of the current block in a different view. This signaling is very minimal and this method has the potential to offer notable reduction in bit rate.

An analysis of the coding gains offered by both illumination compensation and motion skip mode was reported in [13]. A rate reduction of 10% was reported over a wide range of sequences with a maximum reduction of approximately 18%. A sample plot of rate-distortion performance for the Race1 sequence is given in Figure 9. While the gains are considered sizeable, these tools would require block-level changes to the decoding process, which was viewed as undesirable at the time and led to the decision not to adopt such tools into the MVC standard.

View Synthesis Prediction

Another novel macroblock level coding tool that has been explored for improved coding of multiview video is view synthesis prediction. This coding technique predicts a picture in the current view from synthesized references generated from neighboring views.

One approach for view synthesis prediction is to encode depth for each block, which is then used at the decoder to generate the view synthesis data used for prediction, as first described by Martinian, et al. [25] and fully elaborated on by Yea and Vetro [26]. Another approach estimates pixel-level disparities at both the encoder and decoder and encodes only disparity correction values [27].

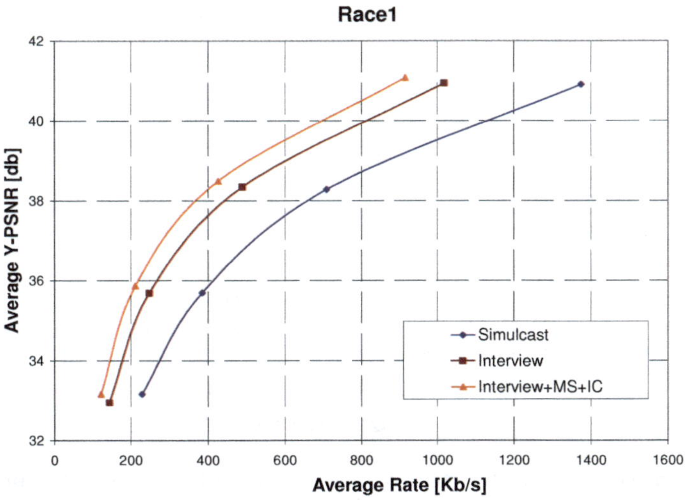

Fig. 9. Illustration of coding efficiency for Race1 sequence demonstrating gains using motion skip (MS) and illumination compensation (IC) coding tools.

As we know, conventional inter-view prediction maps every pixel in a rectangular block in the predicted view to the corresponding pixel in the reference view, where every pixel in the block is displaced by the same amount using a single disparity vector. In contrast, view synthesis prediction maps the matching pixels according to the scene depth a camera configuration; such a mapping could provide better prediction when the matching area in the reference view is non-rectangular or the correspondence between the views is non-translational. Further details on the view synthesis process, the estimation of depth and the coding of side information are available in the cited papers.

As a sample result to illustrate the coding gains, the results from [26] are discussed. Figure 10 shows a rate-distortion curve that compares the performance of disparity-compensated prediction (DCP) with and without view synthesis prediction (VSP). The results in this plot are averaged over all 100 frames of the B-views in the Breakdancer sequence, i.e., views 1, 3 and 5; these which are the views that utilize two spatially neighboring views from different directions in addition to temporal prediction. While the gains are not substantial at the higher bit rates, we do observe notable gains in the middle to low bit rate range that are between 0.2 and 1.5 dB.

4.3 Depth Compression Techniques

As discussed in previous sections, depth information could be used at the receiver to generate additional novel views or used at the encoder to realize more efficient compression with view synthesis prediction schemes. Regardless of the application, maintaining the fidelity of depth information is important since the quality of the view synthesis result is highly dependent on the accuracy of the geometric

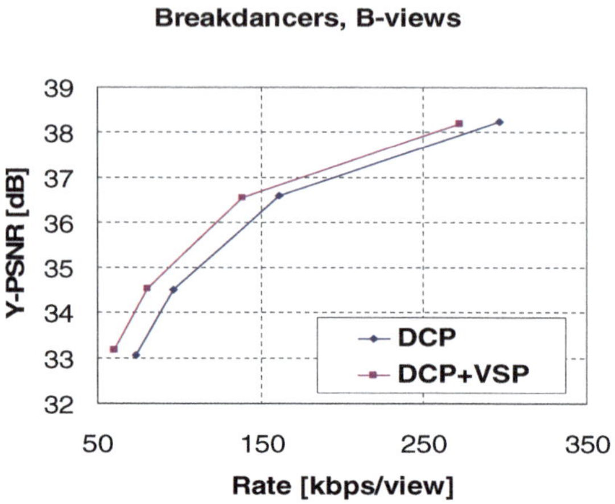

Fig. 10. Comparison of coding performance of disparity-compensated prediction and disparity-compensated prediction with view synthesis prediction for 100 frames of the Breakdancer sequence. Results are computed on B-views only.

information provided by depth. Therefore, it is crucial to strike a good balance between the fidelity of depth data and the associated bandwidth requirement.

As reported in [28], the rate used to code depth video with pixel-level accuracy could be quite high and on the same order as that of the texture video. Experiments were performed to demonstrate how the video synthesis quality varies as a function of bit rate for both the texture and depth videos. It was found that higher bit rates were needed to code the depth data so that the view rendering quality around object boundaries could be maintained.

There have been various approaches considered in the literature to reduce the required rate for coding depth, while maintaining high view synthesis and multiview rendering quality.

One approach is to code a reduced resolution version of the depth using conventional compression techniques. This method could provide substantial rate reductions, but the filtering and reconstruction techniques need to be carefully designed to maximize quality. A set of experiments were performed in [8] using simple averaging and interpolation filters. These results demonstrate effective reduction in rate, but artifacts are introduced in the reconstructed images due to the simple filters. Improved down/up sampling filters were proposed by Oh, et al. in [29]. This work not only achieves very substantial reductions in the bit rate, but also improved rendering quality around the object boundaries.

Another approach is to code the depth based on geometric representation of the data. In [30], Morvan, et al. model depth images using a piece-wise linear function; they referred to this representation as platelets. The image is subdivided using a quadtree decomposition and an appropriate modeling function is selected for each region of the image in order to optimize the overall rate-distortion cost. The benefit of this approach for improved rendering quality relative to JPEG 2000 was clearly shown. In subsequent work, comparisons to AVC intra coding were also made and similar benefits have been shown [31].

A drawback of the platelet-based approach is that it appears difficult to extend this scheme to video. An alternative multi-layered coding scheme for depth was suggested in [32]. In this approach, it was argued that the quality of depth information around object boundaries needs to be maintained with higher fidelity since it as notable impact on subjective visual quality. The proposed scheme guarantees a near-lossless bound on the depth values around the edges by adding an extra enhancement layer. This method effectively improve the visual quality of the synthesized images, and is flexible in the sense that it could incorporate any lossy coder as the base layer thereby making it easily extendible to coding of depth video.

5 Discussion

After many false starts, it now appears that 3D video is forming a positive impression on consumers around the world through high quality digital theater experiences. 3D content production is ramping up and the infrastructure and equipment is being put in place to delivery 3D video to the home. There is still a significant amount of work to be done in the various standards organizations to ensure interoperable 3D services across a wide range of application domains and equipment. One critical issue will be determining suitable 3D data formats among the various options that have been described in this chapter.

Current industry activities are now focused on distribution of stereoscopic video since this is what current display technology, production and delivery infrastructure could practically support. However, there is significant research and standardization initiatives underway that target 3D solutions that would not require glasses. To enable this, a very rich and accurate representation of the scene is needed. The data needs to be coded in an efficient manner and be rendered on devices that are capable of providing an immersive and realistic 3D viewing experience. It is equally important for this to be done in a practical way based on current state of technologies so that the experience could be made available to consumers on a large scale.

References

1. ITU-T Rec. & ISO/IEC 14496-10 AVC: Advanced video coding for generic audio-visual services (2009)
2. SMPTE: Report of SMPTE Task Force on 3D to the Home (2009)
3. Broberg, D.K.: Considerations for stereoscopic 3D video delivery on cable. In: IEEE International Conference on Consumer Electronics, Las Vegas, NV (2010)
4. Konrad, J., Halle, M.: 3D displays and signal processing: An answer to 3D ills? IEEE Signal Processing Magazine 24(6), 97–111 (2007)
5. HDMI Licensing, LLC.: HDMI Specification 1.4 (2009)

6. Markwalter, B., Stockfisch, M.: Enabling 3D content in consumer electronics. In: IEEE International Conference on Consumer Electronics, Las Vegas, NV (2010)
7. Video Group: Text of ISO/IEC 14496-10:2009/FDAM1: Constrained baseline profile, stereo high profile and frame packing arrangement SEI message. ISO/IEC JTC1/SC29/WG11 Doc. N10707, London, UK (2009)
8. Vetro, A., Yea, S., Smolic, A.: Towards a 3D video format for auto-stereoscopic displays. In: SPIE Conference on Applications of Digital Image Processing XXXI (2008)
9. Video and Requirements Group: Vision on 3D Video. ISO/IEC JTC1/SC29/WG11 Doc. N10357, Lausanne, Switzerland (2009)
10. Video and Requirements Group: Applications and Requirements on 3D Video Coding. ISO/IEC JTC1/SC29/WG11 Doc. N10857, London, UK (2009)
11. Merkle, P., Müller, K., Smolic, A., Wiegand, T.: Efficient compression of multiview video exploiting inter-view dependencies based on H.264/AVC. In: IEEE International Conference on Multimedia & Expo, Toronto, Canada (2006)
12. Merkle, P., Smolic, A., Müller, K., Wiegand, T.: Efficient prediction structures for multiview video coding. IEEE Transactions on Circuits and Systems for Video Technology 17(11), 1461–1473 (2007)
13. Tian, D., Pandit, P., Yin, P., Gomila, C.: Study of MVC coding tools. ITU-T & ISO/IEC JTC1/SC29/WG11 Doc. JVT-Y044, Shenzhen, China (2007)
14. Chen, T., Kashiwagi, Y., Lim, C.S., Nishi, T.: Coding performance of Stereo High Profile for movie sequences. ITU-T & ISO/IEC JTC1/SC29/WG11 Doc. JVT-AE022, London, UK (2009)
15. Droese, M., Clemens, C.: Results of CE1-D on multiview video coding. ISO/IEC JTC1/SC29/WG11 Doc. m13247, Montreux, Switzerland (2006)
16. Stelmach, L., Tam, W., Meegan, D., Vincent, A.: Stereo image quality: effects of mixed spatio-temporal resolution. IEEE Transactions on Circuits and Systems for Video Technology 10(2), 188–193 (2000)
17. Fehn, C., Kauff, P., Cho, S., Kwon, H., Hur, N., Kim, J.: Asymmetric coding of stereoscopic video for transmission over T-DMB. In: 3DTV-CON 2007, Kos, Greece (May 2007)
18. Brust, H., Smolic, A., Müller, K., Tech, G., Wiegand, T.: Mixed resolution coding of stereoscopic video for mobile devices. In: 3DTV-CON 2009, Potsdam, Germany (2009)
19. Lee, Y.-L., Hur, J.-H., Lee, Y.-K., Han, K.-H., Cho, S.-H., Hur, N.-H., Kim, J.-W., Kim, J.-H., Lai, P.-L., Ortega, A., Su, Y., Yin, P., Gomila, C.: CE11: Illumination compensation. In: ITU-T & ISO/IEC JTC1/SC29/WG11 Doc. JVT-U052, Hangzhou, China (2006)
20. Hur, J.H., Cho, S., Lee, Y.L.: Adaptive local illumination change compensation method for H.264/AVC-based multiview video codin. IEEE Transactions on Circuits and Systems for Video Technology 17(11), 1496–1505 (2007)
21. Lai, P., Ortega, A., Pandit, P., Yin, P., Gomila, C.: Adaptive reference filtering for MVC. In: ITU-T & ISO/IEC JTC1/SC29/WG11 Doc. JVT-W065, San Jose, CA (2007)
22. Lai, P., Ortega, A., Pandit, P., Yin, P., Gomila, C.: Focus mismatches in multiview systems and efficient adaptive reference filtering for multiview video coding. In: SPIE Conference on Visual Communications and Image Processing, San Jose, CA (2008)
23. Koo, H.S., Jeon, Y.J., Jeon, B.M.: MVC motion skip mode. In: ITU-T & ISO/IEC JTC1/SC29/WG11 Doc. JVT-W081, San Jose, CA (2007)

24. Koo, H.S., Jeon, Y.J., Jeon, B.M.: Motion information inferring scheme for multi-view video coding. IEICE Transactions on Communications E91-B(4), 1247–1250 (2008)
25. Martinian, E., Behrens, A., Xin, J., Vetro, A.: View synthesis for multiview video compression. In: Picture Coding Symposium, Beijing, China (2006)
26. Yea, S., Vetro, A.: View synthesis prediction for multiview video coding. Image Communication 24(1-2), 89–100 (2009)
27. Kitahara, M., Kimata, H., Shimizu, S., Kamikura, K., Yashima, Y., Yamamoto, K., Yendo, T., Fujii, T., Tanimoto, M.: Multi-view video coding using view interpolation and reference picture selection. In: IEEE International Conference on Multimedia & Expo, Toronto, Canada, pp. 97–100 (2006)
28. Merkle, P., Müller, K., Smolic, A., Wiegand, T.: Experiments on coding of multi-view video plus depth. In: ITU-T & ISO/IEC JTC1/SC29/WG11 Doc. JVT-X064, Geneva, Switzerland (2007)
29. Oh, K., Yea, S., Vetro, A., Ho, Y.: Depth reconstruction filter and down/up sampling for depth coding in 3D video 16(9), 747–750 (2009)
30. Morvan, Y., Farin, D., de With, P.H.N.: Depth-image compression based on an R-D optimized quadtree decomposition for the transmission of multiview images. In: IEEE International Conference on Image Processing, San Antonio, TX (2007)
31. Merkle, P., Morvan, Y., Smolic, A., Farin, D.: The effects of multiview depth video compression on multiview rendering. Image Communication 24(1-2), 73–88 (2009)
32. Yea, S., Vetro, A.: Multi-layered coding of depth for virtual view synthesis. In: Picture Coding Symposium, Chicago, IL (2009)

Present and Future Video Coding Standards

Jie Dong and King Ngi Ngan

Department of Electronic Engineering, The Chinese University of Hong Kong
jdong@ee.cuhk.edu.hk, knngan@ee.cuhk.edu.hk

Video coding systems may greatly differ from each other, as they consist of complicated functional modules and each module can be realized by using different techniques. For the interoperability in communication, video coding standards are developed to restrict various video coding systems, such that manufacturers can successfully interwork with each other by producing compliant encoders and decoders, and at the same time still have the freedom to develop competitive and innovative products.

A standard defines a coded representation, syntax, which describes the video in a compressed form. In other words, a standard only specifies the output of the encoder, i.e., the input of the decoder, instead of the codec itself. Although the standard also provides a method of decoding the syntax to reconstruct the video, the manufacturers are free to develop alternative decoders as long as they can decode the syntax and produce the same result as that in the standard.

Video coding standards have been developing for about 20 years, driven by applications and advances in hardware capability. This chapter begins with an introduction of the block-based hybrid coding scheme, which is essentially the core of all the video coding standards. In Section 2, the past video coding standards are briefly reviewed, including H.261, MPEG-1, MPEG-2/H.262, H.263, and MPEG-4. The latest standard, H.264/AVC, is the emphasis of this chapter, which is introduced in Section 3, including the new technical developments and profiles favoring a wide range of applications. The recently finalized amendments on scalable video coding (SVC) and multiview video coding (MVC) are another two highlights. Section 4 presents the Audio and Video Coding Standard (AVS) of China, which has received much attention throughout the world, even though it is a national standard. Finally, Section 5 introduces the current topics intensively studied in the standardization groups, which may become the key techniques in the future coding standards.

1 Block-Based Hybrid Coding Scheme

Hybrid coding refers to the combination of motion-compensated prediction (MCP) and transform coding. The former exploits the temporal correlation of

C.W. Chen et al. (Eds.): Intel. Multimedia Communication: Tech. and Appli., SCI 280, pp. 75–124.
springerlink.com

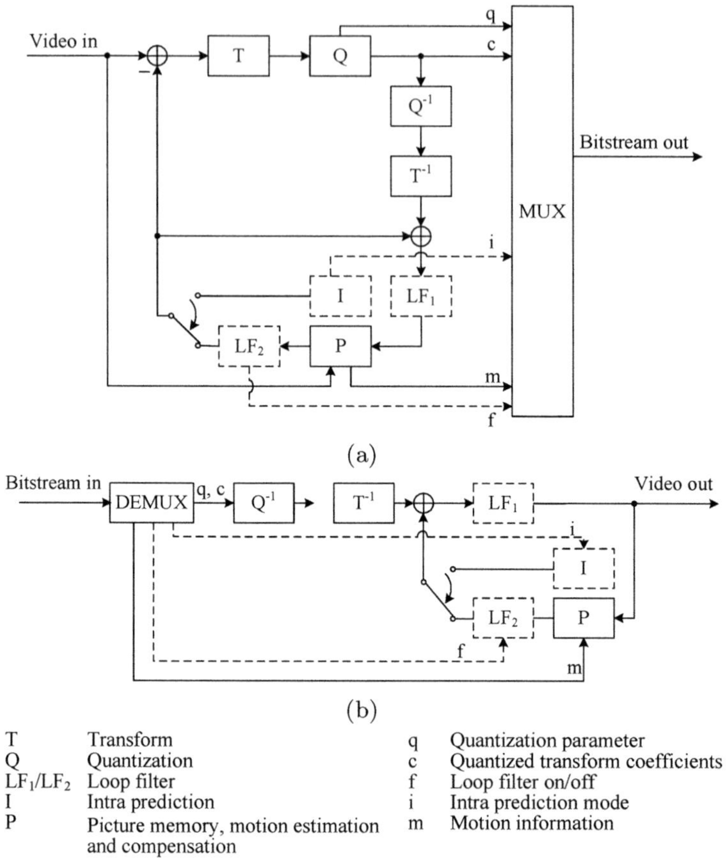

Fig. 1. Block diagrams of the hybrid (a) encoder and (b) decoder

videos, whereas the latter removes the spatial redundancy of the temporal prediction error. The designation "block-based" means each video frame is divided into non-overlapped blocks; each block is the unit where the hybrid coding applies. The phrase "hybrid coding" will refer to block-based hybrid coding scheme here and subsequently in this chapter for convenience.

Fig. 1 shows the generic block diagrams of the hybrid coding scheme for a Macroblock (MB). Note that the modules with dashed boxes are not necessarily included in all standards. Each MB consists of 16×16 luminance pixels and the associated chrominance pixels, depending on the color format, e.g., in 4:2:0 format, Cb and Cr pixels within one MB are both 8×8.

At the encoder (see Fig. 1 (a)), an MB is predicted from a coded frame stored in *P*, known as reference frame. A motion vector (MV) consisting of vertical and horizontal displacements is used to indicate the position of the prediction block, called target block, in the reference frame, and the target block is simply copied for MCP. Besides storing the reference frames, the other function of P is to search the MV, known as motion estimation (ME).

The difference of the current MB and its prediction is input to T, where 2-D transform is applied to compact the prediction error to less coefficients; the coefficients are then quantized in Q. In most standards, the transform is 2-D order-8 Discrete Cosine Transform (DCT) and the quantization is uniform. The quantization stepsize is signaled by the quantization parameter (QP), where a larger QP corresponds to a larger stepsize. The coefficients after quantization are converted to a 1-D array following a zig-zag scan order putting the low frequency coefficients in front of the high frequency ones.

Such hybrid coding process converts the original video to a compressed representation that comprises symbols, called syntax elements (SE). It is followed by a lossless process, known as entropy coding, where all SEs are converted to binary codewords by using variable length coding (VLC) and multiplexed together to a bitstream.

A coded MB belonging to a reference frame should be reconstructed by subsequently applying dequantization and inverse transform and then being compensated by its prediction. A reconstructed reference frame may contain blockiness near the transform block boundaries, which is caused by the quantization and leads to inaccurate MCP. Some standards apply an in-loop deblocking filter on reference frames. The filtering can be enabled before the reconstructed frame is stored in P (see LF_1), e.g., H.264/AVC, or immediately before the target block is used for prediction (see LF_2), e.g., H.261.

At the decoder (see Fig. 1 (b)), the SEs in a bitstream are demultiplexed. The quantized transform coefficients are dequantized, inverse transformed, and compensated by the intra- (optional) or inter-prediction, according to the coding mode. The reference frame used in encoder and decoder should be the same so as to produce identical reconstructed videos.

Using MCP to code an MB is known as inter-mode. It is further classified as P-mode, if the prediction is restricted in the past frame. Otherwise, the inter-mode is known as B-mode, where the prediction is the target block in the past or future frame, or the average of two target blocks from the past and the future frames, respectively. Generally, B-mode can achieve more accurate MCP than P-mode. The coding method without MCP is known as intra-mode, which can be used when inter-mode is not efficient or possible. In early standards, intra-mode directly applies the transform and quantization to original samples. In the recent standards, intra-prediction (see I) is employed in the spatial domain or the transform domain to reduce the spatial redundancy and only the intra-prediction errors are coded. Extending I-, P-, and B-modes to the frame level leads to the concepts of I-, P-, and B-frames.

2 History of Hybrid Coding

The history of hybrid coding is mainly reflected in the evolution of video coding standards, of which the timeline is shown in Fig. 2. The standard series

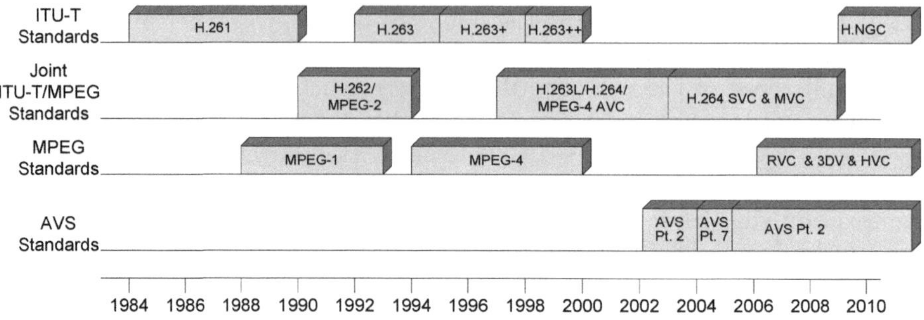

Fig. 2. The timeline of video coding standards

H.26x (x= $1 \cdots 4$) are recommended by the Video Coding Experts Group (VCEG) [1] in ITU-T, and MPEG-x (x= $1, 2, 4$) are developed by the Moving Picture Experts Group (MPEG) [2] in ISO/IEC. A comprehensive survey on the history of MPEG video coding was written by C. Reader [3]. A detailed introduction of the organizations, developments, and techniques of the standards can be found in [4].

2.1 H.261

H.261 [5] was the first standardized hybrid coding scheme published in 1990 by ITU-T, for the interactive video communications over ISDN networks. The supported bit-rate was p×64 kb/s (p= $1 \cdots 30$), depending on the number of ISDN channels used. H.261 was able to operate on QCIF and optionally on CIF, the video formats for visual telephony adopted by ITU-T.

Low complexity was the key factor for the success of H.261, since H.261 was developed at a time when the hardware capability was limited. However, it was soon superseded by H.263, which has higher coding efficiency and greater flexibility, and is still relevant today.

2.2 H.263

H.263 [6] was developed for low bit-rate video communications over analog telephone lines and the Internet and was an evolutionary improvement based on H.261. The initial standard was finished in 1995. Two extensions, nicknamed H.263+ and H.263++, were incorporated into the standard as 21 annexes in 1997 and 2000, respectively.

Baseline H.263 was the basic configuration of the coding algorithm that every compliant H.263 decoder should support. It was distinguished from H.261 by using half-pixel MCP, MV prediction, (level, run, EOB) joint coding for transform coefficients, and so on. Furthermore, H.263 supported more video formats, e.g., sub-QCIF, QCIF, CIF, 4CIF and 16CIF, and a broad range of custom video formats.

Besides baseline H.263, eighteen negotiable coding options (Annexes C-G, I-K, M-V) could be used, either together or separately, subject to certain restrictions. Additional supplemental information (Annexes L and W) might also be included in the bitstream for enhanced display capability and for external usage. A forward error correction method for coded video signal (Annex H) was provided for use, when necessary.

2.3 MPEG-1 Video

MPEG-1 [7] was developed to compress the multimedia content distributed on CD-ROMs at a bit-rate of 1.5 mb/s, catering for the access rate of CD-ROM players. It was conceived to support the "video CD", a format for VHS quality video storage and playback. MPEG-1 typically operated on CIF and SIF videos, and also supported higher rates and resolutions. The functionalities, such as random access, fast forward and rewind, were enabled by the concept of group of pictures (GOP), which starts with an I-frame followed by interleaving P- and B-frames. One can access any GOP without decoding previous GOPs, accomplish fast forward by decoding only I- or I- and P-frames, realize fast rewind by decoding only I-frames in a backward order. Although video CD was not commercially successful, MPEG-1 video had been widely used in many storage and transmission applications.

2.4 MPEG-2 Video

MPEG-2 [8] maintained much similarity to MPEG-1 and also provided the MPEG-1 like functionalities. The main feature distinguishing MPEG-2 from MPEG-1 was the efficient coding of interlaced videos, which was the key enabler of digital storage media and television (TV) broadcast. Two opposite fields in an interlaced frame could be coded in an interleaving manner, known as frame-mode, or separately, known as field-mode. These two modes could be selected once a frame, i.e., picture level adaptive frame/field coding (PAFF), or adapt to the MB level, i.e., MB level adaptive frame/field coding (MBAFF). Coding a field-mode MB basically followed the diagram in Fig. 1, but the reference pictures were fields instead of frames. Other modifications, such as zig-zag scan, were also made.

MPEG-2 mainly operated on the formats specified in ITU-R BT.601 [9] (formerly CCIR 601) with 4:2:0 color format and produced broadcast quality videos at the bit-rates of 4 to 8 mb/s and high quality videos at 10 to 15 mb/s [4]. It also handled HD videos [10] or other color formats at even higher bit-rates. MPEG-2 was a great success with worldwide adoption for digital TV broadcast via cable, satellite and terrestrial channels and for DVD-Video, and finally replaced VHS videotapes.

As a generic coding standard, full MPEG-2 syntax covered many features and parameters to meet a wide spectrum of bit-rates, video formats, quality levels and functionalities. However, certain application required only a subset of the

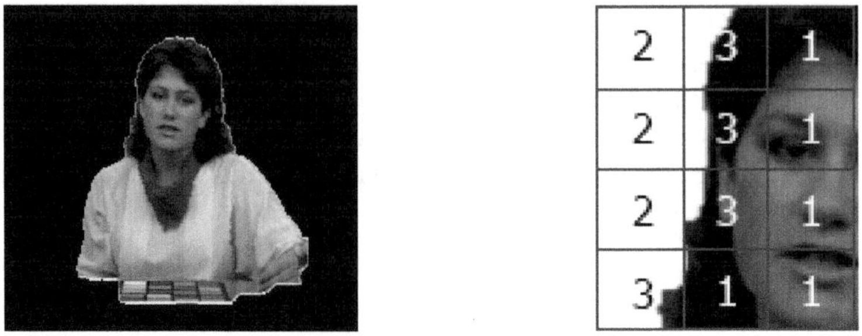

(a) (b)

Fig. 3. Content-based coding scheme for arbitrary-shaped VO. (a) The binary alpha map. (b) The opaque, transparent, and boundary MBs

features and consumed limited processing power. To reduce the implementation cost and keep the interoperability, MPEG-2 is the first introducing the concept of profile and level, such that encoders and decoders compliant to the same profile and level could successfully interwork with each other without supporting the full syntax. A profile describes the necessary features; a level indicates the processing capability by specifying the upper limits of spatial resolution, frame rate, and bit-rate. Among the seven profiles in MPEG-2, the Main Profile at the Main Level was the most widely used for TV broadcast.

2.5 MPEG-4 Visual

MPEG-4 was the first attempt to standardize the content-based coding techniques. Although further improving coding efficiency, MPEG-4 Visual [11] went significantly beyond the pure hybrid coding scheme and provided a rich set of tools to code natural and synthetic objects.

The unit of video content is called video object (VO), e.g., a talking person without background. A VO is taken as a sequence of snapshots of an arbitrary-shaped object; its time instances are called video object planes (VOP). As successive VOPs are highly correlated, MCP is also efficient, but the concepts of I-, P-, and B-frame are herein extended to I-, P-, and B-VOP, respectively.

If the shape of a VOP is a frame, i.e., VO is equivalent to a video sequence, the content-based coding scheme is reduced to the conventional hybrid coding. The Simple Profile of MPEG-4 is compatible with the baseline H.263, but MPEG-4 introduces new techniques to improve the coding efficiency, including AC/DC intra-prediction, horizontal scan, global motion compensation (GMC), GMC based on Sprite, 8×8 motion-compensation (MC) block size, and 1/4-pixel MCP.

If VO is of arbitrary shape and location, the coding scheme is extended by additionally coding the shape, represented by a binary or an 8-bit gray-scaled alpha map. The former (see Fig. 3 (a)) defines whether a pixel belongs to an object; the later describes the transparency information. The alpha map is coded on the MB base. For each opaque and transparent MB, labeled 1 and 2 respectively in Fig 3 (b), the encoder just signals whether the MB is inside or outside the VOP. For an MB containing VOP boundaries, labeled 3 in Fig. 3 (b), the alpha map is coded by a context-based arithmetic coder and the VOP texture inside the boundary MB is transformed by the shape-adaptive DCT.

Meanwhile, MPEG-4 took into account specificities of a wide variety of networks and terminals and allows universal access to multimedia information, by employing techniques for a high degree of scalability and error resilience. A comprehensive overview of MPEG-4 Visual can be found in [12], which includes the nature and synthetic video coding, error resilience tools, and scalable coding.

3 H.264/AVC

H.264/AVC [13] is the latest international video coding standard, jointly developed by VCEG and MPEG. In contrast to MPEG-4 Visual covering the ambitious breadth of functionalities, H.264/AVC returns to the narrower and more traditional focus on two goals, efficient compression and robustness to network environments, for generic rectangular-picture camera-shot video content. A comprehensive introduction of H.264/AVC can be found in [14].

This section first illustrates the history of H.264/AVC in Section 3.1. Then, detailed technical designs to improve coding efficiency and facilitate the network transportation are presented in Sections 3.2 and 3.3, respectively. Section 3.4 explains the profiles in H.264/AVC and the corresponding application areas. The recently finalized amendments on scalable video coding (SVC) and multi-view video coding (MVC) are another two highlights, introduced in Sections 3.5 and 3.6, respectively.

3.1 History

The H.264/AVC standardization project was launched by VCEG in early 1998, initially entitled H.26L (the L stood for "long-term"). It was targeted at doubling the coding efficiency in comparison to any other existing video coding standards for a broad variety of applications. The first draft was created in August 1999. In 2001, MPEG issued a Call for Proposals (CfP) with the similar target of H.26L and then adopted the developing H.26L proposed by VCEG as the starting point. Meanwhile, VCEG and MPEG formed a Joint Video Team (JVT) to jointly develop the new standard. The reference software, continuously integrating adopted coding tools, was named joint model (JM). In 2003, the first phase of the standard was finalized and submitted for formal approval, which was later be adopted by ITU-T under the name of H.264 and by ISO/IEC as MPEG-4 Part 10 Advanced Video Coding (AVC) in the MPEG-4 suite of standards. An overview of the first phase of H.264/AVC was presented in [15].

The first version, focusing on entertainment-quality videos with 4:2:0 color format and 8 bit/sample sources, did not support the highest video resolutions used in the most demanding professional environments. To address the needs of professional applications, a continuation of the joint project was launched to add

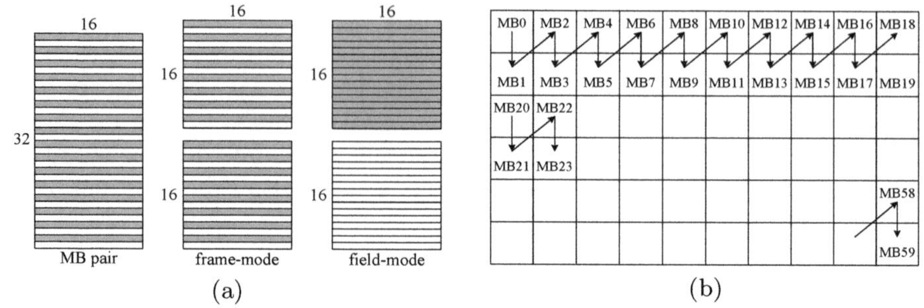

Fig. 4. The concept of MB pair. (a) Coding modes. (b) Coding procedure.

new extensions, named fidelity range extensions (FRExt), to the capabilities of the original standard. FRExt, finalized in 2005, produced a suite of four new profiles, collectively called high profiles. Later in 2007, the breadth of the high profiles were enlarged to eight profiles. In this chapter, the referred H.264/AVC includes the high profiles. An overview of H.264/AVC FRExt is given in [16].

In recent years, SVC and MVC were developed as two amendments to H.264/AVC, driven by the boost of application and network capability. Their histories and backgrounds are introduced in Sections 3.5 and 3.6, respectively.

3.2 Video Coding Layer (VCL)

The VCL design of H.264/AVC essentially follows the hybrid coding scheme, as introduced in Section 1, but significantly outperforms all previous video coding standards. The gain is due to more accurate prediction and more efficient entropy coding, contributed by a plurality of coding elements.

Adaptive Frame/Field Coding

A coded video sequence of H.264/AVC consists of a sequence of coded pictures. In an interlaced video, a coded picture represents either an entire frame or a single field. Two coding modes, PAFF and MBAFF, as in MPEG-2 (see Section 2.4), are also employed. However, as shown in Fig. 4, MBAFF herein makes the frame/field decision at the MB pair level, a 16×32 luminance region, rather than the MB level, so that the basic MB processing structure is kept intact. Fig. 4 (b) illustrates the MBAFF coding process in an interlaced frame.

If an interlaced frame consists of mixed regions, where some regions are moving and others are not, MBAFF coding is efficient to code the non-moving regions in frame-mode and the moving regions in field-mode. However, in the case of rapid global motion and scene change, MBAFF coding is not so efficient as PAFF coding, because one field cannot use the opposite field in the same frame as a reference picture for MCP.

Slice

H.264/AVC supports flexible slice structure, which means the sizes and shapes of slices are not necessarily the multiples of MB rows as in the previous video coding standards. There are five types of slices as explained below, and a coded picture may be composed of different types of slices.

- I-Slice. A slice in which all MBs of the slice are coded as intra-mode.
- P-Slice. In addition to intra-mode, some MBs can be coded using MCP from picture list 0.
- B-Slice. In addition to the coding modes available in a P-slice, some MBs can be coded using bi-directional prediction from picture list 0 and list 1.
- SP-Slice. A so-called switching P-slice that is coded such that efficient switching between different pre-coded pictures becomes possible (see Section 3.3).
- SI-Slice. A so-called switching I-slice that allows an exact match of an MB in an SP-slice for random access and error recovery purposes (see Section 3.3).

Multipicture Reference

H.264/AVC supports multipicture reference, which means more than one previously coded pictures can be used as references for MCP.

Generalized P- and B-Slices

At the decoder, the buffer storing reference pictures is named decoded picture buffer (DPB), which synchronously replicates the buffer of the encoder according to memory management control operations (MMCO), specified in the bitstream. DPB can hold up to 16 pictures, depending on the conformance point and picture size. Any picture type, including B-picture, is allowed to be stored in DPB for future reference, according to the indication in its NAL unit header (see Section 3.3). Thus, the picture type and the referencing capability are decoupled.

Two lists of reference pictures, denoted as list 0 and list 1, can be arbitrarily constructed using the available pictures in the DPB. P-slices use only list 0 pictures for reference, whereas B-slices can use both list 0 and list 1. However, it is not restricted that list 0 and list 1 consist of temporally preceding and succeeding pictures, respectively. Instead, list 0 may contain temporally succeeding pictures and vice versa. It is also possible for list 0 and list 1 to have same pictures. Fig. 5 shows the possible prediction cases for a B-picture, which have never been valid in any previous standards. By this means, the prediction direction and the picture type are decoupled.

Therefore, the only substantial difference between P- and B-slices is that B-slices are coded in a manner that some MC blocks may use a weighted average of two distinct MCP values for building the prediction signal.

Reference Picture Management

The reference pictures in list 0 or list 1 can be classified as short-term and long-term. By default, a short-term picture is a recently coded reference picture, whereas long-term pictures are typically older reference pictures.

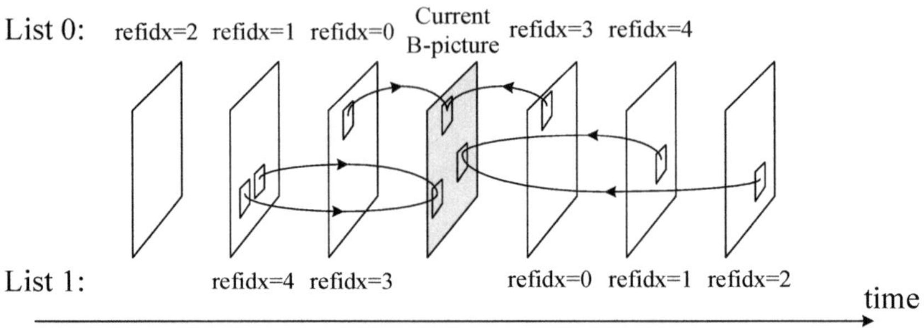

Fig. 5. The construction of list 0 and list 1, and possible bi-directional predictions for B-pictures (reference index is denoted as refidx.)

Each reference picture in a picture list is identified by an index, called reference index, which together with an MV locates a target block for MCP. Among short-term pictures, the one temporally closest to the current picture has reference index equal to 0, and the reference index increases with the temporal distance (see Fig. 5). Each new short-term picture added to the picture list has a reference index equal to 0 and the reference index of each remaining short-term picture is increased by 1. If the number of short-term and long-term pictures reaches the maximum capacity of the picture list, the temporally farthest short-term picture, having the largest reference index, is removed from the DPB.

The operations on long-term pictures, including marking, removing, and replacing, are instructed by MMCO specified in bitstreams. When a short-term picture is marked as a long-term picture, a variable LongTermPicNum is assigned for identification. A long-term picture with smaller LongTermPicNum has a smaller reference index in a picture list and the reference index of long-term pictures starts after all the short-term pictures. Table 1 gives an example of reference picture management for P-pictures, where the size of picture list is five and the current frame number is 100.

The reference picture management introduced in H.264/AVC provides a high degree of flexibility in selecting reference picture, which is constrained only by a total memory capacity bound. This feature is especially efficient for coding videos with periodic transitions, such as interview.

Intra Coding

In contrast to MPEG-1/2/4, where intra-prediction is performed in the transform domain, intra-prediction in H.264/AVC is always conducted in the spatial domain. For the luminance pixels in an MB, three types of partitions are allowed, denoted as INTAR_4×4, INTAR_8×8 and INTAR_16×16; the intra-prediction size for chrominance pixels are fixed to be 8×8.

There are totally nine modes for INTAR_4×4. As shown in Fig. 6, the shaded area is the 4×4 block to predict and the squares labeled with capital letters A-M are the reference pixels in the neighboring coded blocks. Each mode indicates

Table 1. An example of reference picture management for P-pictures

Operation	Reference picture list: reference index 0	1	2	3	4
Initial state	-	-	-	-	-
Encode frame 100	100	-	-	-	-
Encode frame 101	101	100	-	-	-
Encode frame 102	102	101	100	-	-
Encode frame 103	103	102	101	100	-
Assign 101 to LongTermPicNum 0	103	102	100	0	-
Encode frame 104	104	103	102	100	0
Assign 103 to LongTermPicNum 4	104	102	100	0	4
Encode frame 105	105	104	102	0	4
Assign 105 to LongTermPicNum 3	104	102	0	3	4
Encode frame 106	106	104	0	3	4

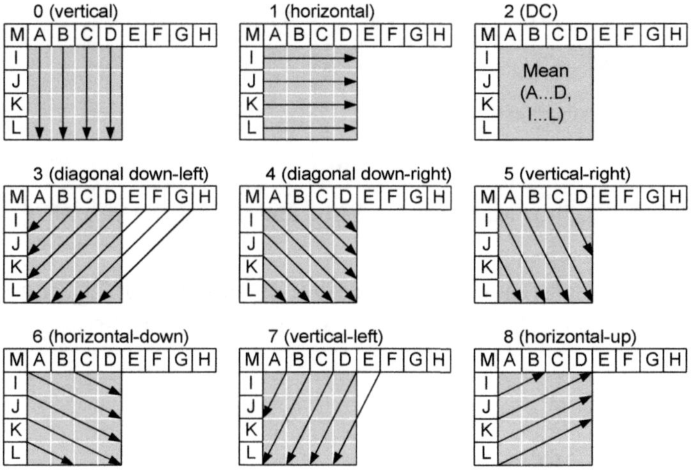

Fig. 6. Nine modes for INTAR_4×4 prediction [14]

a prediction direction except the DC mode, of which the predicted value is the average of all the available reference pixels. INTAR_8×8 for luminance pixels uses basically the same concepts as INTAR_4×4 except the size, but the reference pixels are low pass filtered before used for prediction, in order to reduce prediction error in a larger block.

The intra-prediction with smaller block sizes and more directions is well suited for coding parts of a picture with rich details. On the contrary, for coding very smooth areas, INTAR_16×16 is employed with four modes (see Fig. 7). The 8×8 chrominance pixels within an MB are predicted using a similar prediction as for INTAR_16×16, since chrominance is usually smooth over large areas.

Besides, the standard includes a PCM mode, denoted as I_PCM, with which the values of the samples are sent directly using fixed length coding (FLC).

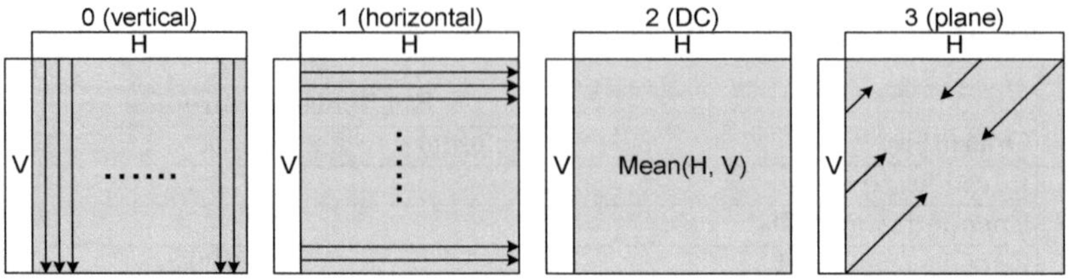

Fig. 7. Four modes for INTAR_16×16 prediction and chrominance intra-prediction [14]

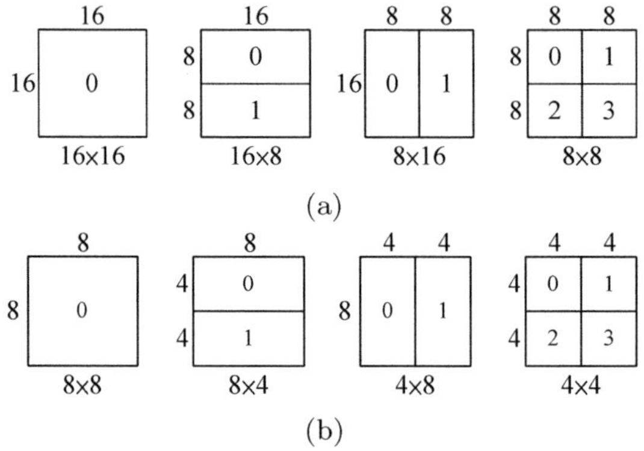

Fig. 8. The partitions of (a) MB and (b) sub-MB for MCP

This mode enables lossless coding for region of interests (ROI) and avoids data expansion, when representing the values in anomalous pictures.

Inter-Coding

Variable MC Block Sizes

In inter-mode, an MB may be partitioned into multiple MC blocks. Fig. 8 (a) shows four partition types for luminance pixels. In case partitions with size 8×8 are chosen, each partition, named sub-MB, can be further partitioned into 8×4, 4×8, or 4×4 MC blocks (see Fig. 8 (b)).

Bigger MC blocks, using less bits to signal the motion information, may result in large MCP errors, and therefore are more appropriate for homogeneous areas of a picture; smaller MC blocks provide more accurate MCP but require more bits for the motion information, so may be beneficial to detailed areas.

MV Prediction

MV in H.264/AVC is predicted from the MVs of the neighboring coded MC blocks and only the prediction error, MVD, is coded. Similar concept has been adopted in previous standards, but the MV prediction in H.264/AVC is improved and more accurate.

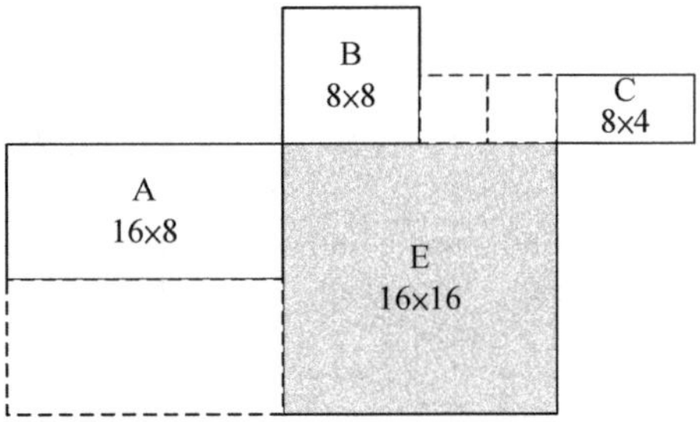

Fig. 9. The adjacent MC blocks involved in MV prediction

As shown in Fig. 9, E is the current block, of which the MV is to predict, and only three adjacent blocks, A, B, and C, are involved to predict the MV of E. The prediction value, MVP, is derived as below.

1. If the size of E is neither 16×8 nor 8×16, the median of the MVs of A, B, and C is used as MVP.
2. For 16×8 MB partition, MVP for the upper 16×8 block is the MV for B and MVP for the lower 16×8 block is the MV for A.
3. For 8×16 MB partition, MVP for the left 8×16 block is the MV for A and MVP for the right 8×16 block is the MV for C.

In case A, B, and C in Fig. 9 are not all available, e.g., outside the current slice or in intra-mode, the derivation of MVP will be modified accordingly.

Interpolation for MCP

For luminance components, the accuracy of MCP is in units of 1/4 distance between pixels and thus the resolution of MV is up to 1/4-pixel. The interpolation for half-pixel samples is separable: a 1-D 6-tap FIR filter [1, -5, 20, 20, -5, 1]/32 is applied for horizontal and vertical directions successively. The samples of 1/4-pixel positions are interpolated using bilinear filter supported by integer pixels and the interpolated half-pixel values. The positions in between chrominance pixels are interpolated by bilinear filtering. Assuming 4:2:0 color format, 1/4-pixel MCP resolution for luminance component corresponds to 1/8-pixel resolution for chrominance components.

Weighted Prediction

In all the previous video coding standards, the target block is directly copied for MCP, or in B-mode, the MCP is the average of the preceding and succeeding target blocks. Such a prediction method is supported by H.264/AVC by default. In addition, H.264/AVC introduces weighted prediction, where the samples of the target blocks are scaled by weighting factors before used for MCP. The weighting factors, w_0 and w_1, can be customized by encoders and transmitted in the slice header, or implicitly derived specially for B-mode in the inverse proportion to the relative temporal distance from the current picture to the reference picture.

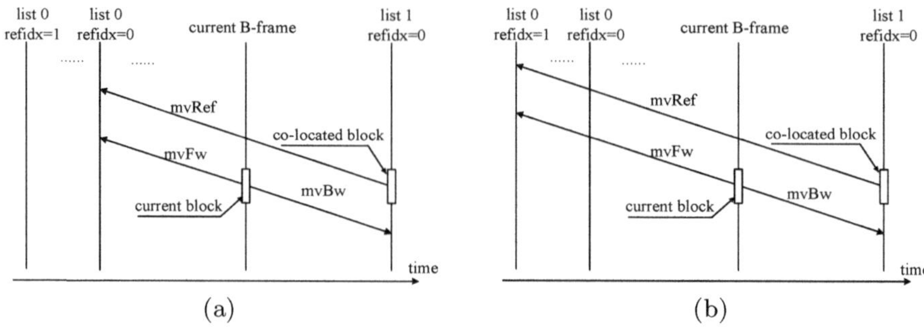

Fig. 10. Derivation of MV and reference index in temporal direct prediction when the MV of the co-located block references the list 0 picture with (a) reference index equal to 0 and (b) reference index equal to 1

Weighted prediction, adaptively controling the relative contributions from target blocks, is especially useful for scene transitions and illumination change.

Improved "Skip" and "Direct" Modes

A skipped MB in previous standards could not move in the scene, which had a detrimental effect for coding video containing global motion. In H.264/AVC, the "Skip" mode for an MB implies no motion in relation to the neighboring MBs rather than absolute rest, such that large areas without change or with constant motion like slow panning can be represented by very few bits.

In P-slices, the "Skip" mode is denoted as P_Skip, with which neither transform coefficients nor motion information are transmitted. The reconstructed MB is exactly the same as its prediction, obtained by using reference index and MVD both equal to 0. In B-slices, the mode similar to P_Skip is called B_Skip, which uses "direct prediction" to infer the MV and reference index. Another mode that also uses direct prediction but requires residual data is known as "Direct" mode, denoted as B_Direct_16×16.

The direct prediction for B_Direct_16×16 and B_Skip modes can be spatial or temporal, as indicated in the slice header. With spatial direct prediction, the MB is predicted as one partition and its MV is derived using the aforementioned MV prediction method. The MV direction can be list 0, list 1, or bi-predictive, depending on the directions of the MVs of the adjacent blocks A, B, and C (see Fig. 9). In some cases, the MV is directly set to zero [13]. In temporal Direct mode, the current MB is partitioned the same way as the co-located MB belonging to the reference picture indexed 0 in list 1. Derivation of the MVs and reference pictures for each partition is illustrated in Fig. 10. The corresponding partitions in the current and co-located MBs use the same list 0 picture. The forward and backward MVs of the current partition are derived by scaling the forward MV of the co-located partition according to the relative temporal distance.

The same spatial and temporal "Direct" concepts can be applied to 8×8 MC blocks, denoted as B_Direct_8×8 mode.

Transform and Quantization

H.264/AVC is the first video coding standard employing an Integer Cosine Ttransform (ICT) rather than DCT, which allows implementation approximations within some specified tolerances [17] (replaced by [18] in 2006). A significant advantage of using ICT is that, with an exact integer inverse transform, the encoder and the decoder perfectly match.

The intra- or inter-prediction error of the luminance pixels in an MB may be split into 4×4 or 8×8 blocks, to which 4×4 and 8×8 ICTs are applied, respectively. The two transforms are adaptively selected for luminance residual data at the MB level, but chrominance pixels are all transformed by the 4×4 ICT. The two transform matrices are given as below.

$$T_4 = \begin{bmatrix} 1 & 1 & 1 & 1 \\ 2 & 1 & -1 & -2 \\ 1 & -1 & -1 & 1 \\ 1 & -2 & 2 & -1 \end{bmatrix} \qquad T_8 = \begin{bmatrix} 8 & 8 & 8 & 8 & 8 & 8 & 8 & 8 \\ 12 & 10 & 6 & 3 & -3 & -6 & -10 & -12 \\ 8 & 4 & -4 & -8 & -8 & -4 & 4 & 8 \\ 10 & -3 & -12 & -6 & 6 & 12 & 3 & -10 \\ 8 & -8 & -8 & 8 & 8 & -8 & -8 & 8 \\ 6 & -12 & 3 & 10 & -10 & -3 & 12 & -6 \\ 4 & -8 & 8 & -4 & -4 & 8 & -8 & 4 \\ 3 & -6 & 10 & -12 & 12 & -10 & 6 & -3 \end{bmatrix}$$

Given 8-bit input video data, the transforms can be easily implemented using 16-bit arithmetic. Especially, the 4×4 transform is so simple that can be implemented using just a few additions and bit shifts.

As mentioned above, INTAR_16×16 prediction modes are intended for coding of smooth areas and chrominance components are usually smooth over large areas. For that reason, the DC coefficients of the sixteen transformed 4×4 luminance blocks in the MB coded by INTAR_16×16 mode are extracted to form a 4×4 block, which is transformed by a secondary Hadamard transform (HT). Similarly, the DC coefficients of chrominance blocks with all MB types are also transformed using a secondary HT. For 4:2:0 color format, this requires a 2×2 HT. Fig. 11 illustrates such a hierarchical transform procedure for 4:2:0 videos, where the index labeled in each block shows the coding order.

After transform, quantization is applied to meet the target bit-rate. By default, the coefficients in one MB are quantized by the same stepsize determined by QP, an integer ranging from 0 to 51. The value of QP is not linearly related to the stepsize as in all previous standards, but influences the stepsize in such a way that the stepsize increases about 16% for one increament and exactly doubles for every six increments. Rather, the value of QP is linearly related to the actual bit-rate.

Besides the default one, H.264/AVC also supports perceptual-based quantization, which had been a mainstay of prior use in MPEG-2. Although the standard defines a default quantization matrix, an encoder is allowed to specify customized quantization matrices and send them at the sequence or picture level. Such perceptual-based quantization typically does not improve objective

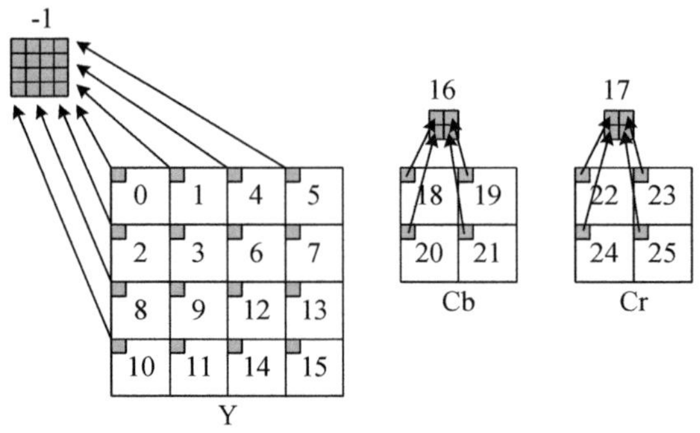

Fig. 11. The hierarchical transform procedure for 4:2:0 videos and the block coding order

quality measured by mean square error (MSE), but it does improve subjective quality, which is eventually the more important criterion.

Entropy Coding

In H.264/AVC, two entropy coding modes are supported. One is a Huffman-like coding, which means each symbol is converted to a binary codeword with integer number of bits, by look-up tables or dynamic codeword construction. The other is the context-adaptive binary arithmetic coding (CABAC). The two modes are switchable at the picture level. The SEs at the higher levels are all coded by FLC or VLC with Exp-Golomb codes. In the mode of Huffman-like coding, the SEs other than the residual data are also coded using Exp-Golomb codes, whereas the residual data are coded using context-adaptive VLC (CAVLC).

Exp-Golomb Codes

Exp-Golomb codes have a generic form of [M zeros][1][INFO] (see Table 2), where INFO is an M-bit field carrying information. Hence, the Exp-Golomb codeword table, which can be constructed in a logical way, is conceptually infinite in length. With Exp-Golomb codes, the standard only defines the mapping from the value of the SE to the look-up index, an unsigned integer denoted as codeNum (see Table 2), according to the probability distribution function (pdf) of the SE.

CAVLC

CAVLC is designed only for coding 4×4 transform blocks. The values and positions of the 1-D 16 coefficients are coded separately, instead of in (level, run) pairs. The coding process is in the reverse zig-zag order, because the trailing non-zero coefficients have high probability to be ±1, whereas the values of leading non-zero coefficients are more random. The events necessary to represent all the information of a block are listed as below. An example of converting the coefficient sequence to events is given in Fig. 12.

Table 2. The Exp-Golomb codes

codeNum	Codeword
0	1
1	010
2	011
3	00100
4	00101
5	00110
6	00111
7	0001000
8	0001001
...	...

		0	1	2	3	4	5	6	7	8	9	10	11	12	13	14	15
Zig-Zag Order		0	1	2	3	4	5	6	7	8	9	10	11	12	13	14	15
Coefficients		0	7	0	0	-4	2	0	1	-1	0	0	1	0	1	0	0
CAVLC	levelCode/T1s		12			7	2		0	T1			T1		T1		
	run_before/zerosLeft		1/1			2/3	0/3		1/4	0/4			2/6		1/7		
	←																
CABAC	significant_coeff_flag	0	1	0	0	1	1	0	1	1	0	0	1	0	1		
	last_significant_coeff_flag		0			0	0		0	0			0		1		
	coeff_abs_level_minus1		6			3	1		0	0			0		0		
	coeff_sign_flag		0			1	0		0	1			0		0		
	→																

Fig. 12. Represent a sequence of transform coefficients by symbols with CAVLC and CABAC

1. TotalCoeff. The number of non-zeros coefficients in the block, jointly coded with T1s as one SE coeff_token. In Fig. 12, TotalCoeff equals 7.
2. Tailing ones (T1s). The number of the successive ±1 coefficients at the end of the coefficient sequence. Note that the maximum value for T1s is 3 and more trailing ±1–if any–will be coded as regular coefficients.
3. trailing_ones_sign_flag. The sign for each T1.
4. levelCode. Levels of regular coefficients, including the magnitude and the sign. If the level is positive, levelCode equals (level-1)<<1. Otherwise, levelCode equals (-level-1)<<1+1. It is represented by two SEs, level_prefix and level_suffix.
5. total_zeros. The number of zeros before the last non-zero coefficient. In Fig. 12, it equals 7.
6. run_before. Number of successive zeros before each coefficient.

In summary, there are six SEs related to CAVLC, coeff_token, trailing_ones_sign_flag, level_prefix, level_suffix, total_zeros, and run_before. For each SE except trailing_ones_sign_flag, which is a 1-bit flag, multiple VLC tables are designed;

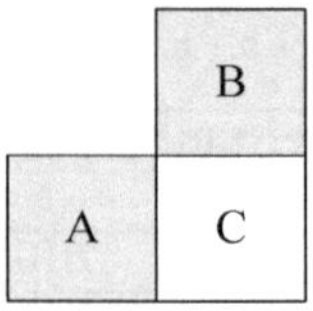

Fig. 13. The context template

each table matches certain conditional statistics. The switching of these tables depends on the local activities, i.e., the context. The meaning of the context for each SE is explained as below.

1. coeff_token. The values of TotalCoeff in neighboring blocks are correlated and can be used as the context. As shown in the context template (see Fig. 13), the table used to code coeff_token in block C is jointly decided by the values of TotalCoeff in A and B.
2. level_prefix and level_suffix. There is no table specified in H.264/AVC for coding levelCode. The codeword for levelCode is the concatenation of the binary representation of level_prefix and level_suffix, both of which have regular forms. The length of the suffix, ranging from 0 to 6, categorizes the codewords into seven conceptual tables, indexed from 0 to 6. The tables in order of ascending indices cater for seven types of contexts from the high frequency coefficients to low frequency coefficients. The switching among the tables is determined by the maximum magnitude of the coded levels in the current block.
3. total_zeros. The choice of the table depends on the TotalCoeff in the current block. Since the summation of total_zeros and TotalCoeff will not exceed 16, large TotalCoeff implies small total_zeros and vice versa.
4. run_before. The choice of the table depends on zerosLeft (see Fig. 12), the number of zeros that have not been processed yet. The range of run_before decreases with zerosLeft.

For coding 8×8 transform blocks, there is no specially designed CAVLC scheme; the 64 coefficients have to be separated into four 4×4 blocks (see Fig. 14), which are successively coded using CAVLC.

CABAC

CABAC is used for coding a broader range of SEs than CAVLC, including slice header, MB header, and residual data. Compared to CAVLC, CABAC typically provides a bit-rate reduction between 5%-15%, although much more computationally complex. The improvement is mainly contributed by three design aspects. Firstly, arithmetic coding allows one to assign a non-integer number of bits per symbol, which is especially beneficial to symbol probabilities greater than 0.5. Secondly, the context modeling allows each SE to have more than one probability models to estimate the conditional probability distributions for different local activities, i.e., different contexts. Thirdly, the probability models can

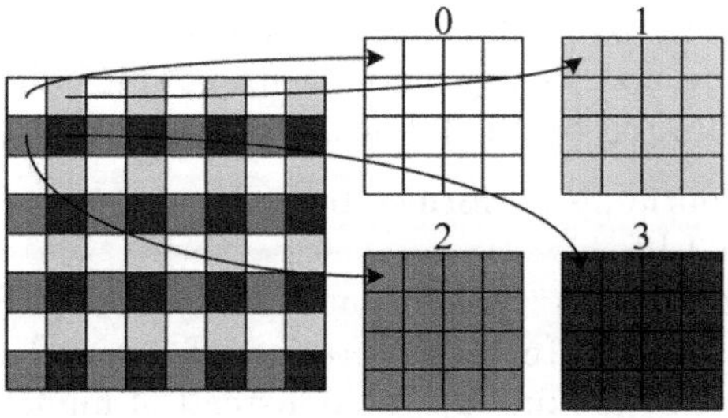

Fig. 14. Separate coefficients of an 8×8 transform block into four 4×4 transform blocks for 8×8 CAVLC

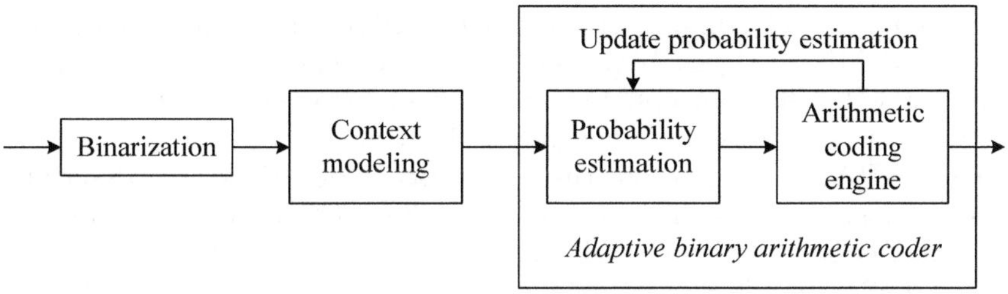

Fig. 15. Generic block diagram of the CABAC entropy coding scheme

be updated to keep track of the actual statistics experienced during the coding process. The diagram of CABAC is depicted in Fig. 15.

The first step is binarization. Since CABAC uses binary arithmetic coding, a non-binary valued SE should first be converted into a binary string. Note that the binary string, still containing much statistical redundancy, is not part of the bitstream. It is only the input to the arithmetic coding engine, of which the output is the compact bitstream.

In the second step of context modeling, each binary decision in a binary string, called "bin", is assigned an appropriate context model, which estimates the probability distribution of the bin. A context model records two types of information. One is the most probable symbol (MPS), which is either 0 or 1. The other is the probability of the least probable symbol (LPS) P_{LPS}, ranging from 0 to 0.5, is discretized into 64 probability states rather than continuous. In case a context model is used for a single bin or is shared by several bins, there is no need to do the context model selection. Otherwise, the selection among available models is made according to the values of the relevant SEs in the context template (see Fig. 13). After context modeling, each bin as well as its probability distribution, described by the associated context model, are input to the arithmetic coding engine.

The third step is the arithmetic coding. Conventional arithmetic coding is based on the principle of recursive interval subdivision, which introduces multiplications. In H.264/AVC, the arithmetic coding is specified as a multiplication-free low complexity method, using only shifts and table look-ups, which is mainly facilitated by the following three properties.

1. Probability estimation is performed by a transition process between 64 probability states for LPS.
2. The range R representing the current state of the arithmetic coder is quantized to pre-set values. In each iteration of interval subdivision, the new range is obtained by looking up table instead of multiplication.
3. SEs with near-uniform probability distributions are coded by a simplified process, where the context modeling is bypassed,

Each context model has an initial probability estimate defined by the standard and is reset at the beginning of every slice. After the coding of each bin, the context model that has just been used is updated depending on the coded bin is an MPS or LPS. Such an update behavior will change the probability state or even cause the transition between LPS and MPS.

When CABAC is used to code the a transform block, the events describing the values and positions of coefficients are quite different from those of CAVLC. First of all, a flag coded_block_flag is signaled to indicate whether the block contains non-zero coefficients. If so, the coding process is then performed in the order of forward zig-zag scan. As the example in Fig. 12, the value of each coefficient is coded by coeff_abs_level_minus1 and coeff_sign_flag, representing the magnitude and the sign, respectively. The positions of these coefficients are located by a significant coefficient map, which is constructed by significant_coeff_flag and last_significant_coeff_flag. Representing a transform block using such a set of events, which are all binary-valued except coeff_abs_level_minus1, can facilitate the use of binary arithmetic coding.

In-loop Deblocking Filter

In-loop deblocking filters have been adopted in previous video coding standards, such as H.261 and H.263 (Annex J), to reduce the blockiness introduced by discontinuity of motion and block-based transform and quantization. The deblocking filter in H.264/AVC is brought within the MCP loop (see LF_1 in Fig. 1), which leads to not only better subjective and objective qualities but also improved capability to predict other pictures. The deblocking filter is applied to the vertical and horizontal boundaries of 8×8 and 4×4 transform blocks. Each filtering operation affects up to three pixels on either side of the boundary. Fig. 16 shows four pixels on either side of a vertical boundary in adjacent blocks P and Q, where the values of $p0$, $p1$, $p2$ and $q0$, $q1$, $q2$ may be changed by filtering. There are three steps for filtering one block.

First of all, the parameter, boundary strength (BS), is calculated for each boundary, ranging from 0 (no filtering) to 4 (strongest filtering). The value of BS implies the possible extent to which the two adjacent blocks suffer from the blockiness, and thus is used to select an appropriate filter. Fig. 17 shows the BS determination procedure used in progressive video coding, which depends on the

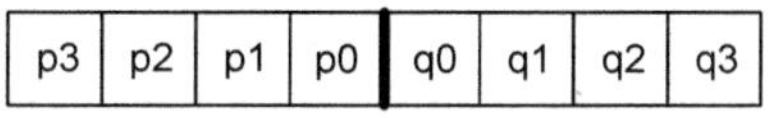

Fig. 16. Pixels on either side of a vertical boundary of adjacent blocks P and Q

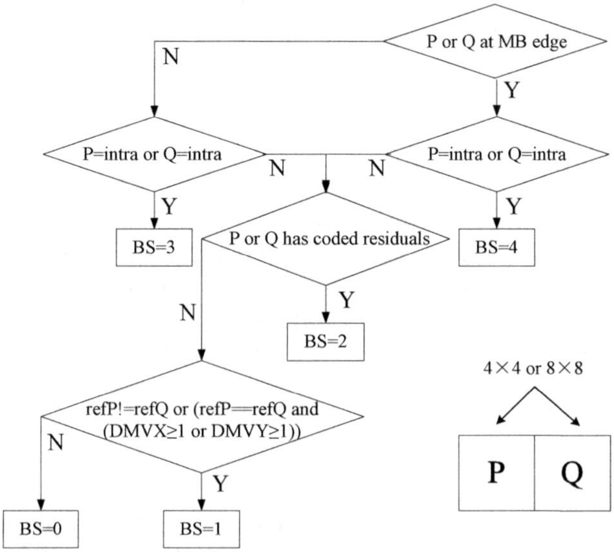

Fig. 17. BS determination procedure used in progressive video coding

boundary location, the coding mode (intra or inter), the existence of non-zero coefficients, and the motion information. In the figure, refP and refQ are the reference information for P and Q, including the direction (list 0, list 1, or bi-prediction) and the reference picture for each direction. "DMVX≥1" means the horizontal distance of the P and Q's target blocks in the same reference picture is greater than or equal to one integer pixel. For coding interlaced videos, the concept is similar, but more logical decisions are involved.

Secondly, while the filter has been selected by BS at the block level, whether to apply such filtering operation is left to be decided for each set of pixels across the boundary (see Fig. 16). The filter decision is made by thresholds $\alpha(x)$ and $\beta(x)$, which decrease with x. The index x is by default the average of the QPs used in P and Q, but can be adjusted by adding an offset transmitted in the slice header. The filtering of $p0$ and $q0$ only takes place if the following condition is true,

$$|p0 - q0| < \alpha(x) \text{ and } |p1 - p0| < \beta(x) \text{ and } |q1 - q0| < \beta(x)$$

where $\beta(x)$ is considerably smaller than $\alpha(x)$. The filtering of $p1$ and $q1$ takes place if the following condition is true.

$$|p2 - p0| < \beta(x) \text{ and } |q2 - q0| < \beta(x)$$

The basic idea is that a relatively large absolute difference between pixels near a block boundary implies blockiness and should therefore be reduced. However, if the magnitude of that difference is so large that it cannot be explained by the coarseness of the quantization, the boundary is more likely to reflect the actual edge in the source picture and should be preserved.

Finally, the filtering operation is applied to where is necessary. The filter varies with BS values, pixel positions, and color components. More details can be found in the standard itself [13].

Designs for 4:4:4 Color Format

As the 4:4:4 color format is usually used in the most demanding video applications, special requirements, such as very high or even lossless fidelity, should be fulfilled. Furthermore, in contrast to 4:2:0 color format, chrominance components are represented in full spatial resolution and thus should be coded using the techniques as powerful as those for luminance components in order to improve the overall performance.

Lossless Coding

The PCM mode in H.264/AVC is developed only to enable the functionality of lossless coding, regardless of coding efficiency. The FRExt includes a transform-bypass lossless mode, where the transform and quantization are bypassed and the intra- or inter-prediction errors are entropy coded in the spatial domain. When a block is coded using intra-prediction with the direction horizontal or vertical, a special method is applied, where the samples along the prediction direction are coded using DPCM. The lossless coding methods introduced herein can also be used for monochrome, 4:2:0, and 4:2:2 videos, but are only enabled in the 4:4:4 profiles (see Section 3.4).

Color Plane Separated Coding

The three color planes in 4:4:4 videos can share the common slice structure and MB coding parameters as for 4:2:0 videos. Alternatively, each color plane can be treated as a separate monochrome video picture and coded individually with its own slice segmentation and mode selections, using the powerful coding tools for luminance components. Furthermore, such a color plane separated coding scheme provides an enhanced parallel processing solution for the encoder by allowing the splitting of the entire encoding processes of the three color planes. The switching of the two coding schemes is indicated by a flag separate_colour_plane_flag in the sequence parameter set.

3.3 Features for Network Transportation

Robustness to data errors and losses and flexibility for operation over a variety of network environments are enabled by a number of new design aspects in H.264/AVC, including the following highlighted features.

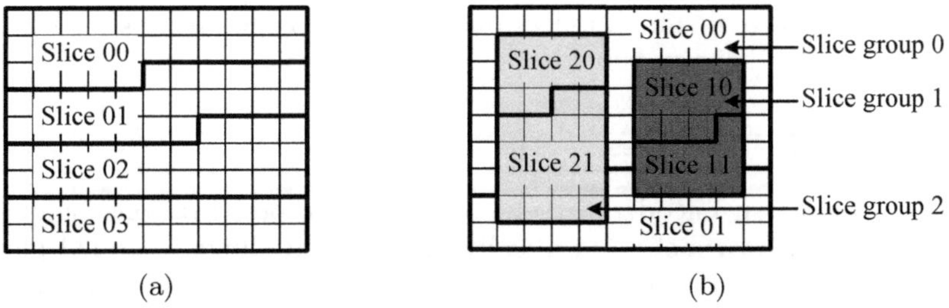

Fig. 18. Subdivision of a QCIF frame into slices (a) without FMO and (b) with FMO.

Parameter Sets

Transmission errors of key information, such as sequence header or picture header, have a severe impact on the decoding process. The parameter set design in H.264/AVC provides robust and efficient conveyance of header information. There are two types of parameter sets, sequence parameter sets (SPS) applied to a coded video sequence and picture parameter sets (PPS) applied to one or more pictures within a coded video sequence. Each VCL NAL unit contains an identifier, referring to the content of the relevant PPS; each PPS contains an identifier, referring to the content of the relevant SPS. In this manner, a small amount of data, i.e., the identifier, can be used to refer to a larger amount of information, i.e., the parameter sets. Separated from the transmission of the video data representation, SPS and PPS can be sent well ahead and repeatedly whenever necessary. It is also possible to convey SPS and PPS "out-of-band" using a more reliable transport mechanism.

Flexible MB Ordering (FMO)

FMO allows a picture to be partitioned into various MB scanning patterns rather than the raster scanning only, by utilizing the concept of slice group. Each slice group is a set of MBs defined by an MB to slice group map, which labels each MB with a particular slice group in the picture. Each slice group can be partitioned into one or more slices, and the MBs in each slice should have the raster scan order inside the belonged slice group. Fig. 18 shows how a QCIF frame is divided into slices. In Fig. 18 (a), where FMO is not used, the whole picture can be considered consisting of a single slice group. In Fig. 18 (b), where FMO is applied, the frame is split into three slice groups, indicated with different shades, and each slice group is further split into two slices. The pattern in Fig. 18 (b) is suitable for coding ROI. Some patterns, such as dispersed and checker-board MB allocations, have been demonstrated to be useful for error concealment in video conferencing applications.

Arbitrary Slice Ordering (ASO)

H.264/AVC enables sending and receiving the slices of a picture in any order relative to each other. This feature, first found in H.263 (Annex K), can improve end-to-end delay in real-time applications, particularly when used in networks having out-of-order delivery behavior, e.g., IP networks.

Redundant Coded Pictures

H.264/AVC allows an encoder to send redundant representations of pictures or regions of pictures, which are typically degraded. Redundant coded pictures have no normative effect on the decoding process and will be decoded only if the relevant primary representation has been lost during data transmission.

Data Partitioning

The concept of data partitioning has been adopted in previous standards, such as H.263 (Annex V) and MPEG-4 Visual. H.264/AVC allows the SEs of each slice to be separated into three different partitions, Partitions A, B, and C, for transmission. Partition A contains the slice header and header data for each MB in the slice; Partitions B and C contain residual data for intra and inter-coded MBs, respectively.

Switching P- (SP-) and Switching I- (SI-) Pictures

In previous video encoding standards, the functionalities of bitstream switching, random access, fast forward/rewind, and error recovery are usually realized by periodically inserting I-pictures. In H.264/AVC, SP- and SI-pictures are developed to provide such functionalities in a more efficient way.

Fig. 19 (a) depicts how to utilize SP-pictures to switch from Bitstream 1 to Bitstream 2. The two bitstreams are assumed to represent the same video sequence at different bit-rates and/or at different temporal resolutions. In each bitstream, primary SP-pictures, such as $S_{1,n}$ and $S_{2,n}$, are periodically inserted to serve as switching points. For each primary SP-picture, the corresponding secondary SP-picture, such as $S_{12,n}$, is generated and sent in response to the switching requirement. A secondary SP-picture has the identical reconstructed values as the relevant primary SP-picture. When switching from Bitstream 1 to Bitstream 2 is required, $S_{12,n}$, predicted from Bitstream 1 but reconstructed as a picture in Bitstream 2, will be sent instead of $S_{1,n}$. Since the reconstructed pictures of $S_{12,n}$ and $S_{2,n}$ are identically the same, the pictures following the switching point, e.g., $P_{2,n+1}$, can be reconstructed without drift, no matter whether the switching has occurred. Note that in fact another secondary SP-picture not shown in the figure, $S_{21,n}$, would be required for switching in the other direction.

Fig. 19 (b) illustrates the switching between bitstreams representing different video sequences by using SI-pictures. SI-pictures are used in the similar way to SP-pictures, except that their relevant secondary pictures are coded in intra-mode. Thus, the switching point enabled by SI-pictures also serves as random access point.

Fig. 19 (c) gives an example of using SP-pictures for error resilience and recovery. When a bitstream is being streamed and there has been a packet loss

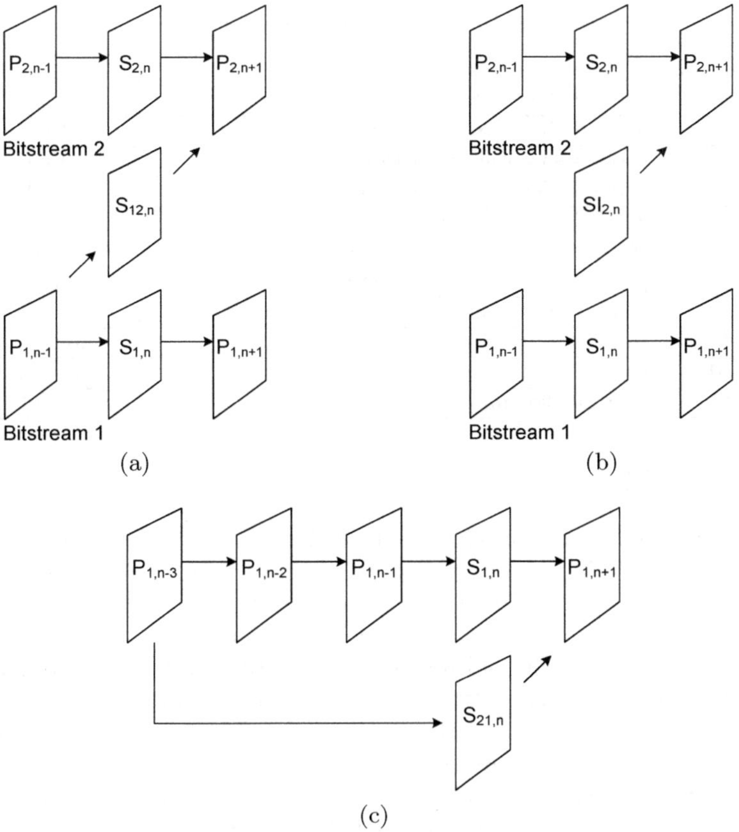

Fig. 19. (a) Switching between bitstreams using SP-frames. (b) Random access using SI-frames. (c) SP-frames in error resilience and recovery.

leading to one or more pictures lost, the client signals the lost pictures to the server, which then responds by sending a secondary SP-picture $S_{21,n}$ of the next primary SP-picture $S_{1,n}$. $S_{21,n}$ references picture $P_{1,n-3}$, which has been correctly decoded. Alternatively, the secondary representation of the next primary SP-picture can be coded as an SI-picture not referencing the previous correctly decoded pictures.

The detailed encoding and decoding processes of SP- and SI-pictures are introduced in [19].

Network Abstraction Layer (NAL)

Rather than forcing a specific bitstream interface to the system as in previous video coding standards, H.264/AVC designs the NAL, which enables simple and effective customization of the method of carrying the video content in a manner appropriate for a broad variety of networks.

Table 3. Types of NAL units

nal_unit_type	Content of NALU	NALU type class
1	Coded slice of a non-IDR picture	VCL
2	Coded slice data partition A	VCL
3	Coded slice data partition B	VCL
4	Coded slice data partition C	VCL
5	Coded slice of an IDR picture	VCL
6	Supplemental enhancement information (SEI)	non-VCL
7	Sequence parameter set (SPS)	non-VCL
8	Picture parameter set (PPS)	non-VCL
9	Access unit delimiter	non-VCL
10	End of sequence	non-VCL
11	End of stream	non-VCL

NAL Unit (NALU)

The coded video data are organized into logical data packets, called NAL units (NALU). Each NALU contains an integer number of bytes, where the first byte is a header and the remaining bytes contain the payload data. The header byte consists of three SEs, forbidden_zero_bit (1 bit), nal_ref_idc (2 bits), and nal_unit_type (5 bits). The nal_unit_type indicates the type of the payload in the NALU. The main NALU types and their corresponding content are shown in Table 3. As can be seen, NALUs are classified into two categories, VCL and non-VCL. The former comprises the data representing the values of the samples in the video pictures; the latter comprises any associated additional information, such as parameter sets and supplemental enhancement information (SEI), which enhances the usability of the decoded video signal but does not affect the normative decoding process.

The nal_ref_idc indicates the reference capability of the payload. If the payload is SPS, PPS, a slice or a slice data partitioning in an IDR or reference picture, which will be referenced for decoding other NALUs, the value of nal_ref_idc is not equal to 0, but no specific value is assigned in the standard. Otherwise, the value of nal_ref_idc equals 0.

NALU Stream

A series of NALUs generated by an encoder is referred to as a NALU stream, which may contain one or more coded video sequences. Each coded video sequence can be decoded independently, given the necessary parameter set information. The NALUs packeting the parameter sets may be conveyed "in-band" or "out-of-band". A coded video sequence consists of a series of access units. The decoding of each access unit results in one decoded picture. At the beginning of a coded video sequence is an instantaneous decoding refresh (IDR) access unit, which represents an I-picture. The presence of an IDR access unit indicates all following coded pictures in decoding order can be decoded without inter-prediction from any picture decoded prior to the IDR picture.

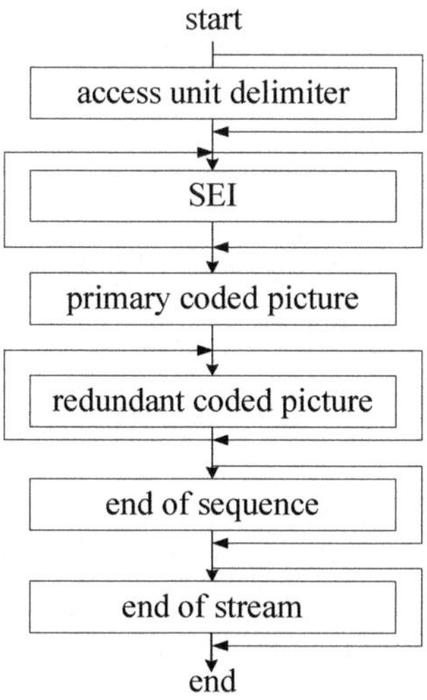

Fig. 20. The structure of an access unit

Fig. 20 shows the format of an access unit. Each access unit contains a primary coded picture, consisting of a set of VCL NALUs. It may also be prefixed with an access unit delimiter to aid in locating the start of the access unit. Some SEI may also precede the primary coded picture. Following the primary coded picture may be some additional VCL NALUs that contain redundant representations of areas of the same video picture, i.e., the aforementioned redundant coded pictures. Finally, if the access unit is the last one of a coded video sequence and/or the entire NALU stream, the NALUs with nal_unit_type equal to 10 and 11 may be presented for indication.

3.4 Profiles and Applications

As introduced in Section 2.4, a profile defines a set of coding tools used in generating a conforming bitstream. Well defined profiles and levels facilitate interoperability between various applications that have similar functional requirements. In the initial version of H.264/AVC finalized in 2003, only three profiles were supported, the Baseline, Main, and Extended Profiles. Fig. 21 shows the coding tools included in these profiles, where the core coding tools cover I-slice, P-slice, intra-prediction, variable MC block sizes, 1/4-pixel MCP, MV prediction, multipicture reference, 4×4 ICT as well as the hierarchical procedure, in-loop deblocking filter, and CAVLC.

The Baseline Profile includes the core coding tools and some error resilience tools. It may be used by conversational services operating typically below 1 mb/s with low latency requirements, such as video telephony, video conferencing, and wireless communications. The Main Profile includes more efficiency-oriented tools to favor entertainment-quality consumer applications operating from 1 to 8 mb/s with moderate latency, such as broadcast, video-on-demand (VOD) via

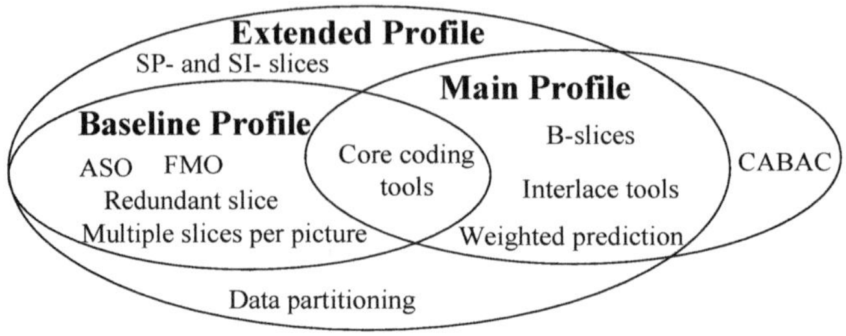

Fig. 21. The coding tools in H.264/AVC profiles

various channels, and high-density storage media. The Extended Profile, a superset of the Baseline Profile, provides improved error resilience and more efficient switching between coded bitstreams. It may be particularly useful for streaming services over the Internet, which typically operates at 50-1500 kb/s and tolerates higher latency.

The union of these three profiles covers all the coding tools in the initial version of H.264/AVC, but excludes those in the FRExt, developed by JVT from 2003 to 2007. The FRExt produces totally eight profiles, collectively named high profiles. Table 4 compares four high profiles, which build upon the design of the Main Profile. As can be seen, the difference in capability among the four high profiles is primarily in terms of the supported bit depths and the chrominance formats. New coding tools further enhancing the coding efficiency are also included, of which some are specially designed for videos with 4:4:4 color format.

The other four high profiles not shown in Table 4 are defined for all-intra coding. Among them, High 10 Intra, High 4:2:2 Intra, and High 4:4:4 Intra Profiles provide the same features and capabilities as High 10, High 4:2:2, and High 4:4:4 Profiles, respectively, except disabling all inter-prediction modes. The last profile, CAVLC 4:4:4 Intra Profile is similar to High 4:4:4 Intra, but the entropy coding is restricted to CAVLC to reduce the complexity.

High profiles were developed to support high quality applications, such as studio camcorders, professional digital video recording and editing systems, digital cinema/large-screen digital imagery, and high fidelity display systems. The four all-intra coding profiles are especially suitable for easy editing of bitstreams. Furthermore, it is anticipated that the High Profile will overtake the Main Profile in the near future as the primary choice for entertainment-quality applications, since the High Profile significantly outperforms the Main Profile without adding much implementation complexity.

3.5 Scalable Video Coding (SVC)

Scalability means parts of the bitstream coded at highest quality and spatial/temporal resolution can be removed in a way that the resulting substream forms another valid bitstream representing the source content with a reduced reconstruction quality of the complete bitstream, e.g., lower fidelity, frame rate, and resolution. This feature is important for bitstreams to adapt to various needs

Table 4. Comparison of the features in four high profiles of H.264/AVC

Features	High	High 10	High 4:2:2	High 4:4:4 Predictive
Main Profile Tools	x	x	x	x
8×8 and 4×4 adaptive transform	x	x	x	x
Perceptual-based quantization	x	x	x	x
Separate Cb and Cr QP control	x	x	x	x
Color plane separated coding				x
Residual color transform				x
Intra residual lossless coding				x
4:2:0 color format	x	x	x	x
4:2:2 color format			x	x
4:4:4 color format				x
8-bit bit depth	x	x	x	x
10-bit bit depth		x	x	x
14-bit bit depth				x

or preferences of end users and to varying terminal capabilities and network conditions. SVC enables this feature by partitioning the data of a bitstream into layers. The base layer is non-scalable and has to be decoded to provide the lowest quality and spatio-temporal resolution, which will be gradually improved with each extra enhancement layer being decoded.

Actually, H.263, MPEG-2, and MPEG-4 have attempted to standardize the technology of SVC. However, the previous designs were not successful in terms of industry adoption, owing to the significant efficiency loss and higher decoding complexity compared with the single-layer coding. In October 2003, MPEG issued a CfP for a new SVC standard, intending to address the shortcoming of the previous designs. The SVC standard was required to enable more flexible transmission in today's highly heterogeneous and time-varying environments and provide comparable rate-distortion (R-D) performance with H.264/AVC single-layer coding. Hence, it is quite natural to build the SVC scheme upon the strong compression foundation of H.264/AVC and employ additional tools to support the required types of scalabilities. In October 2004, the scalability extension of H.264/AVC [20] defeated other proposals and was chosen as the starting point of MPEG's SVC project. In January 2005, MPEG and VCEG agreed to jointly finalize the SVC project as an amendment to H.264/AVC within JVT. The SVC amendment was finally approved in 2007 [21].

The SVC amendment is distinguished from the previous SVC designs by the following three key elements.

- Excellent coding efficiency. For each subset of the scalable bitstream, bit-rate increase at the same fidelity is limited within 10%, compared with the single-layer coding. The detailed R-D performances in a large number of scalability cases can be found in [22].
- Acceptable decoding complexity. A novel approach, known as single MC loop decoding, is adopted, such that the complexity of decoding any subset of the scalable bitstream is comparable with the single-layer decoding.

- Maximal design consistency. The base layer is backward compatible with H.264/AVC, such that the core coding tools of H.264/AVC can be inherited. New tools are added only if necessary for supporting the required types of scalabilities.

Three fundamental types of scalabilities are enabled, i.e., temporal, spatial, and quality scalabilities, which will be discussed respectively in the following sub-sections. An overview of the SVC amendment can be found in [23].

Temporal Scalability

The flexible reference picture management in H.264/AVC (see Section 3.2) enables arbitrary temporal dependencies among successive pictures. Therefore, no change to H.264/AVC is needed to support the temporal scalability with a reasonable number of temporal layers, except the signaling of temporal layers.

Fig. 22 gives three prediction structures, which provide temporal scalability with different temporal dependencies. The picture labeled T_k belongs to the kth temporal layer and T_0 means the base layer. Fig. 22 (a) explains the concept of hierarchical B- or P-pictures, which efficiently provide temporal scalability with dyadic temporal enhancement layers. If the enhancement layer pictures are coded as B-pictures, excellent coding efficiency is achieved. Using P-pictures in enhancement layers can also realize such a temporal coding structure, owing to the decoupling of the picture type and the referencing capability in H.264/AVC. The set of pictures between two successive base layer pictures together with the succeeding base layer picture is referred to as a group of pictures (GOP). Fig. 22 (b) illustrates a non-dyadic hierarchical prediction structure that provides two independently decodable sub-sequences with 1/9-th and 1/3-rd of the full frame rate. Fig. 22 (c) shows a hierarchical prediction structure that does not employ MCP from pictures in the future. This structure provides the same degree of temporal scalability as that in Fig. 22 (a), while its structural delay is reduced to 0, much smaller than 7 pictures for the prediction structure in Fig. 22 (a). However, such a low-delay structure decreases coding efficiency.

Spatial Scalability

The approaches used in previous video coding standards for spatial scalability include intra-layer prediction and inter-layer prediction. With intra-layer prediction, MCP and intra-prediction are employed within each spatial layer as for the single-layer coding. With inter-layer prediction, the reconstructed lower layer signals are upsampled and used as the reference for the MCP of the higher layer. This inter-layer prediction has two disadvantages. Firstly, the statistical dependencies between different layers are not fully exploited, so the R-D efficiency of enhancement layers is relatively low. Secondly, the decoding complexity is high for the enhancement layers, because they cannot be reconstructed until all the depended lower layers are reconstructed.

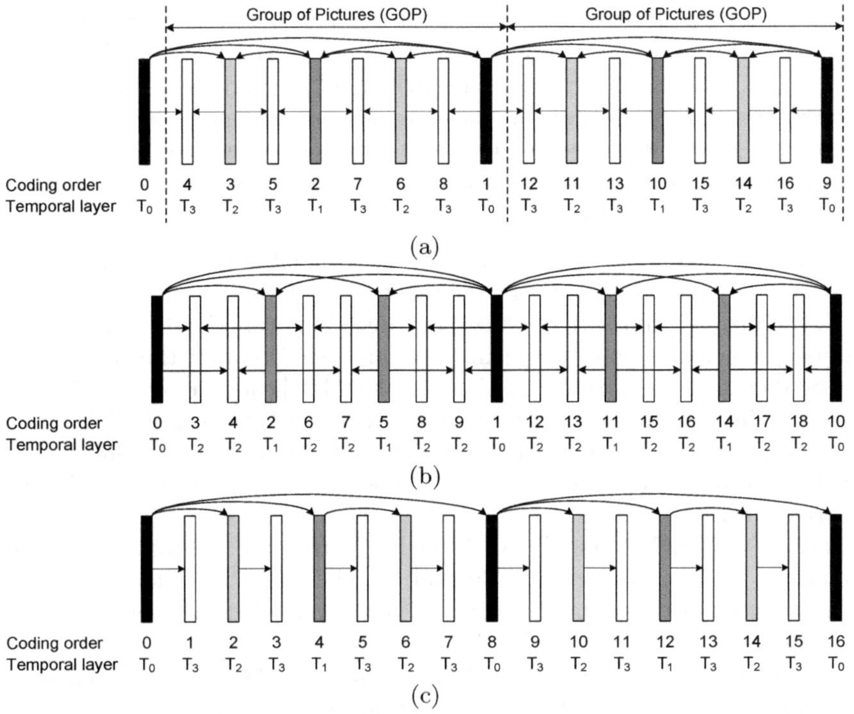

Fig. 22. Hierarchical prediction structures for temporal scalability [23]. (a) Hierarchical B prediction structure. (b) Non-dyadic hierarchical prediction structure. (c) Hierarchical prediction structure with a structural encoding/decoding delay of zero.

In the SVC amendment, the conventional intra-layer prediction is followed by using the tools of H.264/AVC and the inter-layer prediction is improved to overcome the aforementioned two disadvantages. In detail, the SVC amendment adopts two new inter-layer prediction modes, inter-layer motion prediction and inter-layer residual prediction, while also keeping the conventional prediction of using the upsampled reconstructed lower layer signals, named inter-layer intra-prediction. Note that the three modes can be chosen at the MB or sub-MB level.

- Inter-layer motion prediction. When the SE base_mode_flag of the MB to be coded equals 1 and the co-located block in the reference layer is inter-coded, the MB is also inter-coded as for the single-layer H.264/AVC coding, but the motion information, including MC block size, MVs, reference index, is completely derived from the co-located block in the reference layer.
- Inter-layer residual prediction. In the transform domain, the coefficients of the co-located block in the reference layer are upsampled and then subtracted from the coefficients of the block to be coded in the enhancement layer. Only the resulting difference, which often has smaller energy than the original residual data, is encoded using entropy coding as specified in H.264/AVC.

- Inter-layer intra-prediction. When the SE base_mode_flag of the MB to be coded is equal to 1 and the co-located block in the reference layer is intra-coded, the co-located block is reconstructed, upsampled, and used as the prediction for the MB to be coded in the enhancement layer. It is further required that the intra-coded MBs in the reference layer should be coded using constrained intra-prediction, of which the reconstruction is independent of any inter-coded MBs.

In the SVC amendment, each spatial enhancement layer can be decoded without completely reconstructing the reference layer. This feature, called single MC loop decoding, significantly reduces the decoding complexity. The inter-layer residual prediction is performed in the transform domain; the inter-layer motion prediction is performed right after the MB header is decoded; only the inter-layer intra-prediction requires reconstructing the co-located block in the reference layer, which, however, typically accounts for a small proportion and does not require reconstructing any neighboring inter-coded blocks because of constraint intra-prediction.

Similar to MPEG-2 and MPEG-4, the SVC amendment supports spatial scalable coding with arbitrary resolution ratios. The only restriction is that neither the horizontal nor the vertical resolution can decrease from one layer to the next. The SVC design further includes the possibility that an enhancement layer picture represents only a selected rectangular area of its corresponding reference layer picture, using a higher or identical spatial resolution.

Quality Scalability

Quality scalability provides the same spatio-temporal resolution as the complete bitstream with a lower fidelity, where the fidelity is often informally referred to as MSE. In the SVC amendment to H.264/AVC, quality scalability can be supported as a special case of the spatial scalability with the resolution ratio between layers equal to 1, which means the aforementioned three inter-layer prediction modes are employed without using the upsampling. In this case, called coarse-grain quality scalable (CGS) coding, the quality refinement of the enhancement layer is achieved by re-quantizing the residual data by a stepsize smaller than that used for the lower layers, and therefore the number of switching rate points is identical to the number of layers.

To increase the flexibility of bitstream adaptation, medium-grain quality scalable (MGS) coding, a variation of the CGS coding, is developed, which modifies the high level signaling to enable the picture (or access unit) level bit-rate adaptation. In detail, in an access unit, formed by the representations with different fidelities for a given time instant, any NALU containing an enhancement layer slice can be discarded. Furthermore, an enhancement layer slice is not necessarily discarded as a whole, since its transform coefficients, allowed to be further distributed to several slices and then packeted by different NALUs, can be discarded partially. Hence, graceful quality degradation is achieved within one enhancement layer slice. In short, MGS significantly increases the granularity

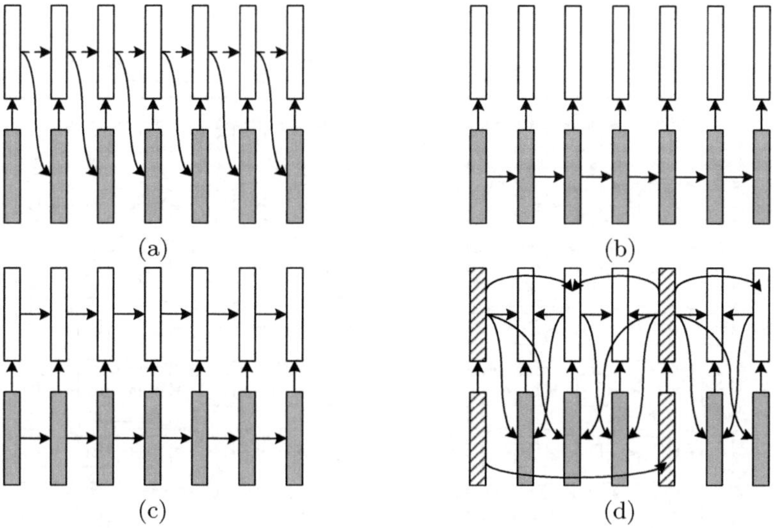

Fig. 23. Concepts for trading off enhancement layer coding efficiency and drift [23]. (a) Enhancement layer control. (b) Base layer control. (c) Two-loop control. (d) Key picture (marked by hatched boxes) concept for hierarchical prediction structures.

of quality scalable coding compared to CGS, although not so flexible as fine-grain quality scalable (FGS) coding, which generates quality scalable bitstreams partially decodable at any bit-rate.

Besides the flexibility of bitstream adaption, the other great concern of quality scalability is the drift, the effect that causes MCP loops at encoder and decoder to lose synchronization. Drift will be inevitably introduced, if the enhancement layer with data loss is used for the MCP of the base layer (see Fig. 23 (a)). Although it is possible to use only the base layer for MCP (see Fig. 23 (b)) such that loss of the enhancement layer does not have any impact on the MCP loop, e.g., the FGS coding in MPEG-4 Visual, the coding efficiency of the enhancement layer is significantly decreased compared to the single-layer coding. The aforementioned CGS coding uses two MC loops as shown in Fig. 23 (c), with which any loss of a quality refinement packet results in a drift for the enhancement layer reconstruction, although the base layer is not influenced. To make a suitable trade-off between drift and enhancement layer coding efficiency, the MGS coding adopts the concept of key picture. Fig. 23 (d) illustrates how the key picture concept is incorporated into hierarchical prediction structures. All pictures of the temporal base layer (marked by hatched boxes) are transmitted as key pictures. In order to limit the decoding overhead, the quality base and enhancement layers of key pictures use exactly the same motion information. The key pictures use the quality base layer as the reference pictures to avoid any drift in the MCP loop of the temporal base layer. In contrast, all pictures in the temporal enhancement layers, i.e., the non-key pictures, typically use the reference with the highest available quality for MCP, which provides a high coding

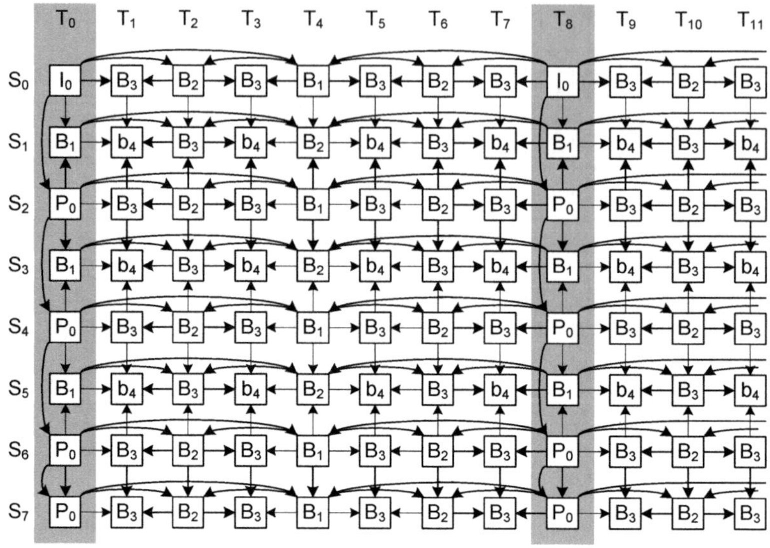

Fig. 24. Joint temporal and inter-view prediction structure for the MVC amendment to H.264/AVC [27]

efficiency. Since the key pictures serve as resynchronization points between encoder and decoder reconstruction, drift propagation is efficiently limited within the GOP. The trade-off between enhancement layer coding efficiency and drift can be adjusted by the GOP size and the number of hierarchy stages.

3.6 Multiview Video Coding (MVC)

Multiview video, captured by synchronized cameras from different viewpoints, provides rich 3-D information of a scene and thus expands viewers' experience beyond what is offered by traditional videos. Thanks to the recent advances in acquisition and display technologies, multiview video is foreseen to be the feasible media in consumer domain, including 3DTV and free viewpoint TV (FTV). 3DTV offers a 3-D depth impression of the observed scenery; FTV allows an interactive selection of viewpoint and direction within a certain operating range. The realization of 3-D vision applications will depend on the complete processing chain, including capturing, compression, transmission, display and interactive presentation. Among them, MVC is one of the most challenging technologies, since a tremendous amount of data, in proportion to the number of cameras, needs to be compressed to such an extent that it can be transmitted within the capability of today's network.

MPEG has been investigating the MVC related topics since 2001 in an ad hoc group (AHG), named 3DAV (3-D audio and visual). In July 2005, MPEG issued a formal CfP [24] to address the requirements in [25], based on the evidence brought forward for MVC technology. The proposals in response to the CfP were all backward compatible with H.264/AVC. After one year's evaluation and

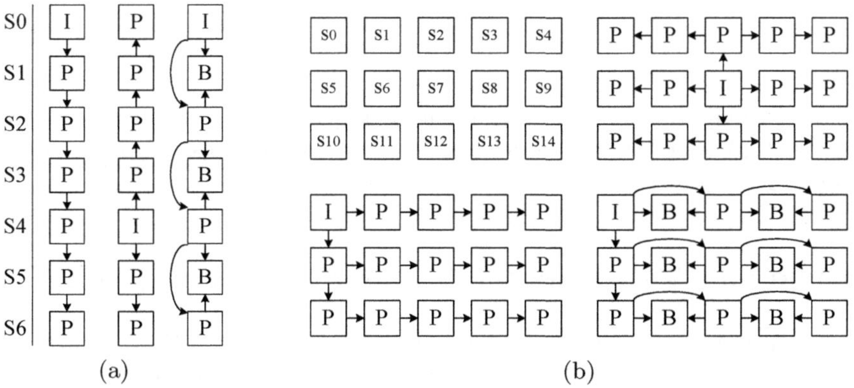

Fig. 25. Inter-view reference dependencies according to camera arrangements [27]

competition, the first MVC model was released. Meanwhile, JVT started to devote part of its effort on the MVC project as another amendment to H.264/AVC. At the time of writing, JVT is at the completion of the MVC amendment [26].

The MVC amendment to H.264/AVC employs inter-view prediction in addition to temporal prediction to remove the inter-view statistical redundancy. Fig. 24 [27] illustrates the joint temporal and inter-view prediction for a multiview video generated by eight linearly arranged cameras. S_n $(n = 0 \cdots 7)$ denotes the individual view sequences and T_n $(n = 0 \cdots 11)$ denotes the consecutive time instances. The first view S_0 employs only temporal prediction as for the single-view coding. More specifically, hierarchical B prediction structure is used for better coding efficiency. As S_0 can be decoded independently of other views, the view scalability is thus provided and S_0 is the base view. For any other view, the temporal prediction structure within the view is the same as that of S_0, whilst inter-view prediction is also enabled, where the pictures from neighboring views at the same time instance are used for MCP. The pictures in the coarsest temporal layer (see the shaded time instances), called key pictures, cannot use temporal prediction, as they provide resynchronization and random access. Therefore, the key pictures in S_0 are all I-pictures and the key pictures in other views can use inter-view prediction. The example in Fig. 24 fully exploits the statistical dependencies, but as a result the video sequences of individual views cannot be processed independently anymore. They have to be either interleaved into one bitstream for sequential processing or signaled and stored in a shared buffer for parallel processing.

Besides the example in Fig. 24, the MVC amendment actually allows a wide variation of joint temporal and inter-view prediction structures, such that the trade-off between the coding efficiency and decoding complexity, including delay and DPB management, can be made. For example, the inter-view prediction can be restricted in key pictures. Furthermore, the inter-view reference dependency can be selected according to the camera arrangement. Fig. 25 (a) shows three possible inter-view reference dependencies for coding the multiview video captured by seven linearly arranged cameras, which result in different complexities. Similarly, Fig. 25 (b) is for a 5×3 camera array.

All the possible joint temporal and inter-view prediction structures are basically the special cases of the SVC amendment. Therefore, no change to SVC is needed except some high level designs, focusing on interface, transport of the MVC bitstreams, and MVC decoder resource management.

4 Audio and Video Coding Standard of China (AVS)

AVS is a suite of standards, including system (Part 1), video (Part 2 and Part 7), audio (Part 3), digital copyright management (DRM) (Part 6), and other supporting standards. The target applications lie in the pivotal fields of Chinese information industry, such as HD digital broadcast, high-density storage media, wireless broad-band multimedia communication, and multimedia streaming. AVS is developed by AVS workgroup, a Chinese standard body established by National Information Industry Ministry in June 2002. This section starts with an introduction of AVS in Section 4.1. Section 4.2 narrows the focus to the technical designs of AVS-Video. The profiles of AVS are presented in Section 4.3.

4.1 Introduction

Video coding standards adopt a large number of patented innovations, which should be licensed at a cost. The standard bodies, e.g., ITU-T and ISO/IEC, deal with such patent issues on a "reasonable and non-discriminatory" (RAND) basis, where the word "reasonable" is very ambiguous. Eventually, the development of necessary patent licensing programs falls to the industry; the manufacturers and end users may risk significantly delayed licensing, unacceptable fees, and complex charging mechanism. For example, the licensing terms for MPEG-4 Visual are so harsh that they have been rejected by most of the market players. In China, the manufacturers produce more than 30 million DVDs and 10 million set-top boxes every year, but make a narrow profit margin because of high license fees. To lower the licensing cost, AVS workgroup was established to develop a suite of national standards that are comprehensive, technically competitive, and affordable for the digital multimedia industry.

The working environment of AVS workgroup is greatly influenced by its Intellectual Property Rights (IPR) policy. An overview of AVS IPR policy is presented in [28]. In short, AVS IPR policy is designed to consider the licensing in parallel with the technical work, which results in two advantages. Firstly, the delay between completing the technical work and the license becoming available is minimized; the licensing should be available soon after the standard is officially approved. Secondly, when deciding which contributions to adopt into the standard, the standard committee does not consider only efficiency and complexity, but also considers the IPR implications, including RAND with royalty-free license, AVS patent pool, and RAND. To avoid high IP cost, some compromises are required, but the benefits of a non-proprietary open standard and the licensing cost savings easily outweigh the small loss in performance. Therefore, AVS IPR policy facilitates early provision of affordable licensing and quick adoption of the standards in markets.

Besides low license fee, AVS also provides competitive performance with succinct design. The coding efficiency of AVS Part 2 is about 2 to 3 times of MPEG-2 and similar to H.264/AVC. In 2006, the Jizhun Profile of AVS Part 2 [29] was officially approved as a Chinese national standard. The standards for audio, system, and DRM have all completed the technical work and are pending for approval.

Nowadays, AVS has been receiving increasing worldwide attention. ITU-T is considering AVS Part 2 as one of the four coding standards for Internet Protocol TV (IPTV). MPEG is also developing the AVS toolbox within the RVC framework (see Section 5.3).

4.2 Highlights of AVS-Video

AVS-Video is essentially similar to H.264/AVC but designed to reduce the implementation cost. Compared to H.264/AVC, some coding tools providing functionalities rather than improving coding efficiency, such as FMO, data partitioning, and SP-/SI- pictures, are not included. To make the design more succinct, the flexibilities of the necessary coding tools, such as intra and inter-prediction, reference picture management, and deblocking filter, are greatly reduced. This section highlights the innovations in AVS-Video, including Part 2 [29]-[31] and Part 7 [32]. A summary of the differences between H.264/AVC and AVS-Video is given in Table 5.

AVS Part 2

Special Reference Pictures

Up to two reference frames are allowed for MCP, which can be used as four fields for field-mode. By default, P-pictures reference the nearest preceding reference pictures, and B-pictures reference the nearest preceding and succeeding reference pictures, both in the display order. To increase the flexibility, three special types of reference pictures are introduced.

- Core picture. Core pictures form the coarsest temporal layer in a bitstream, which has the same concept as the key pictures used for the SVC amendment to H.264/AVC (see Section 3.5).
- Background picture. It is an I-picture, which is expected to be coded at higher quality with noise suppressed for better background prediction. It is stored in the buffer as one of the two reference frames and replaced only upon the reception of the next background frame. This feature is beneficial to video surveillance, where the background is seldom changed, and can be regarded as a special case of long-term picture in H.264/AVC.

Table 5. Comparison of the features in AVS and H.264/AVC

Features	AVS Part 2	AVS Part 7	H.264/AVC
Number of profiles	3	1	11
Color format	4:2:0,4:2:2,4:4:4	4:2:0	4:2:0,4:2:2,4:4:4
Bit depth	8	8	up to 14
Picture type	I,P,B	I,P	I,P,B,SP,SI
Interlace handling	PAFF	N/A	PAFF,MBAFF
Luminance intra prediction	8×8 5 modes	4×4 9 modes	4×4,8×8,16×16 totally 22 modes
No. of reference frame	up to 2	up to 2	up to 16
Reference picture management	background- and core-picture, non-reference P	adaptive sliding window, non-reference P	MMCO-based commands
Prediction in B-picture	forward,backward, Symmetric,Direct	N/A	list0,list1,Direct, bi-predictive
Direct mode	temporal	N/A	spatial, temporal
MV prediction	geometry median	geometry median	median
Interpolation	1/2-pel:[-1,5,5,-1] 1/4-pel:[1,7,7,1]	1/2-pel(Hor.): [-1,4,-12,41,41,-12,4,-1] 1/2-pel(Ver.):[-1,5,5,-1] 1/4-pel: bilinear	1/2-pel: [1,-5,20,20,-5,1] 1/4-pel: bilinear
Minimum MC size	8×8	4×4	4×4
Transform	8×8 PIT	4×4 PIT	4×4/8×8 ICTs
Hierarchical transform	No	No	Yes
Default quantization	64 QPs stepsize doubles for each 8 inc.	64 QPs stepsize doubles for each 8 inc.	52 QPs stepsize doubles for each 6 inc.
Adaptive quantization	perceptual-based	No	perceptual-based
Separate CbCr QP	Yes	No	Yes
Entropy coding	8×8 CA-2D-VLC 8×8 CBAC	4×4 CA-2D-VLC	CAVLC CABAC
Deblocking filter	3 BS levels	2 BS levels	4 BS levels
Lossless coding	No	No	Yes
Weighted prediction	Yes	No	Yes
NAL	No	Yes	Yes
Parameter set	No	Yes	Yes
SEI	No	Yes	Yes
Flexible slice structure	Yes	Yes	Yes
FMO	No	No	Yes
Redundant picture	No	No	Yes
Data partitioning	No	No	Yes

- Non-reference P-picture. P-picture can be marked as a non-reference picture at the picture header, such that temporal scalability is enabled without decoding delay.

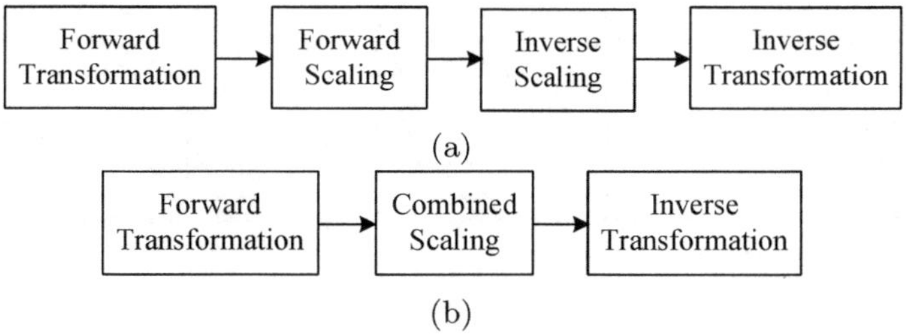

Fig. 26. The flow diagrams of (a) ICT and (b) PIT

"Symmetric" mode in B-pictures

Four prediction modes are used in B-pictures: forward, backward, Direct and Symmetric modes, where the latter two are bi-directional. The Symmetric mode transmits only the forward MV; the backward MV is derived by scaling the forward MV according to the temporal distance. The underlying assumption is that most of the motions in a video sequence are translation and occur within a few neighboring pictures.

Pre-scaled Integer Transform (PIT)

8×8 integer transform is used in order to match the smallest MC block size. The transform matrix given below resembles DCT more than that in H.246/AVC and has higher transform coding gain.

$$T_8 = \begin{bmatrix} 8 & 8 & 8 & 8 & 8 & 8 & 8 & 8 \\ 10 & 9 & 6 & 2 & -2 & -6 & -9 & -10 \\ 10 & 4 & -4 & -10 & -10 & -4 & 4 & 10 \\ 9 & -2 & -10 & -6 & 6 & 10 & 2 & -9 \\ 8 & -8 & -8 & 8 & 8 & -8 & -8 & 8 \\ 6 & -10 & 2 & 9 & -9 & -2 & 10 & -6 \\ 4 & -10 & 10 & -4 & -4 & 10 & -10 & 4 \\ 2 & -6 & 9 & -10 & 10 & -9 & 6 & -2 \end{bmatrix}$$

Compared to the ICT scheme in H.264/AVC, of which the flow diagram is shown in Fig. 26 (a), the transform procedure in AVS is further simplified. As shown in Fig. 26 (b), the inverse scaling is moved from the decoder to the encoder and combined with the forward scaling as one single process, named combined scaling. By this means, the inverse scaling is saved and thus no scaling matrix needs to be stored, whilst the complexity of encoders remains unchanged. The simplified transform is named Pre-scaled Integer Transform (PIT), and interested readers are referred to [33] for in-depth expositions.

Adaptive Quantization

AVS Part 2 supports perceptual-based quantization, but the customization of the quantization matrix is restricted by three quantization patterns, as shown in

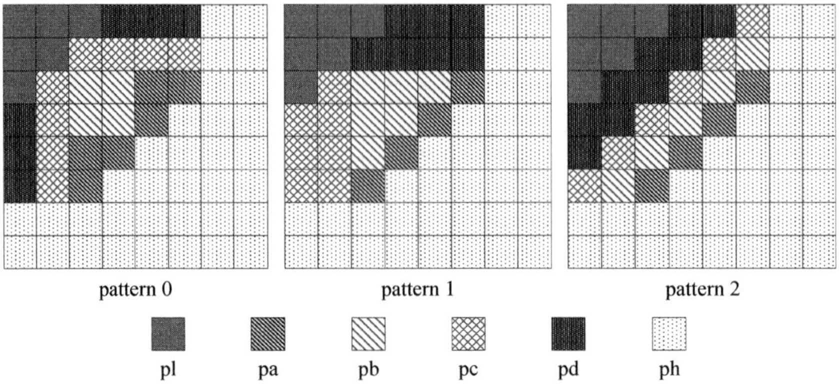

Fig. 27. Three AVS quantization patterns in AVS Part 2

Fig. 27. Each quantization pattern divides a transform block into six specific subbands and the coefficients in one subband use the common quantization weights. There are two default sets of the weights. One set {pl, pa, pb, pc, pd, ph} equal to [135, 143, 143, 160, 160, 213]>>7 is for detail protection; the other set [128, 98, 106, 116, 116, 128]>>7 is for detail reduction. To determine a quantization matrix, three types of information are transmitted in the picture header, the choice of the pattern, the choice of the default set, and the differences between the actual weights and the selected default set if any.

Entropy Coding

Like H.264/AVC, AVS Part 2 supports two entropy coding schemes, Huffman-like coding and arithmetic coding. The former also employs Exp-Golomb codes to code the SEs other than residual data.

In the Huffman-like coding scheme, the residual data are coded using context-adaptive 2-D VLC (CA-2D-VLC), where the levels and runs are jointly coded as (level, run) pairs along the reverse zig-zag scan order and are ended with "EOB". Two main features distinguish CA-2D-VLC from the previous (level, run) coding schemes as in MPEG-2.

Firstly, a plurality of look-up tables is defined to code one transform block. When coding a sequence of (level, run) pairs in one transform block, CA-2D-VLC uses the table indexed 0 for the first pair and switches the tables upon the detection of context changing. The context changing herein means the maximum magnitude of the levels that have been coded in the current block exceeds one of the pre-defined thresholds. There are seven thresholds defined for inter-coded luminance blocks, seven for intra-coded luminance blocks, and five for chrominance blocks, and therefore, totally 19 tables are designed for these three types of blocks.

Secondly, all the codewords in tables are constructed using order-k Exp-Golomb codes, of which the aforementioned Exp-Golomb codes in Section 3.2 are the special cases with k equal to 0 (see Table 6). Exp-Golomb codes with

Table 6. Order-k Exp-Golomb codes

order-k	Codeword	codeNum Range
	1	0
	$01x_0$	1-2
k=0	$001x_1x_0$	3-6
	$0001x_2x_1x_0$	7-14
	...	...
	$1x_0$	0-1
	$01x_1x_0$	2-5
k=1	$001x_2x_1x_0$	6-13
	$0001x_3x_2x_1x_0$	14-29
	...	...
	$1x_1x_0$	0-3
	$01x_2x_1x_0$	4-11
k=2	$001x_3x_2x_1x_0$	12-27
	...	...
	$1x_2x_1x_0$	0-7
	$01x_3x_2x_1x_0$	8-23
k=3	$001x_4x_3x_2x_1x_0$	24-55
	...	...

larger orders favor flatter-shaped pdf and vice versa, whereas the mapping from a (level, run) pair to a codeNum only reflects a relative probability.

The arithmetic coding is named context-based arithmetic coding (CBAC), which basically has the same diagram as CABAC in H.264/AVC (see Fig. 15). The main difference between CBAC and CABAC lies in two aspects. Firstly, the arithmetic coding engine performs the interval subdivisions, i.e., probability update, in the logarithmic domain, where multiplication in the physical domain is equivalent to addition. Therefore, the probability update of the arithmetic coding is approximated by additions and shifts only. Secondly, the transform block is coded as a sequence of (level, run) pairs as in CA-2D-VLC and the context selection is jointly determined by the maximum magnitude of the levels that have been coded in the current block and the position of the current level in the zig-zag order.

AVS Part 7

Flexible Reference Frame

H.264/AVC can handle up to 16 reference frames by complicated MMCO-based commands, and the reference pictures are categorized into short-term and long-term ones. In AVS, only two reference frames are used, so the operation to realize such a categorization is significantly simplified. The concept of adaptive sliding window is introduced, of which the window size can be one or two, selected at the frame level. If the window size equals two, both reference frames will be involved in the first-in-first-out sliding window operation; the temporally farther

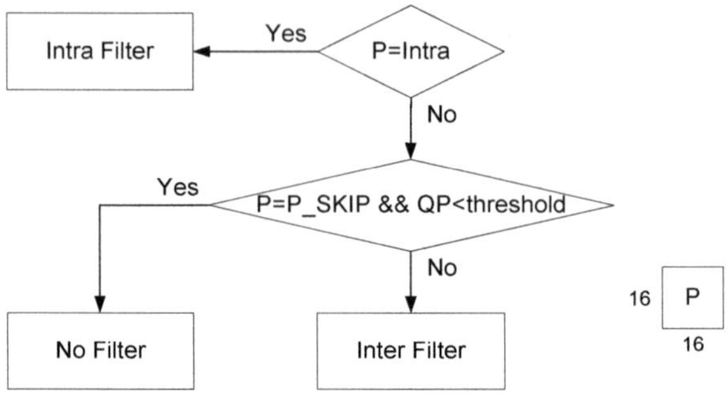

Fig. 28. The deblocking filter mode decision in AVS Part 7

reference frame will be pushed out of the DPB. Otherwise, only the temporally nearer frame is in the sliding window, which will be removed from the DPB by the just arrived reference frame. Obviously, window size equal to one makes the reference frame outside the sliding window act as a long-term frame.

All the high-level designs and error resilience tools in AVS Part 7, including this reference frame management, are discussed in depth in [34].

Transform Matrix

The transform in AVS Part 7 is also PIT, and the transform matrix is given as below.

$$T_4 = \begin{bmatrix} 2 & 2 & 2 & 2 \\ 3 & 1 & -1 & -3 \\ 2 & -2 & -2 & 2 \\ 1 & -3 & 3 & -1 \end{bmatrix}$$

Simplified Deblocking Filter

The deblocking filter in AVS Part 7 is significantly simpler than that in Part 2 or H.264/AVC. Only two filtering modes, decided at the MB level, are used, whereas H.264/AVC uses four modes (BS) decided at the block level. Fig. 28 shows the procedure of determining filtering mode, which involves much less coding information and logical decisions than Fig. 17. The "threshold" in the figure is equal to 40 by default and can be modified by transmitting an offset in the frame header. Furthermore, each filtering operation affects only two pixels on either side of the boundary, and thus the pixels filtered for neighboring boundaries do not overlap. This enables parallel operations during the deblocking process.

Interested readers are referred to [35] for the details of all the low complexity coding tools in AVS.

4.3 Profile and Application

AVS Part 2 contains three profiles, Jizhun (Base) [29], Jiaqiang (Enhanced) [30], and Shenzhan (Extended) [31] Profiles. Fig. 29 shows the coding tools included

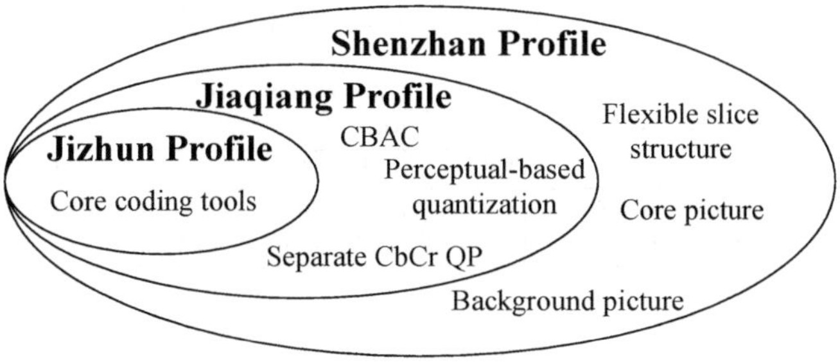

Fig. 29. The coding tools in AVS profiles

in different profiles, where the core coding tools are I-, P-, and B-pictures, 8×8 intra-prediction, variable MC block sizes, 1/4-pixel MCP, MV prediction, two reference frames, 8×8 PIT, deblocking filter, CA-2D-VLC, weighted prediction and PAFF.

Jizhun Profile mainly focuses on digital TV applications, such as terrestrial TV, IPTV, satellite TV, as well as storage media. As the enhancement of Jizhun Profile, Jiaqiang Profile adds CBAC and adaptive quantization, in order to address the requirements of high quality video applications, such as digital cinema. Shenzhan Profile is developed specially for video surveillance applications. Features of flexible slice structure, core picture, and background picture are incorporated, which improve the capability of error resilience, temporal scalability, and friendliness to the surveillance environment.

AVS Part 7 has been developed to meet the needs of mobile video applications, such as interactive storage media, video services on wireless broad-band and packet networks. AVS Part 7 has only one profile, named Jiben (Basic) Profile, which contains all the features in this part.

5 Future Video Coding Standards

At the time of writing, the two standard bodies, MPEG and VCEG, are moving forward to develop the future video coding standards based on the foreseen requirements. This section gives a brief introduction of the requirements and the ongoing standardization efforts.

5.1 High-Performance Video Coding (HVC)

Thanks to the technology evolution, more and more video materials with increased quality and spatio-temporal resolution (see Table 7) will be captured and distributed in the near future [36]. It is anticipated that the bit-rate produced by the current coding technology will go up faster than the increased capacity of the wireless or wired network infrastructure. Therefore, a new generation of video compression technology aiming at sufficiently higher compression capability rather than rich functionalities is required. Both VCEG and MPEG have been studying the feasibility of such a new standard.

Table 7. The video formats to be supported by high-performance video coding

Frame rate	typically 24-60 fps, up to 172 fps
Spatial resolution	VGA/720p/1080p/4K×2K
Color space	YCbCr or RGB
Color format	4:2:0/4:2:2/4:4:4
Bit depth	typically 8 bits, up to 14 bits

H.NGVC in VCEG

The project VCEG is currently working on is named H.NGVC, standing for "Next Generation Video Coding". H.NGVC could mean either an extension of H.264/AVC or a new standard, depending on the technical design. In January 2009, VCEG drafted a requirement for H.NGVC, which was basically consistent with the specification in Table 7. Regarding the coding efficiency, H.NGVC is expected to provide 50% bit-rate reduction at the same subjective quality compared to H.264/AVC.

Actually, VCEG has been seeking evidence regarding the possibility of a major objective performance gain over H.264/AVC ever since 2005. To better evaluate the techniques and retain the progress, key technical area (KTA) is developed as the reference software, which uses JM11 as the baseline and continuously integrates promising coding tools. Here listed are the techniques having been adopted in KTA so far.

- High-resolution MCP. The resolution of MV is increased from commonly used 1/4-pixel to 1/8-pixel, which has been proved especially efficient for low resolution video sequences.
- Adaptive interpolation filter (AIF). The coefficients of AIF are customized at the picture level and coded as the side information. Different AIF techniques make different approximations to the optimal filter that minimizes the prediction error energy, such as reducing the support region, imposing the symmetry constraints, separating the 2-D operation, and quantizing the filter coefficients, so as to reduce the side information and complexity.
- Adaptive quantization matrix selection (AQMS). The quantization matrix is designed on-the-fly or selected from a pre-defined candidate pool at the MB level. The selection is based on the criterion of R-D cost and signaled in the bitstream.
- Adaptive prediction error coding (APEC). With the increased prediction accuracy, the correlation of the residual signals decreases, so sometimes the transform becomes inefficient for energy compaction. APEC allows the residual data to be coded in either transform or spatial domain, decided and signaled down to the transform block level. In spatial coding mode, prediction errors are coded by specially developed quantization and entropy coding, without transform.
- Competition-based MV prediction. Instead of having one single MV predictor as in H.264/AVC, a set of spatial, temporal, and spatio-temporal predictors compete with each other; the predictor resulting in the lowest R-D cost wins.

Such a competition-based scheme is also employed to derive the MV for Skip mode. No matter what the MB type is, the selected MV predictor is signaled in the bitstream.

- Mode-dependent directional transform (MDDT). For intra-prediction modes with strong directionality, e.g., vertical mode and horizontal mode, corresponding MDDTs are derived from KLT to favor the high energy along the directions. The type of MDDT is coupled with the selected intra-prediction mode, so is not explicitly signaled. However, the memory to store all predefined MDDT bases is up to 1.5Kb.
- Extended block sizes for MCP and transform. The MB size is enlarged to 32×32 or 64×64, and the sizes for MCP are scaled accordingly. A 2D order-16 transform are also adopted for the residual blocks produced by MC block larger than or equal to 16×16. The transform matrix is obtained by scaling the transform matrix of 2D order-16 DCT by the factor 128 and rounding. This technique is developed to serve the upcoming applications of ultra-high definition TV (UDTV).

Meanwhile, VCEG keeps seeking the best combination of these coding tools, as their performance gains are not necessarily addictive.

HVC in MPEG

In October 2008, MPEG issued a Call for Evidence [37], which means a more rigorous evaluation phase for potential technologies. The target technology is not necessarily built on the top of the state-of-the-art standard. An absolutely new paradigm is also possible, as long as sufficient performance gain is observed. Both subjective and objective evaluation methodologies will be employed. Positive evaluation result may lead to a formal CfP and the new standard development with a tentative name high-performance video coding (HVC).

5.2 3D Video Coding in MPEG

The MVC amendment to H.264/AVC introduced in Section 3.6 efficiently reduces the bit-rate compared to multiview simulcast, but the bit-rate proportionally increases with the number of views, because the required views at the receiver all need to be coded. Therefore, the supported maximum number of views is limited due to the constraints of transmission and processing capabilities.

MPEG envisions that 3D media applications, providing the viewers more immersed experience, will become reality in the next few years [38]. The viewers will perceive stereoscopic videos at arbitrary viewpoint by using auto-stereoscopic displays, head-mounter displays (HMD) or head tracking. These applications require producing a large number of views for display, which go beyond the capabilities of the MVC amendment to H.264/AVC. In April 2007, MPEG launched an exploration on a new multiview video presentation format, named 3-D video (3DV), which bears explicit 3-D information, i.e., the depth, and a few channels

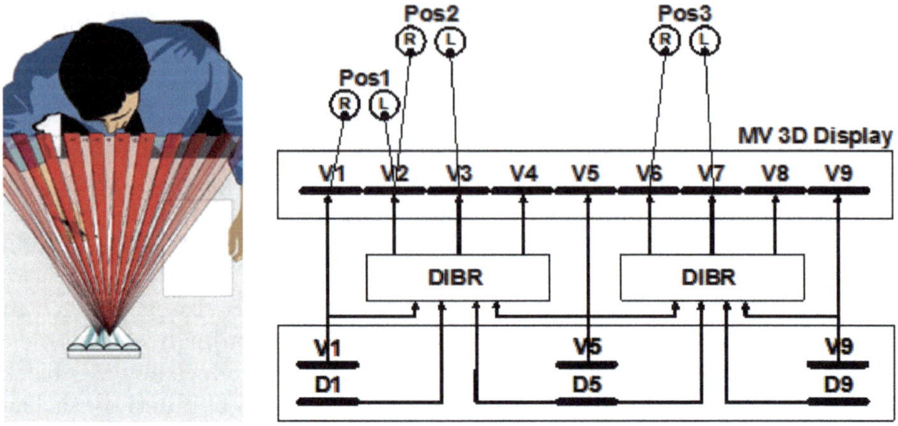

(a) (b)

Fig. 30. Illustration of the 3DV application scenarios [39]. (a) An auto-stereoscopic display. (b) Simultaneously presenting nine views by decoding and rendering.

of videos, and supports synthesizing high-quality views for continuous viewpoints. By this means, the transmission rate, which is almost constant, is decoupled from the number of output views.

Fig. 30 illustrates the 3DV application scenarios [39]. Fig. 30 (a) shows an auto-stereoscopic display presenting nine views simultaneously to the users, where the neighboring views, forming a stereo pair, provide 3-D perception while the user is moving within a narrow angle. As shown in Fig. 30 (b), only three views (V1, V5, V9) with the associated depth data (D1, D5, D9) are received, and the other six views to display are generated by depth image based rendering (DIBR).

In the whole architecture of 3DV communication, three designs are considered by MPEG for standardization, i.e., 3DV data format, decoder, and view interpolation.

- 3DV data format, including multiview video, camera parameters, depth data, and additional information (if necessary to make the system more efficient), is the output of the decoder and the input for view interpolation. 3DV data format should be hardware-independent to ensure wide applicability and interoperability.
- A decoder reconstructs 3DV data format in a normative manner.
- The interpolation module interpolates the views for display, according to the standardized algorithm, the decoded 3DV data, and some parameters from the display module, such as size of image, number of views, viewpoints, and frame rate. However, at the moment, the standardization of view interpolation has been receiving arguments concerning the evolution of rendering technology and unnecessary cost for certain cases.

At the time of writing, the formal standardization activity has not been launched; MPEG is exploring the related techniques by four exploration experiments (EE), focusing on depth estimation/generation, view synthesis, 3DV data format definition, and coding experiments, respectively. Especially, the fourth EE aims at getting an impression of how the depth map coding affects the quality of synthesized views. Compared to the MVC amendment to H.264/AVC, the preliminary results show that utilizing depth information improves the synthesis quality to some extent but rather sequence-dependent. Depth rate ranges from 10% to 50% of the texture rate.

5.3 Reconfigurable Video Coding in MPEG

Having successfully developed many video coding standards, MPEG initiated another standardization activity, named reconfigurable video coding (RVC), motivated by the following considerations [41].

Firstly, nowadays a multimedia device quite often should support not only the latest standard but also several legacy ones, as well as their multiple profiles. The standards are realized case by case without recognition or exploitation of the commonality among their coding tools. Such implementation redundancy can be efficiently reduced by modularizing and reusing the coding tools in a multi-codec device.

Secondly, the traditional standardization procedure, typically taking two to three years, is too slow to satisfy the rapidly changing landscape and requirements of media coding applications. Hence, standardizing new technologies at the coding tool level instead of the codec level offers a faster path to improving MPEG standards and better addressing the growing requirements.

Thirdly, coding tools that have to be implemented by a compliant decoder may not be all used in decoding the bitstream produced by a certain encoder, as encoder can choose the coding tools more freely. The decoder will become more customized, if dedicatedly configured based on the corresponding encoder.

RVC, formally launched by MPEG in January 2006 [40], is actually a high level specification model for direct and efficient codec synthesis. In other words, RVC offers a framework capable of constructing a video codec by configuring its coding tools, thus enabling a dynamic development, implementation and adoption of standardized video coding solutions. The RVC framework, comprising video tool library (VTL) and decoder description (DD) [41], is introduced as below.

A normative VTL consists of function units (FU) drawn from the existing MPEG coding standards, so-called MPEG toolbox. VTL is specified by a textual specification and the corresponding reference software. The reference software is provided using the specification language RVC-CAL, standardized by MPEG, where the data flow components are called actors. An actor encapsulates its own state and thus cannot read or modify the state of any other ones. The only interaction between actors is via messages, known as tokens in RVC-CAL, which are passed from an output of one actor to an input of another. The behavior of an actor is defined in terms of a set of actions, such as reading input tokens,

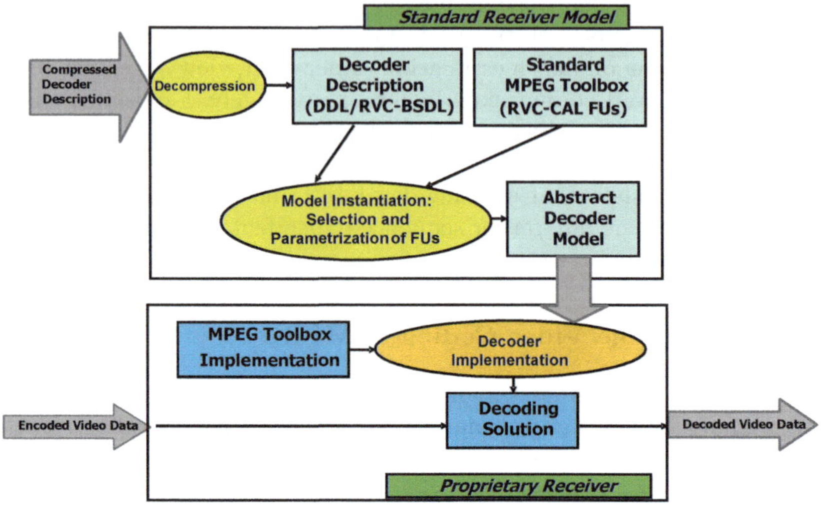

Fig. 31. The conceptual process of deriving an abstract decoder model in the MPEG RVC specification [41]

modifying internal state, producing output tokens, and interacting with the underlying platform on which the actor is running. At most one action is active at any time instance.

A decoder within the RVC framework should implement at least one VTL. At the moment, the defined VTLs include MPEG-2 Simple Profile and Main Profile, MPEG-4 Visual Simple Profile, MPEG-4 AVC Baseline Profile and SVC Baseline Profile. MPEG continues working on defining the remaining coding tools in MPEG toolbox for RVC-oriented FUs. Non-MPEG VTLs, such as AVS coding tools, are also supported by RVC framework, but MPEG will not take the responsibility of specification, conformance, or any technical qualification assessment.

DD is coded and transmitted together with the encoded video data, which carries two types of information. The first type describes how FUs are connected and parameterized. The second type, which describes the bitstream syntax structure by using the profiled Bitstream Syntax Description Language (BSDL), called RVC-BSDL, is used to synthesize a bitstream parser.

With the introduced normative elements, an abstract decoder model (ADM) is constituted (see Fig. 31). ADM, a behavioral model of decoder configuration, is platform-independent, whereas the decoder implementation is platform-dependent, as it is free to substitute proprietary VTLs having standard I/O interface for standard VTLs. Finally, an instantiation of decoding solution is established, which is actually a dedicated decoder to reconstruct the received video bitstream.

References

1. Terms of reference, ITU-T SG16/6 web site, `http://www.itu.int/ITU-T/studygroups/com16`
2. Terms of reference, MPEG home paper, `http://www.chiariglione.org/mpeg/`
3. Reader, C.: History of MPEG video compression. JVT document JVT-E066 (October 2002)
4. Wang, Y., Ostermann, J., Zhang, Y.-Q.: Video Processing and Communications. Prentice-Hall, New Jersey (2002)
5. ITU-T Recommendation H.261, Video codec for audiovisual services at p×64 kbits (1993)
6. ITU-T Recommendation H.263, Video coding for low bit rate communication (1998)
7. ISO/IEC IS 11172-2, Information technology – Coding of moving pictures and associated audio for digital storage media at up to about 1.5 Mbit/s – Part 2: Video (1993)
8. ISO/IEC IS 13818-2, Information technology – Generic coding of moving pictures and associated audio information: Video (1996)
9. ITU-R Recommendation BT.601-5, Studio encoding parameters of digital television for standard 4:3 and wide-screen 16:9 aspect ratios (1995)
10. ITU-R Recommendation BT.709-5, Basic parameter values for the HDTV standard, for the studio and for international program exchange (2002)
11. ISO/IEC IS 14496-2, Information technology – Coding of audio-visual objects – Part 2: Visual (1998)
12. Koenen, R.: MPEG-4 overview. ISO/IEC JTC1/SC29/WG11 document N4668 (March 2002)
13. ITU-T Recommendation H.264, Advanced video coding for generic audiovisual services (2005)
14. Richardson, I.: H.264 and MPEG-4 Video Compression Video Coding for Next-Generation Multimedia. John Wiley & Sons Ltd., England (2003)
15. Wiegand, T., Sullivan, J., Bjontegaard, G., Luthra, A.: Overview of the H.264/AVC video coding standard. IEEE Trans. Circuits Syst. Video Technol. 13, 560–576 (2003)
16. Sullivan, J., Topiwala, P., Luthra, A.: The H.264/AVC advanced video coding standard: overview and introduction to the fidelity range extensions. In: SPIE Conference on Applications of Digital Image Processing XXVII (August 2004)
17. IEEE Standard 1180-1990, IEEE standard specifications for the implementations of 8×8 inverse cosine transform (1990)
18. ISO/IEC IS 23003-1, Information technology – MPEG video technologies – Part 1: Accuracy requirements for implementation of integer-output 8×8 inverse discrete cosine transform (2006)
19. Karczewicz, M., Kurceren, R.: The SP- and SI-frames design for H.264/AVC. IEEE Trans. Circuits Syst. Video Technol. 13, 637–644 (2003)
20. Schwarz, H., Hinz, T., Kirchhoffer, H., Marpe, D., Wiegand, T.: Technical description of the HHI proposal for SVC CE1. ISO/IEC JTC1/SC29/WG11 document M11244 (October 2004)
21. ITU-T Recommendation H.264, Advanced video coding for generic audiovisual services (2007)
22. Wien, M., Schwarz, H., Oelbaum, T.: Performance analysis of SVC. IEEE Trans. Circuits Syst. Video Technol. 17, 1194–1203 (2007)

23. Schwarz, H., Marpe, D., Wiegand, T.: Overview of the scalable video coding extension of the H.264/AVC standard. IEEE Trans. Circuits Syst. Video Technol. 17, 1103–1120 (2007)
24. MPEG, Call for proposals on multi-view video coding. ISO/IEC JTC1/SC29/WG11 document N7327 (July 2005)
25. MPEG, Requirements on multi-view video coding v.4. ISO/IEC JTC1/SC29/WG11 document N7282 (July 2005)
26. Vetro, A., Pandit, P., Kimata, H., Smolic, A., Wang, Y.-K.: Joint draft 8.0 on multiview video coding. JVT document JVT-AB204 (July 2008)
27. Merkle, P., Smolic, A., Muller, K., Wiegand, T.: Efficient prediction structures for multiview video coding. IEEE Trans. Circuits Syst. Video Technol. 17, 1461–1473 (2007)
28. Reader, C.: AVS intellectual property rights (IPR) policy. J. Comput. Sci. & Techol. 21(3), 306–309 (2006)
29. GB/T20090.2, Information technology – Advanced coding of audio and video – Part2: Video (2006)
30. AVS workgroup, Information technology – Advanced coding of audio and video – Part2: Video. AVS document N1572 (December 2008)
31. AVS workgroup, Information technology – Advanced coding of audio and video – Part2: Video. AVS document N1540 (September 2008)
32. AVS workgroup, Information technology – Advanced coding of audio and video – Part7: Mobility video. AVS document N1151 (December 2004)
33. Zhang, C., et al.: The technique of prescaled integer transform: concept, design and applications. IEEE Trans. Circuits Syst. Video Technol. 18, 84–97 (2008)
34. Wang, Y.-K.: AVS-M: from standards to applications. J. Comput. Sci. & Techol. 21(3), 332–344 (2006)
35. Yi, F., Sun, Q.-C., Dong, J., Yu, L.: Low-complexity tools in AVS part 7. J. Comput. Sci. & Techol. 21(3), 345–353 (2006)
36. MPEG, Vision and requirements for high-performance video coding (HVC). ISO/IEC JTC1/SC29/WG11 document N10361 (February 2009)
37. MPEG, Draft call for evidence on high-performance video coding. ISO/IEC JTC1/SC29/WG11 document N10363 (February 2009)
38. MPEG, Vision on 3D video. ISO/IEC JTC1/SC29/WG11 document N10357 (February 2009)
39. MPEG Introduction to 3D video. ISO/IEC JTC1/SC29/WG11 document N9784 (April 2008)
40. MPEG, Draft call for proposals on reconfigurable video coding (RVC). ISO/IEC JTC1/SC29/WG11 document N7776 (January 2006)
41. MPEG, Whitepaper on reconfigurable video coding (RVC). ISO/IEC JTC1/SC29/WG11 document N9586 (January 2008)

AVS Video Coding Standard

Wen Gao[1], Siwei Ma[1], Li Zhang[1], Li Su[2], and Debin Zhao[3]

[1] Peking University, Beijing, China
[2] Gradute University, Chinese Academy of Sciences
[3] Harbin Institute of Technology, Harbin, China

Abstract. The AVS video coding standard developed by the China Audio Video Coding Standard (AVS) Working Group has attracted more and more attentions both from the industries and research institutes. Recently it has been accepted as an option by ITU-TFGIPTV for IPTV applications. As MPEG standards, the AVS standard is composed of several parts, such as system, video, audio, conformance testing, and reference software etc. This chapter will give an overview of AVS standard, which includes Part 2 and Part 7. AVS Part 2 is targeted to high resolution, high bit rate video applications, such as broadcasting and called AVS1-P2 in AVS. AVS Part 7, also called AVS1-P7, is for low resolution, low bit rate video applications, such as streaming, wireless multimedia communication etc. Here AVS1 denotes the fist stage standard activity of AVS working group. The second stage standard activity, called AVS2, which targets to the next generation highly efficient video coding standards, is just started.

In the chapter, we will give an overview of AVS standard including background, technique features, the performance comparison between AVS and H.264/AVC, and the complexity analysis of AVS.

1 Introduction

In the latest few years, the emerging advanced video coding standards or codecs have become the focus of multimedia industries and research institutes, such as H.264/AVC established by JVT (Joint Video Team) 0, VC-1 by Microsoft 0, and AVS 0. The AVS video coding standard is developed by the China Audio Video Coding Standard (AVS) Working Group, which was founded in June 2002. The role of the group is to establish general technical standards for the compression, decoding, processing, and the representation of digital audio-video, thereby enabling digital audio-video equipment and systems with high-efficiency and economical coding/decoding technologies. After seven years, AVS has published a series of standards. So far, AVS standards are becoming more and more internationalized. The AVS video coding standard has been accepted as an option by ITU-TFGIPTV for IPTV applications. This chapter will give an overview of the background and technical features of the AVS standard.

As MPEG standards, the AVS standard is composed of several parts, such as technical specifications for system multiplexing, video coding, audio coding, supporting specifications for conformance testing, and reference software etc. For video coding, AVS video standards include two parts 0. One part is Part 2, called AVS1-P2 in the AVS Working Group, which is mainly targeted to high definition and high quality digital broadcasting, digital storage media. AVS1-P2 (Jizhun Profile) has been approved as

C.W. Chen et al. (Eds.): Intel. Multimedia Communication: Tech. and Appli., SCI 280, pp. 125–166.
springerlink.com

the national standard in 2006 in China. After that, an extended work of AVS1-P2 Jizhun Profile, called Zengqiang profile, was started to further improve the coding efficiency of AVS1-P2. In March 2008, a new profile called Jiaqiang profile is defined based on the partial work of Zengqiang profile; the other one is Part 7 (AVS1-P7), which is targeted to the growing mobile applications, wireless broad-band multimedia communication, and internet broad-band streaming media. Only one profile was defined in AVS1-P7 now. In the chapter, we say AVS meaning AVS1 video coding standards.

Table 1 shows the history of the AVS video coding standard and the development process of the major video coding tools, such as variable block size motion compensation, multiple reference pictures and quarter pixel motion interpolation techniques etc. Compared with H.264/AVC, AVS standards provide a good tradeoff between the performance and complexity for the specific applications, because all coding tools in AVS are selected by jointly considering the coding complexity and performance gain for the target applications. This chapter will give a detail introduction to these coding tools with coding performance and coding complexity analysis.

The rest of this chapter is organized as follows: Section 2 first gives a brief introduction to the framework of AVS video coding standards, and why these coding tools were selected for AVS is described. In Section 3, the performance comparison between AVS standards and H.264/AVC and the complexity issue is discussed. Finally, Section 4 concludes the chapter.

Table 1. History of AVS Standard

Time	Document	Coding tools and comments
Oct. 2002	AVS1-P2 Jizhun Profile WD	Prediction/transform based Hybrid framework is selected
Jul. 2003	AVS1-P2 Jizhun Profile CD	8x8 integer transform, 2D VLC entropy coding, improved B frame direct mode and symmetric mode
Oct. 2003	AVS1-P2 Jizhun Profile FCD	Improved 8x8 integer transform, low complexity loop filter, low complexity quarter pixel interpolation, low complexity intra prediction, motion vector prediction
Dec. 2003	AVS1-P2 Jizhun Profile DS	AVS1-P2 Jizhun Profile was finished
Aug. 2004	AVS1-P7 WD	4x4 block based prediction/transform framework
Nov. 2004	AVS1-P7 CD	Improved quarter pixel interpolation, intra prediction, loop filter, motion vector prediction
Dec. 2004	AVS1-P7 FCD	AVS1-P7 first stage was finished
Sep. 2005	AVS1-P2 Zengqiang Profile WD	New context arithmetic coding, adaptive scan, improved MBAFF interlace coding
Feb. 2006	AVS1-P2 NS	AVS1-P2 was approved as China National Standard
Sep. 2008	AVS1-P2 Jiaqiang Profile FCD	Context Binary Arithmetic Coding (CBAC), Adaptive Weighting Quantization (AWQ)
Notes	WD: working draft; CD: committee draft; FCD: final committee draft; NS: national standard	

2 An Overview of the AVS Standard

As shown in Fig.1 the AVS video encoder is based on the traditional transform/prediction hybrid framework. In the coding process, each input macroblock is predicted with intra prediction or inter prediction. After prediction, the predicted residual is transformed and quantized. Then the quantized coefficients are coded with an entropy coder. At the same time, the quantized coefficients are processed with inverse quantization and inverse transform to produce reconstructed prediction error, and the reconstructed prediction error and prediction sample are added together to get the reconstructed picture. The reconstructed picture is filtered and sent to the frame buffer. In the framework, video coding tools are classified as:

- Intra prediction
- Variable Block-size Motion Compensation (VBMC)
- Multiple reference picture inter prediction
- Quarter pixel motion interpolation
- Improved direct mode for B picture (only in AVS1-P2)
- Symmetric mode for B picture (only in AVS1-P2)
- Integer transform
- Quantization
- Entropy coding
- Loop filter
- Interlace coding tools(only in AVS1-P2)
- Weighted quantization
- Error resilience tools

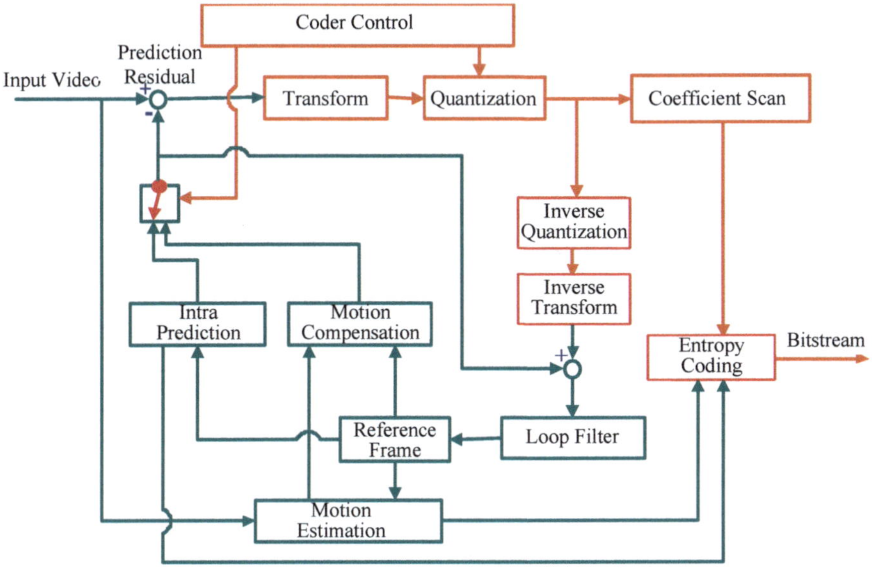

Fig. 1. The block diagram of AVS video encoder

Table 1 has shown the history of the AVS video coding standard and the development process of major video coding tools. We will detail these coding tools and the profile/level definition of AVS1-P2 in the following sections.

2.1 Intra Prediction

Spatial domain intra prediction has been used both in H.264/AVC and AVS to improve the coding efficiency of intra coding. In AVS standards, the intra prediction is 8x8 luma block based for AVS1-P2 and 4×4 luma block based for AVS1-P7, as shown in Fig. 2. That is because AVS1-P2 is mainly targeted to high resolution, such as Standard/High Definition (SD/HD) coding where the smaller block prediction is not efficient as the bigger block prediction for coding performance improvement. On the contrary, AVS1-P7 mainly focuses on low resolution coding, and smaller block prediction is more efficient.

In Fig. 2, when the neighbor samples r[8..16](or c[8..16]) are not available, they are replaced with r[8](or c[8]). Fig. 3 shows five intra prediction modes in AVS1-P2. Mode 0 (vertical prediction) and Mode 1 (horizontal prediction) in AVS1-P2 are the same as that in H.264/AVC, for example, the sample at position $[i, j]$ is predicted with $r[i+1]$ for Mode 0 and $c[j+1]$ for Mode 1. For the other modes, the filtered neighbor samples in the prediction direction will be used as prediction samples. For example, for mode 2 (DC prediction), the sample at [0, 0] position is predicted with:

$$((r[0]+ r[1] \ll 1 + r[2] + 2) \gg 2 + (c[0]+ c[1] \ll 1 + c[2] + 2) \gg 2) \gg 1.$$

For intra prediction in AVS1-P7, the same luma prediction directions are used with little difference on samples selection and mode index number, as shown in Fig. 2.

For chroma prediction, AVS1-P2 has also four chroma modes (vertical, horizontal, DC, and plane mode). However, AVS1-P7 only has three chroma modes (vertical, horizontal, DC) for complexity reduction.

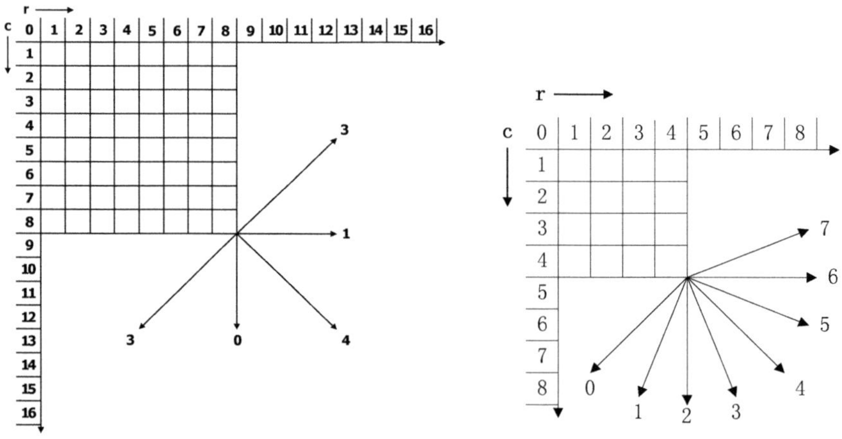

Fig. 2. Neighboring samples used for Intra luma prediction. Left: AVS1-P2, Right: AVS1-P7

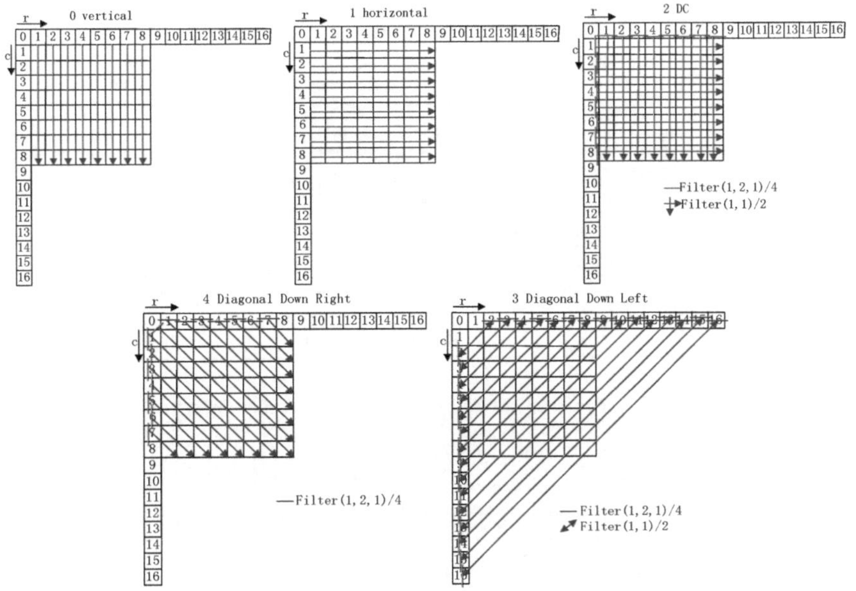

Fig. 3. Five intra luma prediction modes in AVS1-P2

2.2 Variable Block-Size Motion Compensation

Variable Block-size Motion Compensation (VBMC) is very efficient for prediction accuracy and coding efficiency improvement; however it is also the most costly in video encoder due to the high cost of motion estimation and the rate-distortion optimization mode decision. Macroblock partitions used in AVS standard are shown in Fig. 4. In AVS1-P7, the block size varies from 16×16 to 4×4, which is the same as

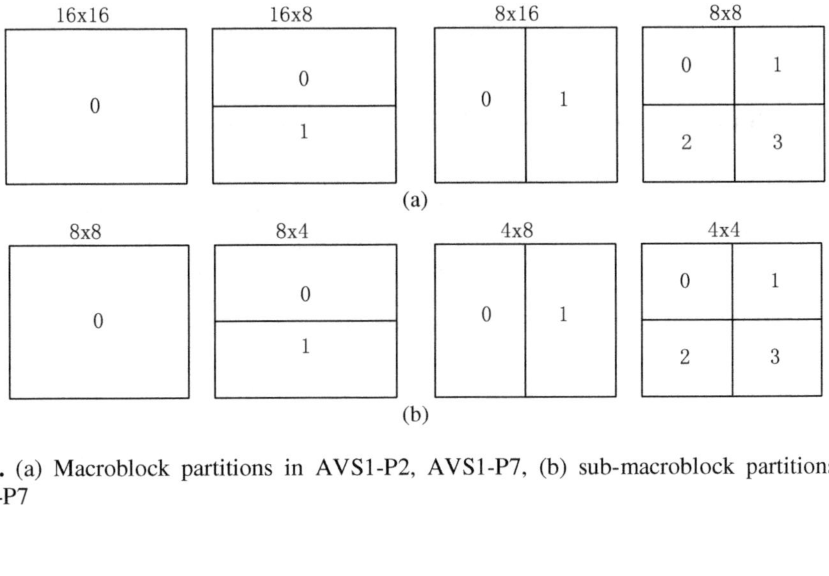

Fig. 4. (a) Macroblock partitions in AVS1-P2, AVS1-P7, (b) sub-macroblock partitions in AVS1-P7

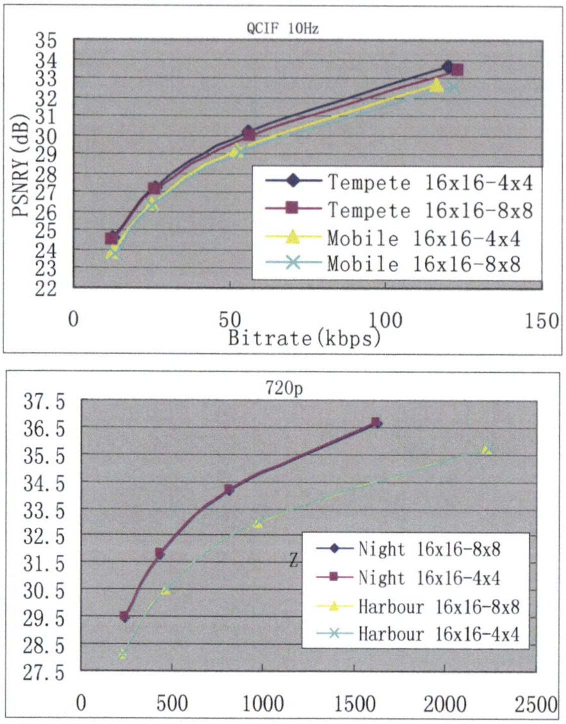

Fig. 5. VBMC performance testing on QCIF and 720p test sequences

that in H.264/AVC. However, in AVS1-P2, the minimum macroblock partition is 8×8 block. This is because the block size less than 8x8 does not give much improvement for high resolution coding. Fig. 5 shows the VBMC performance for QCIF and 720p HD sequences. Compared with the minimum 4×4 block motion compensation, the minimum 8×8 block motion compensation reduces encoder and decoder complexity significantly.

2.3 Multiple Reference Picture Motion Compensation

Multiple reference picture coding can further improve coding efficiency compared to using one reference e.g. MPEG-20. In general, two or three reference pictures give almost the best performance and more reference pictures will not bring significant performance improvement but increase the complexity greatly, as shown in Fig. 6, which shows the multi reference picture performance comparison. AVS1-P2 and AVS1-P7 restrict the maximum number of reference picture to be 2. Actually, for AVS1-P2 setting the maximum number of reference to be 2 does not increase the reference buffer size. In the previous video coding standards, such as MPEG-1 and MPEG-2, although P pictures use only one previous picture to predict the current picture, B pictures use one previous picture and one future picture as references, therefore the reference buffer size in a decoder has to be twice the picture size. Compared

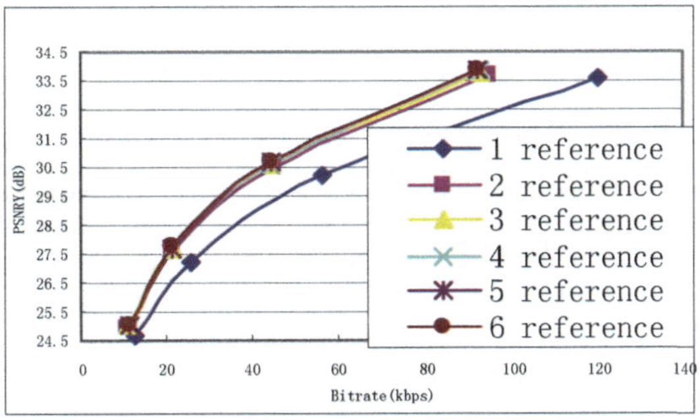

Fig. 6. Multiple reference picture performance testing

with MPEG-1 and MPEG-2, AVS can improve coding efficiency while using the same reference buffer size.

2.4 Quarter-Pixel Interpolation

Fractional pixel motion accuracy has already been used since MPEG-1. Since MPEG-4(Advanced Simple Profile), quarter pixel interpolation has been developed. For fractional pixel interpolation, the interpolation filter does important effect on coding efficiency. In 0, the performance of Wiener filter and bilinear filter is studied. In the development process of H.264/AVC, the filter parameter has been discussed in many proposals 000. In the final H.264/AVC standard, a 6-tap filter is used for half pixel interpolation. But in AVS1-P2, a 4-tap filter is used for half pixel interpolation 0. This is because the 6-tap filter does better for low resolution video, such as QCIF, CIF. But a 4-tap filter can achieve similar performance with 6-tap filter while with much lower computing and memory access complexity. Fig. 7 shows the performance comparison between the 6-tap filter of H.264/AVC and 4-tap filter of AVS on several QCIF and 720p sequences.

For the interpolation process in AVS1-P2, as shown in Fig. 8, the samples at half sample positions labeled *b* and *h* are derived by first calculating intermediate values *b'* and *h'*, respectively by applying the 4-tap filter (-1, 5, 5, -1) to the neighboring integer samples, as follows:

$$b'=(-C+5D+5E-F)$$
$$h'=(-A+5D+5H-K)$$

The final prediction values for locations *b* and *h* are obtained as follows and clipped to the range of 0 to 255:

$$h=(h'+4)>>3$$
$$b=(b'+4)>>3$$

The samples at half sample positions labeled as *j* are obtained by applying the 4-tap filter (-1, 5, 5, -1) to the neighboring half samples as follows:

$$j'=(-bb'+5h'+5m'-cc') \text{ or}$$
$$j'=(-aa'+5b'+5s'-dd'),$$

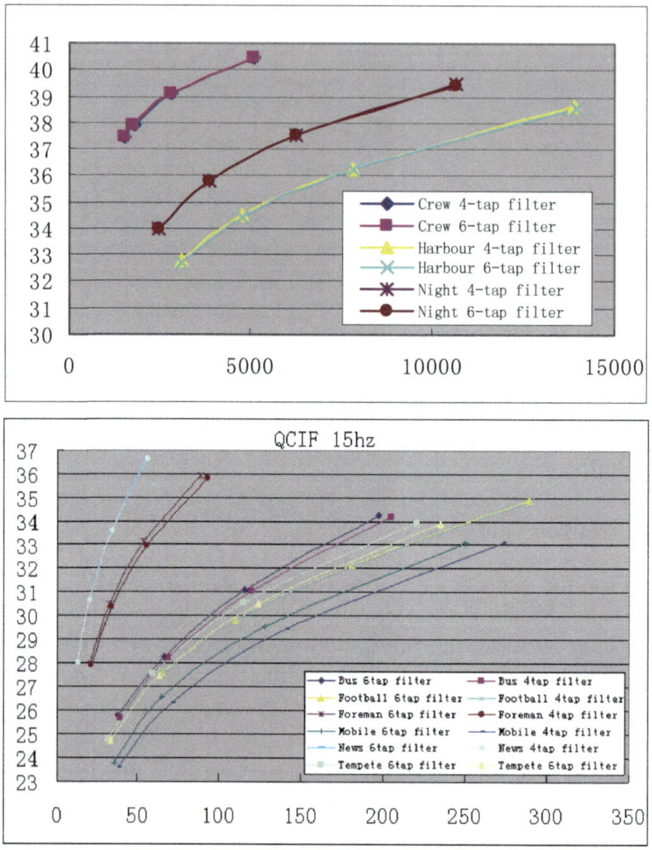

Fig. 7. Interpolation filter performance comparison

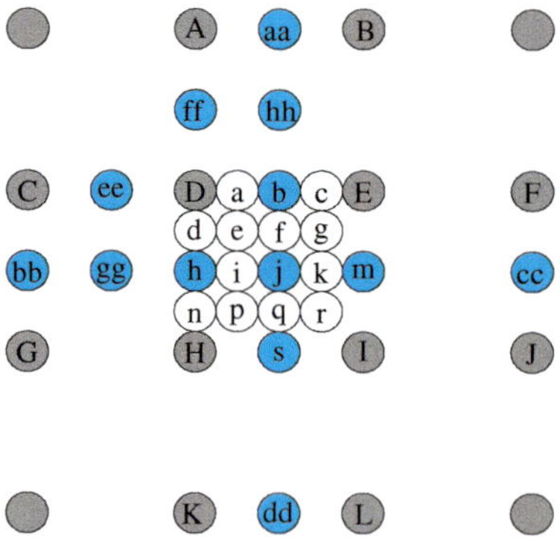

Fig. 8. Filtering for fractional-sample accuracy motion compensation. Upper-case letters indicate samples on the full-sample grid, while lower case samples indicate samples at fractional-sample positions.

here, aa', bb', cc', dd', m', s' is the intermediate value of samples at half sample positions aa, bb, cc, dd, m, s respectively.

The final prediction values for locations j is obtained as follows and clipped to the range of 0 to 255:

$$j=(j'+32)>>6$$

The samples at quarter sample positions labeled as a, i, d, f are obtained by applying the 4-tap filter (1, 7, 7, 1) to the neighboring integer and half sample position samples, e.g.:

$$a'=(ee'+7D'+7b'+E')$$
$$i'=(gg'+7h''+7j'+m'')$$
$$d'=(ff'+7D'+7h'+H')$$
$$f'=(hh'+7b''+7j'+s''),$$

here ee', gg', ff', gg' are the intermediate values of samples at half sample positions. D', E', H' are the scaled integer samples with 8 times. h'', m'', b'', s'' are the half samples scaled with 8 times.

The final prediction values for locations a, i, d, f are obtained as follows and clipped to the range of 0 to 255:

$$a=(a'+64)>>7$$
$$i=(i'+512)>>10$$
$$d=(d'+64)>>7$$
$$f=(f'+512)>>10$$

The samples at quarter sample positions labeled as e, g, p, r are obtained by averaging the neighboring integer position samples and j:

$$e = (D'' + j' + 64) >> 7$$
$$g = (E'' + j' + 64) >> 7$$
$$p = (H'' + j' + 64) >> 7$$
$$r = (I'' + j' + 64) >> 7$$

D'', E'', H'', I'' are the integer samples scaled with 8 times.

In AVS1-P7, an 8-tap filter and a 6-tap filter is used for horizontal and vertical direction interpolation respectively. The samples at half sample positions labeled b and h are derived by first calculating intermediate values b' and h', respectively by applying the filter as follows:

$$b'= (-C+4D-12E+41F+41G-12H+4I-J)$$
$$h'=(-A+5F+5N-S)$$

The final prediction values for locations b and h are obtained as follows and clipped to the range of 0 to 255:

$$h= (h'+4)>>3;$$
$$b= (b'+32)>>6$$

The samples at half sample positions labeled as j are obtained by applying the 4-tap filter as follows:

$$j'=(-aa+5b+5t-hh)$$

here, aa', t', hh' are the intermediate values of samples at half sample positions.

The samples at quarter sample positions labeled as *a, c, d, n* are obtained by averaging the neighboring integer and half position sample, e.g.

$$a=(F+b+1)>>1$$
$$d=(F+h+1)>>1$$

i, k, f, q are obtained by averaging the neighboring half pixel position samples, e.g.:

$$i=(h+j+1)>>1$$
$$f=(b+j+1)>>1$$

The samples at quarter sample positions labeled as *e, g, p, r* are obtained by averaging the neighboring integer sample and half position sample *j*, e.g.:

$$e = (F + j + 1) >> 1$$

2.5 Improved Direct Mode

The direct mode previously existing in H.263+ and MPEG-4 for global motion is improved in H.264/AVC and AVS. In H.264/AVC, it is classified as temporal and spatial direct mode according to the motion vector derivation scheme. The temporal direct mode and spatial direct mode are independent of each other. However, in AVS1-P2, the temporal direct mode and spatial direct mode are combined together. In the prediction process of direct mode, spatial prediction will be used when the co-located macroblock for temporal prediction is intra coded.

Fig. 9 (a) shows the motion vector derivation process for direct mode in AVS1-P2 frame coding. The motion vectors for the current block in B picture can be derived as follows:

$$mvFw\_x = \text{sign}(mvRef\_x)(((\frac{16384}{BlockDistanceRef}) \times (1 + \text{abs}(\frac{mvRef\_x}{BlockDistanceFw})) - 1) >> 14)$$
$$mvFw\_y = \text{sign}(mvRef\_y)(((\frac{16384}{BlockDistanceRef}) \times (1 + \text{abs}(\frac{mvRef\_y}{BlockDistanceFw})) - 1) >> 14)$$
$$mvBw\_x = -\text{sign}(mvRef\_x)(((\frac{16384}{BlockDistanceRef}) \times (1 + \text{abs}(\frac{mvRef\_x}{BlockDistanceBw})) - 1) >> 14)$$
$$mvBw\_y = -\text{sign}(mvRef\_y)(((\frac{16384}{BlockDistanceRef}) \times (1 + \text{abs}(\frac{mvRef\_y}{BlockDistanceBw})) - 1) >> 14)$$

Here (*mvFw_x*, *mvFw_y*), (*mvBw_x*, *mvBw_y*) are the forward and backward motion vector of the current block respectively. (*mvRef_x*, *mvRef_y*) is the motion vector of the co-located block in the backward reference. *BlockDistanceRef* is the distance between the reference picture to which the co-located block belongs and the co-located block's reference picture. *BlockDistanceFw* is the distance between the current picture and the current block's forward reference picture. *BlockDistanceBw* is the distance between the current picture and the current block's backward reference picture.

For interlace coding, with different field parity of the picture where the current block exists and different reference index of the co-located block, different reference pictures will be selected for the current block and the corresponding temporal picture distance will be calculated for motion vector derivation, as shown in Fig. 9 (b)-(e).

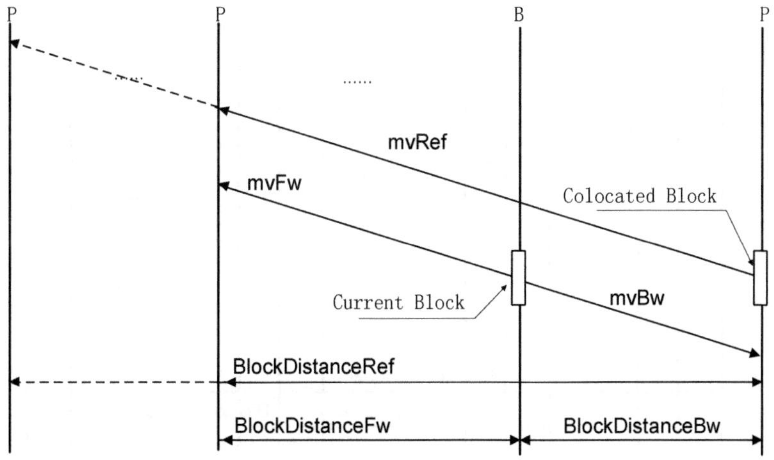

(a) Motion vector derivation for direct mode in frame coding. Co-located block's reference index is 0 (solid line), or 1(dash line).

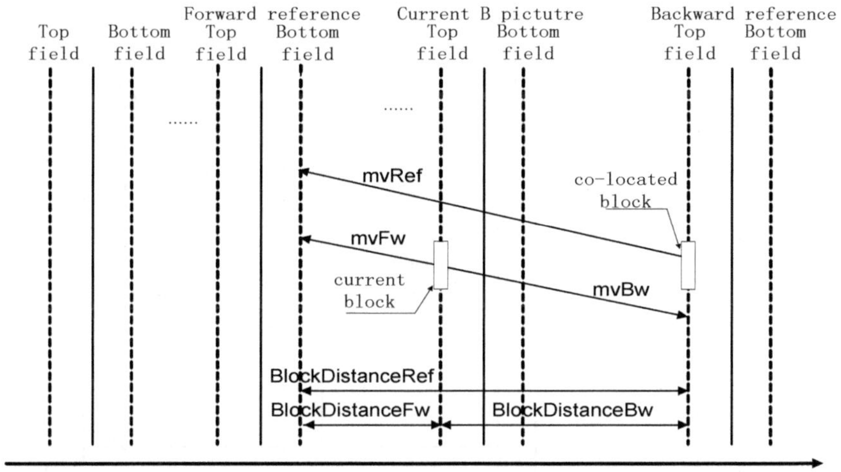

(b) Motion vector derivation for direct mode in top field coding. Co-located block's reference index is 0.

Fig. 9. Temporal direct mode in AVS1-P2

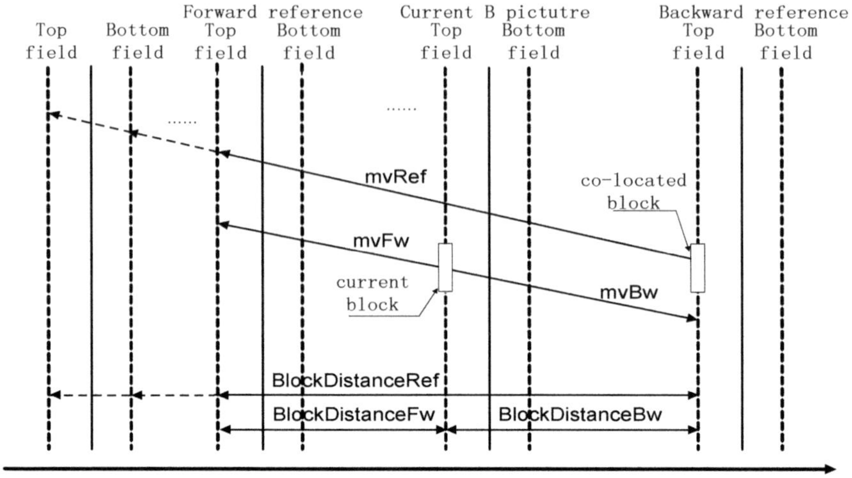

(c) Motion vector derivation for direct mode in top field coding. Co-located block's reference index is 1 (solid line), 2(dashed line pointing to bottom field), or 3(dashed line pointing to top field).

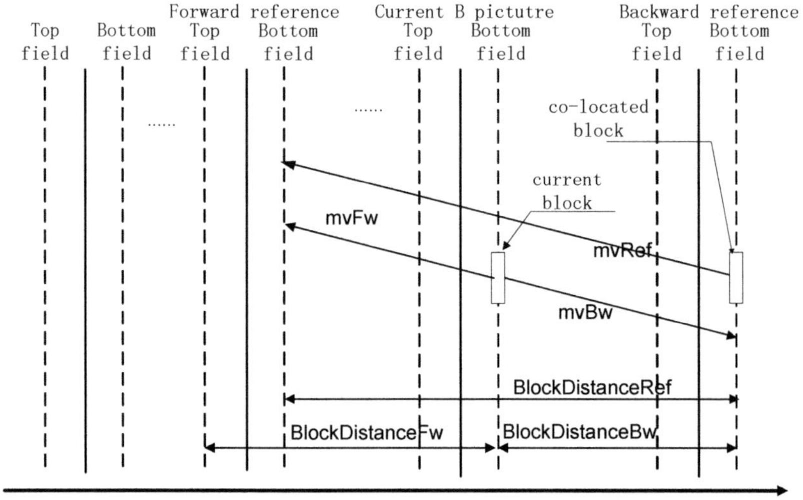

(d) Motion vector derivation for direct mode in top field coding. Co-located block's reference index is 1.

Fig. 9. *(continued)*

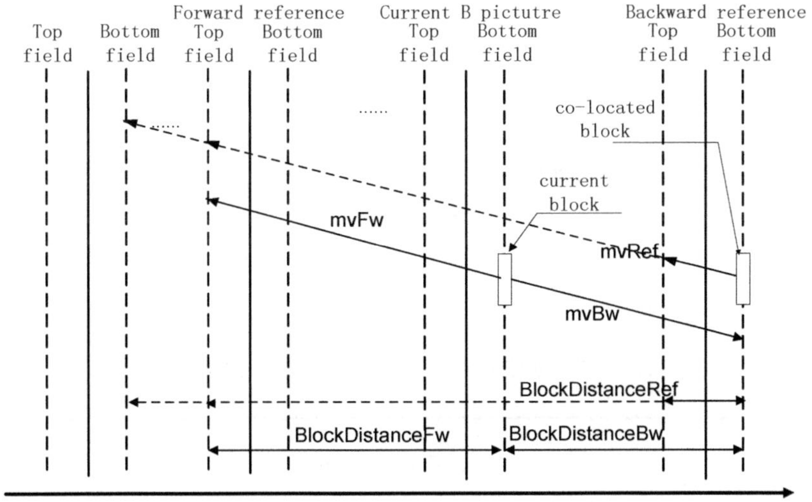

(e) Motion vector derivation for direct mode in top field coding. Co-located block's reference index is 0 (solid line), 2(dashed line pointing to bottom field), or 3(dashed line pointing to top field).

Fig. 9. (*continued*)

2.6 Symmetric Mode

In AVS1-P2, a new symmetrical mode is used for B picture coding to replace the traditional bi-direction coding 0. For the symmetric mode, only a forward motion vector is coded and the backward motion vector is derived from the forward motion vector. That is to say at most one motion vector is coded for a block in a B picture in AVS.

As shown in Fig. 10 (a), the nearest previous and future reference pictures are selected as the forward and backward reference pictures for the current block in frame coding. The forward motion vector is *mvFW*, and the backward motion vector is derived as:

$$mvBW = -\frac{BlockDistanceBw}{BlockDistanceFw} mvFW \cdot$$

Here, *BlockDistanceFw* is the distance between the current picture and the forward reference picture of the current block. *BlockDistanceBw* is the distance between the current picture and the backward reference picture of the current block. For field coding, the forward and backward reference selection is shown as Fig. 10 (b) and Fig. 10 (c).

2.7 Transform and Quantization

In H.264/AVC, the traditional float DCT is replaced with integer transform and the mismatch can be removed 0. In general, a larger size transform has better energy compaction property, while a smaller size transform has the advantages of reducing

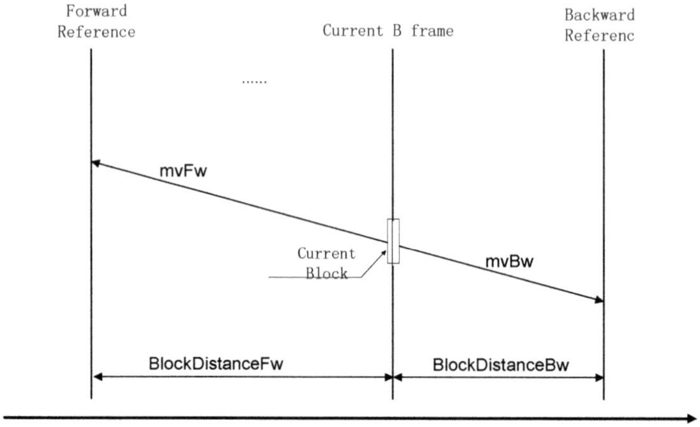

(a) Frame coding

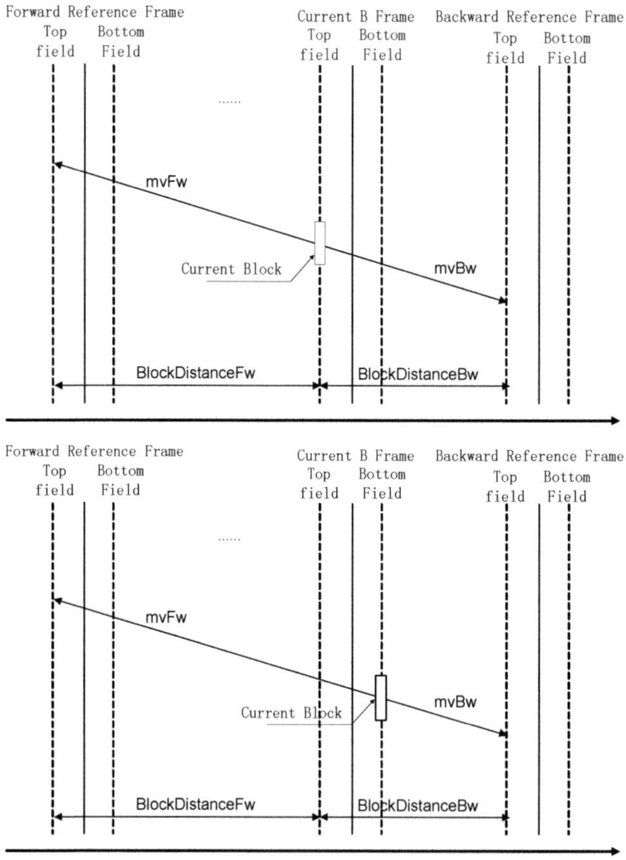

(b) Field coding, forward reference index is 1, backward reference index is 0.

Fig. 10. Motion vector derivation for symmetric mode in AVS1-P2

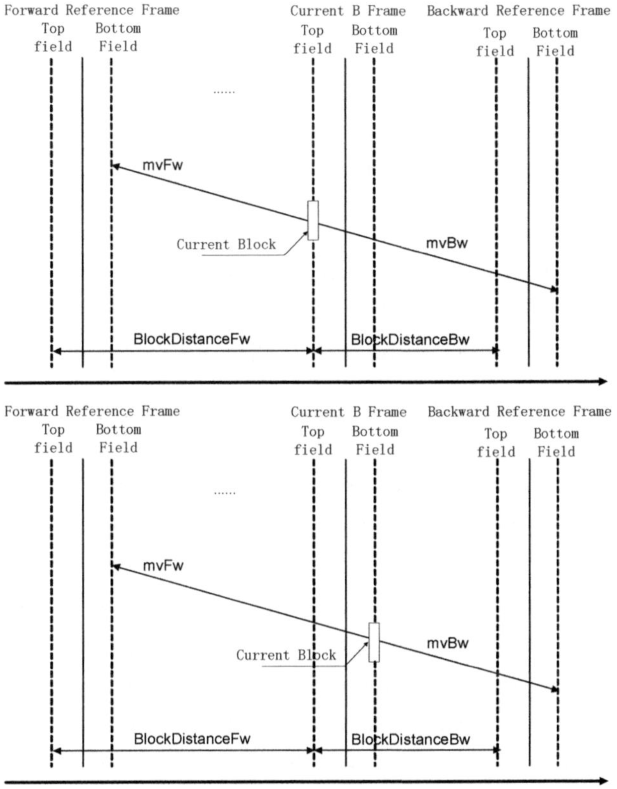

(c) Field coding, forward reference index is 0, backward reference index is 1.

Fig. 10. *(continued)*

ringing artifacts at edges and discontinuities. In the development of H.264/AVC, adaptive block transform (ABT) was proposed and has been adopted into H.264/AVC High Profile 00. As larger transform size is more efficient for high resolution coding, an 8×8 integer transform is used in AVS1-P2, shown as follows:

$$T_8 = \begin{bmatrix} 8 & 10 & 10 & 9 & 8 & 6 & 4 & 2 \\ 8 & 9 & 4 & -2 & -8 & -10 & -10 & -6 \\ 8 & 6 & -4 & -10 & -8 & 2 & 10 & 9 \\ 8 & 2 & -10 & -6 & 8 & 9 & -4 & -10 \\ 8 & -2 & -10 & 6 & 8 & -9 & -4 & 10 \\ 8 & -6 & -4 & 10 & -8 & -2 & 10 & -9 \\ 8 & -9 & 4 & 2 & -8 & 10 & -10 & 6 \\ 8 & -10 & 10 & -9 & 8 & -6 & 4 & -2 \end{bmatrix}$$

The transform has the similar feature as that in WMV9, called Pre-scaled Integer Transform (PIT) in [16]. As the basis of the transform coefficients is very close, the transform normalization can be accounted entirely on the encoder side. The transform in AVS1-P7 has the same feature as that in AVS1-P2, but a 4×4 transform is used:

$$T_4 = \begin{bmatrix} 2 & 2 & 2 & 2 \\ 3 & 1 & -1 & -3 \\ 2 & -2 & -2 & 2 \\ 1 & -2 & 3 & -1 \end{bmatrix}$$

All these transforms can be implemented within16 bits. After transform, the transform coefficient matrix *Y* is normalized with a scale table *ScaleTbl*:

$$Y'_{i,j} = (Y_{i,j} \times ScaleTbl\,[i][j] + 2^{a-1}) >> a$$

In AVS1-P2 reference model, *a* is equal to 19, and the scale table is defined as follows

$$ScaleTbl = \begin{bmatrix} 32768 & 37959 & 36158 & 37958 & 32768 & 37958 & 36158 & 37958 \\ 37958 & 43969 & 41884 & 43969 & 37958 & 43969 & 41884 & 43969 \\ 36158 & 41884 & 39898 & 41884 & 36158 & 41884 & 39898 & 41884 \\ 37958 & 43969 & 41884 & 43969 & 37958 & 43969 & 41884 & 43969 \\ 32768 & 37958 & 36158 & 37958 & 32768 & 37958 & 36158 & 37958 \\ 37958 & 43969 & 41884 & 43969 & 37958 & 43969 & 41884 & 43969 \\ 36158 & 41884 & 39898 & 41884 & 36158 & 41884 & 39898 & 41884 \\ 37958 & 43969 & 41884 & 43969 & 37958 & 43969 & 41884 & 43969 \end{bmatrix}.$$

In AVS1-P7 reference model, *a* is equal to 15, and the scale table is defined as:

$$ScaleTbl = \begin{bmatrix} 32768 & 26214 & 32768 & 26214 \\ 26214 & 20972 & 26214 & 20972 \\ 32768 & 26214 & 32768 & 26214 \\ 26214 & 20972 & 26214 & 20972 \end{bmatrix}$$

As H.264/AVC, quantization/dequantization can be implemented with multiply and right shift. All these operations can be completed within 16 bits. After transform normalization, the quantization is done as:

$$YQ_{i,j} = (Y'_{i,j} \times Q[QP] + 2^{b-1}) >> b$$

Q[*QP*] is the quantization table on encoder set. The dequantization is shown as follows:

$$X_{i,j} = (YQ_{i,j} \times IQ[QP] + 2^{s(QP)-1}) >> s(QP),$$

where *QP* is the quantization parameter, *IQ*(*QP*) is the inverse quantization table and *s*(*QP*) is the varied shift value for inverse quantization. AVS1-P7 uses the same quantization and dequantization table as that in AVS1-P2 except for a little difference on the value of *b*. *b* is 15 for AVS1-P2 reference model and 19 for AVS1-P7 reference model. *Q*(*QP*), *IQ*(*QP*) and *s*(*QP*) is shown as follows:

QP	0	1	2	3	4
Q [QP]	32768	29775	27554	25268	23170
IQ [QP]	32768	36061	38968	42495	46341
s[QP]	14	14	14	14	14
QP	5	6	...	62	63
Q [QP]	21247	19369	...	152	140
IQ [QP]	50535	55437	...	55109	60099
s [QP]	14	14	...	7	7

In order to improve the overall image quality, AVS-P2 Jiaqiang Profile adopts the technique of adaptive weighting quantization (AWQ) [39]. Similar to the adaptive quantization matrix technique in H.264/AVC, users can define the quantization matrices based on the characteristics of test sequences. To achieve this purpose, three default quantization matrix patterns are defined and corresponding six weighting parameters for each pattern can be set by users are defined in AWQ, which are based on the distribution characteristics of transform coefficients as well as the human perception. Fig.11 depicts the three patterns defined in AVS1-P2 Jiaqiang Profile. It can be easily observed that in these three patterns, the quantization matrix is partitioned into six different regions. Each region represents similar characteristics of transform coefficients. The quantization weighting value is set to be the same in different frequency sub-bands of the current region. For each pattern, three pre-defined weighting parameter sets are given in AVS specification. They represent default weighting quantization case, the case for reserving detail texture information and one for removing the detail information of texture. We take the second pattern depicted in Fig.11(b) for an example, the pre-defined weighting parameter sets for this pattern are shown in Fig.12. Therefore, the encoder only needs to send the matrix pattern together with six offsets of weighting values from the pre-defined ones to the decoder to form an optimized quantization matrix. The dequantization process can be derived as:

$$X_{i,j} = ((((YQ_{i,j} \times WQ_{i,j}) >> 3) \times IQ[QP]) >> 4) + 2^{s(QP)-1}) >> s(QP)$$

where $WQ_{i,j}$ indicates the weighting value for each frequency subband.

In AWQ, the quantization matrix can be optimized frame by frame or macroblock by macroblock. For the macroblock-level AWQ, the encoder decides the best quantization matrix pattern index according to its neighbors' types and modes. In addition, the quantization parameter (*QP*) for two chrominance components (UV) in one macroblock can be adjusted and the difference of *QP* for U or V blocks with that of *QP* for luma component is also sent to the decoder.

2.8 Entropy Coding

In H.264/AVC, two entropy coding schemes for transformed coefficients are included: CAVLC and CABAC 0. AVS also has two entropy coding schemes to achieve different level compression efficiency with different coding complexity. One

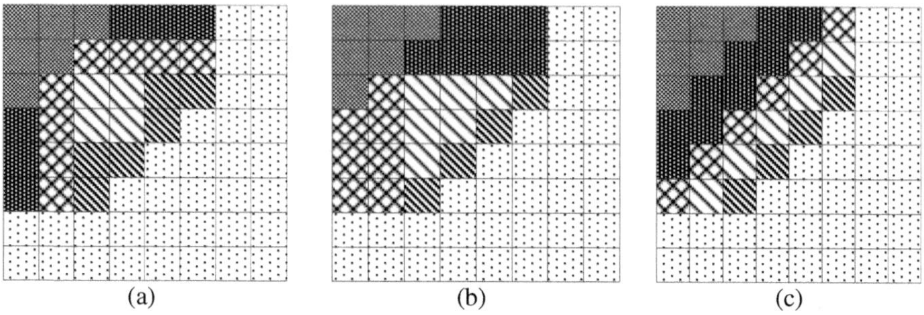

Fig. 11. Quantization matrix patterns in AVS1-P2 Jiaqiang Profile

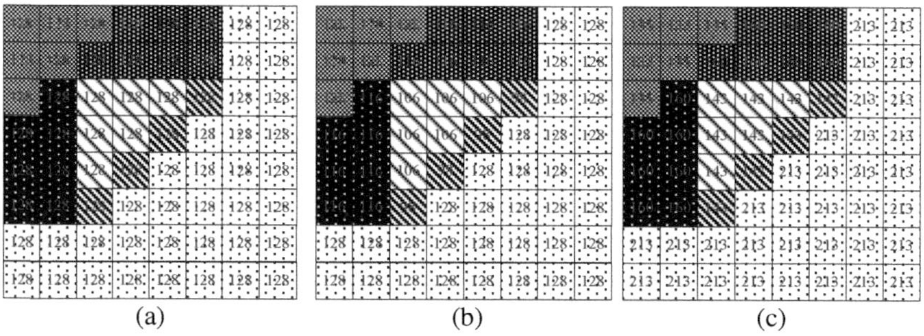

Fig. 12. Pre-defined quantization weighting parameters in AVS1-P2 Jiaqiang Profile, (a) default parameters, (b) parameters for keeping detail information of texture, (c) parameters for removing detail information of texture

is Context-based 2D–VLC (C2DVLC) entropy coding 0, used in AVS1-P2 Jizhun profile and AVS1-P7, and the other one is improved Context-based Adaptive Binary Arithmetic Coding (CBAC) for AVS1-P2 Jiaqiang profile. The two schemes both have lower complexity compared with CAVLC and CABAC in H.264/AVC.

2.8.1 2D Adaptive VLC Coding

2D adaptive VLC coding is used in AVS Jizhun profile. As in H.264/AVC, k^{th}-order Exp-Golomb code (k = 0, 1, 2, 3) is used for entropy coding, defined in Table 2. CBP, macroblock coding mode, motion vectors are coded with 0th-order Exp-Golomp code. The quantized transform coefficients are coded with multiple tables. The difference between AVS and H.264/AVC is that run and level are coded as a pair in AVS and not coded independently as H.264/AVC.

For AVS1-P2 Jizhun Profile, a block of coefficients is coded as follows:

Coefficient scan: As shown in Fig. 13, two scan patterns are used in AVS1-P2. Zigzag scan is used for progressive coding and alternate scan is used for interlace coding. After coefficient scan, the two-dimensional transformed coefficients are organized into one sequence in the form of (*Level*, *Run*) pairs. To indicate the coding end of current block, a special symbol *EOB* is used, that is one (0, 0) pair.

j \ i	0	1	2	3	4	5	6	7
0	0	1	5	6	14	15	27	28
1	2	4	7	13	16	26	29	42
2	3	8	12	17	25	30	41	43
3	9	11	18	24	31	40	44	53
4	10	19	23	32	39	45	52	54
5	20	22	33	38	46	51	55	60
6	21	34	37	47	50	56	59	61
7	35	36	48	49	57	58	62	63

j \ i	0	1	2	3	4	5	6	7
0	0	3	11	16	22	32	38	55
1	1	6	12	20	25	33	42	57
2	2	7	15	21	28	37	43	58
3	4	10	19	27	31	39	47	59
4	5	14	24	30	36	44	50	60
5	8	17	26	35	41	48	52	61
6	9	18	29	40	46	51	54	62
7	13	23	34	45	49	53	56	63

Fig. 13. Coefficient scan in AVS1-P2. Left: zig-zag scan, Right: alternate scan

***Table selection*:** 19 mapping tables are defined in AVS1-P2 to code the codeword after mapping one (*Level*, *Run*) pair into Exp-Golomb code. 7 tables are used for intra luma coefficient coding; 7 tables are used for inter luma coefficient coding; remained 5 tables are used for chroma coding. The (*Level*, *Run*) pairs are coded in reverse scan order, that is to say the last (*Level*, *Run*) pair in scan order will be coded first. Tables are selected adaptively based on the maximum magnitude of all previous *Level*s in coding process, as shown in Fig. 14. Firstly, an initial table is selected for the first non-zero quantized coefficient and its corresponding *Run* value. Afterward, the absolute value of the current coefficient decides which table is used for the next non-zero quantized coefficient following (*Level*, *Run*) pair.

***Run-level mapping and coding*:** The (*Level*, *Run*) pair is converted into a *CodeNum* according to the selected table. An example table is shown in Fig. 15. In the table, only *CodeNum*s for positive *Level* are shown and *CodeNum*+1 is that for the corresponding negative *Level*. Then the *CodeNum* is mapped into a *Code* according to the mapping function of kth-order Exp-Golomb code. If the coefficient is the last non-zero coefficient, an EOB (end of block) symbol is coded, otherwise go to step 2 to code the next (*Level*, *Run*) pair. When the (*Level*, *Run*) pair is not in the table, it will be coded with escape code. An escape code is combined with two codes: one code is 59+*Run*; the other one is *abs*(*Level*)-RefAbsLevel(*Run*). Here *abs*() denotes the absolute function. RefAbsLevel(*Run*) is decided by the selected table. When the *Run* is not in the table, *RefAbsLevel*(*Run*) is 1.

For entropy coding in AVS1-P7, totally 18 tables are defined, including 7 tables for intra luma coefficients and 7 tables for inter luma coefficients and 4 tables for chroma coefficients.

2.8.2 Context-Based Binary Arithmetic Coding

To further improve entropy coding efficiency, a new context-based binary arithmetic coding method is developed in AVS1-P2 enhancement profile. Its complexity is reduced compared with CABAC in H.264/AVC. For CABAC in H.264/AVC, 326 probability models are used for residual data coding. While in AVS, only unary

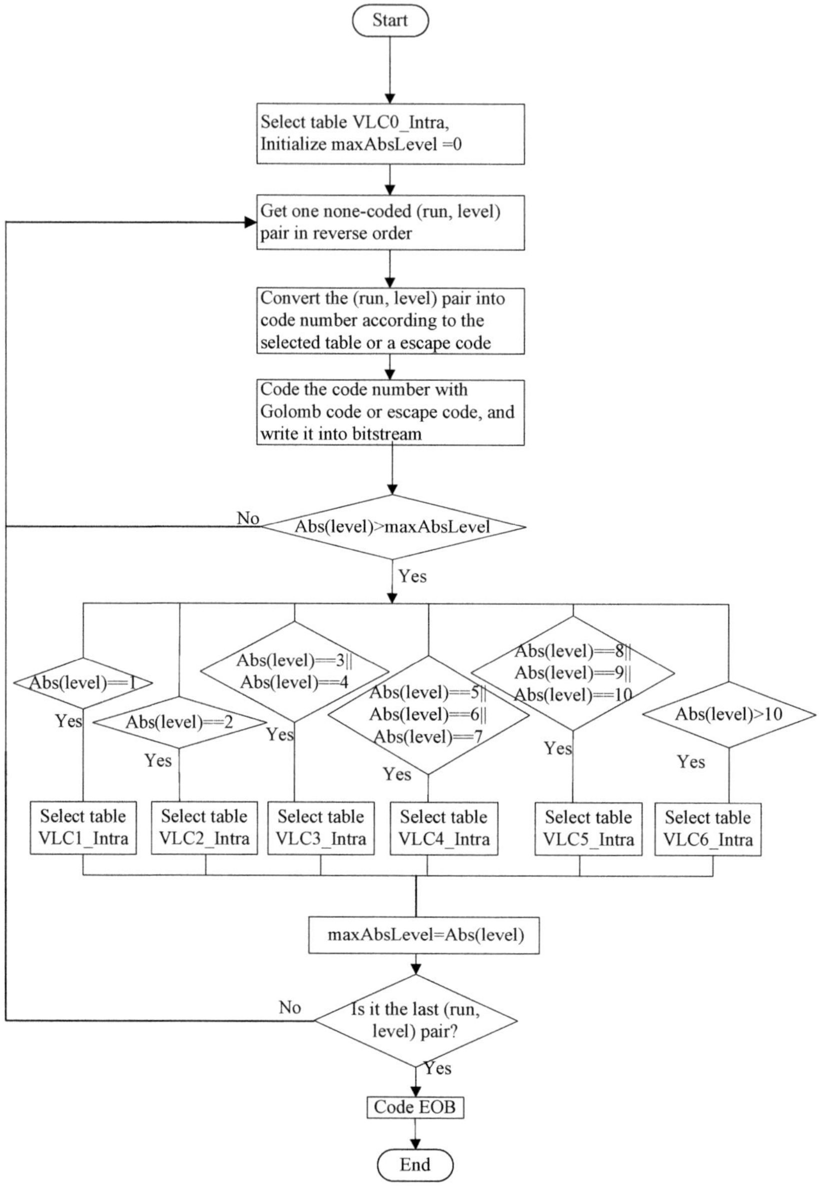

(a) Flowchart of coding one intra luma block

Fig. 14. Coefficient coding process in AVS1-P2 2D VLC entropy coding scheme

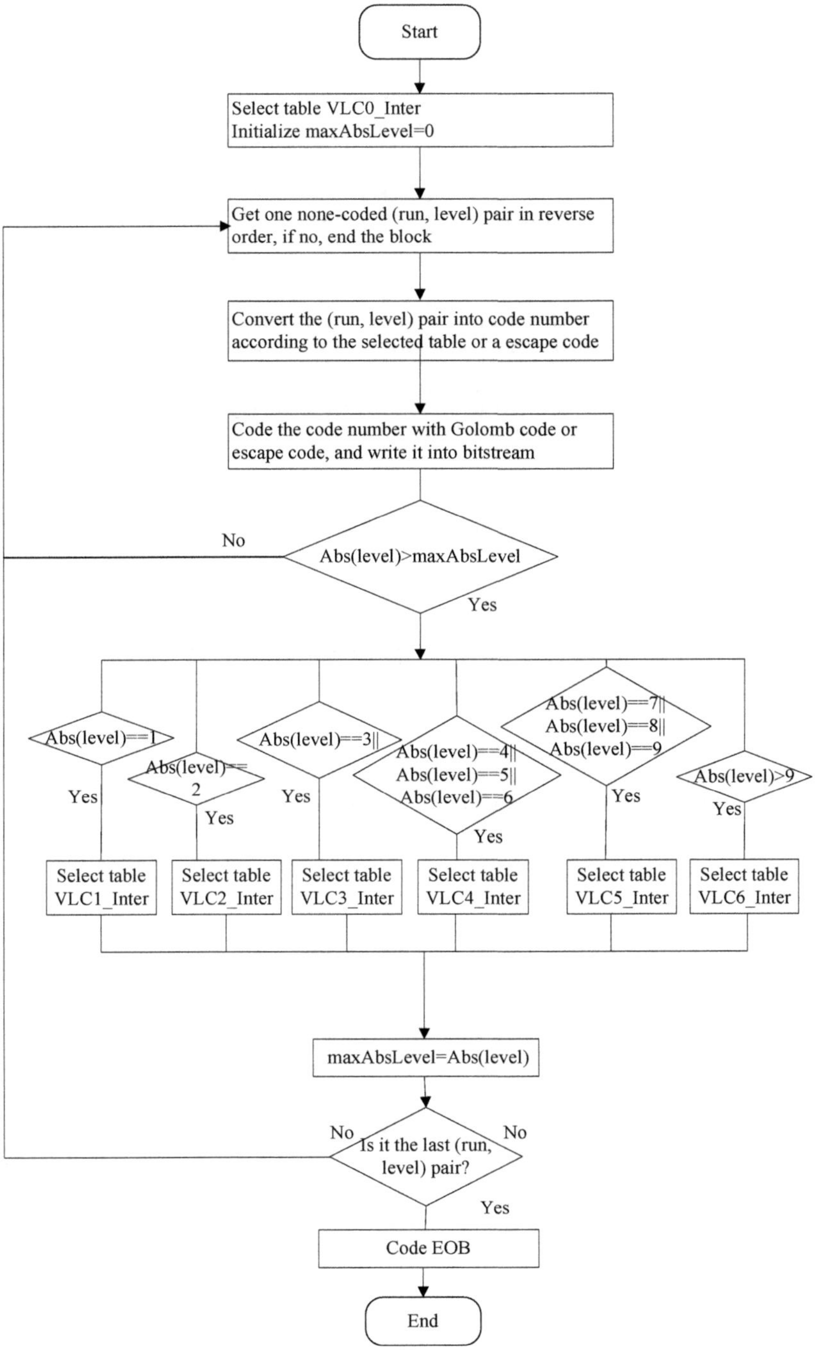

(b) Flowchart of coding one inter luma block

Fig. 14. *(continued)*

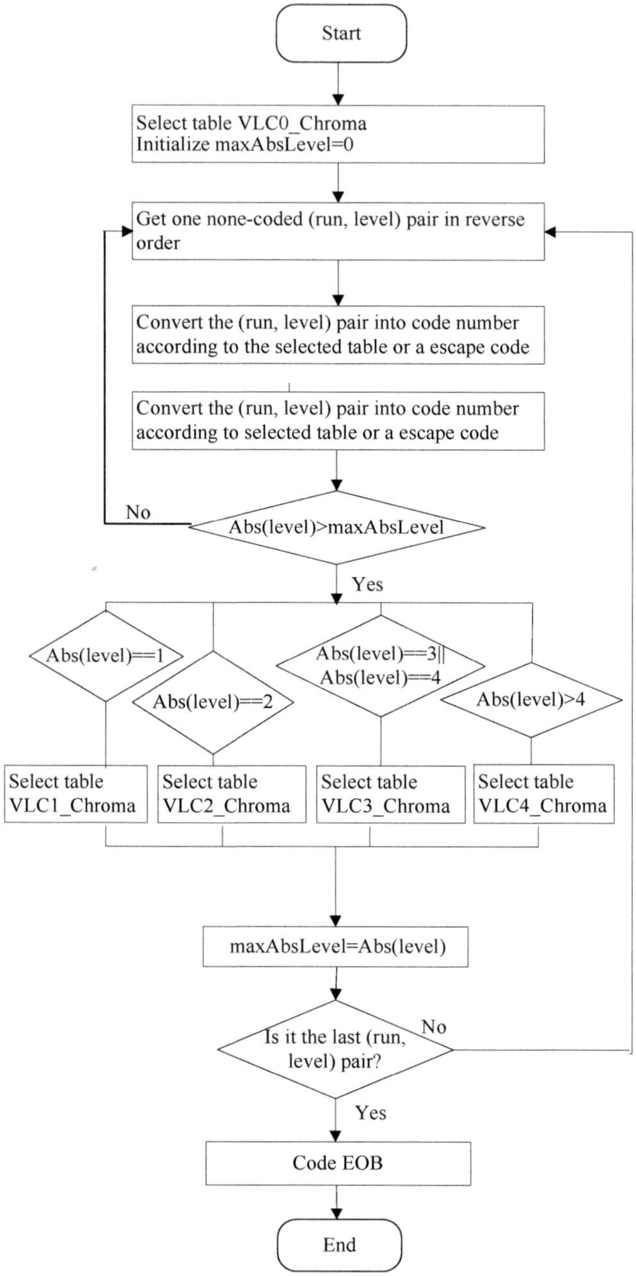

(c) Flowchart of coding one inter chroma block

Fig. 14. (*continued*)

Run	EOB	Level > 0						RefAbsLevel
		1	2	3	4	5	6	
	8	-	-	-	-	-	-	
0	-	0	4	15	27	41	55	7
1	-	2	17	35	-	-	-	4
2	-	6	25	53	-	-	-	4
3	-	9	33	-	-	-	-	3
4	-	11	39	-	-	-	-	3
5	-	13	45	-	-	-	-	3
6	-	19	49	-	-	-	-	3
7	-	21	51	-	-	-	-	3
8	-	23	-	-	-	-	-	2
9	-	29	-	-	-	-	-	2

Fig. 15. An example table in AVS1-P2—VLC1_Intra: from (run, level) to *CodeNum*

binarization and totally 132 context models are used. The coding process of a block of coefficients after zig-zag scan is shown as follows:

***Binarization*:** The *(Level, Run)* pair is in a large range and coding these *Level/Run* values directly by an *m*-ary arithmetic code will have a high computational complexity. So binary arithmetic coding is used in AVS too. Both *Level* and *Run* are binarized using unary binarization scheme. The signed integer *Level* is represented by sign (0/1: +/-) and the unary bits of its magnitude (absLevel: the absolute value of *Level*). *Level* is coded first followed by the *Run*.

***Context model Selection*:** In AVS, a two order context model is used, called as primary context and secondary context. The primary context is defined as the maximum magnitude of all previously coded levels in the current block, denoted with a context variable *Lmax*. *Lmax* will be updated in the coding process. As the dynamic range of the context variable *Lmax* can still be too large, it is reduced by the quantization function defined into five primary contexts. The context quantization function can be defined as

$$\chi_{(Lmax)} = \begin{cases} Lmax & Lmax \in [0, 2] \\ 3 & Lmax \in [3, 4] \\ 4 & otherwise \end{cases}$$

Under each primary context, seven secondary contexts are defined to code *level* and *run*, shown as follows:

0: first bit of *absLevel* (i.e., the EOB symbol).
1: second bit of *absLevel*, if exist.

2: remaining bits of *absLevel*, if exist.
3: first bit of Run, if *absLevel*=1.
4: remaining bits of Run when *absLevel*=1, if exist.
5: first bit of Run when *absLevel*>1
6: remaining bits of Run when *absLevel*>1, if exist.

Here, *absLevel* is the absolute value of the *level*.

In order to further improve compression performance, a context weighting technique is used in AVS entropy coding. Another context variable *ReverseP* is introduced into the context, called accompanying context, which is the position of current non-zero DCT coefficient in the reverse scanning order. For an 8x8 block the range of *ReverseP* is [0, 64], and it is uniformly quantized into 32 accompanying contexts, [0, 31]. In the binary arithmetic coding of the *run* and *level*, each accompanying context created by *ReverseP* will be combined with the same seven secondary contexts as in the case of primary contexts created.

2.9 Loop Filter

Block-based video coding often produces blocking artifacts especially at low bit rates. As in H.264/AVC, AVS also adopts adaptive in-loop deblocking filter to improve the

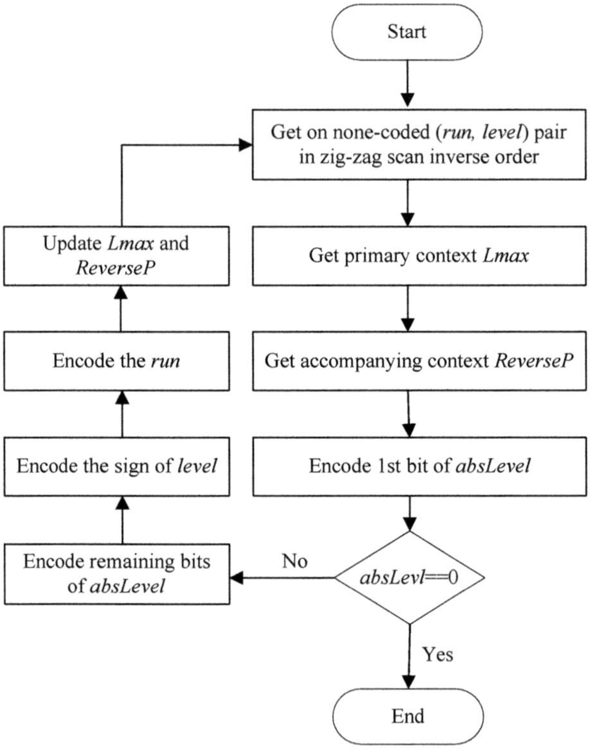

Fig. 16. Coefficient coding process in AVS1-P2 context adaptive arithmetic coding

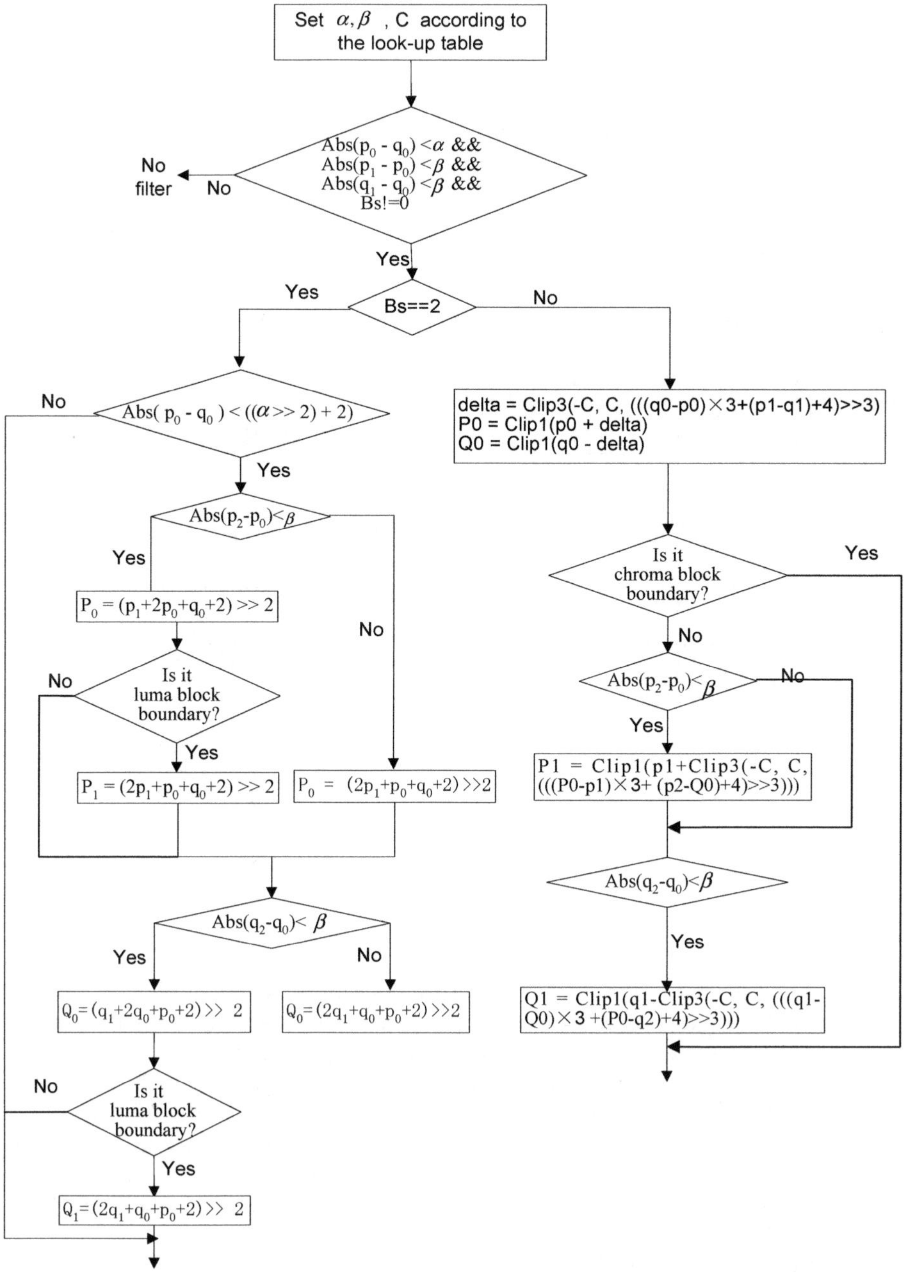

(a) Deblocking filter process in AVS1-P2

Fig. 17. Deblocking filter process in AVS

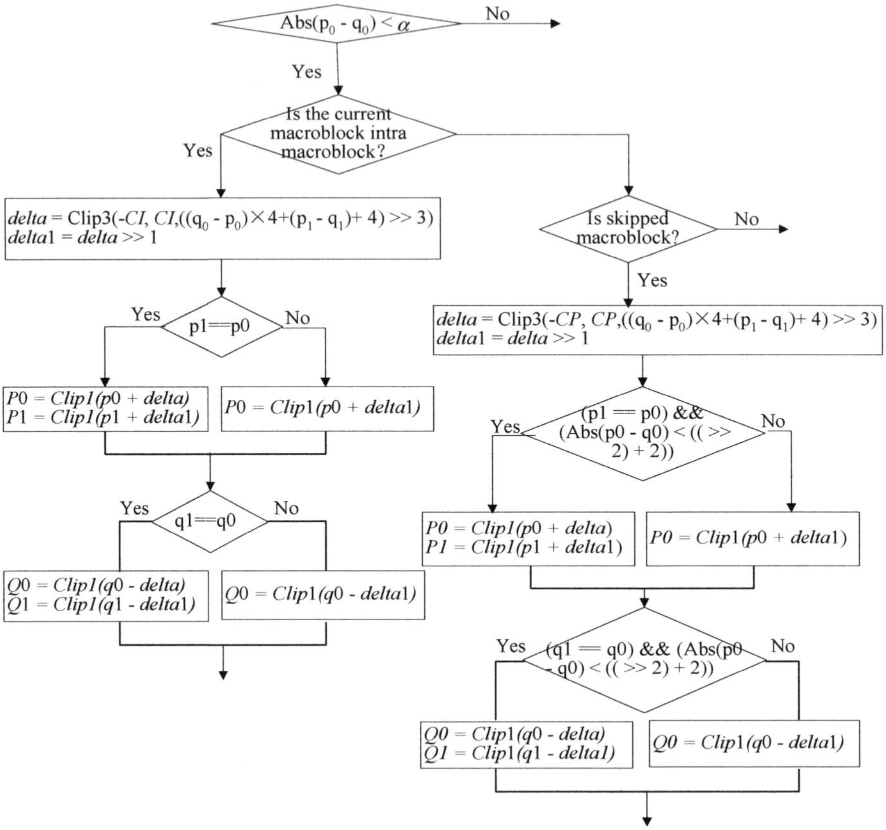

(b) Deblocking filter process in AVS1-P7

Fig. 17. *(continued)*

decoded visual quality 0. For AVS1-P7, the deblock filter is on 4x4 block boundaries, while for AVS1-P2 it is on 8x8 block boundaries. In AVS1-P2, the boundary may be between two 8x8 luma or chroma blocks except for picture or slice boundaries. For each edge, boundary strength (*Bs*) is derived based on the macroblock type, motion vector, reference index etc. Based on the boundary strength *Bs*, the filter process in AVS1-P2 is shown in Fig. 17 (a). In the filtering process, two threshold values α and β are used to control the filtering operation on the block edge. α is cross-block gradient boundary and β is inner block gradient boundary. $|q_0-p_0|<\alpha \&\& |q_0-p_1|<\beta \&\& |q_1-p_0|<\beta$ means the block edge is artifacts and is not an object boundary.

In AVS1-P7, the complexity of deblock filter is also reduced heavily, as shown in Fig. 17 (b) by reducing the number of judgments and the pixels to be filtered.

2.10 Interlace Coding

H.264/AVC supports both macroblock and picture level adaptive frame field coding for interlace coding (MBAFF, PAFF). In general, MBAFF is more complex than

PAFF with little performance improvement, as shown in Fig. 18. In AVS1-P2 Jizhun profile, only picture level frame field coding is used. For PAFF coding, a picture can be coded as either one frame or two fields (a top field and a bottom field). In the current AVS1-P2 Zengqiang profile, an improved MBAFF interlace coding scheme is accepted. As in H.264/AVC, a 32x16 macroblock pair can be coded in progressive or interlace mode. It is shown as NS (None Sampled) macroblock pair and VS (Vertical Sampled macroblock pair in Fig. 19. In H.264/AVC MBAFF coding, macroblock pair is coded one by one in raster scan order, and the top field can not be referenced by the bottom filed in the same frame. However, in AVS1-P2, if a macroblcok pair is coded as NS macroblock pair, the two macroblocks will be coded in order, otherwise, only top VS macroblock is coded. After all macroblock pairs are processed, an interpolated field is used a reference picture to code the left bottom VS macroblocks.

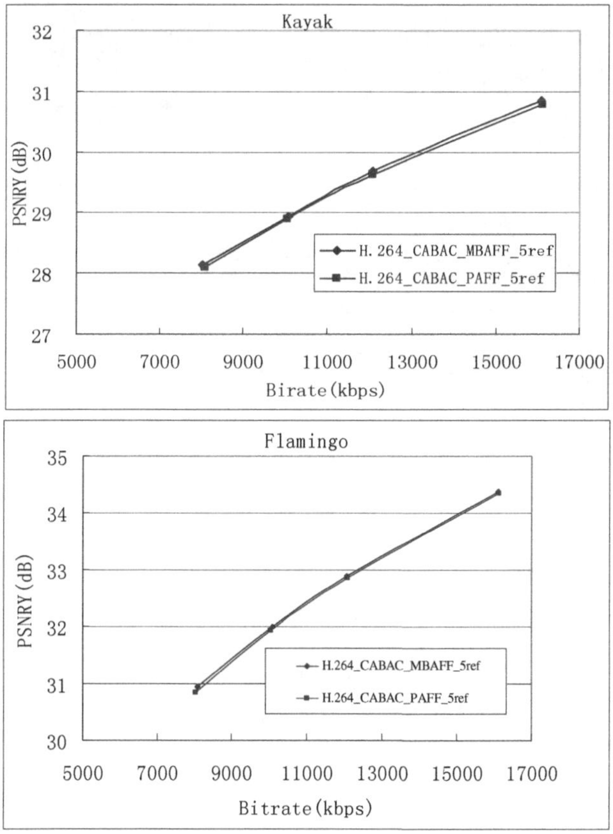

Fig. 18. Performance comparison between MBAFF and PAFF

2.11 Error Resilience Tools

Error resilience is very important for video transport, especially when the network is erroneous. H.264/AVC is very promising for low bit-rate applications such as IP

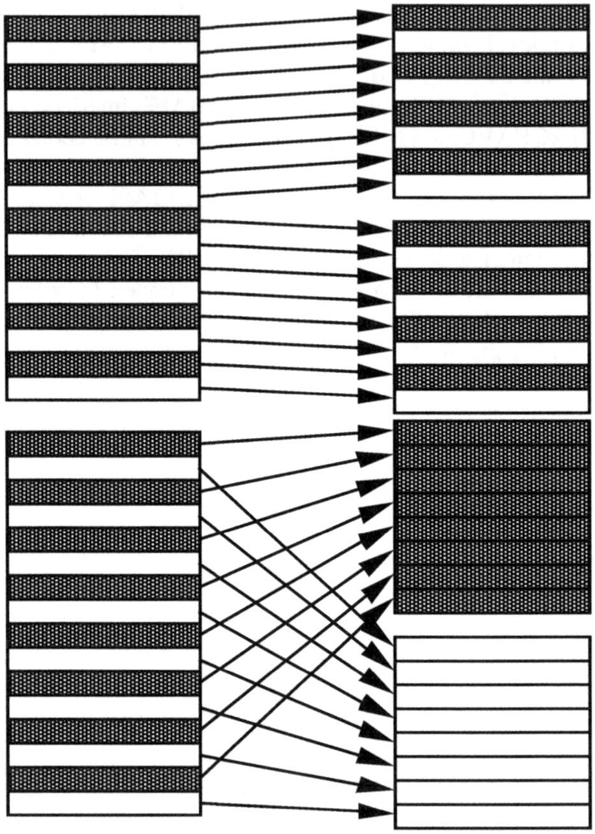

Fig. 19. MB pair in AVS1-P2 Left: None sampled macroblock(NS) Right: Vertical sampled macroblock(VS)

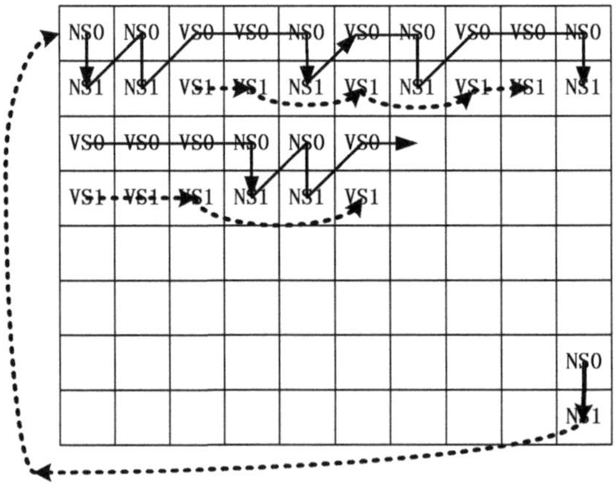

Fig. 20. MBAFF coding process in AVS1-P2

network 0 and wireless communications 0. To adapt these applications, more error resilience tools are introduced into H.264/AVC, such as parameter sets, FMO, redundant slice etc. As AVS1-P2 mainly targets to broadcasting, and the network condition usually is error free, few error resilience tools are defined in AVS1-P2. In AVS1-P2, only slice partition is defined to support error resilience and each slice is composed of one or more rows of macroblocks. However, as AVS1-P7 may be used in wireless network or transport on other erroneous channel, error resilience is very important. So in AVS1-P7, except for the slice partition, flexible reference frame scheme, constrained intra prediction, parameter set, is also introduced as H.264/AVC.

2.12 Profile and Level

In AVS1-P2, two profiles are defined now, named Jizhun profile, and Jiaqiang profile. Jizhun means baseline profile as H.264/AVC in Chinese, and Jiaqiang means an enhancement. There are five levels in Jizhun profile. Each level specifies the upper limits for the picture size, the maximal video bit-rate, the BBV buffer size etc. Level 2.0, 4.0, 4.2, 6.0 and 6.2 targets to streaming video, SD video, SD video with 4:2:2, HD video and HD video with 4:2:2, respectively.

Table 2. Profile definition in AVS1-P2

Coding Tools	AVS1 P2 Jizhun Profile	AVS1 P2 Jiaqiang Profile
Intra prediction	√	√
Variable Block Size Motion Compensation (VBMC)	√	√
Multi-reference prediction	√	√
C2DVLC	√	
CBAC	×	√
Loop filter	√	√
Interlace coding	√	√
B frame	√	√

3 Complexity Analyses and Performance Testing for AVS

This section will give an overview of the complexity and performance comparison between AVS and H.264/AVC. In the software and hardware implementation, coarse complexity estimation for AVS1-P2 is that H.264/AVC main profile encoder/decoder is about 1.5 times more complex than AVS1-P2. The complexity reduction for AVS mainly comes from interpolation, loopfilter and entropy coding. In AVS1-P7, loopfilter and entropy coding complexity is also reduced. The complexity of AVS1-P7 is approximated with H.264/AVC baseline profile.

As referred in 0, it is a challenging work to give an accurate estimation of the encoder and decoder complexity. The practical computation and storage consumption are general platform dependent. In this section, we will give a coarse estimation of complexity reduction in AVS based on the algorithm analysis.

3.1 Encoder Complexity

According to the statistical data, motion estimation and mode decision are the most time consuming parts in the hybrid coding framework. The space and computational consumptions of these two modules increase greatly with more number of prediction modes, more number of reference frames, higher motion vector resolution, and larger motion search range [34]. The main encoding complexity in terms of these key tools is analyzed as following:

3.1.1 Intra Prediction

For intra prediction in AVS, the number of prediction modes is less than H.264/AVC, and only one block size is used for luma and chroma prediction, as shown in Table 3. Therefore, during the rate-distortion optimal mode decision (RDO-MD) module, the maximum number to calculate the RD costs is reduced, as shown in Fig.21.

Table 3. Intra prediction modes in AVS and H.264/AVC

Prediction Type	Prediction Parameters	AVS1-P2 (JiZhun Profile)	AVS1-P7	H.264/AVC (Baseline Profile)	
Luma prediction	Block size	8×8	4×4	4×4	16×16
	Modes number	5	9	9	4
Chroma prediction	Block size	8×8	4×4	4×4	
	Modes number	4	3	4	

3.1.2 Inter Prediction

AVS and H.264/AVC both support the variable block sizes, as shown in Table 4. This affects the access frequency in a linear way.

In AVS1-P7 and H.264/AVC, the minimum block size is 4×4. However, the number of chroma-prediction modes in AVS1-P7 is less than that in H.264/AVC.

In AVS1-P2, the minimum block size is 8×8. Therefore, the number of luma-prediction modes is further reduced. Consequently, in the RDO-MD module, the maximum number to calculate the RD costs is reduced, as shown in Fig.22.

Additionally, for B picture in AVS, the symmetry-based bi-predictive mode only need to code forward motion vector. The backward motion vector is derived from the forward one, and the calculation is very simple via a linear formula [19]. In AVS1-P7, only P picture is adopted.

3.1.3 Multiple Reference Frames

With the increasing number of reference frames for the inter prediction, the maximum number to calculate the RD costs of candidate modes increases in times.

In H.264/AVC, the maximal number of reference frames is 16. While in AVS, the maximal number of reference frames is restricted to 2. Consequently, the storage requirement and the computational complexity are reduced. However, the penalty in performance is acceptable for most test sequences, as shown in Fig. 6.

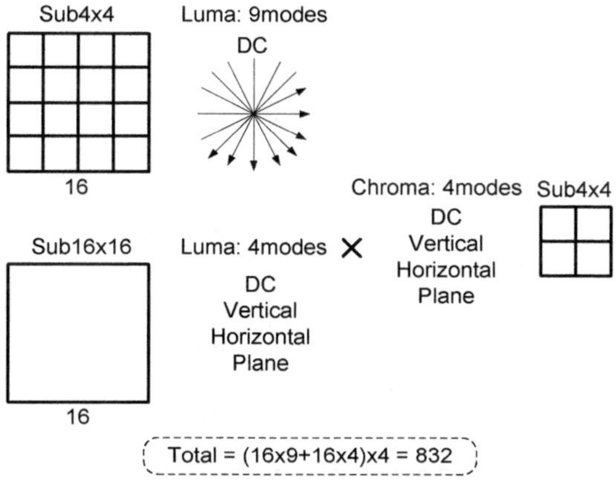

(a) H.264/AVC (basic count unit is 4x4 block)

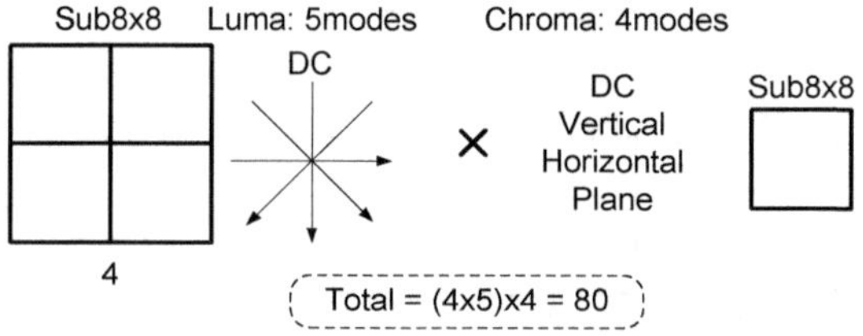

(b) AVS1-P2 (basic count unit is 8x8 block)

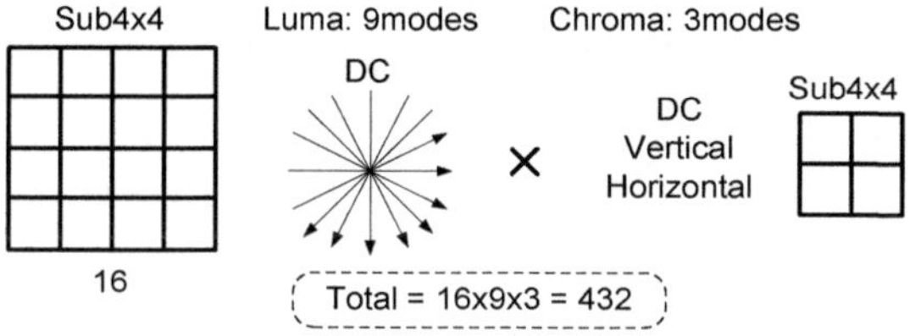

(c)AVS1-P7 (basic count unit is 4x4 block)

Fig. 21. Maximum number of RD costs calculation for intra prediction (with 4:2:0 format)

3.1.4 Sub-pixel Accuracy Motion Vector

The quarter-pixel accuracy motion vector is adopted in both AVS and H.264/AVC. The computational consumption increases mainly due to the additional SAD (Sum of Absolute Difference) calculations for sub-pixel motion compensation. According to the statistic in [34], the encoder prefers 1/2-pel accuracy motion vector instead of 1/4-pel ones, because the 1/4-pel motion search results in 10% increases of access frequency and processing time. However, up to 30% increases of coding efficiency are obtained except for very low bit rates.

The additional sub-pixel interpolations also result in higher space and time consumptions. AVS1-P2 adopts the Two Steps Four Taps (TSFT) interpolation method,

Table 4. Inter prediction modes in AVS and H.264/AVC

Prediction Parameters	AVS1-P2	AVS1-P7	H.264/AVC
Block size	16×16, 16×8, 8×16, 8×8	16×16, 16×8, 8×16, 8×8, 8×4, 4×8, 4×4	16×16, 16×8, 8×16, 8×8, 8×4, 4×8, 4×4
P picture modes	Skip, P16×16, P16×8, P8×16, P8×8	Skip, P16×16, P16×8, P8×16, P8×8, P8×4, P4×8, P4×4	Skip, P16×16, P16v×8, P8×16, P8×8, P8×4, P4×8, P4×4
B picture modes	Skip, Direct, Forward, Backward, Symmetric	----	Skip, Direct, Forward, Backward, Bi-predictive

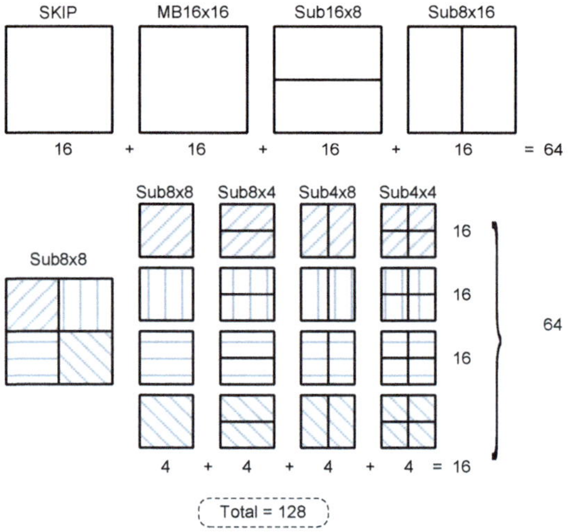

(a) H.264/AVS1-P7 (basic count unit is 4x4 block)

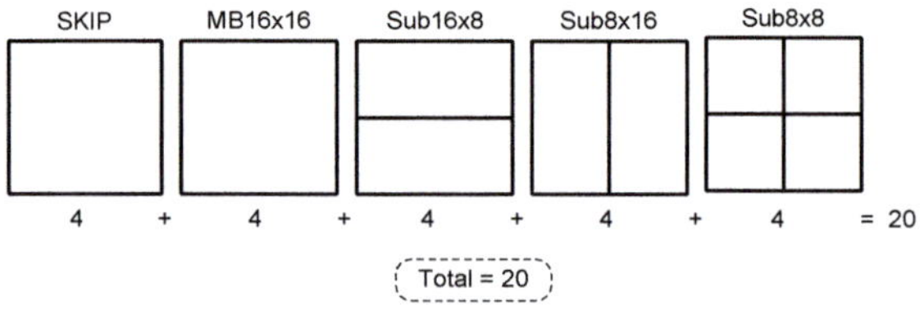

(b) AVS1-P2 (basic count unit is 8x8 block)

Fig. 22. Maximum number of RD costs calculation for inter prediction (with one reference frame) while H.264/AVC 1/4-pel interpolation requires 6-tap filters in horizontal and vertical direction. The 4-tap filter of AVS reduces space complexity compared with the 6-tap filter of H.264/AVC [11]. However, their filter performances are similar on most sequences, as shown in Fig. 7.

3.1.5 Transform and Quantization

H.264/AVC adopts 4x4 and 8x8 Integer Cosine Transform (ICT). ICT obtains equivalent compress efficiency as DCT, but much lower complexity because only additions and shifts operations are needed [37]. In AVS1-P2, the 8x8 ICT with Pre-scaled Integer Transform (PIT) technique is performed [16]. PIT further reduces the space and computational complexity without noticeable penalty in performance by changing the order of inverse scaling [35][38].

As shown in Fig. 23, the scaling operations of ICT are needed on both encoder and decoder side. In order to facilitate the parallel processing, the dequantization matrix is fully expanded to avoid lookup operations. In this case, the 4x4 ICT need allocate a memory of 6×4×4=96 bytes to store the dequantization matrix in decoder. The 8x8 ICT costs 6×8×8=96 bytes instead.

As shown in Fig. 24, the inverse scaling of PIT is moved to the encoder side and combined with forward scaling and quantization as one single process. Therefore, PIT reduces the memory size to only 6 bytes in decoder without increasing computational complexity in encoder.

If the dequantization matrix is not expanded, every transformed coefficient of ICT requires a 3-D lookup operation, which uses QP (Quantization Parameter) and the coordinates of the coefficient in the block, to get the corresponding dequantization coefficient. However, PIT uses QP only and the computational burden is also reduced evidently.

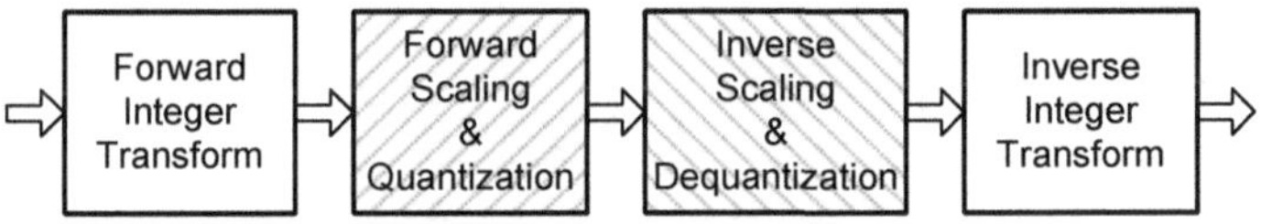

Fig. 23. Block diagram of ICT in H.264/AVC.

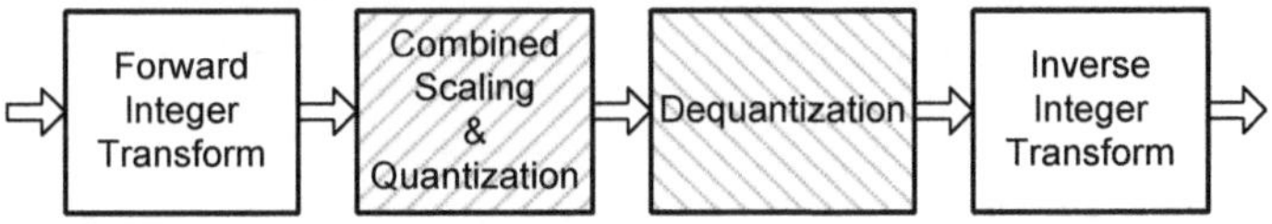

Fig. 24. Block diagram of PIT in AVS1-P2.

3.1.6 Entropy Coding

For entropy coding, 2D VLC coding is adopted in AVS1-P2 Jizhun profile and AVS1-P7. It is much simpler than CABAC in H.264/AVC, especially for hardware design. Compared to the method using a single reversible VLC table, CABAC results in the increases of access frequency up to 12% [34].The higher the bit rate, the higher the access frequency.

Table 5. Test conditions for comparison between AVS1-P2 and H.264/AVC

Coding Tools	H.264/AVC	AVS1-P2
Intra prediction	All Intra 16×16, Intra 4×4 modes	All Intra8×8 modes
Multi-reference frames	2 reference frames (3 reference frames for B frame)	2 reference frames
Variable block-size MC	16×16-4×4	16×16-8×8
Entropy coding	CABAC	VLC(Jizhun Profile); CBAC(Jiaqiang Profile)
RDO	On	On
Loop filter	On	On
Interlace Coding	MBAFF	PAFF
Adaptive block transform	Only 4×4(JM7.6); 8x8 and 4×4(JM10.1)	Only 8×8
Hierarchical B frames	Off	None

Table 6. Test conditions for comparison between AVS1-P7 and H.264/AVC

Coding Tools	H.264/AVC	AVS1-P7
Intra prediction	All Intra 16×16, Intra 4×4 modes	All Intra4×4 modes
Multi-reference frames	2 reference frames	2 reference frames
Variable block-size MC	16×16-4×4	16x16-4x4
Entropy coding	CAVLC	VLC
RDO	On	On
Loop filter	On	On

The improved context adaptive arithmetic coding is adopted in AVS1-P2 Jiaqiang profile. It also has lower complexity compared with CAVLC and CABAC in H.264/AVC [22].

3.2 Decoder Complexity

In 0, it is referred that H.264/AVC decoder complexity is estimated to be 4 times over MPEG-2 decoder and H.264/AVC encoder complexity is about 9 times over MPEG-2 encoder. A detail analysis on the H.264/AVC baseline profile decoder complexity from time complexity and space complexity is provided 0. Based on the analysis it is concluded that H.264/AVC baseline decoder is about 2.5 times more time complex than H.263 baseline decoder.

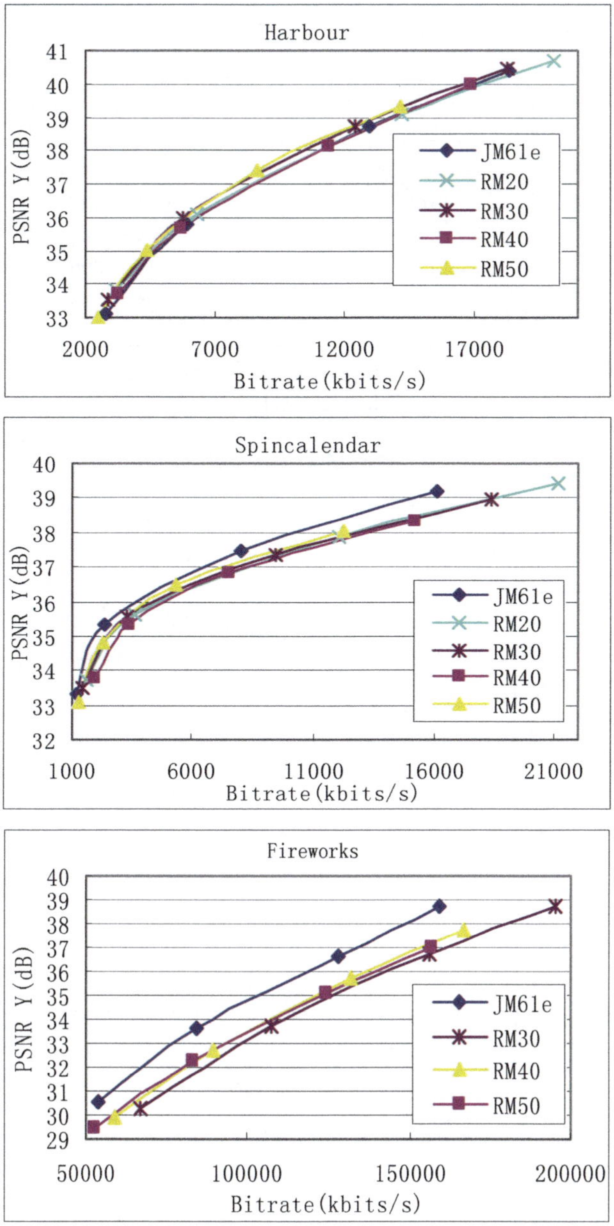

Fig. 25. Experimental results for comparison between H.264/AVC and AVS1-P2 Jizhun Profile

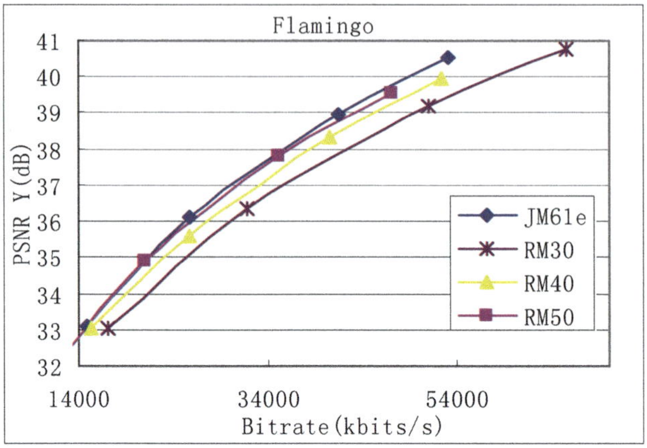

Fig. 25. *(continued)*

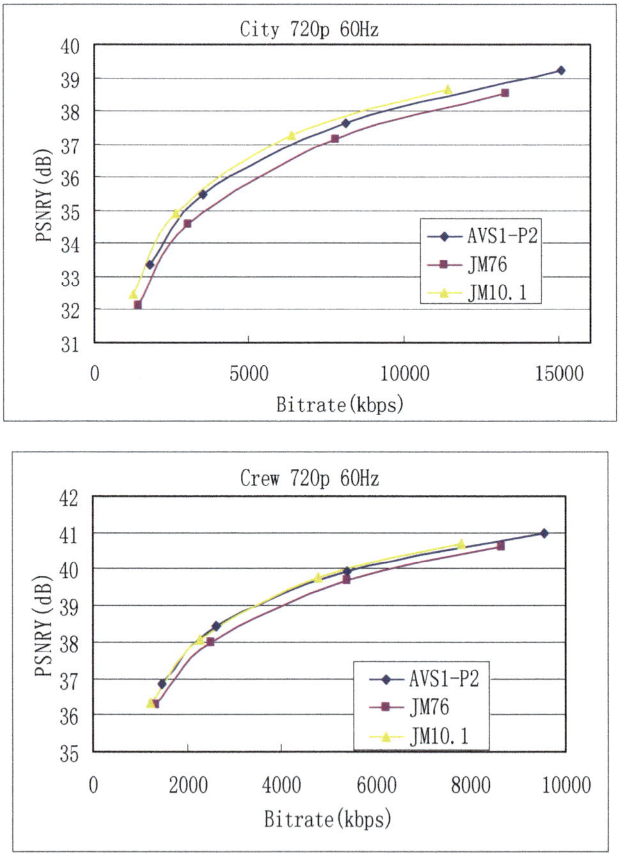

Fig. 26. Experimental results for comparison between H.264/AVC and AVS1-P2 Enhancement Profile (in developing)

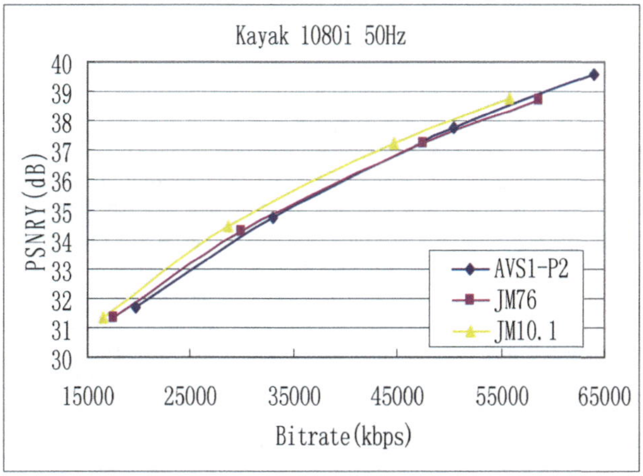

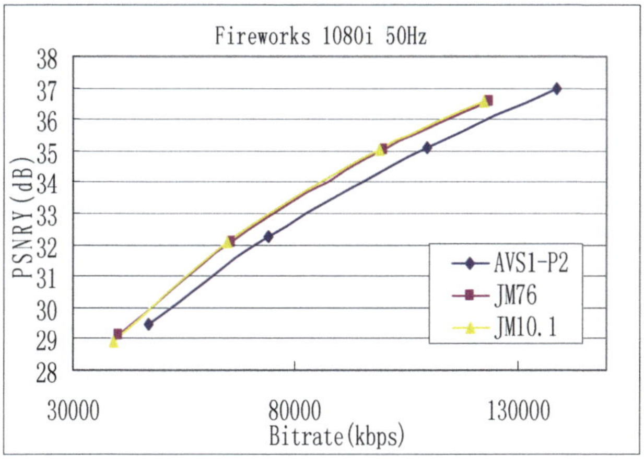

Fig. 26. (*continued*)

So far, some optimized AVS1-P2 software decoder and decoder chips have been released [31][32][33]. According to the statistical data in 00, the most complex two parts in decoder are interpolation and loop filter.

3.2.1 Interpolation

For interpolation in AVS1-P2, the spatial complexity and computation complexity comparison with H.264/AVC is analyzed in 0. AVS1-P2 interpolation method saves 11% memory bandwidth while maintains the similar computational complexity. For high definition video coding, besides the remarkable decreases of storage requirement and access frequency, AVS1-P2 interpolation method averagely achieves 0.0375dB PSNR gain [11][35][36].

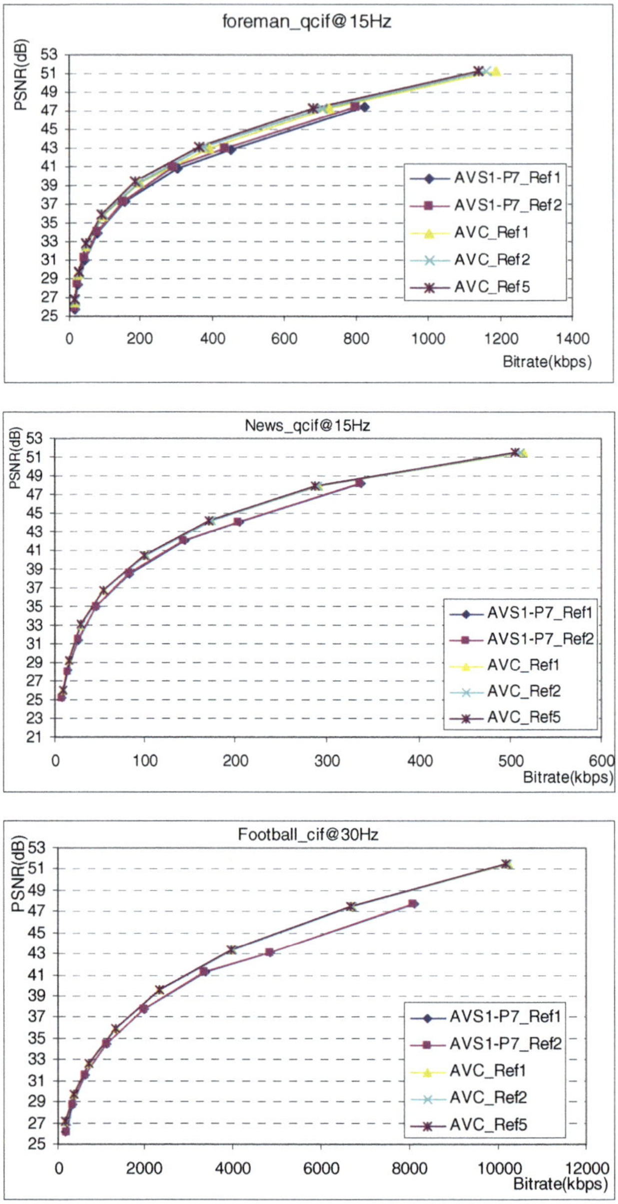

Fig. 27. Experimental results for comparison between H.264/AVC and AVS1-P7

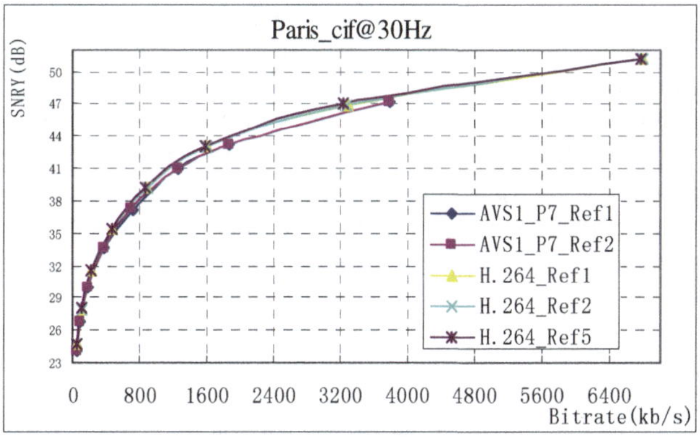

Fig. 27. *(continued)*

3.2.2 Loopfilter

For loopfilter, the most important factor is boundary strength computation and the number of edges to be filtered. In AVS1-P2, the filter is 8×8 block based, while H.264/AVC is 4×4 block based. Therefore, the number of edges to be filtered is reduced heavily. In AVS1-P7, the filter is also 4×4 block based. However, its loopfilter complexity is further reduced by simplifying filter strength decision and reducing the number of pixels to be filtered.

3.3 Performance Comparison

Performance comparison between AVS1-P2 and H.264/AVC Main/High Profile, AVS1-P7 and H.264/AVC Baseline Profile is provided in this part.

Table 5 and Table 6 show the test conditions for the comparison between AVS1-P2 Jizhun Profile and H.264/AVC Main Profile, AVS1-P7 Baseline Profile and H.264/AVC Baseline Profile. For H.264/AVC High profile, JM10.1 reference software is used with adaptive transform on. From the curves in Fig. 25, it can be seen that AVS1-P2 Jizhun Profile shows comparable performance with H.264/AVC for most progressive sequences. For interlace sequence or the sequence with more complex motion or texture, AVS1-P2 does a little worse than H.264/AVC. The loss mainly comes from entropy coding. As shown in Fig. 26, AVS1-P2 enhancement profile can achieve similar or even better performance compared with H.264/AVC High Profile under comparative test conditions, but the complexity of AVS is lower than H.264/AVC. Fig. 27 shows the performance of AVS1-P7 and H.264/AVC Baseline Profile. From the curves, it can be seen AVS1-P7 has similar performance with H.264/AVC Baseline Profile especially at low bitrate, and a little worse at high bitrate.

4 Conclusions

In this chapter, we have given an overview of AVS standard including background and technique features. Performance and complexity comparison between AVS and

H.264/AVC is also presented. Compared with H.264/AVC, AVS standards provide a good tradeoff between the performance and complexity for the specific applications, because all coding tools in AVS, including intra prediction, variable block-size motion compensation, multiple reference frames, interpolation filter, loopfilter and entropy coding, are selected by jointly considering the coding complexity and performance gain for the target applications. Experimental results prove that AVS provides similar performance with H.264/AVC while with lower complexity for the specific applications.

References

[1] Wiegand, T., Sullivan, G., Bjøntegaard, G., Luthra, A.: Overview of the H.264/AVC Video Coding Standard. IEEE Transactions on Circuits and Systems for Video Technology 13(7), 560–576 (2003)

[2] Srinivasan, S., Hsu, P., Holcomb, T., Mukerjee, K., Regunathan, S., Lin, B., Liang, J., Lee, M., Corbera, J.: Windows Media Video: 9 Overview and Applications. Signal Processing: Image Communication 19(9), 851–875 (2004)

[3] AVS Working Group Website, `http://www.avs.org.cn`

[4] AVS Document and Software FTP Site, `ftp://159.226.42.57`

[5] Wiegand T.: Version 3 of H.264/AVC, `http://wftp3.itu.int/av-arch/jvt-site/2004_07_Redmond/JVT-L012d2wcmRelTod1.doc`

[6] Wiegand, T., Sullivan, G., Bjøntegaard, G., Luthra, A.: Long-term Memory Motion-Compensated Prediction. IEEE Transactions on Circuits and Systems for Video Technology 9(1), 70–84 (1999)

[7] Wedi, T., Musmann, H.: Motion- and Aliasing-Compensated Prediction for Hybrid Video Coding. IEEE Transactions on Circuits and Systems for Video Technology 13(7), 577–586 (2003)

[8] Wedi, T.: More Results on Adaptive Interpolation Filter for H.26L, `http://wftp3.itu.int/av-arch/jvt-site/2001_12_Pattaya/VCEG-O28.doc`

[9] Boyce, J., Gomila, C.: 4-Tap Motion Interpolation Filter, `http://wftp3.itu.int/av-arch/jvt-site/2002_05_Fairfax/JVT-C014.doc`

[10] Hallapuro, A., Lainema, J., Karczewicz, M.: 4-tap Filter for Bi-predicted Macroblocks, `http://wftp3.itu.int/av-arch/jvt-site/2002_07_Klagenfurt/JVT-D029.doc`

[11] Wang, R., Huang, C., Li, J., Shen, Y.: Sub-pixel Motion Compensation Interpolation Filter in AVS. In: IEEE International Conference on Multimedia and Expo, vol. 1, pp. 93–96 (2004)

[12] Malvar, H., Hallapuro, A., Karczewicz, M., Kerofsky, L.: Low-Complexity Transform and Quantization in H.264/AVC. IEEE Transaction on Circuits and Systems for Video Technology 13(7), 598–603 (2003)

[13] Wien, M.: Variable Block-size Transforms for H.264/AVC. IEEE Transaction on Circuits and Systems for Video Technology 13(7), 604–613 (2003)

[14] Gordon, S., Marpe, D., Wiegand, T.: Simplified Use of 8x8 Transforms – Updated Proposal and Results, `http://ftp3.itu.ch/av-arch/jvt-site/2004_03_Munich/JVT-K028.doc`

[15] Ma, S., Gao, W., Fan, X.: Low Complexity Integer Transform and High Definition Coding. In: Proceedings of SPIE 49th Annual meeting (2004)
[16] Zhang, C., Lou, J., Yu, L., Dong, J., Cham, W.: The Techniques of Pre-scaled Integer Transform. In: International Symposium on Circuits and Systems, vol. 1, pp. 316–319 (2005)
[17] Flierl, M., Girod, B.: Generallized B Pictures and the Draft H.264/AVC Video-Compression Standard. IEEE Transactions on Circuits and Systems for Video Technology 13(7), 587–597 (2003)
[18] Ji, X., Zhao, D., Gao, W., Lu, Y., Ma, S.: New Scaling Technique for Direct Mode Coding in B Pictures. IEEE International Conference on Image Processing 1, 469–472 (2004)
[19] Ji, X., Zhao, D., Gao, W.: B-picture Coding in AVS Video Compression Standard. Signal Processing: Image Communication 23(1), 31–41 (2008)
[20] Marpe, D., Schwarz, H., Wiegand, T.: Context-Based Adaptive Binary Arithmetic Coding in the H.264/AVC Video Compression Standard. IEEE Transactions on Circuits and Systems for Video Technology 13(7), 620–636 (2003)
[21] Wang, Q., Zhao, D., Gao, W.: Context-Based 2D-VLC Entropy Coder in AVS Video Coing Standard. Journal of Computer Science and Technology 21(3), 315–322 (2006)
[22] Zhang, L., Wang, Q., Zhang, N., Zhao, D., Wu, X., Gao, W.: Context-based entropy coding in AVS video coding standard. Signal Processing: Image Communication 24(4), 263–276 (2009)
[23] List, P., Joch, A., Lainema, J., Bjøntegaard, G., Karczewicz, M.: Adatpive Deblocking Filter. IEEE Transactions on Circuits and Systems for Video Technology 13(7), 614–619 (2003)
[24] Wenger, S.: H.264/AVC over IP. IEEE Transactions on Circuits and Systems for Video Technology 13(7), 645–656 (2003)
[25] Stockhammer, T., Hannuksela, M., Wiegand, T.: H.264/AVC in Wireless Environments. IEEE Transactions on Circuits and Systems for Video Technology 13(7), 657–673 (2003)
[26] Horowitz, M., Joch, A., Kossentini, F., Hallapuro, A.: H.264/AVC Baseline Profile Decoder Complexity Analysis. IEEE Transactions on Circuits and Systems for Video Technology 13(7), 704–716 (2003)
[27] Lappalainen, V., Hallapuro, A., Hämäläinen, T.: Complexity of Optimized H.26L Video Decoder Implementation. IEEE Transactions on Circuits and Systems for Video Technology 13(7), 717–725 (2003)
[28] Puri, A., Chen, X., Luthra, A.: Video Coding Using the H.264/MPEG-4 AVC Compression Standard. Signal Processing: Image Communication 19, 793–849 (2004)
[29] H.264 reference software website, `http://iphome.hhi.de/suehring/tml/`
[30] AHG on AVS1-P7 performance testing: AHG report on AVS1-P7 performance testing. Document AVS_M1635, Shanghai, China (September 2005)
[31] Sheng, B., Gao, W., Xie, D., Wu, D.: An Efficient VLSI Architecture of VLD for AVS HDTV Decoder. IEEE Transactions on Consumer Electronics 52(2), 696–701 (2006)
[32] Sheng, B., Gao, W., Xie, D.: Algorithmic and Architectural Co-Design for Integer Motion Estimation of AVS. IEEE Transactions on Consumer Electronics 52(3), 1092–1098 (2006)
[33] Zheng, J., Deng, L., Zhang, P., Xie, D.: An Efficient VLSI Architecture for Motion Compensation of AVS HDTV Decoder. Journal of Computer Science and Technology 21(3), 370–377 (2006)
[34] Ostermann, J., Bormans, J., List, P., Marpe, D., Narroschke, M., Pereira, F., Stockhammer, T., Wedi, T.: Video Coding with H.264/AVC: Tools, Performance, and Complexity. IEEE Circuits and System Magazine 4(1), 7–28 (2004)

[35] Yu, L., Yi, F., Dong, J., Zhang, C.: Overview of AVS-Video: Tools, Performance and Complexity. SPIE Visual Communications and Image Processing 5960, 679–690 (2005)
[36] Hang, C., Wang, R., Li, J., Shen, Y.: Improvement of Subpixel Interpolation. Document AVS_M1162, Hangzhou, China (October 2003)
[37] Malvar, H., Hallapuro, A., Karczwicz, M., Kerofsky, L.: Low Complexity Transform and Quantization in H.264/AVC. IEEE Transactions on Circuits and Systems for Video Technology 13(7), 598–603 (2003)
[38] Wang, X., Zhao, D.: Performance Comparison of AVS and H.264/AVC Video Coding Standards. Journal of Computer Science and Technology 21(3), 310–314 (2006)
[39] Zheng, J., Zheng, X., Lai, C.: Adaptive Weighting Technology for AVS Jiaqiang Profile. Document AVS_M2427, Tianjin, China (September 2008)

A Resolution Adaptive Video Compression System

Serhan Uslubas[1], Ehsan Maani[2], and Aggelos K. Katsaggelos[3]

[1] Northwestern University, Department of EECS
s-uslubas@northwestern.edu
[2] Northwestern University, Department of EECS
ehssan@northwestern.edu
[3] Northwestern University, Department of EECS
aggk@eecs.northwestern.edu

Modern video encoding systems employ block-based, multi-mode, spatio-temporal prediction methods in order to achieve high compression efficiency. A common practice is to transform, quantize and encode the difference between the prediction and the original along with the system parameters. Obviously, it's crucial to design better prediction and residual encoding methods to obtain higher compression gains. In this work, we examine two such systems which utilize subsampled representations of the sequence and residual data. In the first system, we consider a method for reorganizing, downsampling and interpolating the residual data. In the second system, we propose a new method that employs lower resolution intensity values for spatial and motion-compensated prediction. Both of these methods are macroblock adaptive in the rate-distortion sense. Our experiments show that implementing these methods brings additional compression efficiency compared to the state-of-the-art video encoding standard H.264/AVC.

1 Background and Related Work in Multi-resolution Video Compression Systems

Video encoding systems achieve compression by removing the redundancy in the data, i.e. elements which can be discarded without adversely affecting reproduction fidelity. Video signals take place in time and space; therefore, most video encoding systems exploit both temporal and spatial redundancy present in these signals. Typically, there is high temporal correlation between successive frames. This is also true in the spatial domain for pixels which are within a close proximity of each other. Thus, one can achieve high compression gains by carefully exploiting these spatio-temporal correlations.

In this work, we will consider one of the most widely adopted video coding schemes, namely block-based hybrid video coding. The major video coding standards, such as H.261 [7], H.263 [10], MPEG-2 [13], MPEG-4 Visual [14] and the current state-of-the-art H.264/AVC [11] are based on this model. A block-based

C.W. Chen et al. (Eds.): Intel. Multimedia Communication: Tech. and Appli., SCI 280, pp. 167–194.
springerlink.com

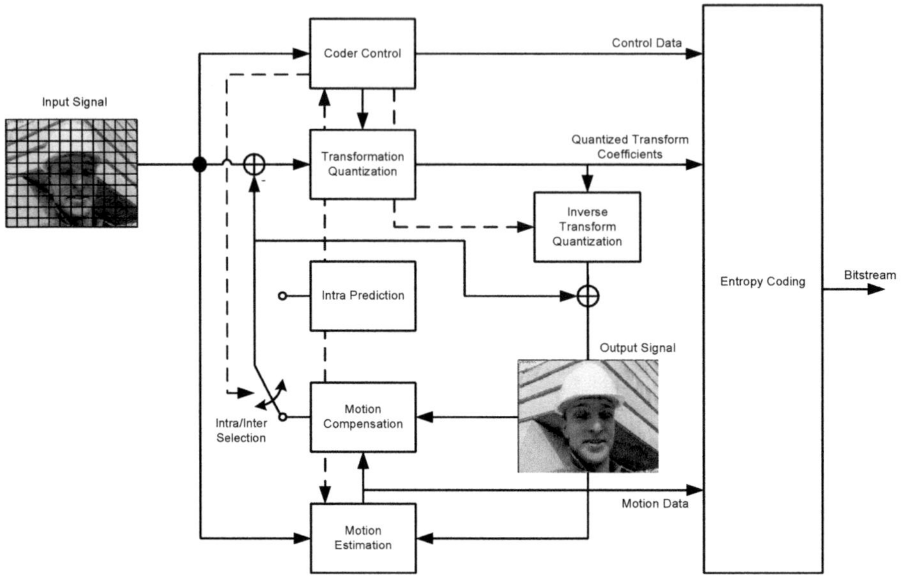

Fig. 1. Block diagram of H.264/AVC encoder

coding approach divides a frame into elemental units called macroblock. For source material in 4:2:0 YUV format, one macroblock encloses a 16×16 region of the luma frame and 8×8 region of the chroma frame. A block-based coding system, such as H.264/AVC, processes each video frame in units of macroblocks. Encoding a macroblock involves a hybrid of three techniques: prediction, transformation, and entropy coding. All luma and chroma samples of a macroblock are predicted spatially or temporally. The difference between the prediction and the original is put through transformation and quantization processes, whose output is encoded using entropy coding methods. Fig. 1 shows a block diagram of an H.264/AVC video encoder which is built on block-based hybrid video coding architecture [25].

An active field of research in the literature is the efficient and quality-rich encoding of high-definition (HD) video. HD video refers to image sequences with spatial resolution exceeding that of standard-definition (SD). Resolutions of 1280x720 (720p) and 1920x1080 (1080p) with 16:9 aspect ratio and frame rates of 60 frames per second (fps) are typical today. Furthermore, several manufacturers have publicly demonstrated LCD panels capable of even higher resolution such as 4096x2160 (4Kx2K or UHD) [17].With these advances in display technology it is evident that HD and UHD video will be the dominant display standard in the near future. However, such extensive use of HD video requires an enormous amount of bandwidth. On the signal processing front, the current state-of-the-art video coding system H.264/AVC [11] offers exceptional video compression efficiency compared to preceding video coding standards. Even with the use of H.264/AVC, the bitrate to represent HD video surpasses what is feasible by today's transmission and storage capabilities, thus creating a bottleneck. This chapter is motivated by these coding efficiency limitations and attempts to improve the performance of H.264/AVC by presenting a mixed resolution video coding architecture.

1.1 Spatial Resolution Scaling

Downsampling (sub-sampling or decimation), as intended for images and image sequences, is defined as representing a version of the original signal with fewer spatial samples. This is achieved by discarding some of the pixels of the original image based on a new sampling grid, as illustrated by Fig. 2. Furthermore, downsampling is considered a compression technique since fewer samples are required to represent the original signal.

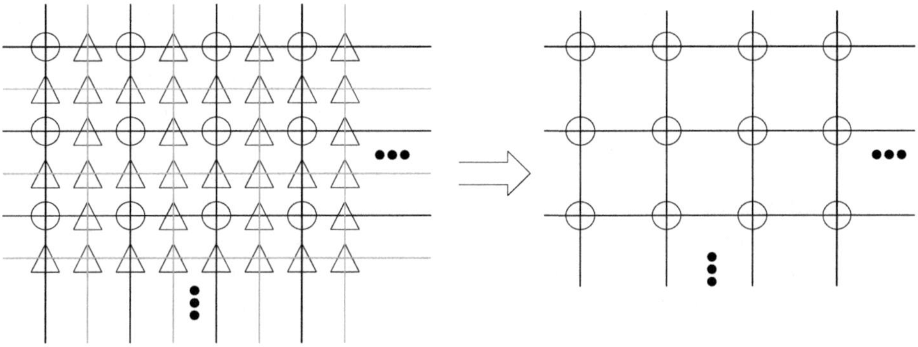

Fig. 2. Downsampling grid

Upsampling, the reverse process of downsampling, is representing a low resolution image in a high resolution grid by calculating the missing samples via interpolation. Figure 3 shows the block diagram of a video coding system, which compresses a low resolution representation of the input for storage and transmission. In order to display the sequence in the original resolution, the decompressed signal is upsampled after decoding. In order to avoid the aliasing problem, a low-pass filter should be utilized before downsampling. This filter causes blur in the detailed regions of the image, which corresponds to the removal of the high frequency content. Therefore, a rate reduction comes at the expense of signal degradation. Detailed image sequences, which contain a lot of high frequency information are more likely to be harmed by a resolution rescaling process than sequences with less details. This observation highlights the importance of better pre-processor and interpolator design in the overall quality of the codec, and it has been addressed by utilizing a localized resolution adaptivity in the proposed system.

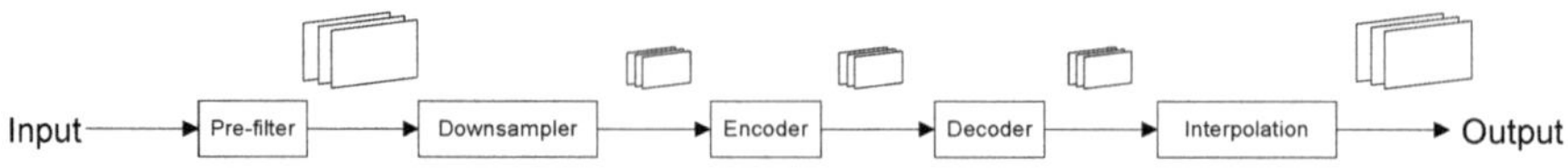

Fig. 3. Low resolution coding system

1.2 Hierarchical Resolution Image Coding

Downsampling prior to compression is an intuitive technique in low bitrate still image compression. The popular image coding standard JPEG [24], uses the Discrete Cosine Transform (DCT) followed by quantization on 8×8 blocks. As the compression factor is increased, the bits-per-pixel ratio is reduced drastically, which causes annoying artifacts and blockiness [12]. One way to restore this ratio is to work with the downsampled representation of the image, e.g. downsample the image, then use compression, and upsample it back to the original resolution after decompression. Such a scheme allows more bit allocation per pixel, which boosts image details. Therefore, downsampling prior to encoding and upsampling after decoding can improve the quality of coded image, especially at low bit rates [5]. This technique is also similar to the hierarchical JPEG encoding mode (also known as pyramidal encoding) in which a lower-resolution image is coded first as the base layer along with the residual difference between its upsampled version and the original as a secondary stream [18]. The decoded base layer is interpolated and added to the residual in order to reconstruct the image at its intended resolution. One can also argue that the Discrete Wavelet Transform (DWT) utilized in the new JPEG 2000 standard [19] inherently provides a similar multi-resolution image encoding property.

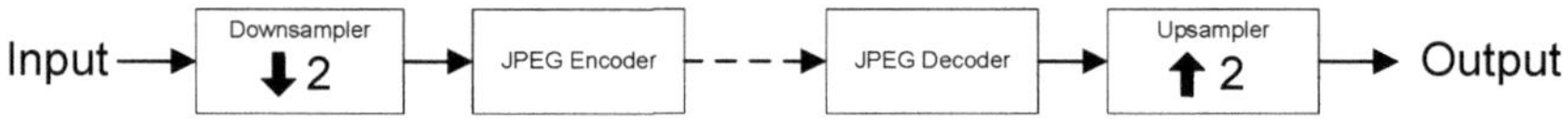

Fig. 4. Low resolution compression system for still images

In order to observe this relation, we conducted a proof-of-concept experiment with a simple downsampling-compression setup whose block diagram is shown in Fig. 4. This experiment was carried out in the Matlab (R) environment using the Image Processing Toolbox functions for resizing and encoding.

In this experiment, we use the first frame of the 1920×1080 HD sequence *Rush Hour* as input to the system. The frame is downsampled by two, in the horizontal and vertical directions, and fed to the JPEG encoder. The encoder bitrate can be adjusted with QP ranging from 0 to 100. This parameter is varied, and the number of bits necessary to encode the low resolution frame is noted. At each step, the decoded JPEG output is interpolated back to the original resolution using a bilinear filter, and the associated peak signal-to-noise ratio is noted. Figure 5 shows PSNR-Bitrate plots of the described and the conventional JPEG encoding schemes. As expected, at low bitrates, the downsampling scheme outperforms the conventional. At 31 Kbytes/frame, the proposed system offers

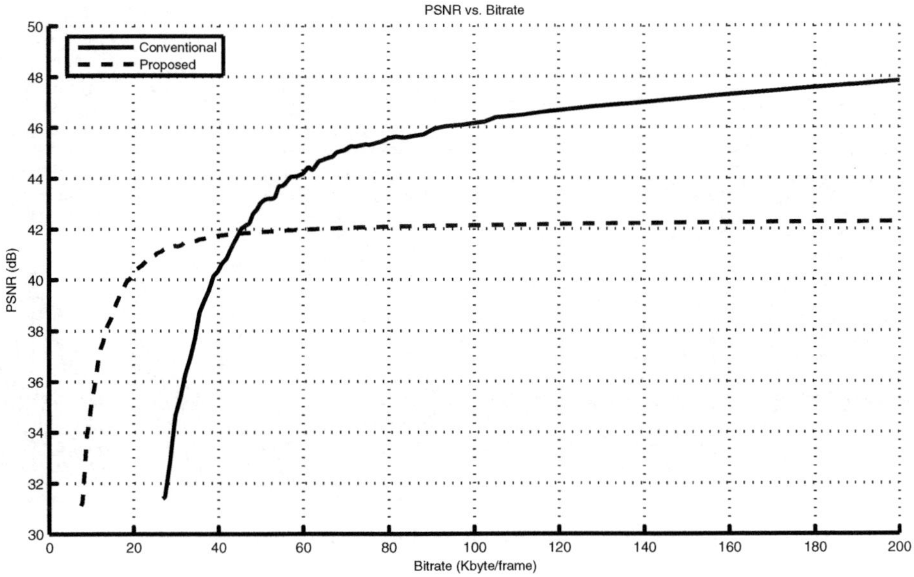

Fig. 5. Comparison of downsampling and the conventional JPEG image compression systems

a quality boost of 6 dB. In the conventional system, blocking artifacts degrade the image quality, while the downsampling system suffer less from such defects, as shown in Figs. 6 and 7, respectively. It should also be noted that the range of bitrates in which the proposed system offers an advantage is limited, and the conventional JPEG encoder rapidly starts to perform better as the number of bits per frame is increased. At these high bitrates, decoded image quality of the proposed system is clipped by the information lost during the resampling process and cannot be increased arbitrarily. Nevertheless, this experiment shows that under certain conditions, downconverting an image prior to compression can introduce substantial coding gain compared to the direct coding.

1.3 Hierarchical Resolution Video Coding

Hierarchical resolution video coding schemes have been investigated primarily in terms of sub-band decomposition [4]. The basic idea behind these schemes is to decompose a video signal into a low-pass subsampled representation and multiple high-frequency bands in a hierarchical manner. Sub-band decomposition can be realized with DWT, as in [23]. The lowest resolution signal is encoded and decoded independently, serving as the base layer, then each frequency band is separately encoded. In order to maximize the compression efficiency, each sub-band can be coded with a different technique, which can be tailored to the properties of the sub-band. On the decoder side, the incoming base layer is upsampled by interpolation and combined with the higher level sub-bands to

Fig. 6. Conventional coding system output

Fig. 7. Downsampling coding system output

accommodate it with high-frequency information. An important shortcoming of this type of scheme is the poor motion compensation performance in the wavelet domain [9].

Another type of coding which utilizes the multi-resolution paradigm is spatial scalability in scalable video coding (SVC). A straightforward approach to achieve spatial scalability is to iteratively downsample the input frame and code each resolution level separately. While this approach is simple, it introduces a very large coding overhead. An improved system employs motion compensated prediction between layers, as it is the case with the scalable extension (SVC) of the H.264/AVC standard [20]. While the purpose of this section is not to describe the many details of SVC, some explanation is necessary to allow the proposed scheme of Sec. 2 distinguish itself from scalable video coding.

The motivation behind the scalable video coding is to address the problem of network heterogeneity, i.e., to adapt the compressed bitstream to a wide range of devices with different processing capabilities and varying transmission conditions. SVC provides three major modes of scalability: temporal scalability, spatial scalability, and quality (SNR) scalability. In this work, the term scalability is referred exclusively to spatial scalability. SVC provides a mechanism for reusing the encoded lower resolution version of an image sequence, the base layer, for coding of the corresponding higher resolution frame, the enhancement layer [22]. An SVC bitstream contains several sub-streams which can be extracted separately and combined to reconstruct a high resolution picture from low resolution base layers.

SVC utilizes a hierarchical resolution decomposition by downsampling the high resolution image sequence. The base layer of this sequence is encoded independently, while the higher resolution enhancement layers are coded relative to the base layer. In order to increase the coding efficiency, enhancement layer samples are predicted from the lower layer pictures (reference layers), which is known as inter-layer prediction. A high resolution macroblock is obtained by upsampling the corresponding region of the reference layer and combining it with the residual information from the enhancement bitstream.

The purpose of having a spatially scalable codec is to enclose multiple resolution representations of an image sequence in a single bitstream, such that each device can extract the necessary information from the same stream and reconstruct the target resolution video without having a need of multicasting. While this property of scalable video codec offers many advantages, it comes with the price of less efficient compression [16]. On the other hand, our proposed method of Sec. 2 utilizes the concept of hierarchical resolution decomposition in order to improve the compression efficiency in a non-scalable, or single layer video coding system.

1.4 Super-Resolution Video Compression

A typical video coding scheme which utilizes multi-resolution representation of an image sequence is described in [15]. The principle idea behind this system is to encode the low resolution representation of the image sequence by downsampling the original and then bring the decoded frame back to the original resolution by utilizing a super-resolution approach [21]. Super-resolution is typically referred to obtaining a high resolution image from a set of low resolution observations [3]. When implemented, this method provides compression efficiency in MPEG-4 Part 2 encoding environments at low bitrates. However at high bitrates, the conventional system outperforms the super-resolution compression scheme, as there is more bits available to the conventional encoder to account for the details in the sequence. On the contrary, it is difficult for the super-resolution scheme to recover the information lost during the resolution rescaling step. Furthermore, it has been shown that the super-resolution process is region dependent, as it does not perform well in textured areas with no motion and provides very little benefit in flat regions [1].

An extension of the previous method has been addressed in [2], where a region-based downsampling and super-resolution scheme is utilized. The mentioned system performs a segmentation algorithm on a group of pictures (GOP) and identifies regions in each frame according to the amount of motion and texture information they contain. This leads to a motion-texture map (MTM) for the GOP with three labels: regions with motion (M), flat regions with no motion (F) and textured regions with no motion (T). On the encoder side, MTM drives the region-based pre-processing and downsampling, while at the decoder side it commands the super-resolution and post-processing. The produced low resolution image sequence is then compressed by an MPEG-4 system, together with the MTM as side information. The decoder reconstructs the low resolution image sequence from the bistream and upsamples it back to the original resolution according to MTM. Regions labeled (F) are upsampled with a bilinear interpolation filter, (M) are super-resolved by [6], and (T) are upsampled according to the downsampling process based on the frame type. Novelties of this scheme, such as high-level content analysis and region-based processing, offer compression efficiency compared to [15]. However, at higher bitrates this system is also surpassed by the conventional MPEG-4 compression. Our proposed method of Sec. 2 tries to overcome these shortcomings by providing a more refined downsampling adaptivity at the macroblock level. Only the macroblocks whose low resolution encoding offers a rate distortion (RD) improvement are downsampled, which makes the proposed system work over a wider range of bitrates.

1.5 Subsampled Residual Video Coding

Low resolution encoding can also be utilized for the residual information of a hybrid video coding system. One method of achieving this scheme is to downconvert the residual macroblock, which is defined as the difference between the prediction and the original, and encode the resulting low resolution representation. Subsequently, the decoded low resolution residual samples can be upsampled and combined with the computed prediction in order to reconstruct the coded macroblock. Subsampled representation of the residual information can be justified by the fact that recent video coding systems, such as H.264/AVC, already provide exceptional prediction for a macroblock through sophisticated multimodal intra and inter methods, leading to a sparse residual macroblock, whose low resolution representation can be brought back to the original resolution without losing much information.

An implementation of the described scheme is given in [8], where the macroblock residual data is downsampled and encoded if the distortion introduced by this rescaling is not too severe. Acknowledged application works by obtaining a 16×16 residual macroblock through the conventional H.264/AVC prediction, which is subsequently downsampled, quantized and transformed. In order to match the information available at the decoder, inverse quantization and transformation is applied on the transmitted coefficients. Next, decoded residual is upsampled and added to the prediction to simulate the reconstructed macroblock. Finally, the resulting RD cost of this method is calculated and compared with the RD cost of the conventional coding. If the RD cost of the low resolution residual coding method is better than the RD cost of H.264/AVC coding, current residual is encoded in reduced resolution. This system can provide compression efficiency compared to H.264/AVC in low bitrates. In Sec. 3, we propose a new a video coding system, which realizes an improved low resolution residual encoding paradigm that outperforms H.264/AVC over a wider range of bitrates.

2 Resolution Adaptive Macroblock Based Video Coding System (RAMB)

The current state-of-the-art video coding standard H.264/AVC provides compression efficiency by predicting a macroblock spatially and temporally in the same spatial resolution as the input frame. While this strategy offers very good prediction performance in the highly detailed and textured areas of a frame, it does not fully exploit the features inherent in constant and smooth regions. Meanwhile, adoption of HD image sequences in video related applications leads to the use of high resolution frames, whereas the dimensions of the processing unit, macroblock, stays the same. This causes the content enclosed by one macroblock to be generally constant. Discussions presented in Sec. 1.2 - Sec. 1.3 suggest that low resolution coding of smooth and invariable regions of a frame increases coding efficiency. While these methods are promising for low bitrate applications, they perform poorly as more bitrate becomes available. This behavior is due to the unrecoverable information loss during the resolution rescaling process. In such systems, it is likely that the extra coding efficiency obtained by low resolution encoding of the smooth regions is exceeded by the loss of quality in textured regions, due to low-pass downsampling and interpolation. Since the whole input frame is subsampled in these systems, it is difficult to suppress the negative effects of low-pass filtering of textured regions. One plausible solution to this problem is to segment the input frame into low-pass and high-pass regions. However, this requires use of complex algorithms, such as principal component analysis, which increases the computational load of the codec unjustifiably.

In this part of the text, we address these problems by proposing a resolution adaptive macroblock based video coding system (RAMBVCS) that adaptively utilizes the low resolution paradigm at the macroblock level.

2.1 System Model and Algorithm

A block diagram of the RAMBVCS encoder is shown in Fig. 8. As the first step, the high resolution input frame is filtered and downsampled by a factor of P to give a low resolution original represention of the source. Ideally, these operations

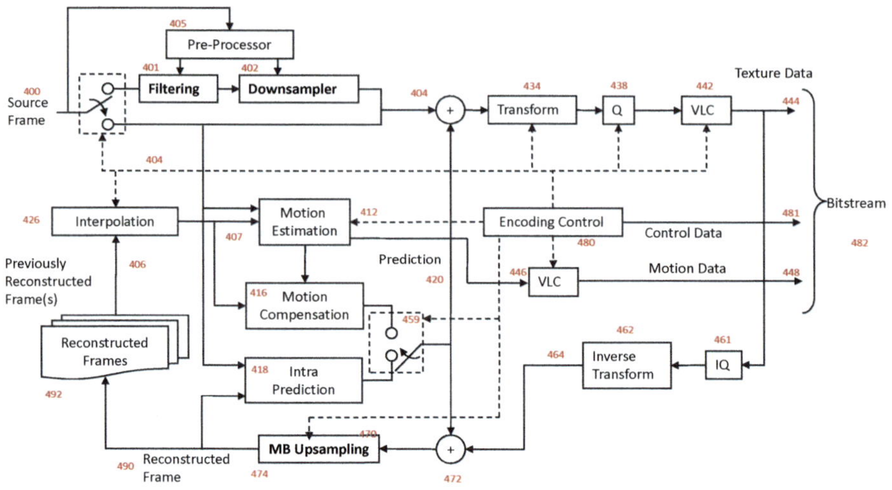

Fig. 8. RAMBVCS encoder

are dictated by the pre-processor, which has knowledge about the state of the encoder and the decoder such that a variable filtering and downsampling strategy can be utilized. Without loss of generality, we shall assume that P is two in both spatial directions and the pre-processing filter is the MPEG-4 dyadic downsampling filter whose kernel H_D is given by

$$H_D = [2 \quad 0 \quad -4 \quad -3 \quad 5 \quad 19 \quad 26 \quad 19 \quad 5 \quad -3 \quad -4 \quad 0 \quad 2]/64. \qquad (1)$$

The downsampling operation takes place by convolving the input frame horizontally and vertically with H and by taking every second pixel in both directions. For an original image size of $M \times N$ this results in a low resolution image of size $M/2 \times N/2$ which is stored in the encoder buffer for future reference. The encoder partitions the high resolution input frame into 16×16 macroblocks which are processed in raster scan order. A macroblock is generally classified by the type of prediction employed: intra or inter. RAMBVCS encoder takes a different coding approach for each type of macroblock.

2.2 Intra Macroblock Encoding

Intra macroblocks refer to block types where only spatial redundancy within a picture is exploited for compression purposes. In previous video coding standards, such as H.263 and MPEG-4, intra prediction has been conducted in the transform domain. Intra prediction in the H.264/AVC, however, is always performed in the spatial domain by referring to neighboring samples of previously coded blocks. In intra prediction, a prediction block is formed based on previously coded and reconstructed blocks before de-blocking. For luma pixels, prediction can take place by considering each 4x4 or 16x16 blocks in various ways.

RAMBVCS can choose to encode a macroblock either in high resolution or in low resolution depending on the final RD cost of each mode. For low resolution prediction and encoding, the system first finds the corresponding low resolution counterpart of the current macroblock. For the 16×16 input macroblock S^{HR} whose vertical and horizontal coordinates are given by (Y, X), corresponding low resolution 8×8 block S^{LR} coordinates are simply $(Y/2, X/2)$. The best low resolution intra prediction mode m^{LR^*} is

$$\begin{aligned} m^{LR^*} &= \underset{m^{LR}}{\operatorname{argmin}} \quad D_I^{LR}(m^{LR}) \\ &= \underset{m^{LR}}{\operatorname{argmin}} \quad \sum_{j,i \in LR} |S_{I_{m^{LR}}}^{LR}(j,i) - S_{LR}(j,i)|^2, \end{aligned} \tag{2}$$

where m^{LR} is the low resolution intra prediction mode, and $S_{I_{m^{LR}}}^{LR}$ is the low resolution prediction of the macroblock for this mode. $S_{I_{m^{LR}}}^{LR}$ is acquired by downsampling the pixel values which have already been encoded and applying intra prediction to the 8×8 block. Subsequently, RD cost of encoding the macroblock in low resolution, C_I^{LR} is computed by

$$C_I^{LR} = D^{LR}(S_{I_{m^{LR*}}}^{LR}, m^{LR^*}) + \lambda_I R_I(m^{LR^*}), \tag{3}$$

where λ_I is the Lagrangian parameter supplied by the rate-controller, $R_I(m^{LR^*})$ is the number of bits required to encode the macroblock in low resolution mode, and $D^{LR}(S_{I_{m^{LR*}}}^{LR}, m^{LR^*})$ is the distortion of the low resolution coding after upsampling of the reconstructed macroblock. $D^{LR}(S_{I_{m^{LR*}}}^{LR}, m^{LR^*})$ is given by

$$D^{LR}(S_{I_{m^{LR*}}}^{LR}, m^{LR^*}) = D\{U(T^{-1}[Q^{-1}\{Q(T[S^{LR} - S_{I_{m^{LR*}}}^{LR}])\}] + S_{I_{m^{LR*}}}^{LR}), S^{HR}\}, \tag{4}$$

where $T(\cdot)$ is integer transform, $Q(\cdot)$ is quantization, $Q(\cdot)^{-1}$ is inverse quantization, and $T(\cdot)^{-1}$ is the inverse transform. Upsampling operation $U(\cdot)$ is performed by projecting the low resolution reconstructed macroblock to a high resolution grid such that every second sample in both directions is presented by the samples of the low resolution input and by computing the missing values using interpolation with H.264/AVC upsampling filter whose kernel is given by

$$H_U = [1 \quad -5 \quad 20 \quad 20 \quad -5 \quad 1]/32. \tag{5}$$

$D(A, B)$ is an SSE metric, computed by $\sum_{j,i \in HR} |A(j,i) - B(j,i)|^2$. RD cost of high resolution encoding, C_I^{HR} is calculated in a similar fashion, except for the intra prediction, which is carried out in the original resolution, and is given by

$$C_I^{HR} = D^{HR}(S_{I_{m^{HR*}}}, m^{HR^*}) + \lambda_I R_I(m^{HR^*}), \tag{6}$$

where

$$D^{HR}(S_{I_{m^{HR*}}}, m^{HR^*}) = D\{T^{-1}[Q^{-1}\{Q(T[S^{HR} - S_{I_{m^{HR*}}}^{HR}])\}] + S_{I_{m^{HR*}}}^{HR}, S^{HR}\}, \tag{7}$$

where, $S_{I_{m^{HR*}}}$ is the best intra prediction of the macroblock in high resolution. Encoder makes a decision on whether to encode the macroblock in high resolution or in low resolution based on RD cost of each option. In other words, macroblock is encoded in low resolution mode if $C_I^{LR} < C_I^{HR}$ or in high resolution mode if $C_I^{HR} < C_I^{LR}$.

Algorithm 1. RAMBVCS intra macroblock encoder

CONTROL PARAMETERS

Denote N_f = number of frames to be encoded
Denote N_{MB} = number of macroblocks in one frame
Denote N_I = number of intra prediction modes
for $k = 0$ to $N_f - 1$ **do**
  Downsample current frame $\rightarrow$ LR input buffer
  for $l = 0$ to $N_{MB} - 1$ **do**
    for $n = 1$ to N_I **do**
      do INTRA prediction m_n^{LR} on LR MB $\rightarrow S^{LR}_{m_n^{LR}}, D^{LR}_{m_n^{LR}}$
      if $D^{LR}_{m_n^{LR}} < D^{LR}_{m^{LR*}}$ **then**
        $m^{LR^*} \leftarrow m_n^{LR}$
        $S^{LR^*} \leftarrow S^{LR}_{m_n^{LR}}$
      end if
      do INTRA prediction m_n^{HR} on HR MB $\rightarrow S^{HR}_{m_n^{LR}}, D^{HR}_{m_n^{HR}}$
      if $D^{HR}_{m_n^{HR}} < D^{HR}_{m^{HR*}}$ **then**
        $m^{HR^*} \leftarrow m_n^{HR}$
        $S^{HR^*} \leftarrow S^{HR}_{m_n^{HR}}$
      end if
    end for
    do Calculate RD costs C_I^{LR} and C_I^{HR}
    if $C_I^{LR} < C_I^{HR}$ **then**
      do Encode current MB in low resolution
    else
      do Encode current MB in high resolution
    end if
  end for
end for

2.3 Inter Macroblock Encoding

Current video compression schemes employ block based motion compensation methods for inter prediction. The macroblock to be processed is compared to available blocks in one or multiple reference frames for the best match. Subsequently, the system encodes the residual difference between the best match and the original block along with the motion vector information associated with it. In this section, we consider temporal prediction methods in a lower spatial resolution. Let $\mathbf{F^{HR}} = \{f_1^{HR}, ...f_K^{HR}\}$ denote the set of high resolution reference frames for inter prediction whose low resolution counterpart $\mathbf{F^{LR}} = \{f_1^{LR}, ...f_K^{LR}\}$ is obtained by filtering and downsampling each frame of $\mathbf{F^{HR}}$. The best low resolution motion predicted block $S^{LR}_{P_{m^{LR*}, \mathbf{v}^{LR*}}}$ can be found by partitioning the current low resolution input block S^{LR} according to the best inter prediction mode m^{LR^*} and displacing it in $\mathbf{F^{LR}}$ by the offset $\mathbf{v}^{LR^*}$. Subsequently, the distortion introduced by the low resolution inter prediction is minimized, resulting in

$$D_P^{LR} = \sum_{j,i \in LR} |S^{LR}(j,i) - f^{LR}_{m^{LR*}}(j + v_y, i + v_x)|^2, \quad (8)$$

where $f^{LR}_{m^{LR*}}$ is the low resolution reference frame, partitioned according to inter mode m^{LR^*}. Following the motion estimation, the RD cost C_P^{LR} for the low resolution inter coding is calculated by

$$C_P^{LR} = D^{LR} + \lambda_P R_P(m^{LR^*}), \quad (9)$$

where

$$D^{LR} = D\{T^{-1}[Q^{-1}\{Q(T[S^{HR} - S^{LR}_{P_{m^{LR*}, \mathbf{v}^{LR*}}}])\}] + S^{LR}_{P_{m^{LR*}, \mathbf{v}^{LR*}}}, S^{HR}\}. \quad (10)$$

In order to calculate the RD cost of the high resolution encoding, the system performs motion prediction in the original resolution and computes

$$C_P^{HR} = D^{HR} + \lambda_P R_P(m^{HR^*}), \quad (11)$$

where m^{HR^*} is the best high resolution inter prediction mode and D^{HR} is calculated the same way as in the previous section. Finally, the encoder makes a decision to encode the macroblock in low resolution mode if $C_P^{LR} < C_P^{HR}$ or in high resolution mode if $C_P^{HR} < C_P^{LR}$.

2.4 RAMBVCS Experimental Results

In this section we present the simulation results obtained with RAMBVCS. The proposed system was implemented by modifying H.264/AVC JM v13.2 software. A baseline profile configuration, which uses CAVLC entropy coder, IPPP...P GOP structure and no loop filter was adopted throughout the experiments. For ME/MC, one reference frame, 32 pixel wide search range, SSE distortion metric and quarter-pel motion vector accuracy options were utilized. One slice per frame with only heterogenous macroblock type was used. For RD control, QP was varied in increments of one, in the range from 14 to 40, which covers a broad range of bitrates and quality levels. Test sequences used in the experiments were 1080p and 720p raw HD video.

Simulation results are based on the widely accepted PSNR vs. Bitrate criterion. In each test case, 50 frames were encoded using the configuration above, whose decoded mean PSNR and bits/frame values are plotted. In order to investigate the improvements attained by the proposed system, we also plot the

Algorithm 2. RAMBVCS inter macroblock encoder

CONTROL PARAMETERS

Denote N_f = number of frames to be encoded
Denote N_{MB} = number of macroblocks in one frame
Denote N_P = number of inter prediction modes
for $k = 0$ to $N_f - 1$ **do**
  Downsample current frame → LR input buffer
  for $l = 0$ to $N_{MB} - 1$ **do**
    for $r = 1$ to N_P **do**
      if $D^{LR}_{m^{LR}_r} < D^{LR}_{m^{LR*}}$ **then**
        $m^{LR^*} \leftarrow m^{LR}_r$
        $S^{LR^*} \leftarrow S^{LR}_{m^{LR}_r}$
        $\mathbf{v}^{\mathbf{LR}^*} \leftarrow \mathbf{v}^{\mathbf{LR}}_{\mathbf{r}}$
      end if
      do INTER prediction m^{HR}_r on LR MB → $S^{HR}_{m^{HR}_r}$, $\mathbf{v}^{\mathbf{HR}}_{\mathbf{r}}$, $D^{HR}_{m^{HR}_r}$
      if $D^{HR}_{m^{HR}_r} < D^{HR}_{m^{HR*}}$ **then**
        $m^{HR^*} \leftarrow m^{HR}_r$
        $S^{HR^*} \leftarrow S^{HR}_{m^{HR}_r}$
        $\mathbf{v}^{\mathbf{HR}^*} \leftarrow \mathbf{v}^{\mathbf{HR}}_{\mathbf{r}}$
      end if
    end for
    do Calculate RD costs C^{LR}_P and C^{HR}_P
    if $C^{LR}_P < C^{HR}_P$ **then**
      do Encode current MB in low resolution
    else
      do Encode current MB in high resolution
    end if
  end for
end for

performance curve of the current state-of-the-art H.264/AVC video codec for each sequence such that a fair comparison is possible.

Figure 9 shows two RD curves obtained by encoding *Pedestrian Area* 1080p HD sequence with the conventional H.264/AVC and RAMBVCS. As it has been anticipated, RAMBVCS encoder provides higher compression gains at low bitrates by using the low resolution encoding option liberally. At these bitrates, bits-per-pixel ratio is very low for the conventional encoder, which causes blocking artifacts, while RAMBVCS can increase the bits-per-pixel ratio by using the downsampled macroblock representation whenever there is an RD benefit. Figure 11 and Fig. 12 illustrate this behavior by providing a close-up view of the top left part of the decoded frames at 86 Kbits/frame for RAMBVCS and H.264/AVC methods, respectively. At this bitrate, H.264/AVC can provide a PSNR value of 34.14 dB, whereas RAMBVCS can support a PSNR of 35.38 dB. At these low bitrates, a PSNR improvement over 1 dB can affect the image quality in a perceivable way, as shown in these figures. One can notice the

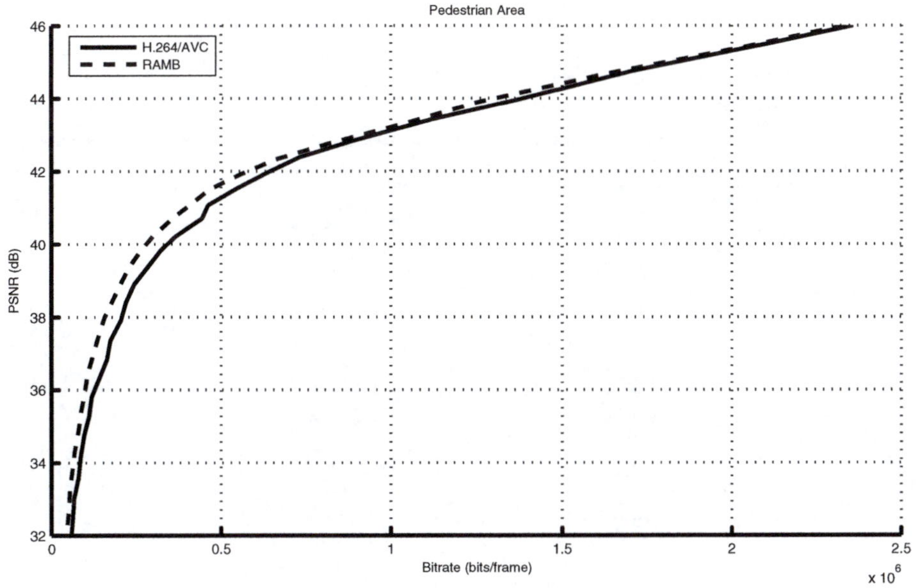

Fig. 9. *Pedestrian Area* RAMBVCS results

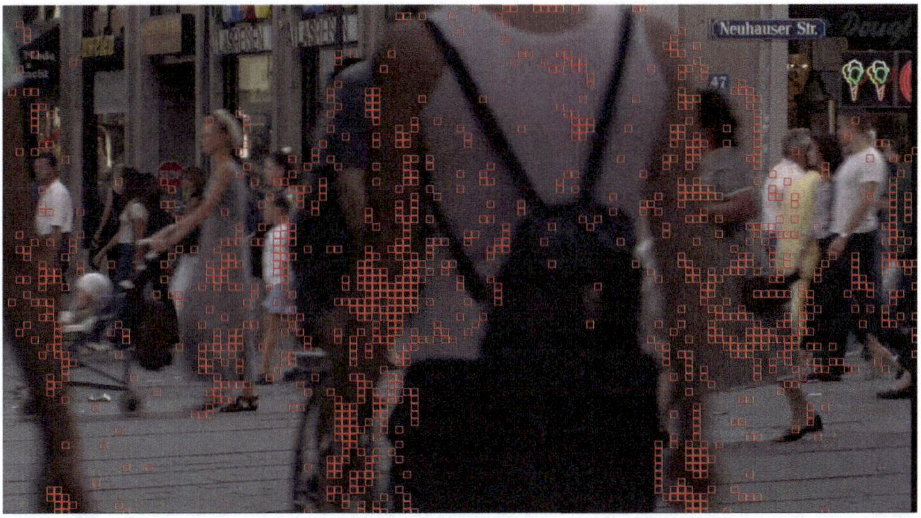

Fig. 10. *Pedestrian Area* RAMBVCS MB distribution

decrease in the amount of the blocking artifact on the woman's face and the increase of texture on the shop signs. Figure 10 shows the distribution of the macroblocks which are encoded in the low resolution. Most of these macroblocks are the smooth parts of the moving objects in the frame, suggesting the fact that

Fig. 11. Close up view of *Pedestrian Area* encoded with RAMBVCS (PSNR = 35.92, Rate = 86 Kbits/frame)

Fig. 12. Close up view of *Pedestrian Area* encoded with H.264/AVC (PSNR = 34.14, Rate = 86 Kbits/frame)

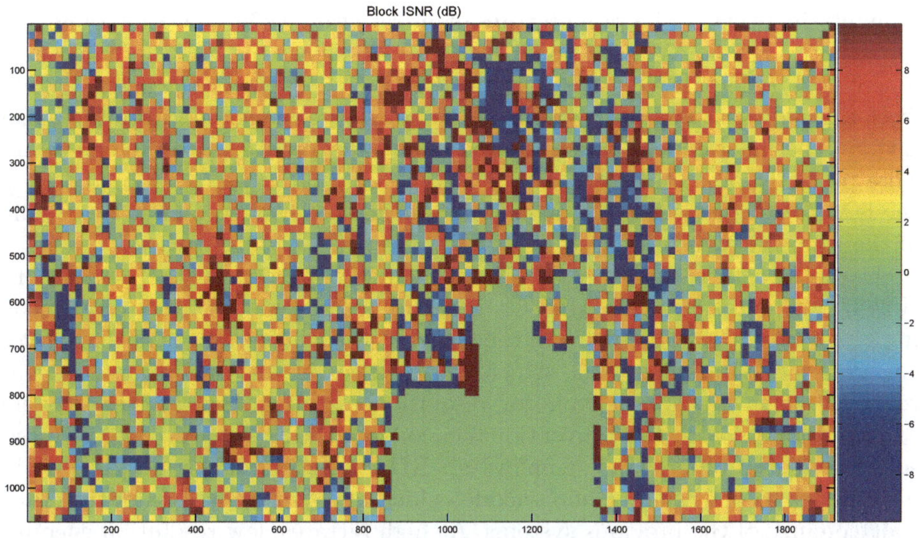

Fig. 13. *Pedestrian Area* RAMBVCS Block ISNR

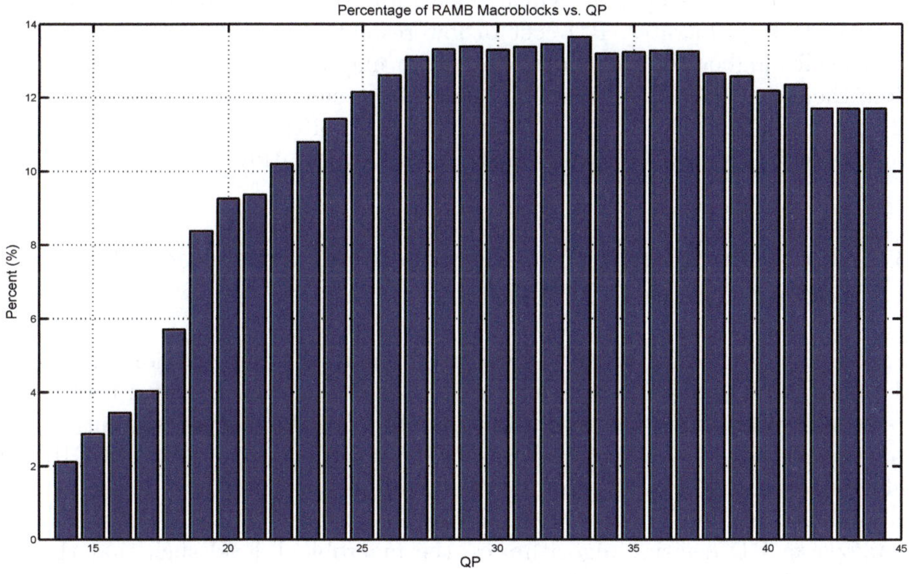

Fig. 14. *Pedestrian Area* RAMBVCS MB distribution vs. QP

the RAMBVCS encoder provides RD benefit in motion compensated low pass macroblocks. These macroblocks are usually blurry due to motion and do not contain a lot of texture; therefore, resolution rescaling does not affect them negatively, while still providing compression efficiency. Bitrate savings from these macroblocks can be used to increase the quality of other macroblocks. Hence, a quality increase at the same bitrate or bitrate savings at the equal quality as provided by H.264/AVC are possible. Figure 13 shows the quality increase per macroblock (Block ISNR) at the equal H.264/AVC rate. As it can be deduced from this figure, the majority of the macroblocks have benefited from the RAMBVCS. The macroblocks which showed quality degradation are the ones which were perfectly predicted by the H.264/AVC and not so perfectly predicted by the RAMBVCS. However, these macroblocks do not affect the subjective quality of the frame as seen in Fig. 10.

As the bitrate is increased, the performance gap between the conventional H.264/AVC and that of RAMBVCS is reduced. Previous low resolution encoding systems 1.2 - 1.3 suffered from the fact that the whole frame is downsampled, encoded and upsampled back, which causes excessive distortion in some macroblocks which are not suitable for resolution rescaling. Hence, these systems cannot compete with the conventional encoders at high bitrates. The proposed solution on the other hand, employs an RD cost criterion for deciding which macroblock to downsample and encode, which makes it exempt from the stated shortcomings of the previous systems. At high bitrates, low resolution encoding system performance is reduced by the loss of information during the resolution scaling process, whereas at low bitrates, codec performance is dominated by the large quantization step size, which makes low resolution encoding a plausible option. At high bitrates, RD cost of low resolution encoding of a macroblock is typically higher than encoding the same macroblock in the original resolution; therefore, RAMBVCS encoder generally prefers to encode the macroblock in high resolution. Figure 14 illustrates the ratio of the low resolution encoding mode selection versus QP, which justifies this expectation.

3 Macroblock Adaptive Hierarchical Resolution Video Coding System (MAHR)

Use of highly effective intra prediction and motion compensation methods in the H.264/AVC codec leads to a sparse residual macroblock, which leaves room for improvement with low resolution encoding. In this chapter we introduce a macroblock adaptive hierarchical resolution video coding system (MAHRVCS) which decomposes the full resolution residual hierarchically into a reduced resolution base layer and enhancement sub-residual layers. Realization of this system employs an RD decision algorithm on the macroblock level such that the best mode can be selected for each macroblock.

3.1 System Model and Algorithm

A block diagram of the MAHR encoder is shown in Fig. 15. The source frame is partitioned into macroblocks of size 16×16, where each macroblock is predicted with one of the two available methods: intra or inter prediction. If the macroblock type is INTRA (I-MB), H.264/AVC intra prediction is utilized, where

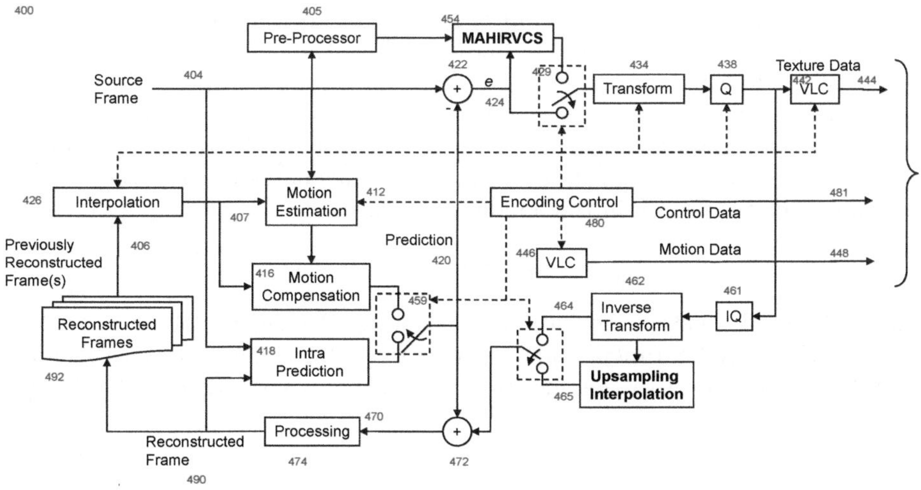

Fig. 15. MAHR encoder block diagram

the encoder tries various spatial prediction modes in order to obtain the best estimate. If macroblock is an INTER (P-MB), encoder employs a motion compensation scheme, where the prediction for the current macroblock is obtained from a composition of blocks in previously encoded reference frames. In either case, a high resolution residual is obtained by taking the difference between the original and the prediction macroblocks. Let $\mathbf{E^{HR}}$ denote this 16×16 original residual whose samples are classified as *a, b, c or d type* according to their coordinates as illustrated in Fig. 16.

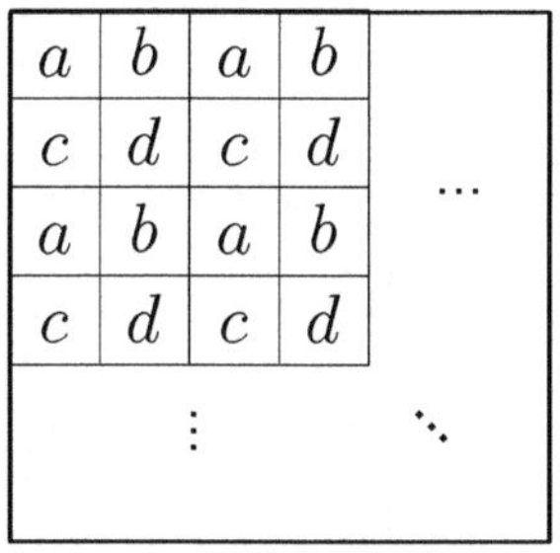

Fig. 16. Pixel classification of $\mathbf{E^{HR}}$

An 8×8 low resolution base layer representation $\mathbf{e_0^{LR}}$ of the original residual can be obtained by convolving $\mathbf{E^{HR}}$ with the MPEG-4 dyadic filter of Sec. 2.1 and retaining only the *a type* samples. MAHRVCS mode m_0^{LR} encodes the output of this operation as the residual by transformation, quantization and entropy coding. On the decoder side, the received low resolution residual is inverse quantized and transformed yielding ${}^{\mathbf{q}}\mathbf{e_0^{LR}}$. Next, a 16×16 residual block $\mathbf{E_0^{LR}}$ is constructed by projecting the samples of ${}^{q}\mathbf{e}_{\mathbf{0}}^{\mathbf{LR}}$ to the high resolution grid and interpolating the missing values as shown in Fig. 17. Towards this end, d_0 values are computed by averaging the nearest four diagonal ${}^{q}a_0$ samples, while b_0 and c_0 are calculated as the mean of the nearest two horizontal and vertical ${}^{q}a_0$ samples, respectively. Subsequently, the encoder adds $\mathbf{E}_{\mathbf{0}}^{\mathbf{LR}}$ to the prediction to obtain the reconstructed macroblock $\mathbf{S}_{\mathbf{0}}^{\mathbf{LR}}$ and calculates the RD cost $\mathbf{C}_{\mathbf{0}}^{\mathbf{LR}}$ of encoding the macroblock in this mode, as given by

$$C_0^{LR} = D\left(\mathbf{S}_{\mathbf{0}}^{\mathbf{LR}}\right) + \lambda R\left(m_0^{LR}\right), \tag{12}$$

where $R\left(m_0^{LR}\right)$ denotes the bitcost of encoding the macroblock with MAHR mode m_0^{LR}, λ is Lagrangian parameter supplied by the rate-controller and $D\left(\mathbf{S}_{\mathbf{0}}^{\mathbf{LR}}\right)$ is the SSE distortion metric of the reconstructed macroblock. $D\left(\mathbf{S}_{\mathbf{0}}^{\mathbf{LR}}\right)$ is given by

$$D\left(\mathbf{S}_{\mathbf{0}}^{\mathbf{LR}}\right) = \sum_{j=0}^{15}\sum_{i=0}^{15}\left|S^{HR}(j,i) - S_0^{LR}(j,i)\right|, \tag{13}$$

where $S^{HR}(j,i)$ denotes the pixel value at j^{th} row and i^{th} column of the original macroblock.

The MAHR encoder has the flexibility of encoding $\mathbf{e}_{\mathbf{0}}^{\mathbf{LR}}$ as the only residual, which provides the highest level of compression but it may also lead to unwanted distortions. Alternatively, the encoder can choose to reinforce $\mathbf{e}_{\mathbf{0}}^{\mathbf{LR}}$ with the enhancement sub-residuals $\mathbf{e}_{\mathbf{1}}^{\mathbf{LR}}$, $\mathbf{e}_{\mathbf{2}}^{\mathbf{LR}}$ and $\mathbf{e}_{\mathbf{3}}^{\mathbf{LR}}$, as associated with the MAHR modes m_1^{LR}, m_2^{LR} and m_3^{LR}, respectively. Inclusion of sub-residuals causes bitcost to increase, while lowering the reconstruction distortion, which offers a gradual RD trade-off mechanism for MAHRVCS.

The 8×8 sub-residual $\mathbf{e}_{\mathbf{1}}^{\mathbf{LR}}$ is defined as the difference between the *d type* values of the original residual $\mathbf{E}^{\mathbf{HR}}$ and the upsampled base layer residual $\mathbf{E}_{\mathbf{0}}^{\mathbf{LR}}$, i.e. $d - d_0$ samples. Following the upsampling of $\mathbf{E}_{\mathbf{0}}^{\mathbf{LR}}$, the encoder extracts

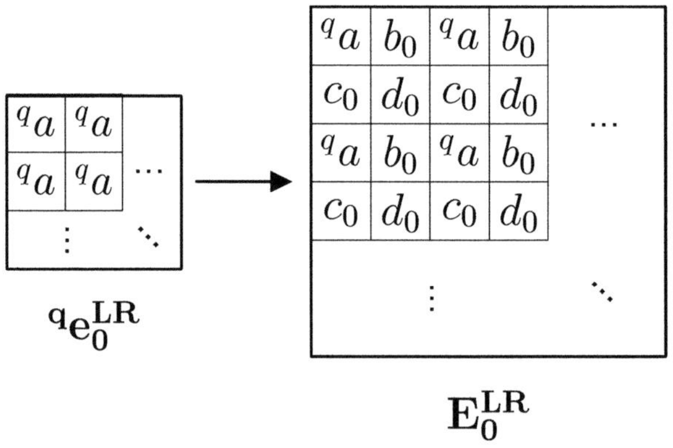

Fig. 17. Upsampling of ${}^{q}\mathbf{e}_{\mathbf{0}}^{\mathbf{LR}}$

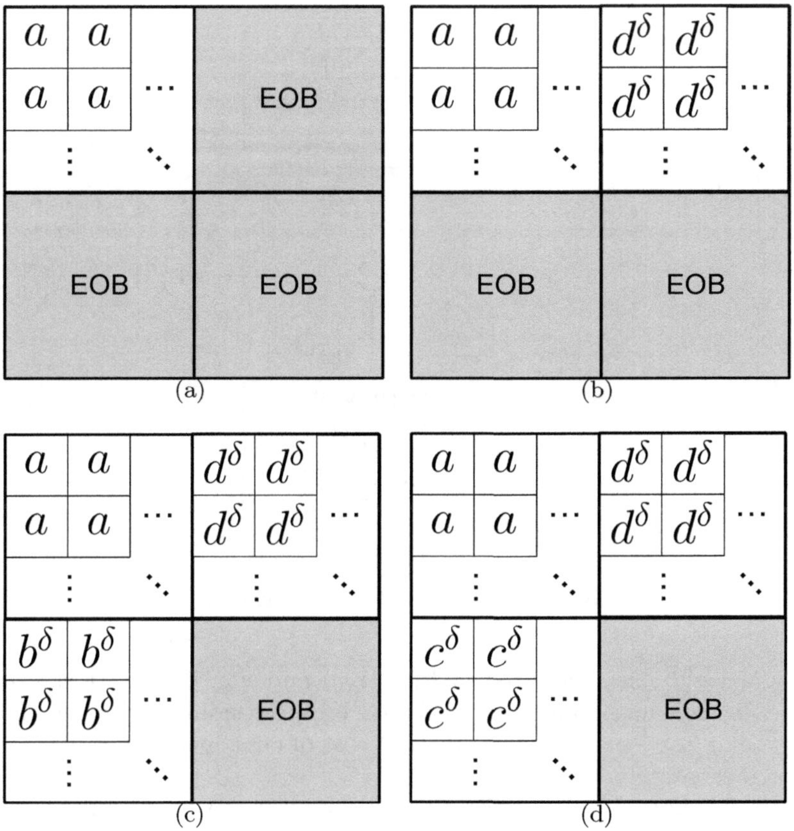

Fig. 18. Residual decomposition for MAHR coding modes. (a)MAHR mode 0 (b) MAHR mode 1 (c) MAHR mode 2 (d) MAHR mode 3

$d^\delta = d - d_0$ values and collects them in a separate 8×8 block, which shall be encoded together with $\mathbf{e_0^{LR}}$, as shown in Fig. 18(b). On the decoder side, received ${}^{\mathbf{q}}\mathbf{e_0^{LR}}$ is first upsampled, as explained in the previous paragraph, yielding $\mathbf{E_0^{LR}}$, whose d_0 samples are refined by the received sub-residual ${}^{\mathbf{q}}\mathbf{e_1^{LR}}$. Associated RD cost C_1^{LR} is also calculated to be used for mode decision later on. Figure 19 illustrates this process.

In order to decrease the distortion of low resolution residual coding even further, MAHR encoder can choose to send an additional sub-residual enhancement block, $\mathbf{e_2^{LR}}$ or $\mathbf{e_3^{LR}}$, which can be used to refine the *b type* or the *c type* residual samples, respectively. The 8×8 sub-residual $\mathbf{e_2^{LR}}$ is computed as the difference between the *b type* values of the original and the upsampled residual $\mathbf{E_1^{LR}}$, namely $b - b_0$. Similarly, $\mathbf{e_3^{LR}}$ can be calculated as the *c type* values of the difference between the two. MAHRVCS encodes $\mathbf{e_2^{LR}}$ for the low resolution mode m_2^{LR} and $\mathbf{e_3^{LR}}$ for m_3^{LR}, in addition to the base layer $\mathbf{e_0^{LR}}$ and previous

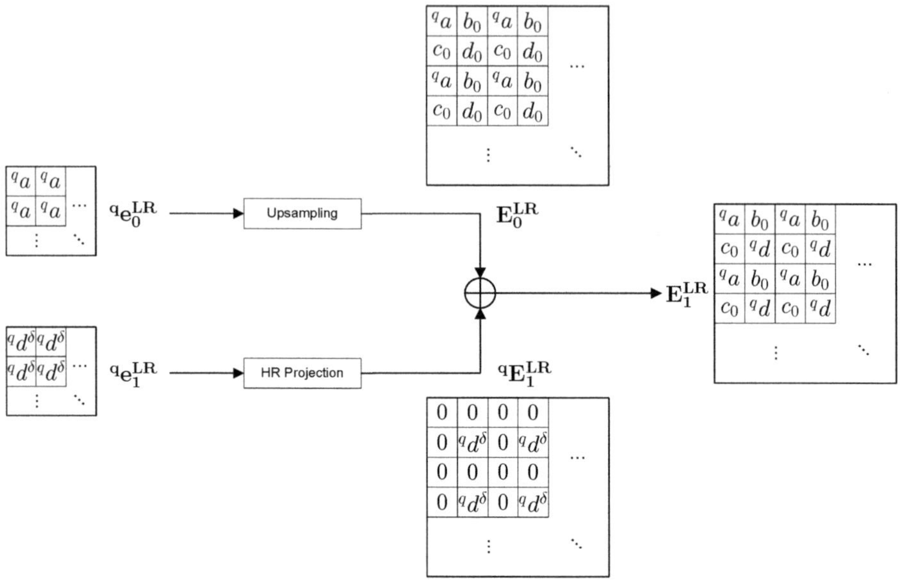

Fig. 19. Sub-residual refinement process

sub-residual $\mathbf{e}_1^{\mathbf{LR}}$, as illustrated in Fig. 18(c) and Fig. 18(d). At the decoder, received ${}^{\mathbf{q}}\mathbf{e}^{\mathbf{BL}}$ is upsampled first, followed by the residual refinement provided by ${}^{\mathbf{q}}\mathbf{e}_1^{\mathbf{LR}}$ and ${}^{\mathbf{q}}\mathbf{e}_2^{\mathbf{LR}}$ or ${}^{\mathbf{q}}\mathbf{e}_3^{\mathbf{LR}}$. Finally, RD cost of encoding with m_2^{LR} or m_3^{LR} is calculated separately.

For the purpose of coding the macroblock with the best available method, the encoder compares the RD cost of the conventional high resolution encoding, C^{HR}, to the RD costs of the introduced MAHR modes, C_0^{LR}, C_1^{LR}, C_2^{LR} and C_3^{LR}. The mode with the lowest RD cost is selected for the current macroblock encoding type and signalled in the bitstream. A flow diagram of the MAHRVCS is shown in Fig. 20.

3.2 MAHRVCS Experimental Results

In this section we present the simulation results obtained with the MAHRVCS. We follow a testing environment similar to the one described in Sec. 2.4. MAHRVCS was realized by modifying H.264/AVC JM v13.2 software. Baseline profile with CAVLC entropy coding, IPPP...P GOP structure, single reference frame motion compensation with SSE distortion metric and quarter-pel motion accuracy options were utilized. QP was varied in increments of one, in the range from 14 to 40 and PSNR-Bitrate point for each was computed.

Figure 21 shows two RD curves obtained by encoding *Shields* sequence with the conventional H.264/AVC and the MAHRVCS methods. As expected, MAHR codec provides compression efficiency at low-to-mid range bitrates, by supplying an optional downsampling of the macroblock residual. MAHRVCS can offer a

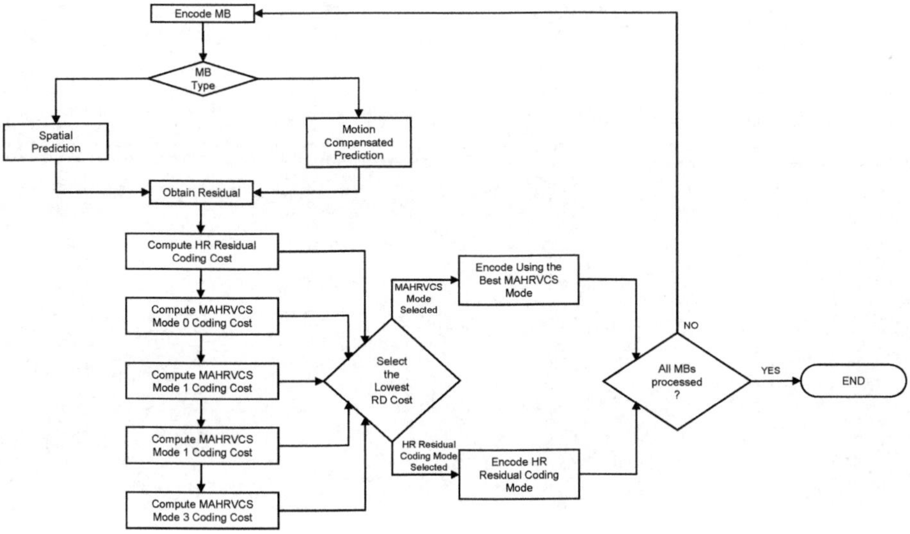

Fig. 20. MAHRVCS encoder flowchart

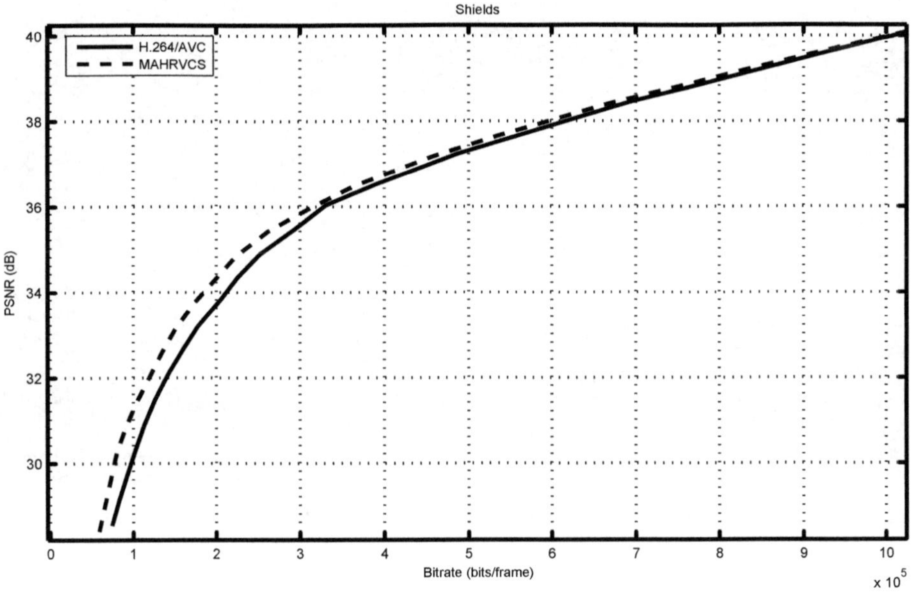

Fig. 21. *Shields* MAHRVCS results

PSNR value of 30.82 dB at 90 Kbits/frame bitrate, while H.264/AVC could only support a PSNR of 29.7 dB at the close but also higher bitrate of 92 Kbits/frame. Fig. 22 and Fig. 23 show the decoded frames at these bitrates for the two systems. Presence of severe blocking artifacts are easily noticeable in H.264/AVC

Fig. 22. *Shields* H.264/AVC encoded frame (PSNR = 29.7 dB, Rate = 92 Kbits/frame)

Fig. 23. *Shields* MAHRVCS encoded frame (PSNR = 30.82 dB, Rate = 92 Kbits/frame)

encoded picture, whereas MAHRVCS presents a better subjective quality with less blockiness. Figure 24 highlights the MAHRVCS encoded macroblocks for the same frame. Ratio of the low resolution residual macroblocks over the whole QP range is presented in Fig. 24. At low bitrates, corresponding to the high QP values, MAHRVCS macroblock ratio is high, which accounts for the observed compression improvement. The ratio starts dropping as the bitrate is increased

Fig. 24. *Shields* MAHRVCS macroblock distribution

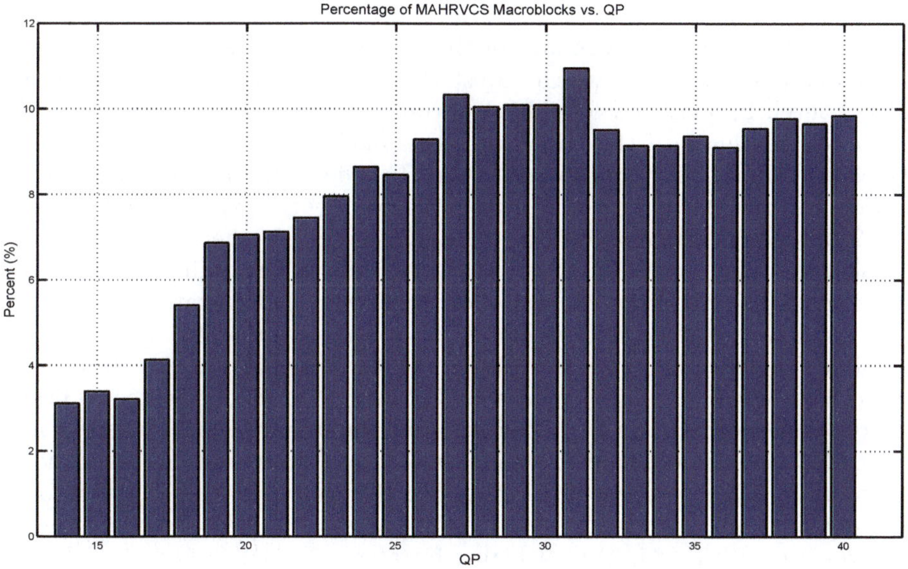

Fig. 25. *Shields* MAHRVCS macroblock ratio vs. QP

because. At high bitrates the conventional system has enough bandwidth allocated to the residual values with small step sizes. Downsampling of these residuals causes information loss which cannot be recovered with interpolation or residual refinement, making the associated RD costs of the MAHRVCS encoding modes higher. Since the encoder decides the downsampling strategy based on the RD cost, ratio of the low resolution residual macroblocks also diminishes and the MAHRVCS coding performance merges with that of H.264/AVC.

4 Conclusion

This work describes a general framework for a macroblock adaptive mixed resolution video encoding system. Towards this end, we started with a general description of a block based motion compensated video coding system. In Sec. 1.3 we provided the principle of low resolution video coding. We established that downsampling prior to compression provides coding efficiency at low bitrates by increasing the allocated bits per pixel ratio. A common shortcoming of the previous methods is the lack of flexibility in the downsampling process. Mentioned systems downsample the whole image frame prior to the compression, which provides coding efficiency only at low bitrates. At high bitrates, the low resolution encoding system performance is clipped by the loss of information during the resolution scaling process, whereas at low bitrates, the codec performance is dominated by the large quantization step size. Therefore, these systems are unable to compete with the conventional compression systems over a wide range of bitrates.

Motivated by the shortcomings of the previous systems, we introduced the RAMB encoder in Sec. 2, which adjust the downsampling strategy on a macroblock level. In another words, RAMB codec chooses to encode a macroblock in the low resolution if the RD cost of this mode is less than the RD cost of the conventional high resolution encoding. This flexibility in the downsampling strategy exposes itself in a positive way, as shown by the experimental results in Sec. 2.4. These results illustrate that the RAMB system offers substantial compression efficiency especially at low bitrates, while still providing a comparable performance to that of the conventional encoders at high bitrates.

In Sec. 3 we described the MAHR video coding system, where the macroblock adaptive downsampling idea was applied to the residual samples. In order to improve the downsampling scheme even further, we introduced three additional sub-residual encoding modes. Realization of these extra modes offers a more refined RD trade-off mechanism among the low resolution residual coding modes. As shown by the experimental results in Sec. 3.2, proposed MAHR video coding system offers coding efficiency especially at low bitrates, while providing on par performance with H.264/AVC at higher bitrates.

As a conclusion, this work addresses the shortcomings of the previous low resolution video encoding systems by providing flexible downsampling capability at the macroblock level. Experimental results of the two proposed systems show that realization of the described algorithms makes the low resolution encoding ideology a plausible option for multimodal video compression systems.

References

1. Barreto, D., Alvarez, L., Abad, J.: Motion estimation techniques in superresolution image reconstruction. A performance evaluation. Virtual Observatory 27, 433–436
2. Barreto, D., Alvarez, L., Molina, R., Katsaggelos, A., Callicó, G.: Region-based super-resolution for compression. Multidimensional Systems and Signal Processing 18(2), 59–81 (2007)
3. Borman, S.: Topics in Multiframe Superresolution Restoration. Ph.D. thesis (2004)
4. Bosveld, F., Lagendijk, R., Biemond, J.: Hierarchical video coding using a spatio-temporal subbanddecomposition. In: IEEE International Conference on Acoustics, Speech, and Signal Processing. ICASSP 1992, vol. 3 (1992)
5. Bruckstein, A., Elad, M., Kimmel, R.: Down-scaling for better transform compression. IEEE Transactions on Image Processing 12(9), 1132–1144 (2003)
6. Callicó, G., Núñez, A., Llopis, R., Sethuraman, R.: Low-Cost and Real-Time Super-Resolution over a Video Encoder IP. In: Fourth international symposium on quality electronic design (ISQED 2003), pp. 79–84 (2003)
7. CCITT S: Recommendation h. 261-video codec for audiovisual services at px64 kbit/s. Tech. rep., COM XV-R37-E, International Telecommunication Union (Augsust 1990)
8. Cheng, H., Kopansky, A., Isnardi, M.: Reduced resolution residual coding for h.264-based compression system. In: Proceedings of IEEE International Symposium on Circuits and Systems. ISCAS 2006 (May 2006)
9. Cheng, P., Li, J., Kuo, C., et al.: Multiscale video compression using wavelet transform and motion compensation. In: Proc. IEEE Int. Conf. Image Processing, pp. 606–609 (1995)
10. Draft I.: Recommendation H. 263, Video Coding for Low Bit Rate Communication. International Telecommunication Union (December 1995)
11. Draft I.: Recommendation and Final Draft International Standard of Joint Video Specification (ITU-T Rec. H. 264| ISO/IEC 14496-10 AVC), Geneva, Switzerland (May 2003)
12. Gandhi, P.: JPEG-based image compression for low-bit-rate coding. In: Proceedings of SPIE, vol. 2669, p. 82. SPIE (1996)
13. IEC I: 13818-1, Information Technology Generic Coding of Moving Pictures and Associated Audio Information: Systems, Part 1
14. IEC I: 14496-2, Information technology coding of audio-visual objects: visual. Committee Draft ISO/IEC JTC1/SC29/WGll, 2202
15. Molina, R., Katsaggelos, A., Alvarez, L., Mateos, J.: Toward a new video compression scheme using super-resolution. In: Proceedings of SPIE, vol. 6077, pp. 67–79 (2006)
16. Ohm, J.: Advances in Scalable Video Coding. Proceedings of the IEEE 93(1), 42–56 (2005)
17. Otani, T.: Sharp presents industry's first 4kx2k direct viewing lcd panel (October 2006)
18. Pennebaker, W., Mitchell, J.: JPEG Still Image Data Compression Standard. Kluwer Academic Publishers, Norwell (1992)
19. Rabbani, M., Joshi, R.: An overview of the JPEG 2000 still image compression standard. Signal Processing: Image Communication 17(1), 3–48 (2002)
20. Schwarz, H., Marpe, D., Wiegand, T.: Overview of the Scalable Video Coding Extension of the H. 264/AVC Standard. IEEE Transactions on Circuits And Systems For Video Technology 17(9), 1103 (2007)

21. Segall, C., Katsaggelos, A., Molina, R., Mateos, J.: Bayesian resolution enhancement of compressed video. IEEE Transactions on Image Processing 13(7), 898–911 (2004)
22. Segall, C.A., Sullivan, G.J.: Spatial Scalability Within the H. 264/AVC Scalable Video Coding Extension. IEEE Transactions on Circuits and Systems For Video Technology 17(9), 1121 (2007)
23. Vishwanath, M., Chou, P.: An efficient algorithm for hierarchical compression of video. In: Proceedings of IEEE International Conference on Image Processing. ICIP 1994, vol. 3 (1994)
24. Wallace, G.: The JPEG still picture compression standard. IEEE Transactions on Consumer Electronics 38(1) (1992)
25. Wiegand, T., Sullivan, G., Bjntegaard, G., Luthra, A.: Overview of the H. 264/AVC video coding standard. IEEE Transactions on Circuits and Systems for Video Technology 13(7), 560–576 (2003)

Peer-to-Peer Streaming Systems

Yifeng He and Ling Guan

Department of Electrical and Computer Engineering,
Ryerson University, Toronto, ON, M5B 2K3 Canada

Summary. With advances in the broadband Internet access technology and the coding techniques, video streaming services have become increasingly popular. Traditionally, video streaming services are deployed in the client/server architecture. However, this centralized architecture cannot provide streaming to a large population of users due to the limited upload capacity from the server. Peer-to-Peer (P2P) technology has recently become a promising approach to provide live streaming services or Video-on-Demand (VoD) services to a huge number of the concurrent users over a global area. P2P streaming systems can be classified into P2P live streaming systems and P2P VoD systems. In a P2P live streaming system, a live video is disseminated to all users in real time. The video playbacks on all users are synchronized. In a P2P VoD system, users can choose any video they like and start to watch it at any time. The playbacks of the same video on different users are not synchronized. Depending on the approaches of overlay constructions, P2P live streaming systems can be categorized into tree-based P2P live streaming systems and mesh-based P2P live streaming systems. In tree-based P2P live streaming systems, the video streams are delivered over a single application-layer tree or multiple application-layer trees. In mesh-based P2P live streaming systems, each peer exchanges the data with a set of neighbors. VoD services provide more flexibility and interactivity to users. Depending on the forwarding approaches, P2P VoD systems can be categorized into buffer-forwarding systems, storage-forwarding systems, and hybrid-forwarding systems. In P2P VoD systems, prefetching scheme enables peers to download the future segments which are away from the current playback position, thus increasing the playback continuity.

1 Introduction

With advances in broadband Internet access technology and the coding techniques, video streaming services have become increasingly popular. For example, YouTube [1], a video-sharing service which streams its videos to users on-demand, has attracted about 20 million views a day [2]. Video traffic is expected to be the dominating traffic on the Internet in the near future.

Conventionally there are several approaches that can be used to deliver video streams to the users. We can roughly categorize the traditional approaches into two categories: unicast-based and multicast-based [3].

C.W. Chen et al. (Eds.): Intel. Multimedia Communication: Tech. and Appli., SCI 280, pp. 195–215.
springerlink.com

In unicast-based approaches, a unicast session is established for every user. There are three unicast approaches to deliver streaming video over the Internet: centralized, proxy, and Content Delivery Network (CDN) [3].

In centralized approach, a powerful server with a high-bandwidth is deployed to serve video streams to the users. This approach is easy to deploy and manage. However, there are drawbacks for the centralized approach. First, there is a risk of single point of failure on the server since all of the users request the services from it. Second, it is not scalable. The server consumes a corresponding amount of bandwidth for each user. When the number of the users is increased to a certain level, the server may not be able to support them due to the bandwidth constraint. For example, if a video is encoded in MPEG-4 format ($\sim$ 1.5 Mbps), a streaming server connected to the Internet through a T3 link ($\sim$ 45 Mb/s) can only support about 30 simultaneous users.

In the proxy approach, proxy servers are deployed near the client domains to cache a fraction of each video. The users can receive a part of the video from the proxy server near them. Therefore, this approach relieves the burden of the streaming server and the traffic load across the Wide Area Network (WAN). However, this approach requires deploying and managing proxies at many locations, which increases the cost.

In CDN approach, the video streams are delivered by the third party, known as the CDN. Content delivery networks, such as Akamai [4], deploy many servers at the edge of the Internet. The user can request the video from the most suitable server. CDN effectively shortens the users' startup delays, reduces the traffic across the WAN. However, the cost for CDN approach is quite high since the CDN operator charges the video provider for every megabyte served.

The multicast-based approaches serve multiple users using the same stream, which is bandwidth efficient. The network-layer multicast establishes a tree over the internal routers with the users as the leaves of the tree. Though network-layer multicast is efficient, it is not widely deployed due to the following reasons. First, network-layer multicast requires routers to maintain per-group state, which introduces high complexity at the IP layer. Second, it is difficult for the transport layer to provide flow control and congestion control in the network-layer multicast. In a word, the lack of incentive to install multicast-capable routers prevents the network-layer multicast from wide deployment [5]. Therefore, network-layer multicast is not a feasible solution for delivering video streams to all the users.

Peer-to-Peer (P2P) technology has recently become a promising approach to build distributed and scalable network applications [6, 7]. The basic design philosophy of P2P is to encourage users to act as both clients and servers, namely as peers. In a P2P network, a peer not only downloads the data from other supplying peers who are buffering or storing them, but also uploads the downloaded data to other requesting peers who are requesting them. The upload bandwidth of end users is efficiently utilized to offload the bandwidth burdens placed on the servers [8].

Since the appearance of Napster [9], a P2P file sharing service, in early 1999, P2P networks have experienced tremendous growth. In 2003, P2P became the most popular web application. At the end of 2004, P2P traffic represented over 60% of the total Internet traffic, dwarfing Web browsing [10]. This rapid success was fueled by file transfer networks which allow users to swap the blocks of a file, despite the large time often necessary to complete a download. Recently P2P technology has been employed to provide video streaming services [6, 7]. Several P2P streaming systems have been deployed to provide Video-on-Demand (VoD) or live video streaming services on the Internet [11, 12, 13]. On January 28, 2006, more than 200,000 simultaneous users watched the live broadcast of Chinese Spring Festival Show on PPLive [11] at bit rates from 400 to 800 kpbs. The aggregate required bandwidth reaches 100 gigabits/second [14].

The P2P network systems can be classified into P2P file-sharing systems and P2P streaming systems. In P2P file-sharing systems, users need to download the whole file before it can be used. A file is divided into multiple blocks. Each user downloads the unavailable blocks from the peers who have those blocks. There are no timing constraints on downloading a block of the file. Rather, the total download time is more important. In P2P video streaming systems, a user plays the video while it is being downloading. Timing constraints are crucial to the streaming service, since a packet arriving after its playback deadline is useless.

P2P streaming systems can be categorized into P2P live streaming systems and P2P VoD systems. In a P2P live streaming system, a live video is disseminated to all users in real time. The video playbacks on all users are synchronized. In other words, all users watch almost the same position of the same video. A typical example of P2P live streaming systems is P2P IPTV, which is used to broadcast the TV programs to all the participating users. In a P2P VoD system, users can choose any video they like and start to watch it at any time. The playbacks of the same video on different users are not synchronized. In other words, each user watches a different position of the same video.

A video file is composed of multiple blocks. The exchanges of the blocks in different P2P network systems are compared in Fig. 1. In a P2P file-sharing system, the file is not played until all of the blocks of the file have been downloaded completely, thus the receiving peer can download any of the unavailable blocks from a supplying peer who is storing them, as shown in Fig. 1(a). The downloaded blocks are stored in the disk and will be served to the other peers. In P2P streaming systems, each peer maintains a buffer to cache the recently downloaded blocks before and after the current playback position. In a P2P live streaming system, users have close playback positions, such that they can exchange the available blocks in their buffer with each other, which is depicted in Fig. 1(b). The small variation of the playback times is caused by the different delivery delays. In a P2P VoD system, different users have different playback positions for the same video. The peer with a earlier playback position can forward the recently played blocks in the buffer to the other peer with a later playback position, as shown in Fig. 1(c).

The rest of the chapter is organized as follows. In Section 2, the P2P live streaming systems are described. In Section 3, the P2P VoD systems are

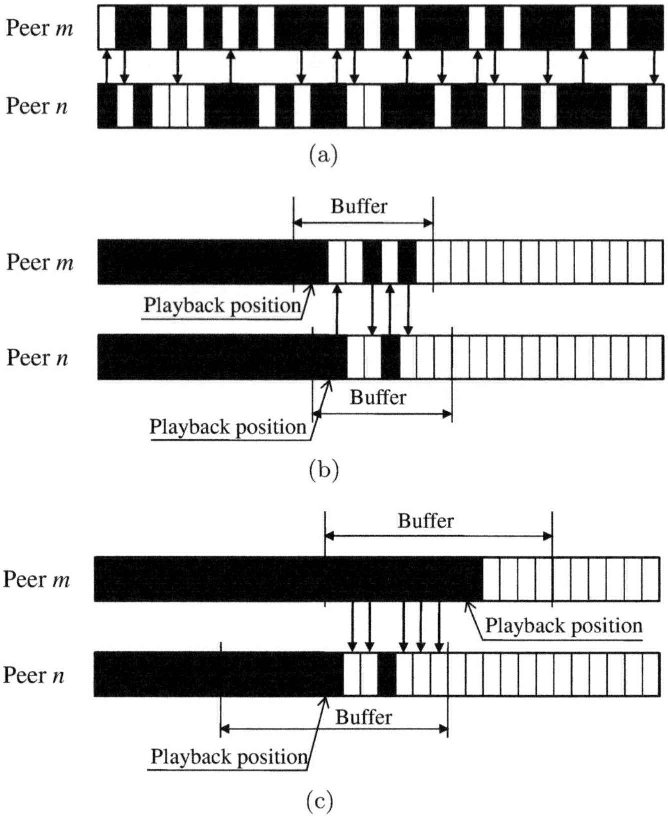

Fig. 1. The exchanges of the blocks between two peers in different P2P systems: (a) P2P file-sharing systems, (b) P2P live streaming systems, and (c) P2P VoD systems

presented. Finally, we discuss the future research directions in Section 4 and give the summary in Section 5, respectively.

2 P2P Live Streaming Systems

Based on the overlay structure, P2P live streaming systems can be broadly classified into tree-based P2P live streaming systems and mesh-based P2P live streaming systems.

2.1 Tree-Based P2P Live Streaming Systems

Though network-layer multicast is efficient in broadcasting a video to all users, it is not widely accepted largely due to the router overhead of managing multicast groups and the complexity of transport control for multicast sessions [5]. Therefore the multicast function is moved from the network layer to the application

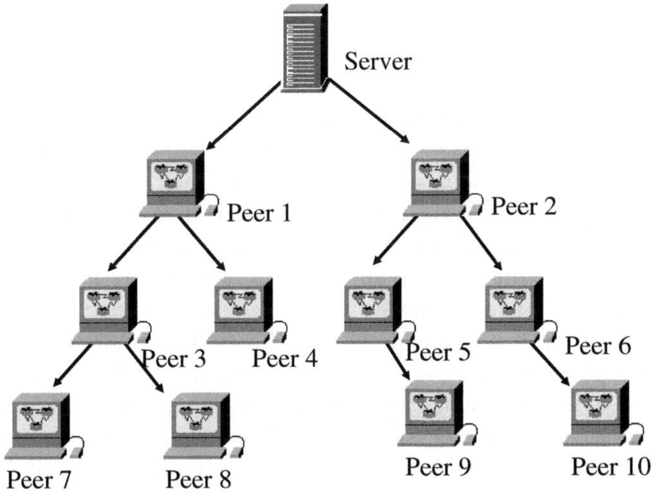

Fig. 2. Illustration of a single-tree based P2P live streaming system

layer. In a tree-Based P2P live streaming systems, a single application-layer tree or multiple application-layer trees are constructed to deliver the video streams.

Single-Tree Based P2P Live Streaming Systems

In a single-tree based P2P live streaming system, users participating in a live video streaming session can form a tree at the application layer. The root of the tree is the server. Each user joins the tree at a certain level. It receives the video from its parent peer at the level above and forward the received video to its child peers at the level below. End System Multicast (ESM) [15] is a typical example of single-tree based P2P live streaming systems.

A single-tree based P2P live streaming system with ten peers is illustrated in Fig. 2. At level 1, there are 2 peers (e.g., Peer 1 and Peer 2) receiving video directly from the server. At level 2, there are 4 peers (e.g., Peers 3-6), receiving video from their parents at level 1 and forwarding the received video to their corresponding child peer(s) at level 3. At level 3, there are 4 peers (e.g., Peers 7-10). The peers without any child peer are called *leaf nodes.* In Fig. 2, the leaf nodes include Peer 4 and Peers 7-10. The peers with at least one child peer are called *internal nodes.*

In the construction of a application-layer tree for live streaming, there are two considerations: the *depth* of the tree and the *fan-out* of the internal nodes. In a single tree, peers receive the video streams from its parent peers at the upper level. To reduce the delays for the peers at the bottom level, a tree with a smaller depth is desired. With a smaller depth, the fan-out of the internal nodes is increased. In other words, the internal nodes need to upload video streams to more child peers in order to reduce the depth of the tree. However, the maximum fan-out degree of a peer is constrained by its uploading bandwidth.

Peers may join or leave a live streaming session at any time. Therefore tree maintenance is important. Tree construction and maintenance can be managed in either a centralized or a distributed fashion. In a centralized solution, a central server controls the tree construction and recovery. When a peer joins the system, it first contacts the server. Based on the existing topology and the characteristics of the newly joined peer, such as its location and network access capacity, the server decides the position of the new peer in the tree and notifies it which parent peer to connect to. A peer might leave the session at any time either gracefully or unexpectedly (e.g., machine crashes). After a peer leaves, all its descendants in the tree get isolated from the server and cannot receive the video. In such a case, the isolated peers can contact the the central server, who then assigns each of the isolated peers to a new parent peer in the tree. The drawback of the centralized management is the performance bottleneck and the single point of failure on the central server. To combat this, various distributed algorithms for tree construction and maintenance have been proposed. In ZIGZAG [16], a peer-join can be accomplished without asking more than $O(\log N)$ existing peers where N is the peer population, and a failure can be recovered quickly and regionally with a constant number of reconnections and no affection on the server.

There are two major drawbacks for single-tree based P2P live streaming systems. First, the departure of a peer causes the isolation of all of its descendants from the video source. Second, all the leaf nodes do not contribute their uploading bandwidths, which degrades the efficiency of the peer bandwidth utilization. A remedy to those drawbacks is a multiple-tree based P2P streaming system.

Multiple-Tree Based P2P Live Streaming Systems

To improve the resiliency of the tree and the bandwidth utilization of the peers, multiple-tree based approaches have been proposed [17]. In multiple-tree based P2P live streaming systems, the video is encoded into multiple sub-streams, and each sub-stream is delivered over one tree. The quality received by a peer depends on the number of sub-streams that it receives. There are two key advantages for the multiple-tree solution. First, if a peer fails or leaves, all its descendants lose the sub-stream delivered from that peer, but they still receive the sub-streams delivered over the other trees. Therefore, all its descendants can receive a coarse video quality in case of a loss of a sub-stream. Second, a peer has different roles in different trees. It might be an internal node in one tree and a leaf node in another tree. When a peer is an internal node in a tree, its upload bandwidth will be utilized to upload the sub-stream delivered over that tree. To achieve high bandwidth utilization, a peer with a high upload bandwidth can supply sub-streams in more trees [5].

Fig. 3 illustrates a multiple-tree based P2P live streaming system with 7 peers. The server encodes the video into 2 sub-streams and then distributes them into tree 1 and tree 2, respectively. Each sub-stream has a bit rate $s/2$, where s is the source rate. Peers 1, 2 and 3 are internal nodes in the left tree and leaf nodes in the right tree, Peers 5, 6 and 7 are internal nodes in the right tree

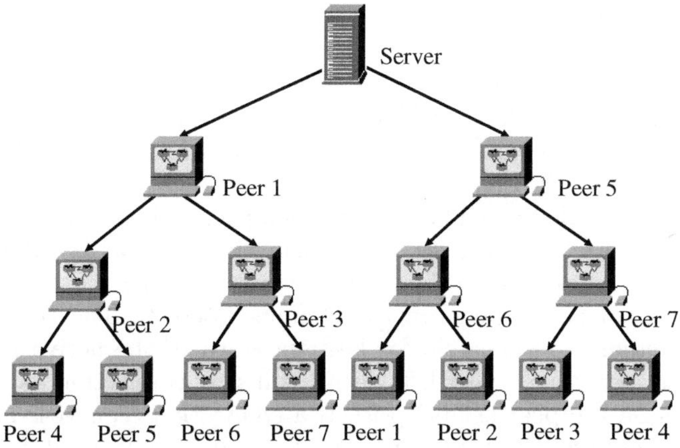

Fig. 3. Illustration of a multiple-tree based P2P live streaming system

and leaf nodes in the left tree. Each of Peers 1, 2, 3, 5, 6, and 7 uploads a sub-stream of rate $s/2$ to two child peers. Peer 4 serves as leaf nodes in both trees. Therefore it contribute zero upload bandwidth to the system. If a peer leaves the system, its descendants can still receive a sub-stream from the other trees. For example, if Peer 1 leaves, the left tree is disconnected from the server. However, the descendants of Peer 1 (e.g., Peers 2-7) can receive a sub-stream delivered over the right tree. They can watch the video at a coarse quality without interruption in case of peer departure.

2.2 Mesh-Based P2P Live Streaming Systems

In single-tree based P2P live streaming systems, a peer receives the streams from a single parent. If the peer's parent leaves, the peer will receive nothing until it is connected to a new parent. In multiple-tree based P2P live streaming systems, a peer receives a sub-stream from the parent in the corresponding tree. If the parent in the tree leaves, the received video quality at the peer will be degraded. Therefore, peer dynamics make tree maintenance a challenging and costly task in tree-based P2P live streaming systems.

To combat the peer dynamics, many recent P2P streaming systems use mesh-based streaming approach [6, 19]. In mesh-based P2P live streaming systems, each peer exchanges the data with a set of neighbors. If one neighbor leaves, the peer can still download the video data from the remaining neighbors. Meanwhile, the peers will add other peers into its neighbor set. Each peer can receive data from multiple supplying peers in mesh-based streaming systems, instead of a single parent in single-tree based streaming systems. Thus mesh-based streaming systems are robust against peer dynamics. The major challenges in mesh-based P2P live streaming systems are neighborhood formation and data scheduling, which will be discussed in details as follows.

Neighborhood Formation

In mesh-based P2P streaming systems, each peer downloads the video data from its neighbors. Let's first look at how a peer finds its neighbors. The process for finding the neighbors consists of 3 steps as shown in Fig. 4. At the beginning, the peer (e.g., peer 1 in Fig. 4) sends out a query message to the channel server to obtain an updated channel list, from which peer 1 chooses a channel to watch (step 1). Next, peer 1 reports its own information, such as IP address and port number, to the tracker, who then provides peer 1 a *peer list* that contains a random subset of active peers in the channel (step 2). The tracker in a P2P streaming system is responsible for keeping track of the active peers for each channel. After receiving a peer list, peer 1 will send connection requests to some remote peers on the peer list (step 3). If a connection request is accepted by a remote peer, peer 1 will add the remote peer into its *neighbor list*. After obtaining sufficient neighbors in the neighbor list, the neighborhood of peer 1 is formed, and then peer 1 starts to exchange video content with its neighbors.

Due to peer dynamics, the neighbor list of a peer may be stale. Therefore, a peer needs to periodically update its neighbor list. A peer can retrieve a new peer list from the tracker periodically, and then find the neighbors from the peer list. It can also exchange the peer list with its neighbors, and then find the new neighbors from the updated peer list.

If a peer leaves the session gracefully, it will notify the tracker and its neighbors of its departure such that its information can be removed from their peer lists immediately. If a peer leaves unexpectedly such as computer crashes, the failure of the peer can be detected by regular heartbeat messages. A peer will remove the

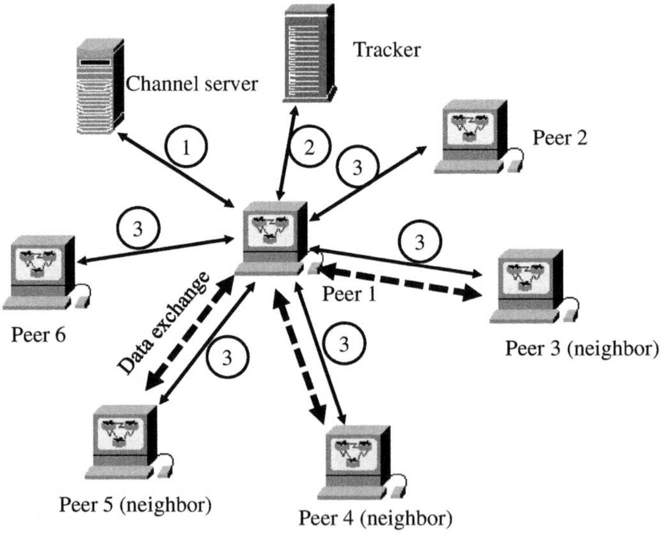

Fig. 4. Neighborhood formation in mesh-based P2P live streaming systems

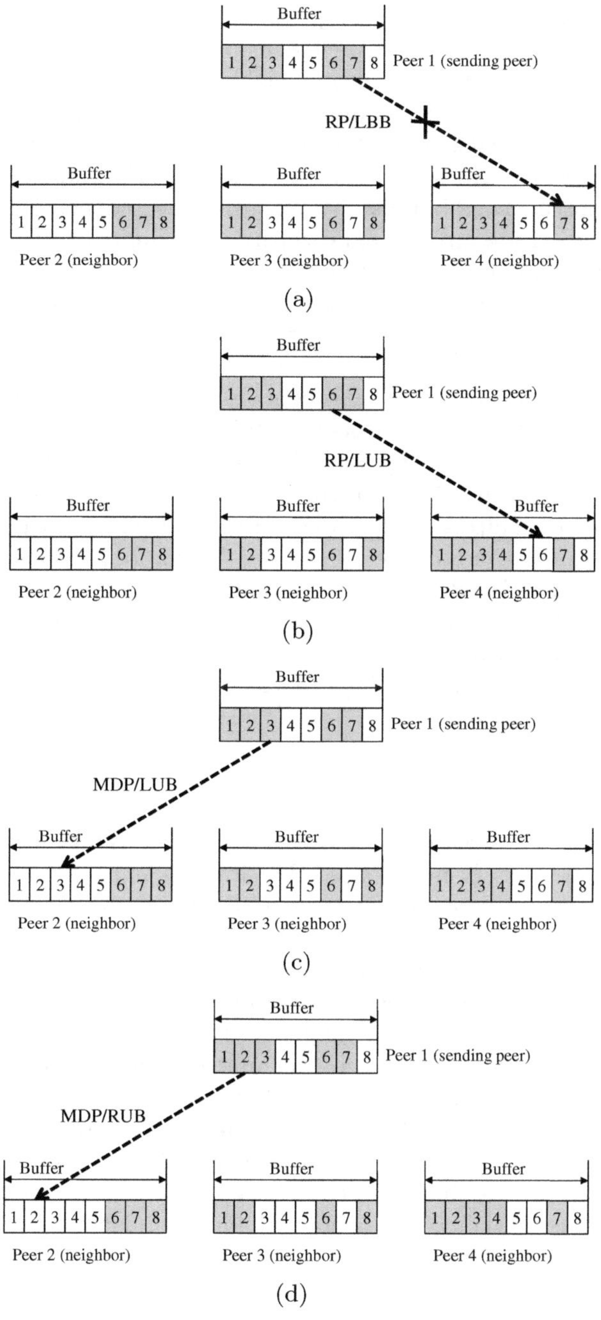

Fig. 5. Different push-based schemes in mesh-based P2P live streaming systems: (a) Random Peer / Latest Blind Block (RP/LBB), (b) Random Peer / Latest Useful Block (RP/LUB), (c) Most Deprived Peer / Latest Useful Block (MDP/LUB), and (d) Most Deprived Peer / Random Useful Block (MDP/RUB)

information of a neighbor from its neighbor list if it has not received a heartbeat message from the neighbor within a timeout period.

If the peer has more neighbors in its neighbor list, it can download the missing blocks from more sources simultaneously, thus increasing the throughput. However, connections with more neighbors introduce a higher communication overhead. In order to achieve a higher streaming performance, a peer can periodically update its neighbor list by rejecting the old neighbors with a low performance (e.g., a low upload bandwidth) and admitting the new neighbors with a higher performance (e.g., a high upload bandwidth). There is a tradeoff on the frequency of neighbor update. If a peer updates its neighbor list frequently, it can achieve a higher video quality at the price of a higher communication overhead.

In neighborhood formation, a peer contacts some peers in its peer list, and then negotiates with the remote peers on establishment of neighbor relationship. It is a challenging issue on neighbor selection. A better neighbor list helps to improve the data exchange between the peer and its neighbors. The decision of the neighbor relationship is made based on the following considerations [5]: 1) the available resources on both peers, such as the number of connections, upload bandwidth, download bandwidth, CPU and memory usage; 2) the quality of the potential overlay link between the two peers, which can be characterized by the transmission delay and packet loss rate; 3) the complementariness of the video content, which means that the video blocks available at a peer are needed by the other peer and vice versa.

Data Scheduling

In tree-based P2P live streaming systems, video streams flow from the source to all peers along the tree. In mesh-based P2P live streaming systems, the concept of video stream becomes invalid. Instead, a peer maintains a buffer to cache the video data. The video player reads the data in the buffer and then plays the video on the screen. A video is divided into multiple blocks. A block is the basic unit for data exchange among peers. Each block is identified with a unique sequence number. A peer downloads the blocks from its neighbors who have them in their buffer, at the same time the peer uploads the buffered blocks to those neighbors who are requesting them. The data scheduling algorithm has a great impact to the performance of P2P live streaming systems. Depending on whether the supplying peer or the receiving peer performs the data scheduling, the data scheduling algorithms can be categorized into *push-based schemes* and *pull-based schemes.* Push-based schemes are more suitable for upload-constrained systems, since the supplying peer can fully utilize its upload capacity to disseminate the blocks to its neighbors. On the other hand, pull-based schemes are more appropriate for download-constrained systems, since the receiving peer can adjust the rate of block requests according to its download capacity.

Let's first look at push-based schemes. Depending on selections of the destination peer and the video block, push-based scheme can be one of the following scheduling schemes [20]:

1. Random Peer / Latest Blind Block (RP/LBB). In this scheme, the supplying peer first chooses uniformly at random a destination peer from the neighbor list. The supplying peer then chooses the most recent block (e.g., the block with the highest sequence number) in its buffer, and pushes it to the destination peer.
2. Random Peer / Latest Useful Block (RP/LUB). In this scheme, the supplying peer first chooses uniformly at random a destination peer from the neighbor list. The supplying peer then chooses the most recent block that is not available in the destination peer, and pushes it to the destination peer.
3. Most Deprived Peer / Latest Useful Block (MDP/LUB). In this scheme, the supplying peer first chooses the destination peer from the neighbor list such that the number of the blocks that are available at the supplying peer but not at the destination peer is maximized. The supplying peer then chooses the most recent block that is not available in the destination peer, and pushes it to the destination peer.
4. Most Deprived Peer / Random Useful Block (MDP/RUB). In this scheme, the supplying peer first chooses the destination peer from the neighbor list such that the number of the blocks that are available at the supplying peer but not at the destination peer is maximized. The supplying peer then chooses uniformly at random a block from the blocks that are available at the supplying peer but not available at the destination peer, and pushes it to the destination peer.

Fig. 5 illustrates different push-based schemes. Peer 1 is the supplying peer, and Peers 2-4 are the neighbors of Peer 1. In the RP/LBB scheme shown in Fig. 5(a), Peer 1 randomly chooses a neighbor (assumed to be Peer 4), and then blindly pushes the most recent block (e.g., Block 7) to Peer 4. However Peer 4 already has Block 7 in its buffer, so the pushed block is redundant thus useless. In the RP/LUB scheme shown in Fig. 5(b), Peer 1 pushes the latest useful block (e.g., Block 6) to the randomly chosen neighbor (e.g., peer 4 in this case). As shown in Fig. 5, the most deprived peer is Peer 2 since it has the maximal number of unavailable blocks that are available at Peer 1. In the MDP/LUB scheme shown in Fig. 5(c), Peer 1 pushes the latest useful block (e.g., Block 3) to Peer 2. In the MDP/RUB scheme shown in Fig. 5(d), the useful block set between Peer 1 and Peer 2 are Blocks 1-3. Peer 1 randomly chooses a block from the useful block set (assumed to be Block 2) and then pushes it to Peer 2.

In pull-based schemes, the receiving peer performs the data scheduling, instead of the supplying peer in push-based schemes. In a mesh-based P2P live streaming system using pull, each peer maintains a *buffer map* to indicate the block availability in its buffer. Each peer broadcasts its buffer map to its neighbors periodically. Therefore each peer knows which block is available at which neighbor. The task of the pull-based scheme is to specify which block should be requested from which neighbor. As shown in Fig. 6, the receiving peer (e.g., Peer 1) is missing Blocks 4-8 in its buffer. After exchanging the buffer map with its neighbors (e.g., Peers 2-4), Peer 1 decides to request Block 4 from Peer 2, Block 5 from Peer 4, and Block 6 from Peer 3.

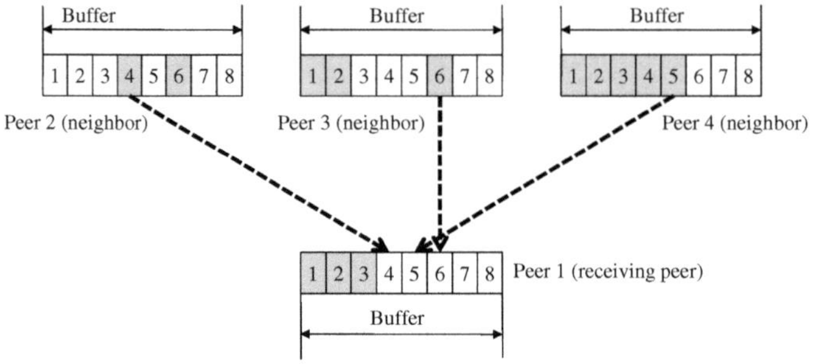

Fig. 6. Illustration of a pull-based scheme in mesh-based P2P live streaming systems

In the example shown in Fig. 6, one may ask a question: why does Peer 1 request Block 4 from Peer 2 instead of Peer 4? To answer the question, we need to consider the upload bandwidth of the neighbor and the available blocks at that neighbor. In the illustrated example, Block 5 is only available at Peer 4. Therefore Peer 1 has to pull Block 5 from Peer 4. Suppose that each neighbor has the identical upload bandwidth. Peer 1 decides to pull Block 4 from Peer 2 instead of Peer 4 since Peer 4 has less remaining upload bandwidth after uploading Block 5. Essentially it is a scheduling optimization problem, which has been investigated in [21]. In [21], each block is assigned a priority, which is a Lagrangian sum of the *rarity factor* and the *emergency factor*. The rarity factor is used to represent the distribution of the blocks in the neighborhood. The rarest block is given the highest rarity factor. The emergency factor is used to represent the time constraints of the blocks. The block closest to the playback deadline is given the highest emergency factor. The scheduling optimization problem is formulated as: to maximize the sum of priorities of all of the currently requested blocks in a mesh-based P2P live streaming system under the bandwidth constraints.

To avoid the delivery of the redundant blocks, the scheduling peer (e.g., the supplying peer in the push-based schemes and the receiving peer in the pull-based schemes) needs to periodically acquire the block availabilities from its neighbors, which introduces more communication overhead and additional delays in block retrieval.

3 P2P VoD Systems

Different from live streaming systems in which each user watches almost the same position of the video, VoD service allows users to watch any point of video at any time. VoD provides more flexibility and interactivity to users, thus attracting more users recently. In a VoD service, each user starts watching the video at a different time, and users may seek to a new position in the video at any time. Therefore different users are watching different positions of the same video at a given instant. Due to the asynchrony of the users, the approaches of overlay construction in P2P VoD systems are different from those in P2P live streaming systems. In P2P live streaming systems, any peer can be a potential supplying peer (e.g., parent in tree-based systems or neighbor in mesh-based systems) of the another peer since they are watching almost the same position. However, a peer in a P2P VoD system has to select a supplying peer from those candidates who have the requested content.

Depending on the forwarding approach, the existing P2P VoD systems can be classified into three categories: buffer-forwarding P2P VoD systems [22, 23], storage-forwarding P2P VoD systems [24, 25], and hybrid-forwarding P2P VoD systems [26]. In buffer-forwarding systems, each peer buffers the recently received content, and forwards it to the child peers. In storage-forwarding systems, the blocks of the video are disseminated over the storage of peers. When a peer wants to watch a video, it first looks for the supplying peers who are storing the content, and then requests the content from them. In hybrid-forwarding systems, a peer may have both buffer-forwarding parents and the storage-forwarding parents, and receive the video content from both kinds of parents.

3.1 Buffer-Forwarding P2P VoD Systems

In buffer-forwarding P2P VoD systems, each peer maintains a buffer to cache the recently received video content around the playback position. The buffered content can be used to serve other peers who have a close viewing point. Fig. 7 shows the buffer at a peer. The start time of the buffer at Peer k is denoted as t_k^S, the end time of the buffer at Peer k is denoted as t_k^E, and the playback time at Peer k is denoted as t_k^P. Suppose that a supplying peer, Peer k, transmits the buffered packets to a receiving peer, Peer j, over an overlay link, Link l. With an initial playback delay T_j^I, Peer j requests the packets with timestamps larger than $(t_j^P + T_j^I)$. Therefore, only the packets within a *transmission sliding window* at Peer k are eligible for transmission to Peer j. In order to be a supplying peer to Peer j, Peer k needs to 1) have a larger playback time than Peer j, and 2) have a buffer overlap with Peer j [27].

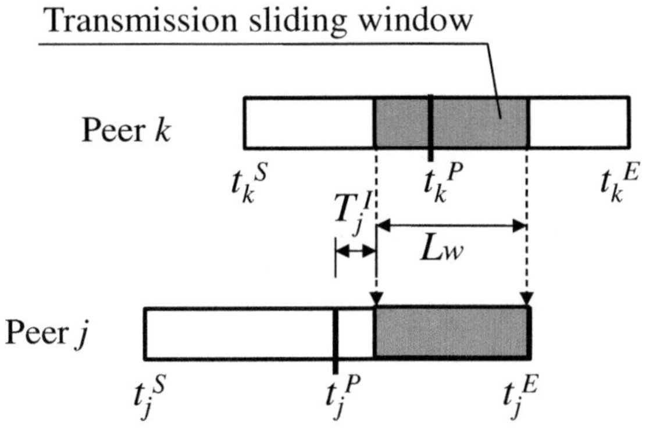

Fig. 7. Buffer overlap between the supplying peer and the receiving peer in buffer-forwarding P2P VoD systems

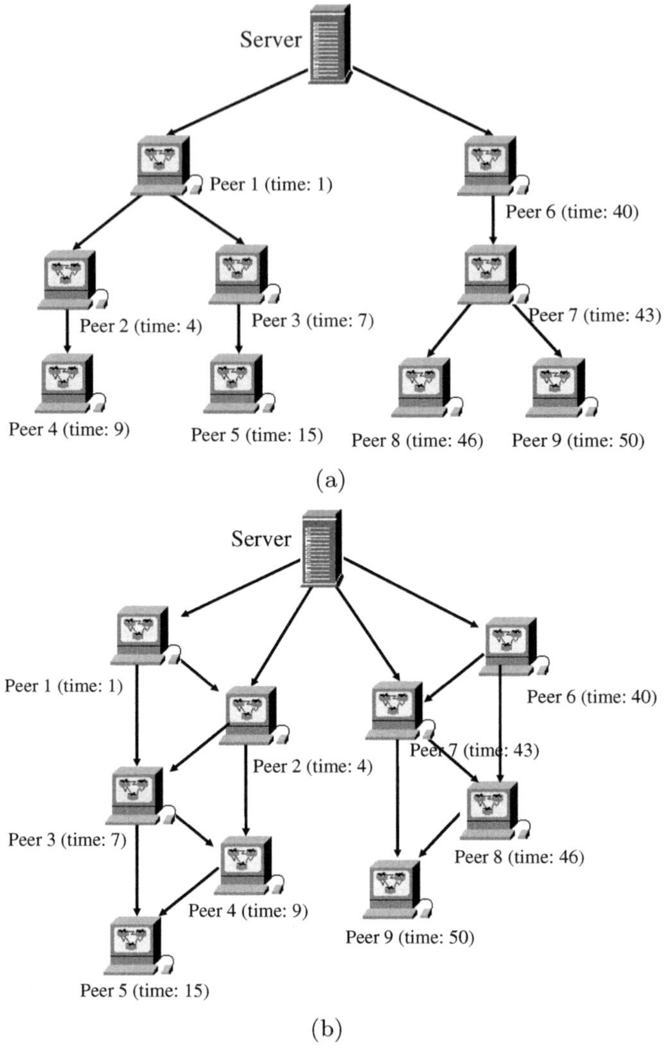

Fig. 8. Overlay constructions in buffer-forwarding P2P VoD systems: (a) tree-based, and (b) mesh-based

Fig. 8 illustrates the overlay constructions in buffer-forwarding P2P VoD systems. Each peer watches the video from the beginning and caches 10 minutes worth of content in its buffer. Fig. 8(a) shows the tree-based overlay construction in which each peer is connected to a single parent (e.g., supplying peer). In Fig. 8(a), Peer 1 arrives at the 1st minute, and it receives the video from the server. Peer 2 and Peer 3 arrive at the 4th and the 7th minute, respectively. Both peers discover that peer 1 is eligible for the parent of them. Therefore they establish overlay links from Peer 1 and receive the streams from it, respectively. Similarly Peer 4 and Peer 5 find its parent in the same way. When Peer 6 joins the system at the 40th minute, it cannot find an eligible supplying peer. Therefore Peer 6 has to ask for the video from the server directly. As shown in Fig. 8(a), a tree-based architecture is formed if each peer is required to be connected to only a single parent.

If each peer is connected to multiple parents, a mesh-based overlay is formed. In the example shown in Fig. 8(b), each peer has two parents except Peer 1 and Peer 6, who have only one parent (e.g., the server). The maximal supported streaming rate in mesh-based architecture is improved compared to the tree-based architecture, since the upload bandwidth of each peer is utilized in a better way. In addition, the mesh-based overlay is more resilient to peer dynamics since each peer can receive the content supplies from multiple sources. However, each peer needs to address content reconciliation in mesh-based overaly, which introduces extra communication overhead.

3.2 Storage-Forwarding P2P VoD Systems

In buffer-forwarding P2P VoD systems, a peer receives the video content from its parents. If the parents jump to other positions in the video, the peer needs to search for new parents again. In storage-forwarding P2P VoD systems, each video is divided into multiple segments, which are distributed among peers. A peer receives the video content from the storage peers who are possessing the requested segment. Each storage peer would not change the video segments that it stores during its lifetime, and hence its activities (i.e., jumping) do not affect its children. VMesh [24] is an example of storage-forwarding P2P VoD systems. In VMesh, a peer randomly stores one or multiple video segments in its local storage. The segment would be served to other peers. If a new peer joins the system, it performs the following procedure [24]:

1. It searches for the segment that it wants to watch using the Distributed Hash Table (DHT).
2. It contacts the returned list of peers.
3. The contacted peers, if they have enough bandwidth, become the peer's parents for supplying the requested segment. Multiple parents are used for fault tolerance purpose.
4. When the peer nearly consumes its current segment, it contacts the other peers who are holding the next segment.
5. If the peer performs a random seek to another position which is not far away from the current position, it can simply follow its forward/backward pointers in the video mesh to contact the new parents. If the new position is too far away, it triggers another DHT search for the destination segment.

3.3 Hybrid-Forwarding P2P VoD Systems

In buffer-forwarding P2P VoD systems, the peers redistribute their buffered content to their child peers, thus offloading the server burden. However, buffer-forwarding systems are less resilient to peer dynamics. If the parents of a peer

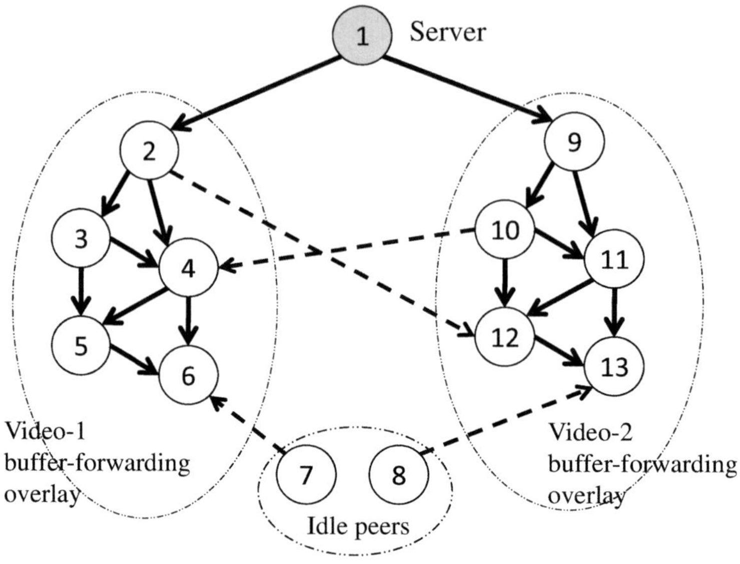

Fig. 9. Illustration of a hybrid-forwarding P2P VoD system

perform random seeks and jump away from the current position, this peer needs to search for the new parents. In storage-forwarding P2P VoD systems, the storage peers do not change the stored segments thus providing a table services. However, the achievable throughput in storage-forwarding systems is low due to the limitation of the total upload capacity.

To overcome these drawbacks, a hybrid-forwarding P2P VoD system is proposed in [26], which improves the throughput by integrating both the buffer-forwarding approach and the storage-forwarding approach. In a hybrid-forwarding system, the peers watching the same video form an overlay using the buffer-forwarding approach. In the example shown in Fig. 9, Peers 1-6 form video-1 buffer-forwarding overlay, and Peers 9-13 and 1 form video-2 buffer-forwarding overlay. Please note that the server (e.g., Peer 1 in the example) participates in all of the video overlays since it contains the contents of all the videos. In the hybrid-forwarding architecture, each peer is encouraged or required to replicate in its storage one or multiple segments of the video that it has watched before. The stored segments can be used to serve other peers. Therefore, the upload bandwidth of the idle peers can be utilized since they can contribute their stored segments to other peers. In the hybrid-forwarding architecture, peers may have both buffer-forwarding links and storage-forwarding links. For example, Peer 10 in Fig. 9 forwards its buffered content to Peer 11 and Peer 12 in video-2 overlay, meanwhile it also forwards its stored content to Peer 4 in video-1 overlay. The two kinds of outgoing links compete for the limited upload bandwidth. By allocating a larger portion of the upload bandwidth to the peers watching a high prioritized video (e.g., paid program) compared to those watching a low prioritized video (e.g., free program), service differentiation among videos can be implemented. Furthermore, the hybrid-forwarding architecture is robust to random seeks because the stored content is a stable source unless the supplier shuts down the P2P application. For example, when the buffer-forwarding parents (e.g., Peers 2 and 3) of Peer 4, both jump to other positions, Peer 4 still has a stream supply from its storage-forwarding parent (e.g., Peer 10), which maintains the continuous playback.

In the overlay construction, the peers watching the same video are first organized into a buffer-forwarding overlay using one of the existing approaches in [22, 23], then each peer locates the storage-forwarding parents either in a centralized way [25] or in a distributed way [24] and connect itself to them.

3.4 Prefetching Schemes

In a P2P VoD system, if a peer has extra bandwidth, it can *prefetch* video content from other peers. *Prefetching* is a behavior that a peer downloads the future segments which are away from the current playback position. Prefetching is an effective way to combat the peer dynamics [28, 23]. In the following, we first describe the prefetching scheme to combat peer churn (e.g., dynamic peer arrivals or departures), and then examine the prefetching scheme to combat peer seeking behaviors.

Prefetching Scheme to Combat Peer Churn

A P2P VoD system is said to be in the *surplus state* if the average upload bandwidth among all the peers is higher than the streaming rate, and in the *deficit state* if the average upload bandwidth is lower than the streaming rate [23]. Due to the dynamic arrivals or departures of the peers, a P2P VoD system may be instantaneously in a surplus or deficit state. When the system enters a deficit state, the server needs to supplement the peers' uploading in order to satisfy the real-time demands. If peers prefetch the future segments when the system is in a surplus state, the server stress can potentially be reduced.

The challenge of the prefeching scheme is how to utilize the upload bandwidth of each peer in a better way. The paper [23] presents a water-leveling prefeching scheme, in which all users' buffer are treated as water tanks, and the buffer with the lowest level is filled first. The water-leveling prefeching scheme can be implemented via the following three steps.

1. The required rate supplied from the server for each peer is determined. The remaining upload bandwidth at each peer is recorded.
2. Let g_i be the *growth rate* of peer i, which represents the extra upload bandwidths assigned to peer i in addition to its real-time demand. Each peer can utilize its remaining upload bandwidth to serve content to the peers with later playback positions. The upload bandwidth assignment is performed from peer $(n-1)$ to peer 1. For peer $(n-1)$, its remaining upload bandwidth is assigned to the later peer (e.g., peer n). For peer $(n-2)$, its remaining upload bandwidth is assigned to the one with the smaller buffer level among the later peers (e.g., peer n and peer $(n-1)$). The process is repeated in a similar way until the remaining upload bandwidth of peer 1 is assigned.

3. If the upload bandwidth is assigned just based on Step 1 and Step 2, a later peer may be allocated with a higher growth rate than an earlier peer such that its buffer point may surpass the earlier peer. Therefore, in order to guarantee the order of the buffer points, we shred off the growth rates of those peers who have already caught up with earlier peers, and re-assign extra bandwidths to later peers.

Prefetching Scheme to Combat Frequent Seeks

In P2P VoD systems, users usually do not play the video successively and passively. Instead, users perform seeks quite frequently [29]. The reasons for these seeking behaviors are: 1) some users may feel that the current segment is boring, so they jump away from it; 2) some users may not have sufficient time to watch the whole video and just want to browse the exciting or interesting segments. The seeking behaviors place a great challenge to the playback continuity.

To deal with the frequent seeks in P2P VoD systems, the concept of *guided seeks* is proposed in [30]. A guided seek is different from a random seek in that it is performed based on the segment access information, which is learned from the seeking statistics in the previous and/or concurrent sessions. Typically, the popular segments are visited more frequently, thus the access count of a segment approximately reflects its popularity. With guidance, users can jump to the desired positions in the video quickly and efficiently. For example, a user watching an on-demand soccer match may want to just watch the segments with goals. However, he or she does not know where the scenes with goals are. In the VoD applications with guidance, the system can provide an indicating bar of the segment popularity in the media player, with a grey scale indicating the degree of the popularity for each segment. This user can just drag the progress bar to the position in dark color since the segment in dark color is popular, thus containing the goal scenes with a higher probability.

In order to provide guidance to users, the *segment access probability* needs to be obtained. The *segment access probability* is a two-dimensional Probability Density Function (PDF), denoted as $P(x, y)$, representing the probability that a user performs a seek from the start segment x to the destination segment y. $P(x, y)$ can be learned from the seeking statistics in the previous and/or concurrent sessions.

In [30], an optimal prefetching framework is proposed for P2P VoD applications with guided seeks. The proposed framework is shown in Fig. 10. It consists of two modules: the seeking statistics aggregation and the optimal prefetching scheme. In the seeking statistics aggregation, the seeking statistics are represented by the *hybrid sketches* [30] to prevent the message size from growing linearly. The peer retrieves a set of random neighbors from the tracker, and then gossips the hybrid sketches with them. Based on the aggregated seeking statistics, the peer estimates the segment access probability $P(x, y)$, which is used for the design of the optimal prefetching scheme, and is also used to guide the user to perform efficient seeks. The module of the optimal prefetching scheme

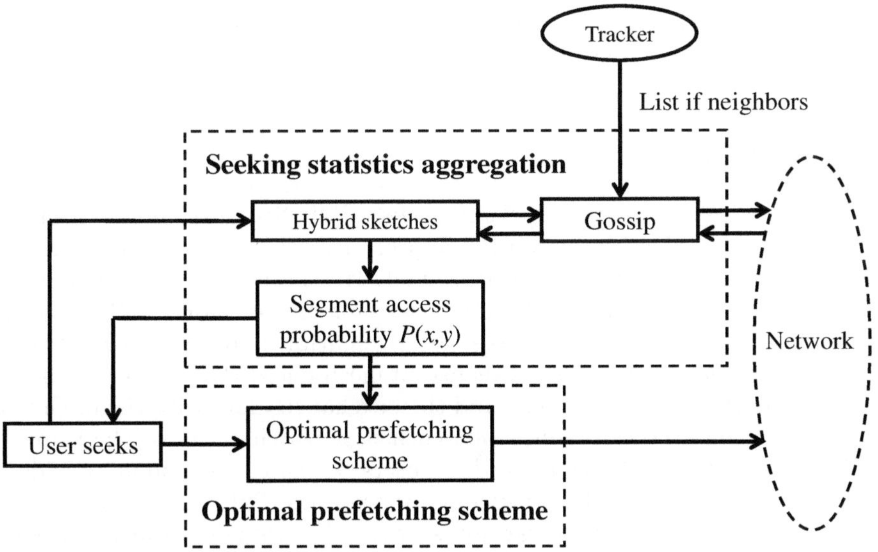

Fig. 10. Optimal prefetching framework in P2P VoD systems with guided seeks

takes the segment access probability $P(x, y)$ as input, and then determines the optimal segments for prefetching and the optimal cache replacement policy.

4 Future Research Directions

Despite a variety of progresses made in P2P streaming systems, there remain a number of important open issues as follows.

1. Users in P2P streaming systems would like to receive the video at a high quality. However, what is the maximal quality that can be achieved in a P2P streaming system is still an open problem. The maximal quality is dependent on the peer bandwidth, the constructed overlay, the source coding scheme and the scheduling algorithm. How to optimize various resources in the P2P streaming system to achieve the maximal quality is quite challenging.
2. A P2P streaming system usually offers multiple channels. A channel is associated with an overlay, formed by the peers watching the channel. Efficiently utilizing the cross-channel resources is expected to improve the system performance. However, how to optimize the resource allocation in a multi-channel P2P streaming system still remains open.
3. The video quality is expected to be improved by jointly optimizing both neighbor selection and data scheduling in a P2P live streaming system. However, this is a challenging task since both neighbor set and data set are time-varying.
4. Peers dynamics place a great challenge in P2P streaming systems. It is a difficult issue for peers to proactively learn the network conditions and then

adaptively take optimal actions to maximize the system utility in a collaborative and distributed manner.

5 Summary

P2P streaming systems can be classified into P2P live streaming systems and P2P VoD systems. Depending on the approaches of overlay constructions, P2P live streaming systems can be categorized into tree-based P2P live streaming systems and mesh-based P2P live streaming systems. In tree-based P2P live streaming systems, a single application-layer tree or multiple application-layer trees are constructed to deliver the video streams. In mesh-based P2P live streaming systems, each peer finds its neighbors and then exchanges data with them. Different from live streaming services in which each user watches almost the same segment of the video, VoD services allow users to watch any point of video at any time. Depending on the forwarding approaches, P2P VoD systems can be categorized into: 1) buffer-forwarding systems, 2) storage-forwarding systems, and 3) hybrid-forwarding systems. To combat peer churn and frequent seeks in P2P VoD systems, prefetching schemes are adopted to increase the playback continuity.

References

1. http://www.youtube.com
2. Gomes, L.: Will all of us get our 15 minutes on a YouTube video? Wall Street Journal (2006)
3. Hefeeda, M., Bhargava, B., Yau, D.: A Hybrid Architecture for Cost -Effective On-Demand Media Streaming. Elsevier Computer Networks 44(3), 353–382 (2004)
4. http://www.akamai.com/
5. Liu, J., Rao, S.G., Li, B., Zhang, H.: Opportunities and Challenges of Peer-to-Peer Internet Video Broadcast. Proceedings of the IEEE 96(1), 11–24 (2008)
6. Zhang, X., Liu, J., Li, B., Yum, T.P.: CoolStreaming/DONet: A Data-Driven Overlay Network for Efficient Live Media Streaming. In: Proc. of IEEE INFOCOM, vol. 3, pp. 2102–2111 (2005)
7. Chi, H., Zhang, Q., Jia, J., Shen, X.: Efficient search and scheduling in P2P-based media-on-demand streaming service. IEEE Journal on Selected Areas in Communications 3, 1467–1472 (2006)
8. Lee, I., Guan, L.: Centralized peer-to-peer streaming with layered video. In: Proc. of IEEE ICME, vol. 1, pp. 513–516 (2003)
9. http://www.napster.com
10. Meeker, M.: The State of the Internet. In: Web 2.0 conference (2006)
11. http://www.pplive.com
12. http://www.uusee.com
13. http://www.sopcast.com
14. Hei, X., Liang, C., Liang, J., Liu, Y., Ross, K.W.: A Measurement Study of a Large-Scale P2P IPTV System. IEEE Transactions on Multimedia 9(8), 1672–1687 (2007)

15. Chu, Y., Rao, S., Seshan, S., Zhang, H.: A Case for End System Multicast. IEEE Journal on Selected Areas in Communications 20(8), 1456–1471 (2002)
16. Tran, D.A., Hua, K.A., Do, T.T.: A peer-to-peer architecture for media streaming. IEEE Journal on Selected Areas in Communications 22(1), 121–133 (2004)
17. Castro, M., Druschel, P., Kermarrec, A.-M., Nandi, A.R.: SplitStream: high-bandwidth multicast in cooperative environments. In: Proc. of ACM SOSP, pp. 298–313 (2003)
18. Liu, Y., Guo, Y., Liang, C.: A survey on peer-to-peer video streaming systems. Springer Journal on Peer-to-Peer Networking and Applications (1), 18–28 (2008)
19. Zhang, M., Luo, J.G., Zhao, L., Yang, S.Q.: A peer-to-peer network for live media streaming using a push-pull approach. In: Proc. of ACM MM, pp. 287–290 (2005)
20. Bonald, T., Massoulie, L., Mathieu, F., Perino, D., Twigg, A.: Epidemic live streaming: optimal performance trade-offs. In: Proc. of ACM SIGMETRICS, pp. 325–336 (2008)
21. Zhang, M., Xiong, Y., Zhang, Q., Sun, L., Yang, S.Q.: Optimizing the Throughput of Data-Driven Peer-to-Peer Streaming. IEEE Transactions on Parallel and Distributed Systems 20(1), 97–110 (2009)
22. Li, Z., Mahanti, A.: A Progressive Flow Auction Approach for Low-Cost On-Demand P2P Media Streaming. In: Proc. of ACM QShine (2006)
23. Huang, C., Li, J., Ross, K.W.: Peer-Assisted VoD: Making Internet Video Distribution Cheap. In: Proc. of IPTPS (2007)
24. Yiu, W.P., Jin, X., Chan, S.H.: VMesh: Distributed segment storage for peer-to-peer interactive video streaming. IEEE Journal on Selected Areas in Communications 25(9), 1717–1731 (2007)
25. Xu, X., Wang, Y., Panwar, S.P., Ross, K.W.: A Peer-to-Peer Video-on-Demand System using Multiple Description Coding and Server Diversity. In: Proc. of IEEE ICIP, vol. 3, pp. 1759–1762 (2004)
26. He, Y., Lee, I., Ling, G.: Distributed throughput maximization in hybrid-forwarding P2P VoD applications. In: Proc. of IEEE ICASSP, pp. 2165–2168 (2008)
27. He, Y., Lee, I., Ling, G.: Distributed throughput maximization in P2P VoD applications. IEEE Transactions on Multimedia 11(3), 509–522 (2009)
28. Shen, Y., Liu, Z., Panwar, S., Ross, K.W., Wang, Y.: On the Design of Prefetching Strategies in a Peer-Driven Video-on-Demand System. In: Proc. of IEEE ICME, pp. 817–820 (2006)
29. Zheng, C., Shen, G., Li, S.: Distributed Prefetching Scheme for Random Seek Support in Peer-to-Peer Streaming Applications. In: Proc. of ACM MM, pp. 29–38 (2005)
30. He, Y., Shen, G., Xiong, Y., Ling, G.: Optimal prefetching scheme in P2P VoD applications with guided seeks. IEEE Transactions on Multimedia 11(1), 138–151 (2009)

Topology Construction and Resource Allocation in P2P Live Streaming

Jacob Chakareski

Ecole Polytechnique Fédérale de Lausanne (EPFL), Lausanne, Switzerland
jakov.cakareski@epfl.ch

Summary. We identify inefficient network topologies and resource allocation mechanisms to be the key obstacles toward adopting present P2P systems as the platform of choice for delivering high-quality live multimedia content. To this end, we propose a comprehensive optimization framework that overcomes these challenges in a systematic manner. In particular, we design a delay-based network construction procedure that creates small-world topologies that provide for efficient data delivery routes and increased data sharing between peers in the network. Furthermore, we equip the nodes with a utility-based packet scheduling technique that maximizes the video quality at a receiving peer subject to available bandwidth resources while allowing for dissemination of less frequently encountered data in the network. Finally, we design an uplink sharing strategy that enables the peers to deal effectively with free-riders in the system. Through theoretical analysis and simulation experiments we demonstrate that the proposed protocols provide for substantial improvements in performance in mesh-pull based P2P live multimedia delivery. Specifically, significant gains are registered over existing solutions in terms of average video quality and decoding rate. The proposed mesh construction procedure provides further gains in performance in terms of reductions in frame-freeze and playback latency relative to the commonly employed approach of random peer neighbour selection. Corresponding gains in video quality for the media presentation are also registered due to the improved continuity of the playback experience.

Keywords: video streaming, multimedia communications, packet scheduling, distributed control, resource allocation, peer-to-peer networks, IPTV.

1 Introduction

Streaming video content over P2P networks is not any longer a distant prospect from the future, but rather a common place frequently encountered on the Internet. Propelled by the steady increase of residential access bandwidth and an audience ever more hungry for an online multimedia experience, a class of applications have emerged that enable sharing data packets with delivery deadlines over a network of peer nodes. Systems like PPLive [1], SopCast [2], PPStream [3],

C.W. Chen et al. (Eds.): Intel. Multimedia Communication: Tech. and Appli., SCI 280, pp. 217–251.
springerlink.com

and Coolstreaming [4], have been successfully deployed and tested for broadcasting/multicasting live events and for streaming pre-encoded content to large audiences in the Internet.

Still, the present P2P multimedia experience is marred by uncontrollable start-up delays, frequent freezes of the multimedia playback, and significant fluctuations of the audio-video quality, as anyone who has used the above applications can witness. Among the reasons for these apparent shortcomings we count the following. First, the design of the existing P2P streaming/multicast applications have been mainly carried over from earlier P2P file/data sharing applications. Hence, as such they are ill equipped to deal with the specificities of multimedia data, such as delivery deadlines and unequal importance for the reconstruction quality. This in turn makes them inefficient in terms of end-to-end performance of the multimedia presentation that they serve. Second, the presence of free-riding in a system also has a negative effect on the overall performance as it counters the main premise of P2P overlay networks, i.e., that the available system bandwidth increases with the number of peers. Specifically, free-riders are peers that want to obtain content from other peers, but that do not want to serve peers with their own content. Hence, effectively this is manifested as a reduction in serving bandwidth to some peers which in turn causes extended delays and variable audio-video quality of the multimedia presentation at these peers. Third and last, the construction of the overlay network over which the data is served and the exchange mechanisms that the peers employ to share their data can also be a contributing factor to poor P2P multimedia performance, if they are not designed and implemented properly.

In this chapter, we present a comprehensive optimization framework that addresses the issues in mesh-pull based P2P streaming described above. Specifically, we design a network construction procedure that creates small-world topologies that feature peer neighbourhoods exhibiting similar delivery delays across its member nodes relative to the broadcast media server. This increases the likelihood of data availability in a neighbourhood thereby reducing the playback latency and the frame freeze frequency of the media presentation at the peers. These improvements also result into a corresponding gain in video quality for the reconstructed presentation. In addition, we design a receiver-driven algorithm for requesting media packets from neighbouring peers that takes into consideration the packets' utility for efficient use of the bandwidth resources. The utility of a packet is based on the packet's delivery deadline and its importance for the reconstruction quality of the media presentation, as well as on its rarity within the neighbourhood. This is done in order to facilitate the delivery of less frequently encountered data units in the network. A peer equipped with our algorithm can thus request the media packets that maximize the performance of its media presentation while contributing simultaneously toward the same goal at neighbouring peers. Therefore, a globally optimized performance of the presentation over the whole peer population is achieved. Finally, we design a bandwidth sharing procedure that targets free-riders. In particular, a sending peers distributes its upload bandwidth among its requesting neighbours in proportion to their own contributions to this peer in terms of data rate. Hence, a free-rider is effectively shut down from receiving any useful data from its neighbours, as its rate contribution to them would typically be non-existing. In addition, the bandwidth sharing procedure is augmented with a periodic search for new neighbours willing to contribute more rate to a peer where the newly selected neighbour replaces the least contributing peer from the original neighbourhood.

Through theoretical analysis we verify the small-world property of the network topologies obtained using the delay-based construction procedure. Similarly, we analytically quantify the advantages of our procedure over the commonly employed approach of constructing P2P network topologies via random peer neighbour selection. Furthermore, through simulation experiments we demonstrate that the proposed framework provides for efficient video broadcast in P2P systems by substantially outperforming existing mesh-pull based solutions over several performance indices. Due to the novel overlay construction procedure, we register significant reductions in frame freeze and playback latency relative to the common random neighbour selection method. In addition, the efficient utility-based scheduling and resource allocation outperform by a wide margin baseline mesh-pull based approaches in terms of average video quality and useful decoding data rate. At the same time, the system remains resilient to free-riders even when we increase their presence/percentage in the network, as enabled by the proportional upload bandwidth sharing strategy.

The rest of the chapter is organized as follows. First, in Section 2, we describe our mesh construction procedure. Then, Section 3 provides a formal analysis of the proposed procedure from two different perspectives. Next, we describe the utility-based packet scheduling technique in Section 4, while the methods for neighbourhood maintenance and uplink bandwidth sharing are presented subsequently in Section 5. Specific performance aspects of the whole system are examined in Section 6 followed by a discussion of related work in Section 7. Finally, we end the chapter with concluding remarks in Section 8.

2 Delay-Based Overlay Construction

2.1 Background

To provide motivation for our approach to overlay construction consider the following example shown in Figure 1. It illustrates the scenario of a new node joining an existing overlay of peers. The new node is represented as a blue (gray) circle in the figure, while three of the peers already present in the overlay are denoted with white circles marked with the numbers one, two, and three in Figure 1. For clarity of presentation, the rest of the nodes and their connections in the overlay are not shown in Figure 1. The relative vertical position of the existing nodes in the overlay represents their respective latencies relative to the origin media server. Specifically, going top to bottom in the figure the delay in receiving media packets originally sent/issued by the media server becomes

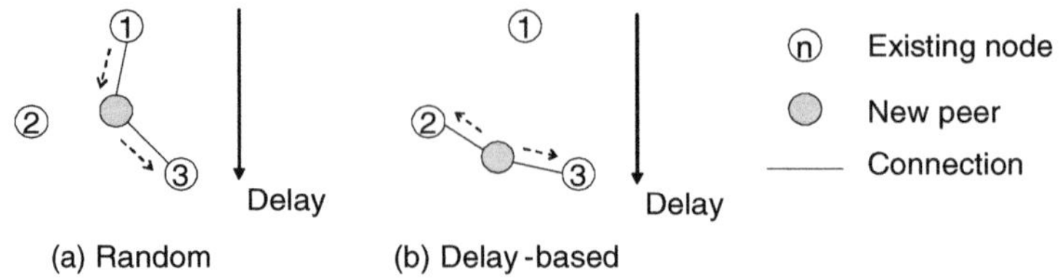

Fig. 1. New node joins: possible neighbourhood configurations with existing peers.

respectively larger at each subsequent node, as represented by the delay axis shown in Figure 1.

Now, assume that in one case the new peer selects Node 1 and Node 3 as its neigbours. Then, the delay configuration between these three nodes may be as the one illustrated in Figure 1(a). Specifically, the new node exhibits longer latency relative to Node 1, which is expected, as it is connected to the media server through this node. For the same reason, the new node may exhibit shorter latency relative to Node 3. Hence, such a configuration will typically create a unidirectional flow of media packets from Node 1 through the new node to Node 3 in the end.

On the other hand, if the new peer selects Node 2 and Node 3 as its neighbour then the prospective latency disposition of the three nodes may be as the one shown in Figure 1(b). Here, Node 2 and Node 3 exhibit similar delays relative to the origin server and the newly joined node exhibits latency that lies somewhere in between. Such a delay configuration may contribute to media packets being exchanged between Node 1 and Node 3 through the new peer in both directions, as illustrated in Figure 1(b). In addition to contributing to a more balanced distribution of data flows in a neighbourhood there in another advantage of the configuration from Figure 1(b) in our opinion. It provides for a more consistent delivery of media packets in the advent of node departures. In particular, in the case examined in Figure 1(a) a sudden departure of Node 1 may disrupt (albeit temporarily) the timely delivery of media packets to the rest of the nodes in the neighbourhood (in this case the new peer and Node 3). On the other hand, if Node 2 decides to leave the newly joined peer can continue to receive new media packets from Node 3 in a timely fashion due to the similar latencies that are involved in the scenario from Figure 1(b).

The configuration from Figure 1(a) can occur if neighbours are selected at random from an existing population of nodes. This is a common method for constructing overlay network graphs nowadays. Though it exhibits some appealing properties, such as strong network connectivity [5], random neighbour selection appears inappropriate for the context of data delivery with deadlines considered herein. Therefore, we propose instead to select peer neighbours based on their delay characteristics relative to the media server, as suggested through the example studied in Figure 1. This brings us to the key idea behind the overlay construction technique proposed in this chapter.

In our approach peers are organized into neighbourhoods that feature similar delays from the media server for all their members. In other words, we aim to construct a neighbourhood such that every peer in it will exhibit a roughly equal latency in receiving the media packets sent by the server originally. This will increase the likelihood of having the peers in that neighbourhood being interested in obtaining content from one another. That is because the sliding windows used to exchange content with other peers (see Section 4.2) will be positioned more or less at the same location relative to the time line of the media presentation. At the same time, this feature will contribute to a lower likelihood of a frame freeze during playback. This is the situation where a video frame is not received/decoded by a peer by the time it is due to be displayed. Hence, the peer freezes the last displayed frame on the screen in order to conceal the loss of this (present) frame. As the peers have bigger prospects for querying data from their neighbours now, the chance of encountering frame freeze during playback decreases. Finally, another advantage of such a neighbourhood is shorter playback start of the media presentation at a peer. In particular, as data sharing among the members of a neighbourhood is increased, the time that it takes for a peer to acquire the necessary amount of data in order for playback to start is reduced. In the next section, we describe formally the goal that the proposed overlay construction technique attempts to achieve.

2.2 Formal Description

Let $P = \{p_1, \ldots, p_N\}$ be a set of nodes through which a joining peer denoted p can connect to an existing overlay. Let $\Delta = \{\delta_1, \ldots, \delta_N\}$ be the corresponding latencies of these nodes relative to the origin server, as described earlier. Similarly, let $\Delta_p = \{\delta_{p \to p_1}, \ldots, \delta_{p \to p_N}\}$ be the corresponding network delays between the new peer and the nodes in P. Finally, let $\Delta_{p \to S} = \{\delta_{p \to p_1} + \delta_1, \ldots, \delta_{p \to p_N} + \delta_N\}$ denote the relative latencies of the new peer. This setting is illustrated in Figure 2 where for clarity the existing peers in the overlay have been denoted solely with their index i.

The new peer is interested in finding the subset of peers $Z \in P$ of size $K < N$ such that the distribution of latencies seen by the peer through these nodes relative to the origin server exhibits the smallest variance. In particular, let $Z = \{p_{l(1)}, \ldots, p_{l(K)}\}$ be a subset of peers from P of size K, where for the mapping $l(j)$ it holds $l(j) \in \{1, \ldots, N\}$ for $j = 1, \ldots, K$. Then, let $\delta_{\min}^Z = \min \Delta_{p \to S}^Z$ and $\delta_{\max}^Z = \max \Delta_{p \to S}^Z$ denote respectively the minimum and maximum latencies that p sees to the server through the peers in Z. Finally, the spread of latencies characterizing the subset Z can be expressed as $|\delta|^Z = \delta_{\max}^Z - \delta_{\min}^Z$.

Hence, the new peer would like to find the subset Z that exhibits the smallest latency spread relative to the origin server, i.e.,

$$Z^* = \arg \min_{Z \in P} |\delta|^Z. \tag{1}$$

It can be shown that (1) can be efficiently solved by performing the following three steps. First, we sort the delays in $\Delta_{p \to S}$ in increasing order. Let $\Delta_{p \to S}^{\text{sort}}$

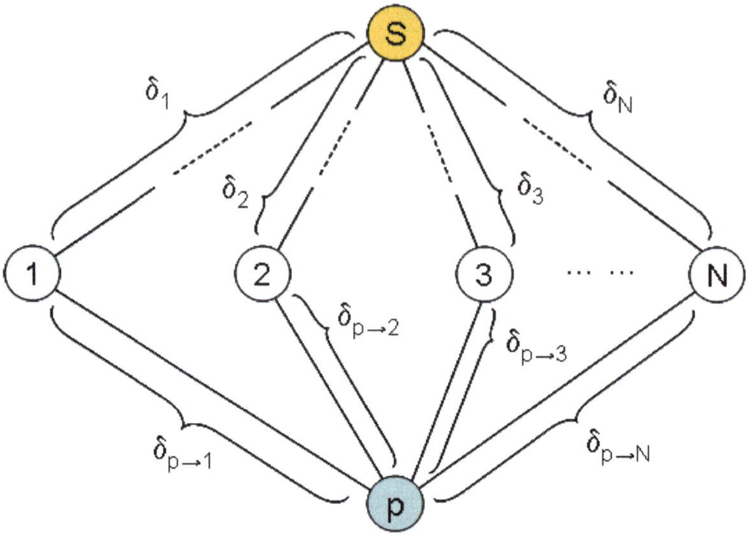

Fig. 2. Peer p experiences different latencies via different existing nodes.

denote the sorted array of delays and let P^{sort} denote the corresponding set of 'sorted' nodes. That is, the index of node p_j, for $j = 1, \ldots, N$, in P^{sort} is equal to the location/position of $(\delta_{p\to p_j} + \delta_j)$ in $\Delta^{\text{sort}}_{p\to S}$. Next, for $m = 1, \ldots, N-K+1$ we create the sets $Z_m \in P^{\text{sort}}$ where the mapping function $l(j) = j + m - 1$, for $j = 1, \ldots, K$. For each set Z_m we compute the corresponding delay spread $|\delta|^{Z_m}$. Finally, we find the index m for which the latency spread is the smallest over all sets Z_m. Let this index be denoted m^*. The corresponding set Z_{m^*} represents the solution to Equation (1). In Figure 3, we summarize the algorithmic steps of the procedure described above for computing the solution in (1). In Section 3, we formally analyze our algorithm from two different perspectives in order to establish its distinguishing properties. Next, we describe how we incorporated this algorithm within an actual overlay construction procedure.

2.3 Implementation Aspects

There is a registry server that keeps track of the peers already in the network. For each peer the server maintains an entry comprising the peer's IP address and the minimum delay that this peer measures with respect to the media server. The nature of this second quantity will become clear as we proceed with the discussion here. The registry server maintains a sorted list of the registered peers in increasing order based on their minimum delays.

A connect procedure for a peer joining the network comprises the following steps. The peer contacts the registry server providing it its IP address. In return, the server provides the sorted list of peers that the connecting peer can use to establish its own neighbourhood in the network. Specifically, the peer begins to

Given p, P, $\Delta_{p\to S}$

(0) Sort $\Delta_{p\to S}$ in increasing order.
$\Rightarrow \overset{\text{sort}}{\Delta_{p\to S}}, \overset{\text{sort}}{P}$.

(1) For $m = 1, \ldots, N - K + 1$ do
For $j = 1, \ldots, K$ and $l(j) = j + m - 1$ do
Construct set $Z_m \in \overset{\text{sort}}{P}$.
Compute latency spread $|\delta|^{Z_m}$.

(2) Find $m^* = \min_m |\delta|^{Z_m}$

(3) Return: Neighbourhood of p is Z_{m^*}.

Fig. 3. Finding the neighbourhood with the smallest latency variation.

contact the nodes in the network starting from the head of the list, checking for the following two quantities. The peer wants to know if the contacted node can accept a new neighbour[1]. At the same time, the peer measures the network delay from the node, based on the data rate that the node is willing to spend on sending packets to the peer. Once the peer has a sufficient number of nodes willing to accept it as their neighbour, the peer creates its own neighbourhood with these nodes. At the same time, the peer provides to the registry server its own minimum delay entry which the peer computes as follows. The peers adds the delay it has measured from a neighbour to the minimum delay value for that neighbour from the sorted list provided by the registry server. The smallest of these values for the neighbourhood of the peer is returned to the registry server[2]. This concludes the connect procedure for a peer.

Note here that for the peers that first join the network, the delays that they measure and store with the registry server actually represent the latency at which they receive data directly from the media server. Subsequent peers however cannot directly access the media server any longer, otherwise the server would be overwhelmed with requesting connections. These peers register their network delay from the media server indirectly, through nodes to which they connect to the network when establishing their own neighbourhoods.

In Figure 4, we show an illustration of a mesh constructed using our procedure. Going radially from the center, where the media server is located, we can see that the delay that the nodes experience in reference to the server increases. Still, nodes within a neighbourhood, denoted as dashed clouds in Figure 4, exhibit similar latencies in receiving media packets issued by the server earlier, which in turn increases their collaboration in delivering these packets to one another,

[1] The nodes usually maintain a bound on the number of neighbours that they are willing to accept in order not to overwhelm their resources by the demanding/requesting neighbours.

[2] We assume that all the latency between two nodes in the network occurs at their access points. Hence, a joining node always measures a smaller overall delay to the media server through a node higher in the sorted list.

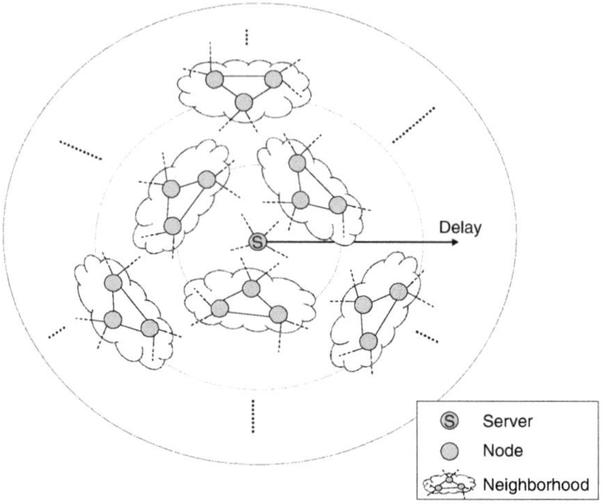

Fig. 4. An example of a minimum delay mesh construction.

as argued previously. Please note that for clarity we do not include in Figure 4 every connection between the nodes in the network.

Periodically, the nodes can check if their minimum delays have changed. This can happen as the data rates at which they receive data from their neighbours can change over time. Hence, a node can update then its minimum delay value stored with the registry server. At the same time, the node informs its neighbours that its own minimum delay has changed so that they can revaluate in turn and if needed the minimum delay values for their respective neighbourhoods. Note that the case of departure of one or multiple neighbours is covered with the above consideration as then the data rate(s) of the departed neighbour(s) to a node would also change (become zero). Lastly, a departing node can inform the registry server of its decision so that its registry entry can be removed.

3 Analysis

3.1 Small World Property

In this section, we study the proposed mesh construction procedure from a graph theoretic perspective. In particular, we show that overlay networks constructed using our procedure exhibit small world properties [6]. This means that such networks feature clusters of highly interconnected nodes and some connections between nodes of different clusters. The benefit of such a node configuration in the network is that once a packet reaches a cluster it can be readily obtained from the cluster members. Furthermore, the smaller number of connections between the dense clusters ensures short overall hop counts between different nodes, which in part can control the end-to-end latency. Small world networks are a class of graphs that exhibit density of edges that is much higher than that of random graphs. In the following, we show that this is also the case for overlays created using the delay-based technique from the present chapter.

Let v be a vertex in a graph and let $V = \{v_1, \ldots, v_n\}$ denote the set of neighbouring vertices to v in the graph. These are the vertices in the graph that have direct connections (edges) to v. Let E denote the set of vertices in the graph between the nodes in V. Then, the coefficient of clustering for vertex v is defined as the ratio of the number of edges between the nodes in V and the number of possible edges between them, i.e.,

$$\mu_c = \frac{|E|}{\binom{n}{2}}. \tag{2}$$

In other words, μ_c describes the density of edges in the neighbourhood of v. The average value of μ_c over all nodes in a graph represents the clustering coefficient for that graph. This quantity is typically much larger for small world networks than for random graphs of the same size, i.e., same number of vertices. In order to provide a lower bound on μ_c in our case, we approximate the total population of peers in the overlay as belonging to two classes. The first one comprises nodes that are mostly embedded in local dense clusters. And the second class features nodes that mostly serve as conduits, i.e., interconnects between different clusters. We compute the lower bound for μ_c as the statistical average of the clustering coefficients for each of the two classes. Finally, we show experimentally that the actual clustering coefficient established in simulations follows the computed bound closely.

Now, let N_1 and N_2 denote respectively the minimum and the maximum sizes of a neighbourhood in the overlay. These correspond respectively to the initial number of neighbours that a node can be associated with in the network and to the maximum number of neighbours that the node can possibly accept during his association with the overlay. In steady state, we model each peer from the first class described above as featuring dense local clustering of edges among N_1 of its neighbours, while the remaining $N_2 - N_1$ neighbours correspond to connections to other dense clusters through the remaining $N_2 - N_1$ neighbours that also feature a densely clustered set of nodes, though rather smaller than the former one. Therefore, the clustering coefficient $\mu_c^{(1)}$ for a node from the first class should be at least as large as

$$\mu_c^{(1)} \geq \frac{\binom{N_1}{2} + \binom{N_2 - N_1}{2}}{\binom{N_2}{2}}. \tag{3}$$

Furthermore, for nodes in the second class it holds that they can be connected to as many as $N_2/(N_2 - N_1)$ different small dense cluster conduits featuring $N_2 - N_1$ neighbours each. Hence, a lower bound on the clustering coefficient $\mu^{(2)}$ for nodes in the second class can be computed in this case as

$$\mu_c^{(2)} \geq \frac{\frac{N_2}{N2 - N_1} \cdot \binom{N_2 - N_1}{2}}{\binom{N_2}{2}}. \tag{4}$$

Finally, in order to be able to compute the average clustering coefficient for the entire network we need to know the number of nodes in each class. We explain how we determine these in the following. Let N be the total number of peers in the overlay. There are roughly $\lfloor N/N_1 \rfloor$ dense clusters of nodes of the first class comprising N_1 such nodes each. Hence, the overall number of nodes of the first class in the overlay is $N_1 \cdot \lfloor N/N_1 \rfloor$, while consequently the fraction of these nodes in the overlay can be computed as

$$p_1 = \frac{N_1 \cdot \lfloor N/N_1 \rfloor}{N}. \tag{5}$$

Given our model, the rest of the nodes in the network $N - N_1 \cdot \lfloor N/N_1 \rfloor$ belong to the second class and hence their fraction p_2 relative to the overall number of vertices in the overlay is computed as

$$p_2 = \frac{N - N_1 \cdot \lfloor N/N_1 \rfloor}{N}. \tag{6}$$

Therefore, a lower bound on the overall clustering coefficient for the entire network can be computed as the statistical average of lower bounds for the clustering coefficients for the nodes in the two classes, given in (3) and (4), using the probabilities (5) and (6). That is,

$$\mu_c = p_1 \cdot \mu_c^{(1)} + p_2 \cdot \mu_c^{(2)}. \tag{7}$$

In Figure 5 below, we examine the bound for μ_c from (7) computed for different N and N_1 values together with the actual values for the clustering coefficient determined in experiments for the same network and neighbourhood sizes. In particular, in Figure 5a we show the bound and the actual value for μ_c as a function of the network size N for $N_1 = 8$ and $N_2 = N_1 + 6$ fixed. It can be seen the actual value of μ_c follows closely the computed bound which justifies the modeling employed in our analysis for computing the bound in (7). Similar behaviour can be observed from Figure 5b where we compute the bound and the actual value for μ_c as a function of N_1, while keeping the network size $N = 10^3$ fixed. Here, we employed the maximum size N_2 of a neighbourhood to be six nodes larger than the initial size N_1 as in the setting for Figure 5a. It can be seen from Figure 5b that again the actual value of μ_c follows closely the computed bound which gives justification to the approximation that lead to the model for computing the bound for μ_c, as already mentioned above.

We conclude the section by comparing the values of μ_c computed for an overlay constructed using the proposed approach and for a random graph of the same vertex size. Specifically, we set the number of peers in the network N to 10^3 and the maximum neighbourhoood size N_2 for a node to be 14 peers. As in the experiments related to Figure 5, we choose the initial neighbourhood size N_1 to be $N_2 - 6$. The corresponding clustering coefficient for this parameter configuration was computed to be 0.472. The corresponding value for μ_c in the case of the random graph is only 0.0122. The significantly larger clustering coefficient in the former case confirms that the proposed delay-based technique creates overlays that indeed belong to the class of small-world networks.

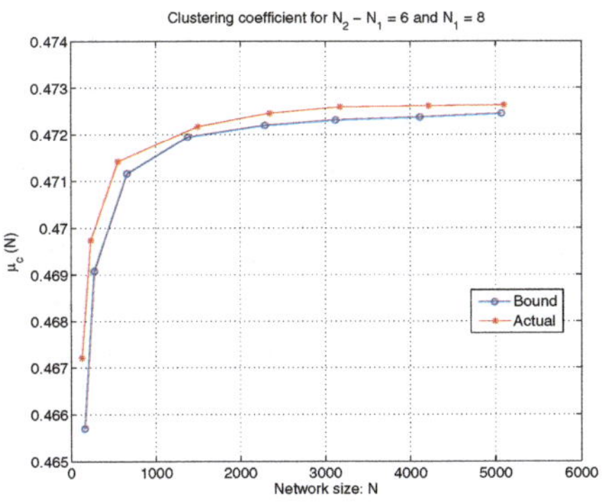

(a) $\mu_c(N)$

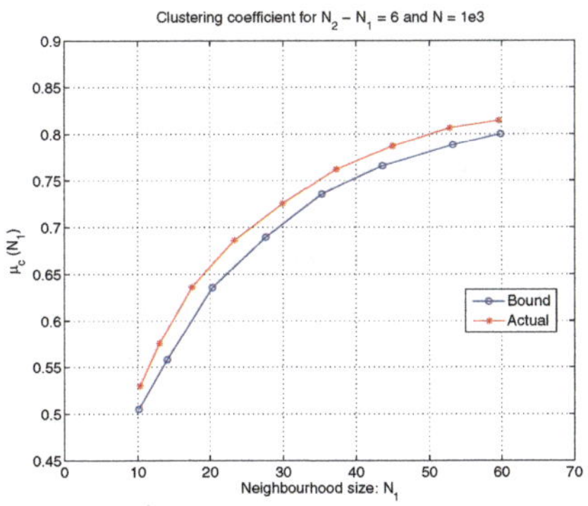

(b) $\mu_c(N_1)$

Fig. 5. Clustering coefficient μ_c as a function of (a) Network size N, for $N_1 = 8$, and $N_2 = 14$, and (b) Neighbourhood size N_1, for $N_2 = N_1 + 6$ and $N = 10^3$.

3.2 Increased Likelihood of Data Sharing

Here, we provide a statistical characterization of the proposed algorithm. Specifically, we analyze the latency distribution that a joining client experiences through its selected neighbours. We study the statistical properties of this distribution, such as mean and variance, and we compare it to the corresponding distribution for the case when neighbours are selected at random. In summary, we show that the former distribution has a tighter/narrower support and features smaller mean and variance values related to the random selection case. These properties are expected given the optimization goal that the proposed techniques aims to achieve, as described in Section 2.2. Moreover, they contribute to increased likelihood of data sharing among neighbours, as also argued in Section 2.2. This in turn improves the overall performance of the system, as demonstrated in Section 6.5 where experimental results are presented.

Now, let $P = \{p_1, \ldots, p_N\}$ denote the set of all peers in the overlay that a joining node can prospectively select as neighbours. Furthermore, let $X = \{x_1, \ldots, x_N\}$ denote the corresponding set of latency values that the joining node experiences relative to the origin server. As explained earlier, these are delays that the new node observes in receiving media packets originally issued by the server at some point in the past. It is a reasonable assumption to consider that the distribution of the delays in X is bounded, i.e., it exhibits lower (x_{min}) and upper (x_{max}) limits on the values it can produce. Specifically, the lower bound can be related to the data rate at which packets are sent to the direct neighbours of the server. The upper bound in turn can be related to the maximum distance or depth (in terms of number of hops) between the origin server and a node on the "periphery" of the overlay.

We consider X to represent a sample set of realizations of N random variables independently and identically distributed over the range described by the two bounds, i.e., $[x_{min}, x_{max}]$. The proposed algorithm sorts X in increasing order which corresponds to creating an order statistics of the original N random variables $X_1, X_2, \ldots, X_N$ denoted as $X_{(1)}, X_{(2)}, \ldots, X_{(N)}$ [7]. Let K represent the size of the neighbourhood in number of peers. As explained in Section 2.2, the algorithm then creates the sets Z_m of size K by selecting in a sliding window fashion a subset of peers from P that correspond to a range of samples from the sorted array of delays X^{sort}. For each of these sets Z_m, for $m = 1, \ldots, N-K+1$, the algorithm computes the latency spread $|\delta|^{Z_m}$ as described in Section 2.2. Given the formalism of order statistics introduced above, this procedure corresponds in fact to computing the realization of the random variable $W_{rs}^{(m)}$ defined as

$$W_{rs}^{(m)} = X_{(s)} - X_{(r)}, \tag{8}$$

where the realizations $x_{(r)}$ and $x_{(s)}$ of the corresponding random variables in (8) denote the respective latency values for the first and last entries (member peers) in the set Z_m, as selected from the sorted array of delays X^{sort}. Note that given the size of the set Z_m, the difference between the indices s and r is always equal to $K-1$.

Next, let $f(x)$ and $F(x)$ denote respectively the probability density function (pdf) and the cumulative distribution function (cdf) for the random variables X_i, $i = 1, \ldots, N$, with respective realizations in X, as explained earlier. Then, it can be shown that the r-th order statistics $X_{(r)}$ is characterized with the following pdf:

$$\begin{aligned} f_r(x) &= \frac{N!}{(r-1)!(N-r)!} F^{r-1}(x)\,(1-F(x))^{N-r} f(x) \\ &= \frac{1}{B(r, N-r+1)} F^{r-1}\,(1-F(x))^{N-r} f(x)\,. \end{aligned} \tag{9}$$

Furthermore, given (9) we can derive the pdf for the random variable introduced in (8). Specifically, first it can be shown that the joint pdf $f_{rs}(x, y)$ of two random variables $X_{(s)}, X_{(r)}$, for $1 \le r < s \le N$ and $x \le y$, from the order statistics can be written as [7]

$$f_{rs}(x, y) = C_{rs}\, F^{r-1}(x)\, f(x)\, (F(y) - F(x))^{s-r-1} f(y)\, (1-F(y))^{N-s}\,, \tag{10}$$

where the constant term C_{rs} is given as

$$C_{rs} = \frac{N!}{(r-1)!(s-r-1)!(N-s)!}\,.$$

Then, using (10) and the transformation of variables $(x, y) \to (x, w_{rs})$[3], where $w_{rs} = y - x$, we can obtain the pdf of $W_{rs}^{(m)}$ by integrating out over x the pdf $f_{rs}(x, w_{rs})$, i.e.,

$$f(w_{rs}) = C_{rs} \int_{-\infty}^{\infty} f_{rs}(x, w_{rs})\, dx. \tag{11}$$

[3] The Jacobian of this transformation has a unity modulus.

Finally, using the previous assumption about $f(x)$ being uniform in the interval $[x_{min}, x_{max}]$ we can obtain more concrete forms for the expressions in (10) and (11). In particular, $f_{rs}(x, y)$ becomes

$$f_{rs}(x,y) = \begin{cases} C_{rs}\, x^{r-1}(y-x)^{s-r-1}(1-y)^{N-s} & : \; x_{min} \leq x \leq y \leq x_{max}, \\ 0 & : \; \text{otherwise.} \end{cases} \tag{12}$$

Similarly, with some work and by noting that $f(y) = f(x + w_{rs}) = 0$ for $x \geq x_{max} - w_{rs}$, $f(w_{rs})$ can be succinctly written as

$$f(w_{rs}) = C_{rs} \int_{x_{min}}^{x_{max}-w_{rs}} x^{r-1} w_{rs}^{s-r-1} (1 - x - w_{rs})^{N-s}\, dx. \tag{13}$$

Now, the integral in (13) can be solved by employing the transformation of variables $x = y(1 - w_{rs})$ in which case one obtains the following expression

$$f(w_{rs}) = I(w_{rs}) w_{rs}^{s-r-1} (1 - w_{rs})^{N-s+r}, \tag{14}$$

for $0 \leq w_{rs} \leq x_{max}$ and where the term $I(w_{rs})$ is given as

$$I(w_{rs}) = C_{rs} \int_{\frac{x_{min}}{1-w_{rs}}}^{\frac{x_{max}-w_{rs}}{1-w_{rs}}} y^{r-1}(1-y)^{N-s} dy\,. \tag{15}$$

The result in (14) shows that the distribution of $W_{rs}^{(m)}$ depends in this case only on the difference $s - r$ and not on s and r individually.

The proposed overlay construction technique computes the realizations of the variables $W_{rs}^{(m)}$, for every set Z_m and $m = 1, \ldots, N-K+1$, as explained earlier. Therefore, it produces an array of values corresponding to these realizations, i.e., $(w_{rs}^{(1)}, \ldots, w_{rs}^{(N-K+1)})$. The algorithm then selects the smallest of these values, i.e., its corresponding index m^*, in order to determine the neighbourhood of peers Z_{m^*} to which the new node will join. This procedure corresponds in fact to finding the order statistics for the variables $W_{rs}^{(m)}$, i.e., $W_{rs(1)}, \ldots, W_{rs(N-K+1)}$ and selecting the neighbourhood Z_m that corresponds to the first order statistics $W_{rs(1)}$. Note that the variables $W_{rs}^{(m)}$ are identically distributed with a pdf given in (14). Therefore, the pdf of $W_{rs(1)}$ can be computed from (9) by using $r = 1$ and $F(w_{rs}) = \int f(w_{rs})$.

In the following, for ease of presentation we normalize the interval $[x_{min}, x_{max}]$ and respectively the samples in X to the unit range $[0, 1]$. Consequently, this restricts the support of the functions $f(x)$ and $F(x)$ to the latter range, which in turn allows for simplification of some of the expressions developed thus far. In particular, the integral $I(w_{rs})$ in (15) simplifies to

$$\begin{aligned} I(w_{rs}) &= C_{rs} \int_0^1 y^{r-1}(1-y)^{N-s} dy \\ &= C_{rs}\, B(r, N-s+1) \\ &= \frac{1}{B(s-r, N-s+r+1)}, \end{aligned} \tag{16}$$

where the constant $B(a,b)$ replacing the integral in the first line of (16) is also known as the Beta function [8], i.e.,

$$B(a,b) = \int_0^1 x^{a-1}(1-x)^{b-1}dx$$

and for which it holds

$$\binom{n}{k} = \frac{1}{(n+1)B(n-k+1,k+1)},$$

a property that we used earlier in Equation (9). Consequently, $f(w_{rs})$ in (14) becomes the density of a beta $\beta(s-r, N-s+1+1)$ variate. Finally, given the above the pdf of the first order statistics $W_{rs(1)}$ obtains the following form:

$$f_{W_{rs(1)}}(w) = \frac{1}{B(1,N)}\left(1 - I_w(s-r, N-s+r+1)\right)^{N-1} f(w_{rs}), \qquad (17)$$

where $I_x(a,b)$ in (17) denotes the incomplete beta function [8] that is sometimes also known as the regularized (due to the denominator term $B(a,b)$) incomplete beta function. Its expression is provided below

$$I_x(a,b) = \frac{1}{B(a,b)}\int_0^x t^{a-1}(1-t)^{b-1}dt\,.$$

Almost surprisingly, the above integral can be worked out, in the case of integer a, b, for example by using integration by parts. This is exactly what we have in (17). Therefore, for completeness we include the solution here

$$I_x(a,b) = \sum_{j=a}^{a+b-1} \frac{(a+b-1)!}{j!\,(a+b-1-j)!}\, x^j\,(1-x)^{a+b-1-j}.$$

Now, given the statistical characterizations of $W_{rs(1)}$ and $W_{rs}^{(m)}$ derived thus far, one can compute their properties such as mean and standard deviation and compare them against the corresponding quantities when peer neighbours are selected at random. Specifically, in this latter case K peers are selected at random from P, the set of available peers in the overlay introduced earlier. Therefore, given the prior exposition this procedure can be modeled as choosing randomly two indices r and s such that $1 \leq r < s < N$ and $s - r \geq K - 1$ that are used afterward to compute the realization of the random variable W_{rs} from the array of sorted latencies X^{sort}. In other words, out of the K peers selected at random, there are two that respectively exhibit the largest and the smallest delay in this subset of nodes. The corresponding indices of these latencies in X^{sort} are s and r, respectively.

The random nature of the difference $s - r$ in this case[4] necessitates employing the pdf of W_{rs} from (14) only as conditioned on the actual realizations of the

[4] That is every time a neighbourhood is selected a different value for $s-r$ is generated. This is in contrast to the approach proposed herein where $s - r = K - 1$ always.

random variables[5] r and s, i.e., $f(w_{rs}|r,s)$. Then, the joint distribution of r, s, and W_{rs} can be obtained from the law of conditional probability as

$$f(w_{rs},r,s) = f(w_{rs}|r,s)f(r,s)\,,$$

where $f(r,s)$ is in fact the joint probability mass function of r and s (as they are discrete random variables) that can be written as follows

$$f(r,s) = \frac{1}{(N-K+1)(N-K-r+2)}\,, \tag{18}$$

for $r \in \{1,\ldots,N-K+1\}$ and $s \in \{r+K-1,\ldots,N\}$. Finally, $f(w_{rs})$ can be obtained from $f(w_{rs},r,s)$ and (18) by integrating out r and s, i.e., $f(w_{rs}) = \iint f(w_{rs},r,s)\,dr\,ds$.

Let μ_u and σ_u represent respectively the mean and standard deviation of the uniformly distributed latency samples in X. Furthermore, let μ and σ represent the corresponding quantities for the distribution of the latency spread in a neigbourhood created using one of the three methods that we compare here. In particular, the first method denoted *Random* corresponds to random peer selection, as explained earlier. The second approach is denoted by the random variables that represents it, i.e., $W_{rs(1)}$, and it corresponds to the solution of (1). Finally, the last technique under comparison is denoted $W_{rs}^{(m)}$ and corresponds in fact to selecting any $m \in \{1,\ldots,N-K+1\}$ in the formalism of the algorithm presented in Section 2.2. One can think of this technique as an intermediate step/solution of the full optimization approach corresponding to $W_{rs(1)}$. As shown before the variables $W_{rs}^{(m)}$ in (8), for $m = 1,\ldots,N-K+1$ will be identically distributed. Hence, the latency spread of a neighbourhood in this case will exhibit the same statistics irrespective of the choice of specific m.

In Table 1a, we express μ and σ in the case of each method as percentages of the corresponding quantities for the underlying delay distribution in X. It can be seen from Table 1a that selecting neighbours at random contributes to a 52% higher mean value for the latency spread of the neighbourhood relative

Table 1. Mean (μ) and standard deviation (σ) for different neighbour selection strategies in percent of (μ,σ) for (a) a Uniform[0,1] r.v., (b) the Random method. (Neighbourhood size $K = 8$).

	μ (%)	σ (%)
Random	152.64	45.50
$W_{rs}^{(m)}$	13.61	8.77
$W_{rs(1)}$	4.90	2.14

(a)

	μ (%)	σ (%)
$W_{rs}^{(m)}$	8.91	19.28
$W_{rs(1)}$	3.21	4.70

(b)

[5] We allow to abuse notation here by employing the same symbols for the random variables and their corresponding realizations. We do that for ease of notation.

to μ_u. On the other hand, the standard deviation of the latency spread is approximately 55% smaller than σ_u. Furthermore, it is encouraging that the other two approaches create neighbourhoods characterized with latency spread statistics that are much smaller than those for the origin delay distribution. Specifically, $W_{rs}^{(m)}$ exhibits mean and standard deviation that represent fractions of 14% and 9% only, relative to μ_u and σ_u, respectively. Further improvement, as expected, is provided by $W_{rs(1)}$ that is characterized with μ and σ representing very small portions of 5% and 2%, respectively, of the corresponding statistics for the original distribution, as seen from Table 1a.

Next, in Table 1b we examine μ and σ for $W_{rs}^{(m)}$ and $W_{rs(1)}$ again expressed in percent but this time relative to those for random node selection (denoted Random in Table 1a). It can be seen that both methods significantly improve performance in terms of latency deviation within a neighbourhood relative to Random. In particular, (μ, σ) for $W_{rs}^{(m)}$ are only 9% and 19% fractions, respectively, of the corresponding quantities for Random. In the case of $W_{rs(1)}$ an even more significant reduction of the magnitude of the statistics of the latency spread is registered, as seen from Table 1b. The sizes of (μ, σ) for $W_{rs(1)}$ represent 3% and 5% fractions of (μ, σ) for Random, respectively.

Finally, in Figure 6 we illustrate the cumulative distribution functions of the latency spread in the case of Random, $W_{rs}^{(m)}$, and $W_{rs(1)}$. As corroborated by the previous results from Table 1, we can see from Figure 6 that the support of the cdf and the mean and standard deviation values for $W_{rs(1)}$ are the smallest, while those for Random are respectively the largest. For example, the support of the cdf for Random spreads over the whole range $[0, 1]$, while the supports of the cumulative distribution functions for $W_{rs(1)}$ and $W_{rs}^{(m)}$ cover respectively only the lower 5% and 20% of the possible range $[0, 1]$. As a last remark in this section, note from Figure 6 that though $W_{rs}^{(m)}$ performs relatively worse relative to $W_{rs(1)}$ it still provides a substantial improvement in latency divergence

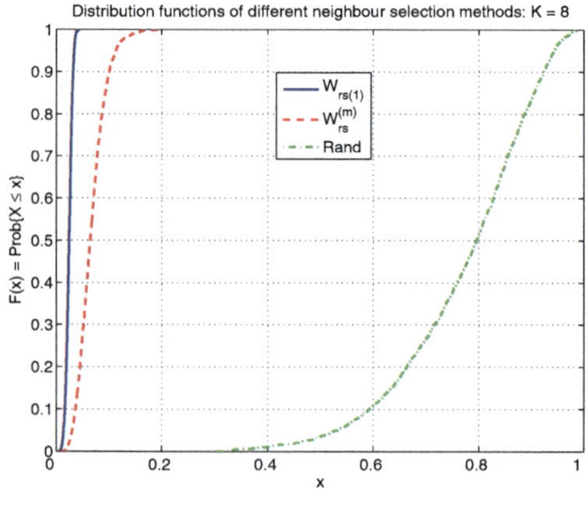

Fig. 6. CDF of different neighbour selection strategies for $K = 8$.

performance relative to Random. Moreover, given its relative simplicity[6], it represent an appealing complexity versus performance compromise that still allows for taking advantage of the overlay construction approach proposed here.

4 Utility-Based Scheduling

In our system, each peers requests video data from a subset of other peers that form its neighbourhood. Since video packets have typically very different importance for the reconstruction quality of the media presentation [9], an effective usage of the bandwidth resources can be achieved by transmitting the most important packets first. Here, we propose a utility-based scheduling algorithm, where the utility of a packet is driven by its rate-distortion characteristics and its popularity within the neighbourhood.

4.1 Rate-Distortion Characteristics

A media presentation comprises data units that are output by an encoding algorithm when the content is originally compressed. The encoding process creates dependencies between the data units that can be abstracted as a directed acyclic graph, as illustrated in Figure 7. Each node in the graph represents a data unit, and an arrow from data unit l to data unit l' in the graph signifies that for decoding data unit l, data unit l' must be decoded first.

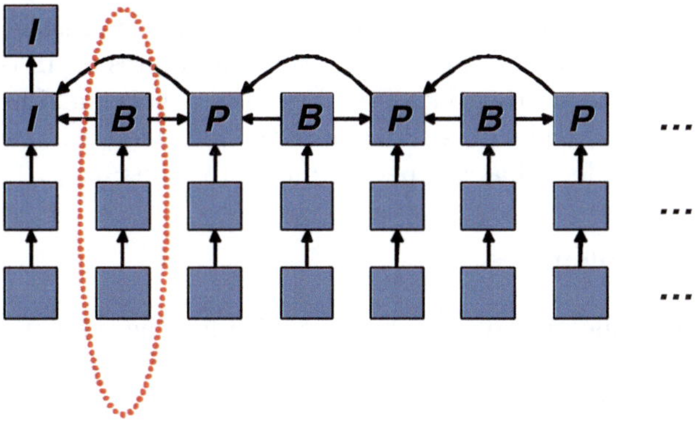

Fig. 7. Dependency graph for data units of a media presentation.

Each data unit in the presentation is characterized with the following quantities. B_l is the size of data unit l in bytes and $t_{d,l}$ is its *delivery deadline*. This is the time by which data unit l needs to be received in order to be usefully decoded[7]. Packets arriving after the delivery deadlines of the respective data units

[6] We only need to maintain an ordered list of peers in the network according to their respective latencies from the origin media server.

[7] In MPEG terminology this is the so called decoder time-stamp.

that they contain are discarded. Furthermore, $\mathcal{N}_c^{(l)} = \{1, \ldots, l\}$ is the set of data units that the receiver considers for error concealment in case data unit l is not decodable by the receiver on time. Finally, $\Delta d_l^{(l_1)}$, for $l_1 \in \mathcal{N}_c^{(l)}$, is the reduction in reconstruction error (distortion) for the media presentation, when data unit l is not decodable but is concealed with data unit l_1 that is received and decoded on time. Note that $\Delta d_l^{(l)}$ denotes simply the reduction in reconstruction error when data unit l is in fact decodable. The media source model presented here has been adopted from [10], which in turn represents a generalized version (accounting for decoder error concealment) of the model originally introduced in [9].

4.2 Receiver-Driven Packet Requests

Each peer maintains a sliding window of data units that periodically advances. A peer buffers the already received data units from this window, while it seeks to request the rest of them from its neighbours. Peers periodically exchange maps describing the presence/absence of data units from their respective windows. In this way, a peer can discover at its neighbours the availability of data units presently missing in its window.

More formally, let W denote the set of data units in the current sliding window at peer p and let $M \in W$ denote the subset of missing data units from W that the peer can request from its neighbours. The peer is interested in requesting these data units such that it maximizes its own utility of them. In particular, a peer may experience different reconstruction qualities of the media presentation played at its end commensurate to the media packets received in response to different request schedules. Therefore, peer n is interested in computing the optimal schedule for requesting data from neighbours that maximizes the reconstruction quality of its media presentation. In the following, we describe in detail the optimization problem that the peer is interested in solving.

Optimization Problem

Let $\boldsymbol{\pi}$ denote the collection of request schedules for the data units in M, i.e., $\boldsymbol{\pi} = \{\pi_1, \ldots, \pi_{|M|}\}$, where π_{l_m} is the request schedule for data unit $l_m \in M$, for $m = 1, \ldots, |M|$. Each request schedule π_m represents a matrix of size $|P| \times N$, where P denotes the set of neighbouring nodes of peer p. The row index n and the column index i of π_m correspond respectively to the neighbour $p_n \in P$ from which data unit l_m can be requested at time slot t_{i-1}, for $n = 1, \ldots, |P|$ and $i = 1, \ldots, N$. In particular, $t_0, t_1, \ldots, t_{N-1}$ represents a horizon of N time instances for requesting data units starting from the present time t_0. Given the above π_m comprises binary entries $a_{ni} \in \{0, 1\}$ that signify whether data unit l_m is requested at time slot t_{i-1} from neighbour p_n ($a_{ni} = 1$) or the opposite is true, i.e., $a_{ni} = 0$. It should be noted that a data unit is not requested beyond its delivery deadline. Therefore, we set $a_{ni} = 0$ for $n = 1, \ldots, N$ and time slots $t_{i-1} \geq t_{d,l_m}$, for every data unit $l_m \in M$. Similarly, we set $a_{ni} = 0$ for

$i = 1, \ldots, N$ for every neighbour p_n that does not have available data unit l_m in its current sliding window. Finally, let $\epsilon(\pi_{l_m})$ denote the *expected error* or the probability of not receiving data unit l_m on time given its request schedule π_{l_m}.

Following the approach in [11], the expected distortion of the media presentation at peer p as a function of the request schedule $\boldsymbol{\pi}$ can be expressed as

$$D(\boldsymbol{\pi}) = D_0 - \sum_{l_m \in M} \sum_{l_1 \in \mathcal{N}_c^{(l_m)}} \Delta d_{l_m}^{(l_1)} \prod_{j \in \mathcal{A}(l_1)} (1 - \epsilon(\pi_j)) \times \prod_{l_2 \in \mathcal{C}(l_m, l_1)} \left(1 - \prod_{l_3 \in \mathcal{A}(l_2) \setminus \mathcal{A}(l_1)} (1 - \epsilon(\pi_{l_3})) \right), \quad (19)$$

where D_0 is the expected reconstruction error for the presentation if no data units are received. $\mathcal{A}(l_1)$ is the set of ancestors of l_1, including l_1. $\mathcal{C}(l_m, l_1)$ is the set of data units $j \in \mathcal{N}_c^{(l_m)} : j > l_1$ that are not mutual descendants, i.e., for $j, k \in \mathcal{C}(l, l_1) : j \notin \mathcal{D}(k), k \notin \mathcal{D}(j)$, where $\mathcal{D}(j)$ is the set of descendants of data unit j including data unit j itself. Finally, "$\setminus$" denotes the operator "set difference". We refer the reader to [11] for details on the derivation of the expression in (19).

Now, a request schedule will also induce a certain data rate on the downlink of peer p. This is the expected amount of data that the neighbours in P will send in response to $\boldsymbol{\pi}$. This quantity can be computed as

$$R(\boldsymbol{\pi}) = \sum_{l_m \in M} B_{l_m} \rho(\pi_{l_m}), \quad (20)$$

where B_{l_m} is the size of data unit l_m in bytes, as introduced before, and $\rho(\pi_{l_m})$ is the *expected cost* or redundancy of requesting data unit l_m under policy π_{l_m}. Precisely, $\rho(\pi_{l_m})$ denotes the expected number of bytes sent per source byte of l_m on the downlink of peer p.

Finally, the peer is interested in minimizing the expected distortion $D(\boldsymbol{\pi})$ such that its downlink capacity $C^{(d)}$ is not exceed as a result. In other words, peer p is interested in computing the optimal policy $\boldsymbol{\pi}^*$ given as

$$\boldsymbol{\pi}^* = \arg\min_{\boldsymbol{\pi}} D(\boldsymbol{\pi}), \qquad \text{s.t. } R(\boldsymbol{\pi}) \leq C^{(d)}. \quad (21)$$

Using the method of Lagrange multipliers, we can reformulate (21) as an unconstrained optimization problem. That is, we seek the policy vector $\boldsymbol{\pi}$ that minimizes the expected Lagrangian $J(\boldsymbol{\pi}) = D(\boldsymbol{\pi}) + \lambda R(\boldsymbol{\pi})$ for some Lagrange multipliers $\lambda > 0$, and therefore achieves a point on the lower convex hull of the set of all achievable distortion-rate pairs $(D(\boldsymbol{\pi}), R(\boldsymbol{\pi}))$. As shown in [11], a policy vector $\boldsymbol{\pi}$ that minimizes the expected Lagrangian $J(\boldsymbol{\pi})$ can be computed using an iterative descent algorithm called Iterative Sensitivity Adjustment (ISA) [9]. However, due to the complexity of such an approach, we propose instead an approximation to the solution in (21) that is more suitable for being incorporated as a part of an actual system. We describe this low complexity approach in the next section.

Practical Solution

For each data unit $l_m \in M$, we define S_{l_m} to be the sensitivity of the media presentation to not receiving data unit l_m on time. This quantity can be computed as the overall increase in distortion affecting the media presentation by the absence of l_m at decoding, i.e.,

$$S_{l_m} = \sum_{j \in \mathcal{D}(l_m)} \Delta d_j^{(j)}, \tag{22}$$

where $\mathcal{D}(l_m)$ is the set of descendants[8] of data unit l_m and $\Delta d_j^{(j)}$ is the reduction of reconstruction distortion associated with data unit j. Both of these quantities were introduced earlier. Furthermore, we define I_{l_m} to be the current importance of data unit l_m for the overall quality of the reconstructed presentation. Using (22) we compute this quantity as

$$\mathcal{I}_{l_m} = \frac{S_{l_m}}{B_{l_m}} \cdot P_{l_m}(k, |P|) \cdot U(t, t_{d,l_m}). \tag{23}$$

We explain each of the multiplicative factors in (23) in the following. The term S_{l_m}/B_{l_m} represents the sensitivity of the media presentation per source byte of data unit l_m. In other words, S_{l_m}/B_{l_m} describes the distortion-rate tradeoff for the media presentation associated with requesting data unit l_m or not. We denote the second term $P_{l_m}(k, |P|)$ the *popularity factor* for data unit l_m in the neighbourhood of peer p. This quantity describes how often this data unit is encountered among the peer nodes in P. Specifically, based on the number of replicas k of data unit l_m found in P and the size of the neighbourhood $|P|$, the popularity factor returns a number that is inversely proportional to the ratio $k/|P|$. When the frequency of coming across l_m in P increases, the popularity factor decreases and vice versa. The motivation behind using such a factor is to alleviate the dissemination of data units less frequently encountered among nodes in the overlay. Finally, the last multiplying factor in (23) accounts for the various delivery deadlines that different data units may have relative to the present time t. In particular, the *urgency factor* $U(t, t_{d,l_m})$ provides a measure of relative urgency of data unit l_m with respect to t and among the data units in M. As the deadline of a data unit approaches t, its urgency factor increases. Conversely, for data units with delivery deadlines far into the future, this factor should exhibit respectively smaller values. The idea for employing an urgency factor when evaluating the present importance of the data units in M is to be able to give preference to data units that need to be received sooner by peer p due to their more pressing delivery deadlines.

The proposed light weight optimization algorithm for computing the request schedule for the data units in M operates as follows. First, the current importance values for data unit $l_m \in M$ and $m = 1, \ldots, |M|$ are computed using (23).

[8] For example, for the first "B" data unit in Figure 7, this is the collection of data units encircled in dotted red.

These quantities are then sorted in decreasing order. Let M^{sort} denote the corresponding set of 'sorted' data units. That is the index of data unit l_m in M^{sort}, for $m = 1, \ldots, |M|$, corresponds to the location/position of its current importance $\mathcal{I}_{l_m}$ in the sorted list of these values. Next, starting from the first element of M^{sort} and moving toward its last one, we compute for each entry in M^{sort} the likelihood of receiving this data unit at p before its delivery deadline. In particular, let $l_{m_j} \in M^{\text{sort}}$, for $j = 1, \ldots, |M|$, be the data unit considered in the algorithm presently. Furthermore, let $P_{(l_{m_j})} \subset P$ denote the subset of neighbours of p that have data unit l_{m_j} available in their sliding windows at present. Then, for every node $p_{n_k} \in P_{(l_{m_j})}$ we compute the probability that data unit l_{m_j} will arrive at peer p no later than $t + t_{d,l_{m_j}}$ in response to a request sent by p to node p_{n_k} at present, i.e., at time t. In other words, this is the probability of experiencing a delay shorter than $t_{d,l_{m_j}}$ between the events of sending the request on the forward channel $p \rightarrow p_{n_k}$ and receiving the data unit on the backward channel $p_{n_k} \rightarrow p$. In the terminology of computer networks this delay is called the round-trip time and we denote it here $RTT_{(p,p_{n_k})}$. Hence, we compute $Prob\{RTT_{(p,p_{n_k})} < t_{d,l_{m_j}}\}$ for $p_{n_k} \in P_{(l_{m_j})}$ and $k = 1, \ldots, |P_{(l_{m_j})}|$. The algorithm selects to send a request for l_{m_j} to the node p_{n_k} that exhibits the highest nonzero $Prob\{RTT_{(p,p_{n_k})} < t_{d,l_{m_j}}\}$. Otherwise, if there is no such value[9], the data unit is not requested and the algorithm proceeds to the next element of M^{sort}. Finally, once p goes through all data units in the 'sorted' set, it sends the computed requests to the appropriate nodes in P. The major computational steps of the algorithm are summarized in Figure 8 below. Next, we describe the procedure for computing the probabilities $Prob\{RTT_{(p,p_{n_k})} < t_{d,l_{m_j}}\}$.

Computing $Prob\{RTT_{(p,p_{n_k})} < t_{d,l_{m_j}}\}$

It should be mentioned that for computing $Prob\{RTT_{(p,p_{n_k})} < t_{d,l_{m_j}}\}$, the algorithm takes into account (i) the statistics of the communication channel from node p_{n_k} to peer p, (ii) any previous (pending) requests to this sender for which peer p has not received yet the corresponding data units[10], and (iii) the estimated transmission bandwidth of the channel $p_{n_k} \rightarrow p$. In particular, requesting a data unit comprises sending a small control packet to a designated neighbour. Moreover, the frequency of sending such packets is typically much smaller than the rate at which the corresponding data units are returned in response. That is because multiple data units can be requested with a single request packet. Hence, requesting data units typically consumes a very small fraction of the transmission bandwidth between two peers. Thus, it is reasonable to assume that the network effects in terms of delay and packet loss that requests experience on the

[9] That is the probability of receiving this data unit on time from any of the prospective senders is zero.

[10] This includes any data units $l_{m_i} \in M^{\text{sort}}$, for $i = 1, \ldots, j-1$, that are going to be requested in this round prior to l_{m_j} from the same sender p_{n_k}.

Given M, P, t

(0) Initialize schedules :
  NodeSchedule = {}, DataUnitSchedule = {}.
(1) For $l_m \in M$ and $m = 1, \ldots, |M|$
  Compute $\mathcal{I}_{l_m}$.
(2) Sort $\{\mathcal{I}_{l_m}\}$ in decreasing order.
  $\Rightarrow M^{\text{sort}}$.
(3) For $l_{m_j} \in M^{\text{sort}}$ and $j = 1, \ldots, |M|$ do
  For $p_{n_k} \in P_{(l_{m_j})}$ and $k = 1, \ldots, |P_{(l_{m_j})}|$ do
    Compute $Prob\{RTT_{(p,p_{n_k})} < t_{d,l_{m_j}}\}$.
  Find $MAX = \max_{p_{n_k}}\{\, Prob\{RTT_{(p,p_{n_k})} < t_{d,l_{m_j}}\}\}$.
  If $MAX > 0$
    Update schedules :
      Find $p^*_{n_k} = \arg\max_{p_{n_k}}\{\, Prob\{RTT_{(p,p_{n_k})} < t_{d,l_{m_j}}\}\}$.
      NodeSchedule = {NodeSchedule, $p^*_{n_k}$}.
      DataUnitSchedule = {DataUnitSchedule, l_{m_j}}.
(4) Execute schedules :
  For $n = 1, \ldots, |\text{NodeSchedule}|$ do
    p_n = NodeSchedule(n); l_n = DataUnitSchedule(n).
    Send request to node p_n for data unit l_n.

Fig. 8. Computing the optimal policy for requesting data units from neighbours.

forward channel $p \rightarrow p_{n_k}$ are quite marginal and can be ignored for practical purposes. This is the approach that we follow here as we associate the overall delay $RTT_{(p,p_{n_k})}$ in receiving a requested data unit to the characteristics of the backward channel $p_{n_k} \rightarrow p$ only.

Now, in order to be able to compute $Prob\{RTT_{(p,p_{n_k})} < t_{d,l_{m_j}}\}$ we need a statistical characterization for the backward channel. Here, we model $p_{n_k} \rightarrow p$ as a packet erasure channel with random transmission delays [9]. Specifically, packets carrying requested data units sent on this channel are either lost with a probability ϵ_B or otherwise they experience a random transmission delay y generated according to a certain probability distribution $f(y)$. Then, $Prob\{RTT_{(p,p_{n_k})} < t_{d,l_{m_j}}\}$ can be written as

$$Prob\{RTT_{(p,p_{n_k})} < t_{d,l_{m_j}}\} = (1 - \epsilon_B) \int_{y < t_{d,l_{m_j}}} f(y)dy. \tag{24}$$

In our case, we characterize the delay as exponentially distributed with a right shift of κ. This means that the delay y comprises a constant component associated with κ and a random component x exhibiting a exponential distribution with a parameter θ. Thus, $f(y)$ can be written as

$$f(y) = \begin{cases} \theta e^{-\theta(y-\kappa)} & : \quad y \geq \kappa, \\ 0 & : \quad \text{otherwise.} \end{cases} \tag{25}$$

We attribute the existence of κ to the prospective backlog of previously requested data units from p_{n_k} that has not been received yet by p and in addition to the required amount of time to empty out data unit l_{m_j} itself from the transmission buffer of node p_{n_k}. Furthermore, we relate the random component of the delay x to transient bandwidth variations of the network links comprising the channel $p_{n_k} \rightarrow p$ which in turn are caused by random occurrences of cross traffic on these links. The requesting peer estimates ϵ_B based on gaps in sequence numbers of arriving data units from p_{n_k} and similarly it estimates the parameter θ based on the jitter of the inter-arrival times of these data units. Finally, let $\tilde{r}_{(p_{n_k},p)}$ denote the present estimate of the download rate from node p_{n_k} that peer p has[11] and let $\mathcal{DU}$ denote the set of data units previously requested from p_{n_k} that has not been received yet. Then, p computes κ as

$$\kappa = \frac{\sum_{l \in \mathcal{DU}} B_l + B_{l_{m_j}}}{\tilde{r}_{(p_{n_k},p)}} \tag{26}$$

Once the peer has values for κ, θ, and ϵ_B, it can compute $Prob\{RTT_{(p,p_{n_k})} < t_{d,l_{m_j}}\}$ using (24) and (25) as

$$Prob\{RTT_{(p,p_{n_k})} < t_{d,l_{m_j}}\} = (1 - \epsilon_B) \int_{\kappa}^{t_{d,l_{m_j}}} \theta\, e^{-\theta(y-\kappa)} dy\ . \tag{27}$$

5 Neighbourhood Maintenance and Uplink Sharing

In order to deal with random node departures and uncooperating peers a neighbourhood needs to be continuously updated over time. We design two efficient techniques for achieving this goal that simultaneously maximize the small world nature of a neighbourhood, i.e., the amount of content exchanged between its member peers. The first technique allows a peer to estimate the rate contributions of its neighbours while the second one provides a mechanism for sharing its uplink capacity across its neighbours. We explain these two techniques next.

5.1 Download Rate Estimation and Peer Replacement

A peer periodically estimates the respective download rates from its neighbours. This is done by computing the total amount of data received from each neighbour since the last time the download rate was computed. For example, let $\mathcal{DU}^{(p_k)}$ represents the set of data units that peer p has received from its neighbour p_k within the last download rate estimation period T. Then, p computes the the received rate contribution from p_k as

[11] In the next section, we explain how p computes $\tilde{r}_{(p_{n_k},p)}$ periodically.

$$\tilde{r}_{(p_k,p)} = \frac{\sum_{l \in \mathcal{DU}^{(p_k)}} B_l}{T} . \quad (28)$$

In this way, a peer can sort its neighbours based on their send rate contributions to this peer. Then, the peer can periodically replace the least contributing neighbour with a new peer selected using the procedure from Section 2[12]. Furthermore, if the peer experiences multiple neighbour nodes with no rate contribution, it will simultaneously replace all of them with newly selected neighbours. The replaced nodes in this latter case will typically represent free-riders that are not interested in sharing their resources with other nodes in the network.

5.2 Sender Upload Bandwidth Sharing

The algorithm for sharing the upload bandwidth of a peer among its requesting neighbours operates as follows. Let $C^{(u)}$ be the upload bandwidth of peer p, and let PR denote the subset of neighbours from which p has pending requests at present. Then, to every node $p_k \in PR$, peer p allocates a share of its upload bandwidth computed as

$$r_{(p,p_k)} = \frac{\tilde{r}_{(p_k,p)}}{\sum_{p_k \in PR} \tilde{r}_{(p_k,p)}} \cdot C^{(u)}, \quad (29)$$

where $\tilde{r}_{p_k,p}$ denotes the present estimate of the sending rate from node p_k to peer p. Hence, nodes that contribute more of their sending rate to peer p will receive in return a larger share of its own upload bandwidth, as provided through Equation (29).

6 Experiments

6.1 Setup

In this section, we examine the performance of the proposed framework for streaming actual video content. In the simulations, we employ the common test video sequence *Foreman* in CIF image size encoded at 30 fps using a codec based on the scalable extension (SVC) of the H.264 standard [12]. The content is encoded into four SNR-scalable layers, with data rates of 455 kbps, 640 kbps, 877 kbps, and 1212 kbps, respectively. The corresponding video quality of the layers is 36.5 dB, 37.8 dB, 39.1 dB, and 40.5 dB, respectively, measured as the average luminance (Y) PSNR of the encoded video frames. The group of pictures (GOP) size of the compressed content is 30 frames, comprising the following frame type pattern IBBPBBP..., i.e., there are two B-frames between every two P frames or P and I frames. The 300 frames of the encoded sequence are concatenated multiple times in order to create a 900 second long video clip that is used afterwards in our simulations.

[12] In particular, the new peer is selected such that the updated neighbourhood again exhibits minimum latency spread across its member nodes.

The P2P network in the experiments comprises 1000 peers, out of which 5% are free-riders, while we distribute the rest in two categories: cable/dsl peers and ethernet peers, in the ratio 7:2.5. The upload bandwidth for ethernet and cable/dsl peers is 1000 and 300 kbps, respectively, while the corresponding download bandwidth values for these two peer type categories are 1500 kbps and 750 kbps. The downlink data rate for free-riders is set to 1000 kbps. The uplink data rate of free-riders is irrelevant for the investigation here. In the simulations, we measure performance as the average Y-PSNR (dB) of the reconstructed video frames at each peer. The content is originally stored at a media server with an upload bandwidth of 6 Mbps. The play-out delay for the presentation is set by the peers to 15 seconds. This is the initial amount of data that each peer needs to accumulate in its buffer before starting the playback of the presentation. The size of the sliding window introduced in Section 4.2 for keeping track of data units at each peer is 30 seconds of data. Sending requests to its neighbours is considered by a peer at intervals of 1 sec. The contribution of each sending peer in terms of data rate is measured by the receiving peer every 30 seconds of time. The exclusion of the least contributing peer in a neighbourhood and the consecutive selection of a new replacement neighbour is done by a peer every 30 sec. Initially each peer selects 8 other peers as its neighbours. The size of a neighbourhood for a peer can grow subsequently to contain up to 14 other peers.

6.2 Influence of Urgency and Popularity Factors

In this section, we examine the influence of the urgency factor $U(t, t_{d,l_m})$ and the popularity factor $P_{l_m}(k, |P|)$ on the overall performance of our system. As introduced in Section 4.2, these quantities convey respectively the relative temporal and the relative spatial importance of a data unit. That is, how soon the data unit is due to expire and how often the data unit is encountered in a neighbourhood. In the following, we describe our specific choices for $U(t, t_{d,l_m})$ and $P_{l_m}(k, |P|)$.

We model these two factors as simple polynomials composed of a single term that satisfy the functional requirements on $U(t, t_{d,l_m})$ and $P_{l_m}(k, |P|)$ described in Section 4.2. Specifically, for data unit l_m they are computed at time t as

$$U(t, t_{d,l_m}) = \left(\frac{t}{t_{d,l_m}}\right)^{\alpha}, \qquad (30)$$

$$P_{l_m}(k, |P|) = \left(\frac{|P|}{k}\right)^{\beta},$$

where the parameters $\alpha, \beta \geq 0$ are the powers of the polynomials for $U(t, t_{d,l_m})$ and $P_{l_m}(k, |P|)$, respectively. Using the formulation in (31) for these factors allows for a simple implementation that at the same time provides a lot of flexibility in terms of the range of values that can be covered by $U(t, t_{d,l_m})$ and $P_{l_m}(k, |P|)$ as a function of the parameters α and β. In the following, we consider how the specific forms of the polynomials for the urgency factor and the popularity factor, i.e., their respective polynomial power parameters, influence

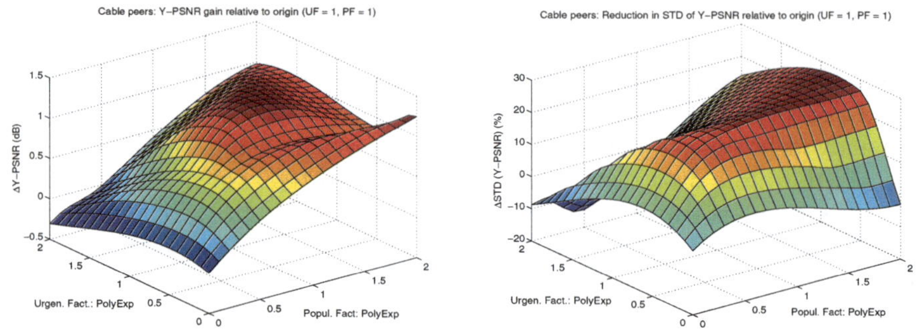

Fig. 9. Gain in Y-PSNR (dB) (left) and reduction (in %) of the standard deviation of Y-PSNR (right) when using the popularity and urgency factors in the case of cable peers.

the quality of the reconstructed media presentation at each peer. In particular, we conduct experiments where we vary α and β in the range [0,2] and measure the corresponding average video quality and its standard deviation for the populations of cable/dsl peers and ethernet peers, respectively.

In Figure 9, we show these quantities in the case of cable/dsl peers. In particular, in Figure 9 (left) we show the gain in dB of the average video quality (Y-PSNR) relative to the case when the urgency and the popularity factors are not employed, i.e., $\alpha, \beta = 0$. It can be seen that the gain can reach as high as 1.3 dB when the power parameters are in the range [1,1.5]. As the range of variations for the Y-PSNR gain in this range ($\alpha \in [1, 1.5], \beta \in [1, 1.5]$) is quite small, to maximize performance the values for these parameters can be selected to correspond anywhere in this plateau. Hence, for ease of implementation we opted to select the same value for both of them, and that is one, in the rest of the experiments in this chapter. Finally, we can see from Figure 9 (right) that the optimum range for α and β in the case of Y-PSNR gain also corresponds to the biggest reduction in standard deviation of this quantity relative to the case when $U(t, t_{d,l_m})$ and $P_{l_m}(k, |P|)$ are not employed. Specifically, for $\alpha \in [1, 1.5], \beta \in [1, 1.5]$ we observe a plateau of maximum standard deviation reduction of 25% in Figure 9 (right). It should be mentioned that similar observations and conclusions can be made for the case of ethernet peers. These results are not included here due to space considerations.

In the following two sections, we study exclusively the performances of our utility-based scheduling and bandwidth sharing techniques from Section 5.2 and Section 4.2, respectively. Therefore, in the related experiments we opted to employ the commonly employed technique of random peer selection for constructing and maintaining the overlay. Then, in Section 6.5, we replace the random mesh construction with our delay-based technique from Section 2 in order to examine in isolation the additional advantages that this method brings to the whole framework.

6.3 Resilience to Free-Riders

In Figure 10, we show the cumulative distribution function of the average video quality for each peer type. It can be seen from the figure that free-riders experience the media presentation at a very low quality. This is actually desirable as these peers do not contribute their upload bandwidth resources to serving data to other peers in the network, as explained earlier. Hence, the degraded video quality that they receive may in fact contribute to them changing their bandwidth sharing policy when they connect to the network next time. On the other hand, cable peers and ethernet peers exhibit distributions of video quality that are quite narrow in range and steep in slope, and most importantly of much higher amplitude relative to that of free-riders, as also seen from Figure 10. Furthermore, the cumulative distributions of video quality for cable peers and ethernet peers are commensurate to their bandwidth capabilities, as ethernet peers can receive more video quality layers from their neighbours and correspondingly serve more layers to them in return. Hence, ethernet peers exhibit video quality that is on the average 2 dB higher than that for cable peers. In particular, the average video quality for most of the cable peers ranges between 37 dB and 38 dB, while the average video quality for most of the ethernet peers is in the range 39 - 40.5 dB, as shown in Figure 10. Note that the distributions of video quality in Figure 10 map to similar relative distributions of the corresponding video decode rate for the three peer classes. These results are not included here due to space constraints.

To examine further the resilience of the proposed framework to free-riding in the system, we conducted the following experiment. We increased the percentage of free-riders in the overall population to 10, 15 and 20 percent, and we measured the corresponding average video quality for the three peer types. In Figure 11, we show these results for ethernet peers, together with the corresponding performance for ethernet peers in the case when sending peers share their upload resources uniformly. In other words, in this latter case, peers send data to their neighbours at same outgoing data rates. It can be seen from Figure 11 that when

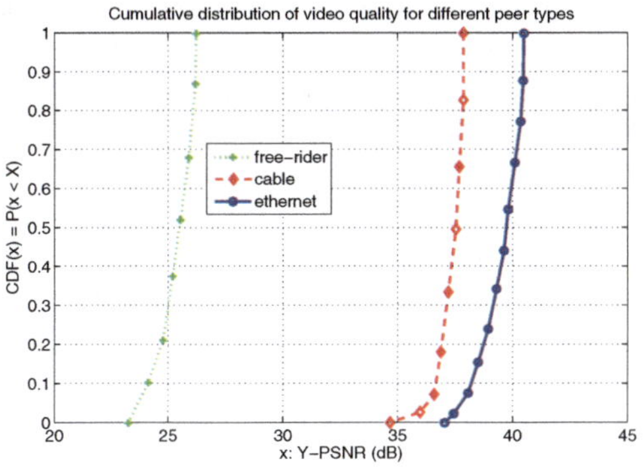

Fig. 10. CDF of average video quality (Y-PSNR) for different peer types.

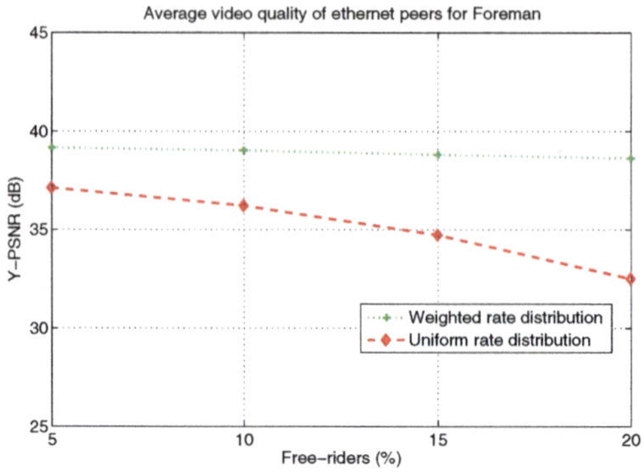

Fig. 11. Influence of free-riders for different upload bandwidth sharing strategies.

our framework is employed, the average performance of the video presentation for the ethernet peers does not vary substantially, as the number of free-riders in the network is increased. However, in the case of uniform send-rate distribution we can see that the average video quality of the ethernet peer population degrades substantially, as more and more resources in the network are consumed by the non-responding free-rider peers. For example, even at 20% free-riders in the network the reduction in average video quality for ethernet peers does not exceed 0.1 - 0.2 dB under the weighted send-rate allocation of our framework, while it reaches around 6 dB in the case of uniform allocation, as evident from Figure 11. Note that similar observations can be drawn when comparing the corresponding results for the cable peer type. These results are not included here for space considerations.

6.4 Utility-Based Scheduling

Here, we explore the performance advantages of utility-based scheduling, henceforth denoted *Utility*, over two commonly employed technique for requesting packets from peer neighbours. With *EDF* we denote the first of these technique which requests packets based on their delivery deadline. In particular, earlier expiring packets are requested first, hence the name Earliest Deadline First (EDF). The second technique used for comparison simply requests packets from neighbours without taking into consideration any specific packet information, hence the name *Random*. In Figure 12, we examine the distribution of average video quality for cable/dsl peers in the case of each of the three scheduling techniques.

It can be seen that *Utility* significantly outperforms the other two techniques. In particular, the average gain over *EDF* is around 4 dB, while the average gain over *Random* is even larger, i.e., around 8 dB, as denoted in Figure 12. The relative gains of utility-based scheduling are due to the fact that this technique takes into account the importance of each packet for the overall reconstruction quality at any given time, which is overlooked in *EDF* and *Random*. Furthermore, it is expected that *EDF* outperforms *Random* as the former at least takes into account the timing information associated with the video packets when requesting

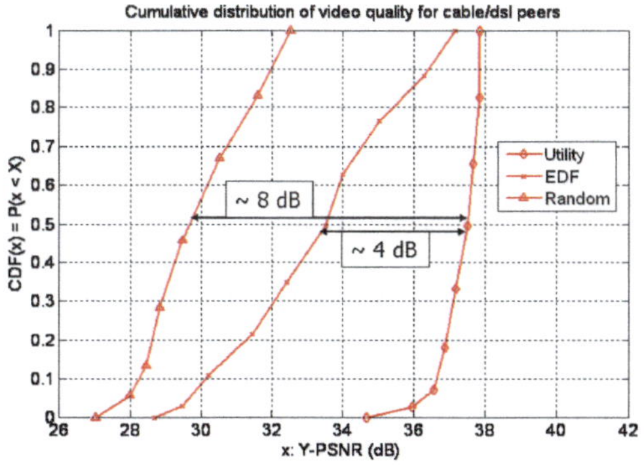

Fig. 12. CDF of average video quality (Y-PSNR) for different scheduling methods in the case of cable/dsl peers.

them. Finally, it should be mentioned that analogous results were obtained for the case of ethernet peers that are not included here due to space constraints.

6.5 Advantages of Delay-Based Mesh Construction

Here, we present the relative improvements in performance when the delay-based algorithm for mesh construction from Section 2 is employed, instead of the standard Random mesh construction technique where each peer selects its neighbours at random. The metrics over which we measure performance in this section are (i) frame freeze duration and (ii) normalized play-out time. We describe them subsequently. Frame freeze duration is the percentage of time relative to the duration of the whole presentation during which a peer experiences frozen video content on its display. Remember that this happens whenever a video frame is not received and decoded by the peer by its decoding/delivery deadline. In order to compensate for this, the peer conceals this frame with the last decodable frame that it has in its buffer. The content of this latter frame is kept on the screen (hence the name freeze frame for this concealment method) until a subsequent frame is decodable and therefore ready to be displayed next.

Next, recall from earlier that a peer has a parameter denoted play-out delay that is set ahead of time. As described previously, this parameter corresponds to the amount of data that the peer needs to buffer from the initial part of the presentation before the playback actually starts at the peer. Typically, it would take a peer a longer period of time than the actual value of its play-out delay parameter to gather the necessary amount of data for the playback to start. Furthermore, one can compute the absolute minimum of this quantity based on the hop distance of a peer from the media server and the data rate at which the server is streaming the presentation (typically the encoding rate of the presentation). Hence, we define normalized play-out time as the ratio of the actual time that a peer requires to fill up its play-out buffer initially and the minimum value of this quantity, as described above.

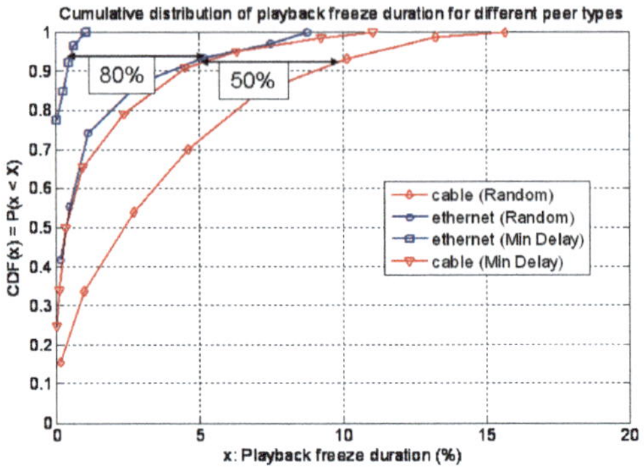

Fig. 13. Frame freeze duration (%) for different peer types and mesh constructions.

First, in Figure 13 we examine the difference in frame freeze duration experienced by the peers when each of the two mesh construction algorithms is used. It can be seen that the proposed technique provides substantial gains over simply selecting peer neighbours at random. For example, 90% of cable/dsl peers experience frame freeze not more than 10% of the time during their presentation in the case of random mesh construction. This percentage is reduced to 5% for the same cable/dsl peer population in the case of delay-based construction. Therefore, a 50% reduction in frame freeze time is achieved, as denoted in Figure 13. Similarly, in the case of ethernet peers, 90% of the population experiences frame freeze not longer than 5% of the time for random mesh construction relative to 1% of the time when peer neighbourhoods are constructed according to their latency spread. Hence, an 80% reduction in frame freeze duration is achieved, as also denoted in Figure 13. In general cable/dsl peers experience longer frame freeze durations relative to ethernet peers, as evident from Figure 13. This is expected and is due to the different uplink/downlink bandwidth capabilities associated with the two peer types. Finally, note that the reductions in frame freeze time result into corresponding gains in average video quality when delay-based mesh construction is employed. These results are not included here due to space limitations.

Next, we study the differences between random and delay-based construction in terms of normalized play-out time for a peer (cable/dsl or ethernet). In Figure 14, we show the CDF of the normalized play-out time for a peer for each of the two mesh construction algorithms. As expected, we can see from Figure 14 that when our algorithm is employed the peers observe much shorter play-out times which in turn improves their audio-visual experience of the media presentation. For example, for 90% of the peers the playback of the presentation can start no longer than twice the preset play-out delay in the case of the proposed technique, compared to about six times the preset play-out delay for the same percentage of peers for random mesh construction. Hence, the former approach provides 67% savings in play-out time or alternatively stated a three times shorter play-out time is achieved, as denoted in Figure 14. Another

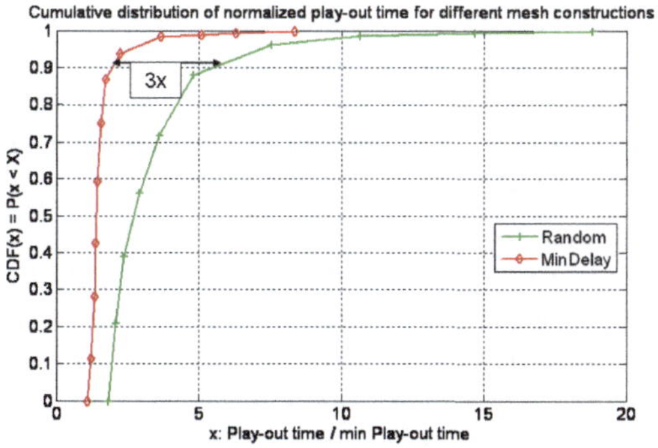

Fig. 14. CDF of normalized play-out time for different mesh constructions.

interpretation of these gains is as a reduction in required buffer size at a peer for smooth playback of the presentation in the case of delay-based construction.

7 Related Work

Due to its promise as a novel technology for delivering multimedia over the Internet at lower cost, P2P streaming has been studied considerably thus far. The solutions are generally built on either tree-based or mesh-based organization of the peers. Despite the plethora of prior work on tree-push based P2P streaming, e.g., [13, 14, 15], we still have to wait to witness an actual deployment of such a system on the Internet. Virtually all such systems to date are rather mesh-pull based[13], and we have therefore focused on this type of solution, which generally offers increased robustness to the dynamics of a P2P system.

From the plethora of prior work, we describe first the most relevant ones from a system-wise perspective. Specifically, in [17], the authors design a global pattern for content delivery in mesh-based overlays that can utilize the upload bandwidth of most of the peers. In addition, a sweet range for the peers' degree is identified that maximizes the delivered quality to the individual peers in the scenario under consideration. A similar approach to overlay architecture construction is taken in [18] where higher bandwidth peers are connected directly to the media server in order to facilitate their capabilities for pushing the content subsequently to many lower bandwidth peers located one layer farther in the overlay network hierarchy. The work in [19] presents a method to monitor the network-wide quality of the media presentation, based on the buffer maps constructed by peers in mesh-based overlays in order to facilitate exchange of data with their neighbours. Furthermore, in [20] video packets are requested from neighbours in prioritized order based on their video layer index and the probability of serving a neighbour in return is commensurate to the rate contribution received from this neighbour. Finally, the work in [21] considers streaming

[13] An interesting overview of mesh-pull based P2P video delivery and its commercial success for IPTV applications can be found in [16].

video multiple descriptions over an overlay mesh created via a gossip-based protocol where the probability of selecting a node for a neighbour is proportional to the size of its existing neighbourhood. A peer in the system employs a packet requesting scheme that maximizes its video quality subject to avoiding buffer underflow events of its video decoder.

There are several important differences relative to the prior works described above. First, we design a delay-based overlay construction procedure that substantially outperforms the random neighbour selection approach employed in [20, 21] for example. In addition, the present work considers content created using generic video encoding that may not necessarily be layered [20] or multiple description encoded [21]. Yet another advantage of our approach is that a more sophisticated utility-based model is proposed in order to determine the packets' relative importance when requesting data from neighbours. Such differentiating characteristics between the packets in terms of their value for the overall reconstruction quality may be overlooked if only the index of the video layer [20] or the video description [21] to which they belong is considered. Lastly, we design a deterministic algorithm for upload bandwidth distribution over the requesting peers which consistently, i.e., all of the time, rewards contributing peers instead of doing that on the average, as in [20].

Another body of prior work that needs to be mentioned in the context of this chapter is on exploiting locality information for improving the efficiency of P2P systems. Specifically, random peer selection combined with flooding-based approaches for content search in overlay networks contribute to excessive amounts of traffic even in moderate size networks [22, 23]. Most of this traffic is unnecessary and is caused by the mismatch between the logical topology of the overlay and the actual underlying physical network. The problem is exacerbated by blindly flooding message on multiple paths that in effect may have the same destination in the end. Therefore, such networks can suffer from scalability issues [24]. Another study that illuminates this mismatch problem is [25] which shows that only a small percentage of the overall number of connections in a Gnutella [26] overlay link peers within a single autonomous system. In order to account for this, there have been proposals for constructing more efficient overlay topologies in recent years, such as [27, 28] that cluster peers based on the closeness of their IP addresses, [29, 30] that cluster peers based on their proximity relative to the underlying network topology, [31][36] that determines the closeness between two peers based on their latency to common landmark servers, and [32] that selects neighbours based on estimates of delay distances between peers. The works outlined above exhibit an analogy with our overlay construction procedure in the sense that they all attempt to cluster peers based on some notion of distance. Still, in the present paper we create neighbourhoods exhibiting min latency spread across their member nodes relative to the origin server. This does not necessarily mean that two neighbours will be close to another according to the distance metrics employed above.

Finally, from the perspective of packet scheduling in P2P systems, the present paper is related to prior works on optimized video streaming in tree-based overlays such as [33] that proposes to employ priority-based mechanisms for allocating resources across different packets at a node in a multicast tree. Differently, the present paper considers utility based scheduling for efficient video streaming in mesh-pull based P2P overlays. Maximizing a utility function of other parameters in a P2P streaming system, such as the playback buffer content reserve or the processor task scheduling at a peer, has been investigated previously in [34] and [35], respectively.

8 Conclusions

The chapter develops a comprehensive optimization framework for live mesh-pull based P2P video delivery. The framework comprises three major building blocks that in concert provide for significant performance advances. In particular, through an overlay construction featuring uniform latency neighbourhoods an effective small-world topology is created that enables efficient data delivery routes and increased data sharing between peers in the network. Furthermore, through utility-based packet scheduling the video quality of the presentation at a peer is maximized for the given bandwidth resources while promoting the same goal at other peers in the overlay. Finally, by employing our uplink sharing strategy that rewards data sharing among neighbours free-riding in the system is effectively canceled out. Our theoretical analysis and simulation experiments make evident the substantial performance improvements in live P2P streaming that the proposed techniques provide. Significant gains in video quality and decoding rate are shown relative to commonly deployed solutions. The delay-based mesh construction allows for additional gains expressed as reductions in frame-freeze duration and playback latency. The improved playback continuity in this case results into further improvement in video quality for the media presentation.

Acknowledgement

I would like to dedicate this chapter to Kelly for always being there for me.

References

1. PPLive Homepage, http://www.pplive.com/
2. SopCast Homepage, http://www.sopcast.com/
3. PPStream Homepage, http://www.ppstream.com/
4. Zhang, X., Liu, J., Li, B., Yum, T.S.: CoolStreaming/DONet: A data-driven overlay network for efficient live media streaming. In: Proc. Conf. on Computer Communications (INFOCOM), vol. 3, pp. 2102–2111. IEEE, Miami (2005)
5. Bollobás, B.: Random Graphs, 2nd edn. Cambridge University Press, New York City (2001)
6. Watts, D.J., Strogatz, S.H.: Collective dynamics of 'small-world' networks. Nature 393(6684), 440–442 (1998)

7. David, H.A., Nagaraja, H.N.: Order Statistics, 3rd edn. Wiley Interscience, Hoboken (2003)
8. Casella, G., Berger, R.L.: Statistical Inference, 2nd edn. Duxbury Press, Boston (2001)
9. Chou, P.A., Miao, Z.: Rate-distortion optimized streaming of packetized media. Tech. Rep. MSR-TR-2001-35, Microsoft Research, Redmond, WA (2001)
10. Chakareski, J., Girod, B.: Rate-distortion optimized packet scheduling and routing for media streaming with path diversity. In: Proc. Data Compression Conference, pp. 203–212. IEEE Computer Society, Snowbird (2003)
11. Chakareski, J., Girod, B.: Server diversity in rate-distortion optimized streaming of multimedia. In: Proc. Int'l Conf. Image Processing, vol. 3, pp. 645–648. IEEE, Barcelona (2003)
12. ITU-T and ISO/IEC JTC 1: Advanced video coding for generic audiovisual services, amendment 3: Scalable video coding. Draft ITU-T Recommendation H.264 - ISO/IEC 14496-10 (AVC) (2005)
13. Castro, M., Druschel, P., Kermarrec, A.M., Nandi, A., Rowstron, A., Singh, A.: Splitstream: High-bandwidth multicast in a cooperative environment. In: Proc. Symp. Operating Systems Principles, pp. 298–313. ACM, Bolton Landing (2003)
14. Padmanabhan, V., Wang, H., Chou, P.: Resilient peer-to-peer streaming. In: Proc. Int'l Conf. on Network Protocols, pp. 16–17. IEEE, Atlanta (2003)
15. Liao, X., Jin, H., Liu, Y., Ni, L.M., Deng, D.: Anysee: Peer-to-peer live streaming. In: Proc. Conf. on Computer Communications (INFOCOM), pp. 1–10. IEEE, Barcelona (2006)
16. Hei, X., Liu, Y., Ross, K.: IPTV over P2P streaming networks: the mesh-pull approach. IEEE Communications Magazine 46(2), 86–92 (2008)
17. Magharei, N., Rejaie, R.: Understanding mesh-based peer-to-peer streaming. In: Proc. Int'l Workshop on Network and Operating Systems Support for Digital Audio and Video, pp. 56–61. ACM, Newport (2006)
18. Liang, C., Liu, Y.: Enabling broadcast of user-generated live video without servers. Tech. rep., Polytechnic Institute of NYU (2009), `http://eeweb.poly.edu/faculty/yongliu/docs/mm09.pdf`
19. Hei, X., Liu, Y., Ross, K.: Inferring network-wide quality in p2p live streaming systems. IEEE J. Selected Areas in Communications 25(10), 1640–1654 (2007)
20. Liu, Z., Shen, Y., Panwar, S., Ross, K., Wang, Y.: Using Layered Video to Provide Incentives in P2P Live Streaming. In: Proc. Workshop on Peer-to-Peer Streaming and IP-TV, pp. 311–316. ACM SIGCOMM, Kyoto (2007)
21. Lu, M.T., Wu, J.C., Peng, K.J., Huang, P., Yao, J., Chen, H.: Design and evaluation of a P2P IPTV system for heterogeneous networks. IEEE Trans. Multimedia 9(8), 1568–1579 (2007)
22. Saroiu, S., Gummadi, K.P., Dunn, R.J., Gribble, S.D., Levy, H.M.: An analysis of internet content delivery systems. In: Proc. 5th Symp. Operating Systems Design and Implementation. USENIX, Boston (2002)
23. Sen, S., Wang, J.: Analyzing peer-to-peer traffic across large networks. IEEE/ACM Trans. Networking 12(2), 219–232 (2004)
24. Ritter, J.: Why Gnutella can't scale. No, really (2001), `http://www.darkridge.com/~jpr5/doc/gnutella.html`
25. Ripeanu, M., Foster, I., Iamnitchi, A.: Mapping the Gnutella Network: Properties of Large-Scale Peer-to-Peer Systems and Implications for System Design. IEEE Internet Computing Journal 6(1), 50–57 (2002)
26. "Gnutella", `http://en.wikipedia.org/wiki/Gnutella`

27. Krishnamurthy, B., Wang, J.: Topology modeling via cluster graphs. In: Proc. 1st Workshop on Internet Measurement, pp. 19–23. ACM SIGCOMM, San Francisco (2001)
28. Padmanabhan, V.N., Subramanian, L.: An investigation of geographic mapping techniques for internet hosts. In: Proc. Data Communication, Ann. Conf. Series, pp. 173–185. ACM, San Diego (2001)
29. Hefeeda, M., Habib, A., Botev, B., Xu, D.: Bhargava: PROMISE: Peer-to-Peer Media Streaming Using CollectCast. In: Proc. Int'l Conf. Multimedia, pp. 45–54. ACM, Berkeley (2003)
30. Cha, M., Rodriguez, P., Moon, S., Crowcroft, J.: On Next-Generation Telco-Managed P2P TV Architectures. In: Proc. Int't Workshop Peer-to-Peer Systems (IPTPS), Tampa Bay, FL, USA (2008)
31. Xu, Z., Tang, C., Zhang, Z.: Building topology-aware overlays using global soft-state. In: Proc. 23rd Int'l Conf. Distributed Computing Systems, pp. 500–508. IEEE Computer Society, Providence (2003)
32. Liu, Y., Xiao, L., Liu, X., Ni, L.M., Zhang, X.: Location awareness in unstructured peer-to-peer systems. IEEE Trans. Parallel and Distributed Systems 16(2), 163–174 (2005)
33. Chakareski, J., Frossard, P.: Adaptive p2p video streaming via packet labeling. In: Proc. Conf. on Visual Communications and Image Processing. SPIE, San Jose (2007)
34. Li, Z., Huang, J., Katsaggelos, A.K.: Content reserve utility based video segment transmission scheduling for peer-to-peer live video streaming system. In: Proc. 45th Allerton Conf. on Communications, Control, and Computing, pp. 563–567. IEEE, Monticello (2007)
35. Chen, F.: A utility-based approach to scheduling multimedia streams in peer-to-peer systems. In: Proc. 18th Int'l Symp. Parallel and Distributed Processing. IEEE Computer Society, Santa Fe (2004)

Intelligent Video Network Engineering with Distributed Optimization: Two Case Studies

Ying Li[1], Zhu Li[2], Mung Chiang[1], and A. Robert Calderbank[1]

[1] Electrical Engineering Department, Princeton University, NJ 08540, USA
{yingli,chiangm,calderbk}@princeton.edu

[2] epartment of Computing, Hong Kong Polytechnic University, Hong Kong
zhu.li@ieee.org

Summary. Video is becoming the dominant traffic over the Internet. To provide better Quality of Service (QoS) to the end users, while also achieve network resource efficiency, is an important problem for both network operators, content providers and consumers. In this work, we present intelligent video networking solutions for IPTV and Peer-to-Peer (P2P) systems that optimizes the users' QoS experiences while under network resource constraints.

Given the limited network bandwidth resources, how to provide Internet users with good video playback Quality of Service (QoS) is a key problem. For IPTV systems video clips competing bandwidth, we propose an approach of Content-Aware distortion-Fair (CAF) video delivery scheme, which is aware of the characteristics of video frames and ensures max-min distortion fair sharing among video flows. Different from bandwidth fair sharing, CAF targets end-to-end video playback quality fairness among users when bandwidth is insufficient, based on the fact that users directly care about video quality rather than bandwidth. The proposed CAF approach does not require rate-distortion modeling of the source, which is difficult to estimate, but instead, it exploits the temporal prediction structure of the video sequences along with a frame drop distortion metric to guide resource allocations and coordination. Experimental results show that the proposed approach operates with limited overhead in computation and communication, and yields better QoS, especially when the network is congested.

For Internet based video broadcasting applications such as IPTV, the Peer-to-Peer (P2P) streaming scheme has been found to be an effective solution. An important issue in live broadcasting is to avoid playback buffer underflow. How to utilize the playback buffer and upload bandwidth of peers to minimize the freeze-ups in playback, is the problem we try to solve. We propose a successive water-filling (SWaF) algorithm for the video transmission scheduling in P2P live streaming system, to minimize the playback freeze-ups among peers. SWaF algorithm only needs each peer to optimally transmit (within its uploading bandwidth) part of its available video segments in the buffer to other peers requiring the content and pass small amount message to some other peers. Moreover, SWaF has low complexity and provable optimality. Numerical results demonstrated the effectiveness of the proposed algorithm.

Keywords: Content-aware, video streaming, peer-to-peer, scheduling, water-filling.

C.W. Chen et al. (Eds.): Intel. Multimedia Communication: Tech. and Appli., SCI 280, pp. 253–290.
springerlink.com

1 Introduction

The rapid advances in computing, communication and storage technologies have heralded a new age of explosive growth in multimedia applications, such as online video repository, mobile TV and IPTV, multimedia based social interaction, etc. Multimedia traffic has dominated Internet traffic in volume. These applications open up new opportunities and present new challenges to the technologies in the area of audio/visual information processing, communication and networking, etc. We focus on developing novel and effective approaches in delivering multimedia content.

The goal of this chapter is to enable users (especially with multimedia traffic) of networks to experience better end-to-end quality of service, and to help network operators run networks more efficiently, especially on how efficiently the network resources are utilized. Meanwhile, during the process of achieving such goal, this chapter aims to develop new mathematical methodologies of optimization for network engineering.

1.1 Challenges of Multimedia Traffic

There are many open problems in terms of how to efficiently provision complicated quality of service requirements for mobile users who need triple play where TV (video), voice, and data are all available from one service provider on one service medium.

One particular challenging problem is multimedia streaming, where demand for better playback quality and small transmission delays needs to be reconciled with the limited and often time-varying communication resources.

The main technical difficulties for media streaming include the following:

1. Media streaming usually has large volume of data and the volume of data may vary over time. For instance, the sources for most video streaming applications are typically sequences with relative high bit rates. However, wired/wireless networks have limited bandwidth. In order to support media streaming over capacity limited networks, the high rate media sources need to be adapted through a variety of schemes, such as scalable video stream extraction, compression, and summarization, before they can be accommodated by the links.

2. Media streaming usually has stringent delay requirements. Media streaming cannot tolerate much variation in delay (jitter) if the buffer starves, because once playout starts, it needs to keep playing. Media streaming sometimes cannot tolerate much delay, such as for interactive applications (e.g., VoIP and online gaming).

3. Media streaming can have certain level of packet loss tolerance, i.e, losing certain number of packets will not dramatically affect the receiving end play back quality. On the other hand, video packets have loss/decoding dependence. Certain packet loss will result in whole segment of video not decodable. This is especially true when layered video coding [60] is present.

4. Different media content segments have different rate-distortion characteristics, e.g., some segment may be part of an action movie and requires a lot of

bits to encode, while some others may be part of a video with less motion such as a scene of news anchors talking that requires relatively less bits to encode. In a system supporting many users, this type of multi-user content diversity in content rate-distortion characteristics need also be exploited.

5. The resource consumptions of video users are typically discrete, i.e., measured in frames instead of in bits. As a result, their QoS metrics as functions of allocated resources are discrete as well, and typically do not have close form characterizations.

Therefore most of previous work on resource allocation for elastic data traffic does not directly apply here, and a new optimization framework is needed. Most of the above difficulties are not well-solved in the network design. In this work, we come up with new theory and new design to deliver multimedia traffic over networks, such that users can have better end-to-end quality of service while the networks being efficient, stable, and robust. For this purpose, we develop and apply the design philosophy for network resource allocation explained in the following section.

1.2 Optimization in Network Resource Allocation

Network Utility Maximization

Since the publication of the seminal paper [36] in 1998, the framework of Network Utility Maximization (NUM) has found many applications in network resource allocation algorithms and the design of protocol stacks. Consider a communication network with L logical links, wired or wireless, each with a fixed capacity of c_l bps, and S sources, each transmitting at a rate of x_s bps. Each source s emits one flow, using a fixed set $L(s)$ of links in its path, and has a utility function $U_s(x_s)$. Each link l is shared by a set $S(l)$ of sources. NUM, in its basic version, is the following problem of maximizing the network utility $\sum_s U_s(x_s)$, over the source rates $\mathbf{x}$, subject to linear flow constraints $\sum_{s\in S(l)} x_s \le c_l$ for all links:

$$\begin{aligned} &\text{maximize } \textstyle\sum_s U_s(x_s) \\ &\text{subject to } \textstyle\sum_{s\in S(l)} x_s \le c_l,\ \forall l \\ &\qquad\qquad\ \mathbf{x} \succeq 0. \end{aligned} \tag{1}$$

The basic NUM framework has been widely extended (e.g., see [21] and the references therein). The NUM framework has shown many applications in *forward engineering*, which means given the optimization problem, we design network protocol stacks. Many network resource allocation algorithms have been proposed for different purposes. Following the pioneer work on optimal flow control [37, 36], an extensive study of network flow control based on optimization method and game theory has been carried out, e.g, [6, 4, 5, 27, 55, 56, 81, 73], and there are a lot of extensions to the protocol design across layers, e.g., [39, 41, 29, 42, 45, 49, 43, 44, 48, 79, 80, 52].

Along another thread of *reverse engineering*, which means given the protocols, we find the mathematic framework behind them, it has been found that the mathematics behind Transmission Control Protocol (TCP) and Active Queue Management (AQM) matches the basic NUM framework [54]. This again backs up the forward engineering foundation of NUM framework.

In this work, we design network resource allocation and protocols for multimedia traffic by extending NUM framework.

Utility Function

Utility function basically is a measure of user's happiness of end-to-end quality of service [67].

In basic NUM, the utility function is assumed to be strictly concave in rate, which has an economic interpretation that when the rate is larger, the increase of user's happiness about the service slows down.

Considering different QoS requirement, different application can have different utility functions. For instance, for the end user with online video application, the end user may mainly care about the video distortion of playback quality, hence the video distortion can be a measure of the end user's happiness, the less distortion, the happier. For another instance, in peer-to-peer video network, the end user may mainly care about the playback smoothness, hence the length of the content to be played in the playback buffer can be a measure of the end user's happiness, the more content to be played in the buffer, the happier.

Utility function basically is a measure of user's happiness of end-to-end quality of service .

Optimality in Which Sense?

Efficiency and fairness are two main concerns for resource allocation when many users are sharing the network resources. Here we mainly consider efficiency in utility and fairness in utility, where for some cases the efficiency in utility can be coincident with the efficiency and/or the fairness in bandwidth.

- Efficiency in utility. We consider the objective as to maximize the total utility,

$$\text{maximize} \textstyle\sum_s U_s. \tag{2}$$

  The total optimality means social efficiency in utility. Note that for special cases of utility function in rate only, maximizing total utility in basic NUM gives bandwidth efficiency or bandwidth fairness. For example, for $U_s(x_s) = x_s$, it gives the efficiency in bandwidth, and for $U_s(x_s) = \log x_s$ [37,36,38], it gives proportional fairness in bandwidth, and $U_s(x_s) = \frac{x_s^{1-\alpha}}{1-\alpha}$ gives α-fairness in bandwidth [59], where α is a positive constant. When $\alpha = 1$, it gives bandwidth proportional fairness, and for α going to infinity, it converges to bandwidth max-min fairness. More discussions on bandwidth fairness and efficiency can be found in [13,14,57].
- We also consider utility max-min fair sharing, which is to maximize the minimum utility,

$$\text{maximize} \min U_s. \tag{3}$$

The utility max-min fair was discussed in [17] and some references therein. There are more discussions on utility fairness such as proportional utility fairness [74]. Here we focus on the max-min utility fairness.

Degrees of Freedom: Network Operations over Layers

Here our resource mainly means the link bandwidth. Given the limited link bandwidth and utility objective functions, how to achieve the optimality? What are the design spaces? We use a suite of network operations as the design knobs, such that users with multimedia traffic in the network can have satisfied quality of service.

A suite of network operations in this work includes the operational or functional modules ('*who does what*') and the connections among them ('*how to connect them*') [21]. The operational modules include operations over different layers, such as multimedia source coding at the Application (APP) layer, rate adaptation or congestion control of users at the Transport layer [72]), scheduling at MAC layer. The optimal resource allocation will shape the network operations to operate at the optimal points by adjusting the design knobs - the network operations.

The design knobs may need to be jointly adapted, rather than separately. For instance, the multimedia source coding can be adapted jointly with the congestion control in networks. By joint adaptation, if there is congestion on the path, or if the wireless link is in bad condition, the source can send the most important segment within the limited room on the path, rather than less important segment, such that the receiver can experience good multimedia playback quality.

Distributed Algorithm

In a network, the distributed algorithm is favorable since it is difficult to run a large network such as Internet if everything is controlled by a central server.

Fortunately, in the basic NUM, convexity and separability properties of the optimization problem readily lead to distributed algorithms that converge to the globally optimal rate allocation, by applying primal/dual decomposition [36, 10, 11, 15, 62].

Such distributed algorithms have the economic interpretation that every source solves its best rate for the net utility which is the utility deducting the cost (price per unit rate multiplying the rate), every link updates the price which indicates how congested the link is, based on the current load of the link, and through the network there is massage passing which gives the source the end-to-end price and gives the link the aggregate load [36]. The distributed algorithm matches the Internet TCP/AQM mechanism [54, 70].

In this work, distributed algorithms are proposed for the optimal resource allocation. Since our utility functions have their spatiality because of the application scenarios and they lose the clean structure of the basic NUM, the decomposition becomes difficult. Despite the difficulty, we develop distributed algorithms based on network pricing. In contrast to standard pricing-based rate control algorithms for the basic NUM, in which each link communicates the same congestion price to each of its users and/or each user communicates its willingness to pay for rate allocation to the network, in our algorithms each link provides a possibly different congestion price to users of different traffic type and each user provides its willingness to pay for the reduction of delay/loss to the network.

Underlying Methodology

Layered architectures form one of the most fundamental structures of network design, but there is little quantitative understanding to guide a systematic, rather than an ad hoc, process of designing layered protocol stack for wired and wireless networks. Our research contributes to such understanding.

The underlying mathematical framework of this work is 'Layering as Optimization Decomposition', which has two cornerstones 'Networks as Optimizers' and 'Layering as Decomposition' [21]. The framework views the network as the optimizer itself, puts the end user application needs as the optimization objective, and offers a common set of methodologies to design modularized and distributed solutions.

Our framework provides a top-down approach to design layered protocol stacks from first principles. The design is driven by applications where we focus on multimedia.

To deliver better end-to-end media quality and achieve more efficient communication resource sharing, by distributed optimization in media content adaptation and resource allocation, we adopt two directions, the vertical direction and the horizontal direction.

In the *vertical* direction, the requirement from the application layer motivates and shapes the adaptation of the lower layers. For instance, the content awareness can be derived from media sequences to define richer media quality metrics and utility, which will help intelligent media coding and communication decisions across application, networking and link layers.

In the *horizontal* direction, media communication rarely happens in a single-user and single-hop situation. Conflicts in communication resource sharing need to be addressed. We are therefore looking at efficient ways to exploit multi-user diversity in channel states and content utility-resource tradeoffs. We develop distributed and collaborative solutions for resource allocation, as well as source media adaptation. In some designs, pricing approaches from Economics are applied.

1.3 Intelligent Network Engineering for IPTV and P2P Systems

By applying the optimization in network resource allocation introduced above, in this work, we present intelligent video streaming over congested networks and intelligent video scheduling for video delivery in peer-to-peer networks.

For video clips competing bandwidth in networks, we propose an approach of Content-Aware distortion-Fair (CAF) video delivery scheme, which is aware of the characteristics of video frames and ensures max-min distortion fair sharing among video flows. The optimality goes to the utility fairness among users where the utility is measured by the video distortion. Different from bandwidth fair sharing, CAF targets end-to-end video playback quality fairness among users when bandwidth is insufficient, based on the fact that users directly care about video quality rather than bandwidth. The proposed CAF approach does not require rate-distortion modeling of the source, which is difficult to estimate, but instead, it exploits the temporal prediction structure of the video sequences along with a frame drop distortion metric to guide resource allocations and coordination. Experimental results show that the proposed approach operates with limited overhead in computation and communication, and yields better QoS, especially when the network is congested. Part of the work is presented in [46].

For Internet based video broadcasting applications such as IPTV, the Peer-to-Peer (P2P) streaming scheme has been found to be an effective solution. An important issue in live broadcasting is to avoid playback buffer underflow. How to utilize the playback buffer and upload bandwidth of peers to minimize the freeze-ups in playback, is the problem we try to solve. Here for each peer the utility (happiness) is measured by the length of the content to be played in the playback buffer and the optimality is to optimize the social utility (efficiency in utility). We propose a successive water-filling (SWaF) algorithm for the video transmission scheduling in P2P live streaming system, to minimize the playback freeze-ups among peers. SWaF algorithm only needs each peer to optimally transmit (within its uploading bandwidth) part of its available video segments in the buffer to other peers requiring the content and pass small amount message to some other peers. Moreover, SWaF has low complexity and provable optimality. Numerical results demonstrated the effectiveness of the proposed algorithm. Part of the work is presented in [44].

2 Content-Aware Distortion-Fair (CAF) Video Streaming

2.1 Introduction

Motivation

The recent advances in transmission, display, and storage capabilities in the Internet and end-user devices have ushered an unprecedented growth of video traffic over Internet. Popular video repository solutions like Youtube are driving up this video traffic demand on the network everyday. Unlike web and email traffic, video is characterized by its large bandwidth requirement and stringent quality of service (QoS) parameters, especially delays. This presents a challenge to network engineering different from text and voice dominated traffic [76].

In video traffic, certain packets can be dropped if the network is congested, with controllable decoding quality degradation. Transcoding solutions [78] can offer good flexibilities in rate-distortion shaping and the scalable video coding [77,60,65] can offer temporal, spatial, and signal-noise-ratio (SNR) scalability. All these technologies carry out the adaptive video source coding at the application layer, providing the opportunity of smartly dropping frames or packets when the communication links are congested or bandwidth is limited.

For the problem of transporting video traffic over a network, the design principal is to coordinate network layer resource allocation with the application layer video adaptation schemes, such that a graceful trade-off between resource consumed and quality of service delivered is achieved. Advances in video signal processing and coding have made a rich set of video coding and adaptation tools available to content delivery network engineering.

Traditionally the content generation and the network resource engineering are designed separately. Network engineering traditionally is not content-aware, and the network resource allocation does not allocate the resource adaptively catching up timely the contents. This is appropriate for data traffic which requires every packet to be received successfully. But for video traffic, different frames may have different importance and can contribute differently in terms of the distortion perceived by human eyes, and dropping some less important frames may only give us some little influence on the perceptual video distortion, if the important frames could have better guarantee to be transmitted successfully, or, if the limited bandwidth is used for those important frames while dropping out some less important frames when the link is congested, end users may experience better QoS. The content-awareness can be utilized to improve the QoS provision. In this work, we study video streaming over Internet problem, and develop a content-aware solution for multi-hop video communications.

For network engineering, an important issue is the congestion control at the transport layer. What is interesting and extensively researched in the last decade is that the problem of Internet congestion control can be cast as an optimization problem with economics interpretation [55, 54, 36], the network utility maximization (NUM), which aims to maximize the aggregate utility of all the users, subject to the constraint that the total flow on each link should be no greater than the link capacity. Here, the utility of each user s with a transmission rate x_s is a strictly concave, continuously differentiable non-decreasing function $U_s(x_s)$, measuring the user's 'happiness' when allocated rate x_s. By decomposition method with a pricing interpretation, the optimization problem can be solved via distributed algorithms. Since the publication of the seminal paper [36] in 1998, the framework of NUM has found many applications in network resource allocation algorithms and design of protocol stacks [21]. It has become a mathematical foundation of many cross-layer designs.

Cross-layer design has shown its advantage for QoS provision and resource allocation. The NUM framework can be extended to deliver video, where the utility function can be the opposite of the distortion function (measuring the QoS) of transmission rate. There are existing works in cross-layer optimization in video delivery, e.g. [75, 31, 58, 20, 19, 68, 83]. These works are mainly focusing on the application layer, MAC layer, physical layer, and other layers.

Particularly, a branch of cross-layer design, the content-aware networking, brings the intelligence of video source coding at the application layer and the network resource engineering together. A lot of works utilize content-awareness in different scenarios, such as in video summarization [51], packetized video delivery [22, 66], in P2P streaming [8], in wireless networks [47, 32], in content delivery networks [7], and in image authentication [85]. In [51, 22, 66, 8, 32], the content-awareness lies in the importance of the frames or packets based on the distortion and the dependency of decoding. In [47], the content-awareness is particularly a metric of the motion characteristics of different scenes. In [7], the content-awareness lies in the knowledge of descriptions of the multiple description source coding. In all these works, the content-awareness is utilized adaptively fitting in the delivery requirement. For instance, through content analysis and optimization, video summarization schemes such as in [51] [53] select a subset of frames from the video sequence to form a concise representation of the sequence, to deliver good visual quality at low bit rates. The work for multi-access wireless video streaming [32] develops a network utility maximization with a resource pricing solution via dual decomposition and the resource price in turn, drives content adaptation solutions at each user in a distributed fashion. In this work, we focus on the similar content-awareness as in [51, 32]. Rather than one-hop multi-access networks in [32], we study in the setting of general multi-hop Internet, where we have to deal with the dependency of users and links in terms of who sharing which link.

This work presents a step towards matching content processing with network engineering so as to maximize the user perceptional utility. We propose a framework for resource allocation, including the joint source adaptation at the application layer and link congestion control at the transport layer.

Another theme of this work is about the end-to-end user utility (measuring user's happiness about QoS) fairness. The NUM framework mentioned above is for the maximum aggregate utility, which may not mean utility fairness. A problem of fairness in QoS is very important in network resource allocation and it is of our interest here. Since motion-rich video clip may require more bandwidth than video clip with less motion, when such users are sharing the resource, the bandwidth fairness among users which results in the same bandwidth allocated to different users may not be a good metric. The end users care about the video quality rather than the bandwidth, hence we focus on max-min fair utility provision among uses. The utility can be the opposite of the distortion and we focus on the end-to-end min-max distortion fairness among users.

Related Work

There are existing works on max-min QoS fair sharing, such as [17] [74]. Authors of [17] [74] assume an explicit utility function of rate and the algorithms are based on the utility function, not really based on content. Studies in [17] [74] are not specifically for video and they do not take into account the special characteristics of video traffic. For video, the utility function is time-varying and it is hard to estimate the utility function accurately and on the fly. In addition, the distributed algorithm in [74] may need a lot of iterations and the rate may fluctuate dramatically such that the video quality fluctuates and the perceptual quality may be bad. Different from [17] [74], we do not assume any explicit utility function, but instead we use the utility of every frame which can be easily and explicitly calculated.

In this work, we propose a cross-layer optimization scheme in network engineering for video content delivery. The source video adaptation and traffic shaping along with distortion metrics in the application layer are driving the congestion control and resource coordination at transport layer. The resulting Content-Aware distortion-Fair (CAF) video delivery algorithm, in which the users and links can distributively cooperate with others such that the end-to-end distortion fairness as well as good video quality are achieved among users.

Our work and many studies such as [35,9,64,82,26,18,34] and works mentioned above on the cross-layer video delivery and content-aware networking are all in favor of the optimal rate-distortion tradeoff, but our work is more on the metric of min-max distortion fairness for many users sharing a general multi-hop network, while others do not have such emphasis.

Our work has rich granularity in how to drop frames when the link is congested. In [9], the authors show that the priority dropping frames of layered video can have gain over uniform dropping. The dropping is based on the importance of the layers of the video source coding, while in our work the frame dropping is based on the importance of frame, not limited to layers.

Motivated by that a lot of video in the Internet is stored, our work is mainly for stored video, where the side information of the number of bits and importance of the frames can be calculated in advance. Some existing works, such as [35], are for real time video. In [35], the authors propose an adaptive congestion control scheme for real time packet video transport, where the video rate-distortion needs to be estimated in real time. Our work can be extended and applied to deal with real-time video by allowing some computational complexity to derive the importance and length of frames.

In the proposed algorithm, we have per-flow performance guarantee. Different from Intserv [16] architecture which also offers per-flow performance guarantees, our scheme has content awareness which is signaled over links. Our work is also different from DiffServ [12] which manages resources at the granularity of traffic classes.

The main contributions of this work are as follows:

1. For multi-hop multi-user video delivery networks, we propose a framework of content-aware distortion-fair networking with joint video source adaptation and network resource allocation, by combining the intelligence of the temporal scalability of the video source coding at the application layer with the congestion control at the transport layer.
2. The framework makes users drop less important frames when the network is congested. Architecturally we illustrate that the approach of end users dropping less important frames is in general more bandwidth-efficient than the approach of links dropping frames when the network is congested.

3. Based on the framework, we propose an algorithm with a provable small number of iterations for users and links to cooperate, to achieve the min-max distortion fairness.
4. By simulations, we show the performance improvement achieved by our design.

The rest of this section is organized as follows. Section 2.2 explains the video content adaptation schemes to shape video traffic and to achieve rate and distortion tradeoffs. In Section 2.3 we give the Content Aware distortion-Fair (CAF) video networking formulations and in Section 2.4 we derive a solution. The effectiveness of our solution with simulation results is demonstrated in Section 2.5. Section 2.6 presents conclusions and future work.

2.2 Video Adaptation and Traffic Shaping

Video traffic is different from other Internet traffic like data, email and web. Video traffic is in general elastic in the sense that certain packet losses may incur distortions in playback but not catastrophic. Therefore the utility from end users' point of view, is not how many bits are delivered, rather the received sequence playback quality. Advances in video signal processing and coding have made available a rich set of coding [71] and adaptation tools, as well as quality metrics that allow applications to trade off video traffic rate with the play back quality. Transcoding [78] based solution offers the most flexibility and rate efficiency for this task, as it in essence re-encodes the sequence to suit specific limits in traffic profiles. However the cost of transcoding in computation is considerable and it is not suitable for multicast video sessions with large number of receivers. Scalable Video Coding (SVC) [60] offers the benefit of flexibility in video traffic shaping without transcoding and is well suited for the Internet video delivery. Scalability can be achieved in temporal, frame size, and SNR quality of video.

In this work, we target the temporal scalability of video traffic only, as more advanced scalability features like frame size and SNR scalability are not widely adopted yet, and there exists a large legacy video content coded in MPEG-1/2/4 [1] that has this temporal scalability built in. Also exploiting temporal scalability only requires minimum software changes on consumer side devices. At the time of congestion over the Internet links, certain frames in the video stream can be dropped with limited loss in playback quality. How to quantify this loss and have a unified metric that can allow multiple video sessions to have a quality fair streaming is the goal.

Not all video frames are created equal in terms of visual quality impact if lost, as well as the number of bits required to encode. The MPEG [28] [1] coding scheme supports bi-directional prediction frames, or B-frames, which can be dropped without affecting subsequent frame decoding, while if an intra-coded I-frame or predictive coded P-frame is lost, the subsequent frames with prediction on them will not be able to correctly decode.

This is illustrated in Fig. 1 below, for a group of pictures (GoP) in an MPEG sequence. Consider the GoP with frames $20 \sim 28$, I-frame f_{20} is independently

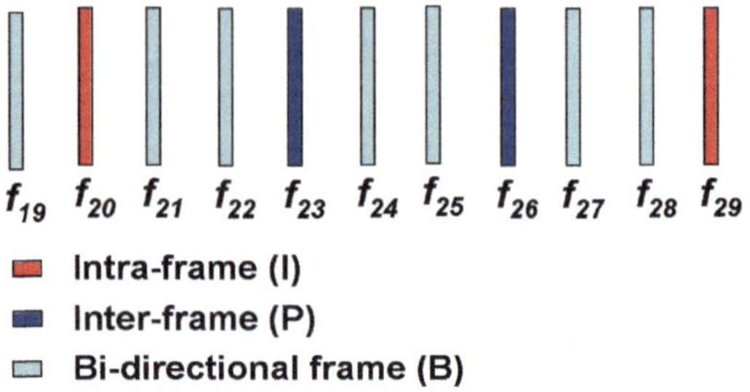

Fig. 1. Group of pictures.

encoded. Its decoding does not depend on any other frames. Inter-frames f_{23} is decoding dependent upon correct decoding of f_{20}, while f_{26} is dependent on f_{23}. For B-frames, they are decoding dependent on correct decoding of its nearest neighboring I-frames and P-frames, e.g, f_{22}'s decoding depends on f_{20} and f_{23}. Therefore, if an I or P-frame is lost, all frames thereafter in the GoP become un-decodable. So if the system decides to drop an I or P frame, it is equivalent to dropping all frames thereafter to the next GoP start.

To characterize the frame loss distortion, in our previous work [51] [53], we developed a frame drop distortion metric for video rate-distortion video summarization. It measures the distortion to the sequence playback as the difference between the original sequence and the zero-th hold playback of the received frame with frame losses. Given an n-frame sequence $V = \{f_1, f_2, \cdots, f_n\}$, denote the set of frame losses in the sequence as Λ,

$$\Lambda = \{\lambda_1, \lambda_2, \cdots, \lambda_m\}, \quad \lambda_k < \lambda_{k+1}, \tag{4}$$

where λ_k is the index of the lost frame. Let the playback sequence be $V = \{f'_1, f'_2, \cdots, f'_n\}$, where playback frame is either the original frame, or the nearest frame that is not lost in the sequence (the latest frame which is not lost prior to the set Λ),

$$f'_k = f_t, \quad t = \arg\max_{j \notin \Lambda} j \leq k, \tag{5}$$

then the sequence distortion can be characterized as the distortions incurred by the frame losses during play the playback,

$$D(k) = d(f_k, f'_k)$$

where the frame loss distortion metric $d(f_k, f'_k)$ is computed as the principle component analysis (PCA) space distance between scaled frames [51]. An example for "foreman" sequence with frame losses are plotted in Fig. 2, where the upper sub-figure shows the frame-by-frame distortion $d(f_k, f_{k-1})$, which gives an indication of activity levels in the video sequence, with the stems in the upper sub-figure marking the remaining frames, and the lower sub-figure shows the resulted distortion $D(k)$ due to the frame losses.

Given the set of frame losses as Λ in (4), the resulting video traffic rate reduction because of the frame losses is characterized as,

$$R(\Lambda) = \sum_{k=1}^{m} R(f_{\lambda_k}), \tag{6}$$

where $R(f_{\lambda_k})$ is the rate of frame f_{λ_k}. The resulted distortion associated with set Λ can be characterized as,

$$D(\Lambda) = \sum_{k} d(f_k, f'_k), \tag{7}$$

where f'_k is given by (5).

Given the decoding dependency of the frames in the GoP in MPEG sequence, dropping different types of frame can result in different set of frame losses according to the decoding dependency. If the GoP anchor I-frame is lost, then the whole group of pictures is lost. If a P-frame is lost, the frames thereafter in the GoP are useless. If only a B-frame is lost, the rate reduction and the resulted distortion are only limited to the B-frame. For the example in Fig. 1, if I-frame f_{20} is lost, the whole GoP with frames $f_{20} \sim f_{28}$ are lost, if P-frame f_{26} is lost, frames $f_{26} \sim f_{28}$ should be expunged, and if B-frame f_{27} is lost, only itself is lost. Hence we have the following frame dropping set associated with the frame taking into account the decoding dependency,

$$\Lambda(f_k) = \left\{ \begin{array}{ll} \{f_k\}, & \text{B frame} \\ \{f_j | j = k-1, \cdots, k+n-1\}, & \text{P frame} \\ \{f_j | j = k, \cdots, k+n-1\}, & \text{I frame} \end{array} \right\}. \tag{8}$$

Therefore, for frame f_k, the associated rate reduction is $R(\Lambda(f_k))$ and the resulted distortion is $D(\Lambda(f_k))$.

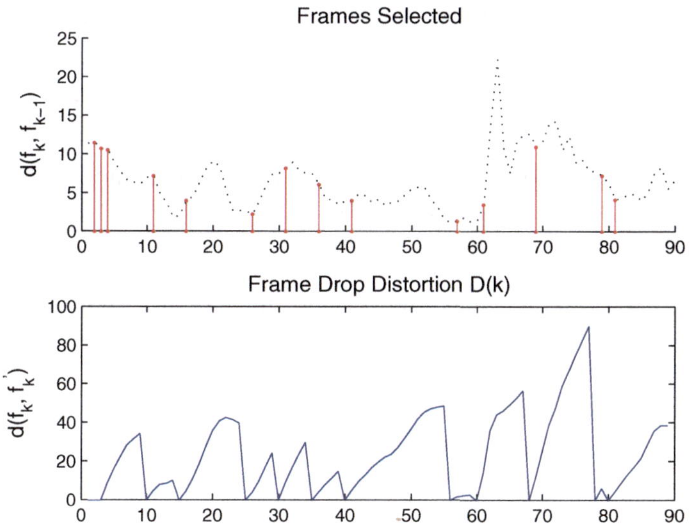

Fig. 2. Distortion: An example for "foreman" sequence with frame losses.

Obviously, different types and locations of frames in the GoP will have different rate and distortion implications if lost. From Information Theory we know that a variety of sources have convex rate-distortion (R-D) functions [23]. This convexity holds true for most practical video source and its operational R-D functions. In practice, there would be a combination of B, P and I frame drops. Some example of operational R-D curves of the sum distortion of $D(k)$ and loss in rates are plotted in Fig. 3. They are indeed exhibiting convexity.

Imagine at certain bottleneck link in the network, the video queues need to be reduced. Armed with the knowledge of R-D operational pair for each frame, we could have a quick sorting of the distortion D, or distortion/rate ratio, D/R, and come up with optimized solutions for either distortion fair (QoS fair), or total distortion minimization. In a network with multiple hops and multiple flows, we develop a distortion fair solution in the following sections.

2.3 Content-Aware Distortion-Fair Video Streaming

Max-min Utility Fair Share

NUM, its basic version 1 is the problem of maximizing the total utility over the source rates $\mathbf{x}$, subject to linear flow constraints $\sum_{s \in S(l)} x_s \leq c_l$, for every link l. The dual decomposition solutions of the basic NUM [36] give us a distributed solution that iterates on source rate control and link adaptation. The objective of the basic NUM is to achieve social efficiency in utility, but efficiency may not mean fairness, which is very important in resource allocation.

In this work, we look at a different problem, that is to achieve distortion fairness among video flows, subject to link capacity constraints. Can we have a similar distributed, iterative solution?

The problem we address in this work, the utility max-min fair sharing, is problem 1 with objective function as 3 instead.

Utility and Distortion

In this work we focus on the problem of stored video streaming and exploit the IBPB GoP structures in the MPEG coded video. We base our video adaptation and traffic shaping on frame drop mechanisms and distortion metrics discussed in Section 2.2.

For a video session, the video adaptation and traffic shaping are achieved through frame pruning, dropping frames with frame indices $\Lambda = \{\lambda_1, \lambda_2, \cdots, \lambda_m\}$ when there is congestion. The resulting video traffic rate reduction is characterized in (6) and the distortion is characterized in (7).

For a given video segment with n frames, there are total $\sum_{k=1}^{n} \binom{n}{k}$ different combination of frame drop options. Because of the MPEG coding structure, some frames have dependency in decoding. If an I or P-frame is lost, then the frames from P frame to the end of the GoP are not decodable, therefore shall be pruned also. This is given by (8).

We define the *importance of a frame* f_k, or the *utility* of a frame f_k, as the resulting distortion summation if this frame is lost, taking into account the decoding dependency,

$$D(\Lambda(f_k))$$

where function D is given by (7) and $\Lambda(f_k)$ given by (8).

Consider a typical 15-frame GoP structure $I_1B_2P_3B_4P_5B_6P_7\cdots\cdots B_{14}P_{15}$. If frame P_7 is lost, the frame drops in effect are $\Lambda(P_7) = \{B_6, P_7, B_8, P_9, \cdots, B_{14}, P_{15}\}$. While if a B frame is lost, the resulting pruning is limited to that frame only, i.e, $\Lambda(B_k) = \{B_k\}$.

In this work, instead of optimization over all combinations of frame drops, we only consider the following drop sequences for a given GoP to approximate its rate-distortion curve. Sort all individual frames by its importance, or utility, as resulting distortion sum if lost, then at the time of link adaptation, we start with an empty frame drop set Λ, and keep adding frames to Λ in the increasing order of its distortion $D(\Lambda(f_j))$. This iterative process of traffic shaping is given as,

$$\begin{aligned} &\Lambda^0 = \varnothing \\ &\Lambda^i = \Lambda^{i-1} \cup \{\Lambda(f_{j=\arg\min_{f_j \notin \lambda^{i-1}} D(\Lambda(f_j))})\} \end{aligned} \tag{9}$$

This way we can approximate R-D curve represented by this distortion incurred and bits saved curve, plotted as examples in the Fig. 3 below. For the "foreman", "paris", "bond", and "akiyo" sequences, we look at a window of 2 sec and operate the frame drop adaptation according to (9). For a bit rate reduction in range of 0 to 20kbps, the resulting distortion incurred are plotted.

Note that in Fig. 3, the resulted distortion actually reflects the importance or the utility of the dropped frames. The shape of the curves in Fig. 3 is the opposite of the traditional R-D curves, because we have the rate as the transmitted rate in the traditional R-D curve while in Fig. 3 we have the rate reduction as the horizontal axis. The distortion in traditional R-D curve in some sense can be regarded as the opposite of the resulted distortion or the utility/importance of the dropped frames.

Min-max Distortion Fair Share

For aggregate network utility maximization (NUM) work [36], the dual decomposition gives it a distributed solution that allows source to adapt its rate based on congestion prices at links, in an iterative fashion. In this work, we look at the quality fair formulation, i.e., to give everyone the same quality of service as possible, given the network resource constraint. This is given as,

$$\begin{aligned} \text{minimize}\ \ & \max D_s(\Lambda_s^l, l \in L(s)) \\ \text{subject to}\ \ & \textstyle\sum_{s\in S(l)}(R_s^0 - R(\Lambda_s^l)) \le c_l,\ \forall l \\ & \Lambda_s^l \subseteq \Lambda_s^k,\ \forall s, \forall l, k \in L(s), k\text{: the next hop of } l \\ \text{variables}\ \ & \Lambda_s^l, \forall s, \forall l \in L(s). \end{aligned} \tag{10}$$

where Λ_s^l is the frame drops at link l for source s, $L(s)$ is the set of the links that source s uses, $S(l)$ is the set of users who use link l, and R_s^0 is the original data

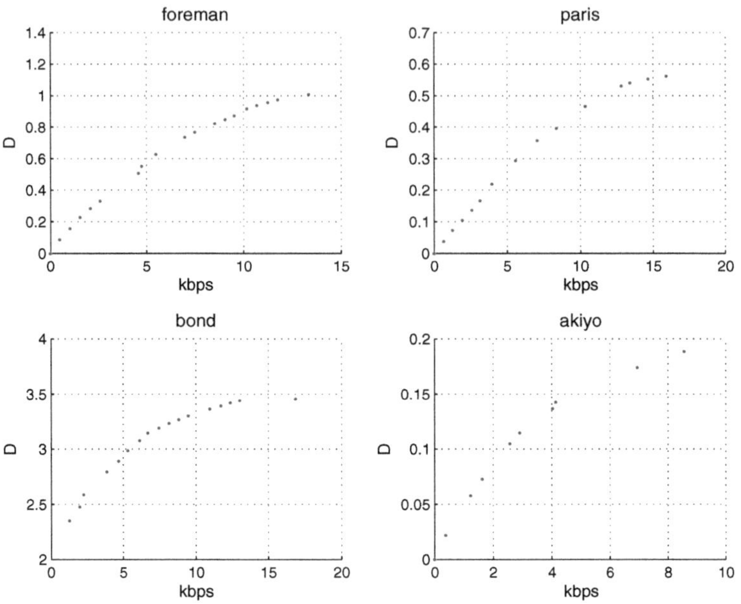

Fig. 3. Rate-saved v.s. resulted distortion curve. Each iteration gives us a set of frames dropped, the resulted distortion of the drops (the importance or the utility of the dropped frames), and rate reduction.

rate for source s without frame drops. In this problem, the first constraint states the flow should not exceed the link capacity, and the second constraint states that for each flow, the set of the frames dropped in the next hop should contain such set at the current hop. The variables are the sets of the frames dropped for each flow at each link it uses.

In this work, since instead of optimization over all combinations of frame drops, we only consider the drop sequences as described in (9) for a given GoP to approximate its rate-distortion curve. The min-max distortion fair sharing problem (10) can be simplified with the way we do traffic shaping as in (9) and the curve of the utility (importance) of the dropped frames v.s. reduced rate for video sequences, such as Fig. 3. Denote the resulted distortion in Fig. 3 as z_s for video flow s. The min-max distortion fair sharing can be stated as to maximize the minimum importance level of z_s among all flows, where the frames with importance less than z_s are pruned for each flow s, and the remaining total flow after pruning of each flow over link l should be no greater than the capacity.

Hence the problem becomes to find out the threshold z_s of dropping frames for each flow s, where the frames with less importance than the threshold are pruned, such that the link capacities are not exceeded after such pruning. This gives us the solution that is based on the thresholding of the distortion levels.

Note that the traffic shaping as in (9) is based on the importance of frames. Although we draw the R-D curves in Fig. 3, for congestion control, we do not need to send links such curves to let them decide how to drop frames to restrict

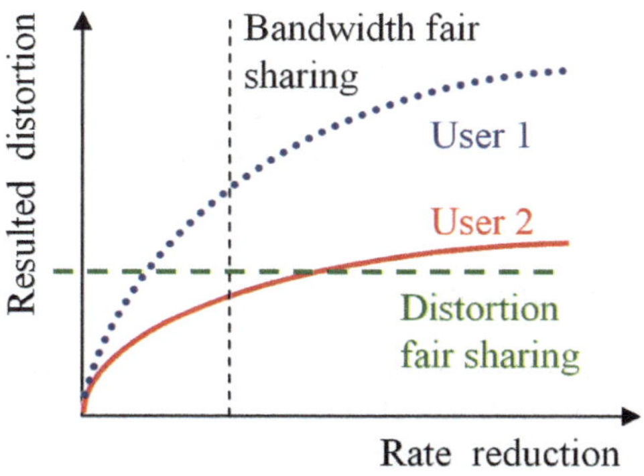

Fig. 4. Illustration of distortion fair sharing.

the flow less than the link capacity, but instead, the links only need to know the importance of the frames and the length of the frames, because the information of the frame importance and length has contained the R-D curve. So for the link adaptation, we only need to signal the links the side information of the importance and number of bits of the frames.

Content-Aware Distortion Fair Share: Users Help Each Other

There exists fairness in terms of the bandwidth sharing. Why do we need distortion fairness, not the bandwidth fairness? Users care about utility (QoS) rather than bandwidth, but bandwidth fair sharing may not always give us utility fairness because the same utility may need different bandwidth according to different content.

Consider two users, each sending a video clip over a common link. When they compete for the bandwidth, by the bandwidth fair sharing, each user gets half of the link capacity. But they may not have the equal happiness. Say, User 1 is transmitting a video with a lot of motion, like commercial, User 2 is transmitting a video with less motion, like a person sitting there reading news. The rate-distortion is illustrated in Fig. 4. If the link is congested and some frames of the video should be dropped, perceptually User 1 is unhappy about the dropping because the content is sensitive to the loss, but User 2 can be happy even if a lot of frames are dropped because the content is not loss sensitive. With distortion fairness, User 1 can get more bandwidth than User 2, but perceptually both users are equally happy. Hence the sharing should be content-aware and distortion fair, with the advantage that flows with less motion help motion-rich flows. Such advantage can be readily viewed through Fig. 3 from the motion-rich video 'foreman' (as User 1) and video 'akiyo' (as User 2) which is with less motions.

Details of our scheme are in the next section.

2.4 Frame Dropping for Max-Min Utility Fair Share

When there is congestion on the link, we use the video traffic shaping technique discussed in Section 2.2 and Section 2.3 to prune the video queues. Considering that frames may have different importance, we propose a smart frame dropping scheme.

A Smart Frame Dropping Scheme

The strategy is to send side information indicating the utility (importance) and length of each frame to the links, then links and users cooperatively decide the optimal threshold of dropping frames for each user, where the frames with less importance than the threshold will be dropped. We define the utility of each frame based on the distortion induced if the frame is lost, while taking into account of the prediction structures of the frames.

Min-max distortion fair (max-min utility fair) bandwidth allocation means at the optimal distortion, the distortion of user s cannot be decreased while still maintaining bandwidth feasibility, without increasing the distortion of some other user with a higher distortion than user s.

For the case of multiple users sharing only one link, the link decides a common threshold (distortion level) of dropping frames for all users, such that the total traffic through the link is less than the link capacity. The same threshold of dropping frames makes users experience a min-max fair distortion.

In a general network, the end-to-end path of a user may consist of several links, each of which may be shared by different users. If each link decides a common threshold of dropping frames for its users, since links are heterogeneous, on its end-to-end path a user (flow) may see the links each with a different threshold of dropping frames, and naturally the distortion level of dropping frames for a user should be decided by the most strict one, say, the highest distortion level over all the links on its path. If a link contains some user who is bottlenecked at other links, the link should know this user's frame dropping level, reduce its capacity by this user's flow, and re-allocate a common threshold of dropping frames for its other users with the remaining capacity, leading a lower common distortion threshold, or equivalently, getting through more traffic for its other users. In this way, each user gets distortion as low as possible and fairly, and the distortion of the user who experiences most stringent bottleneck is minimized, achieving the min-max distortion fair share.

Who drop frames, links or users?

In our approach, the users drop frames based on the threshold, which is found by the signaling over links and users in a control plane. In this way, the rate of the flows is controlled, and the link bandwidth is efficiently utilized.

The other option could be that the links drop frames, say, the link can look at its buffer and drop the less important frames when it is congested. Such scheme looks fine if the bottlenecks are at the beginning of the end-to-end paths, but if

Table 1. Comparison of designs

	Who drops	Content-awareness	Signaling
CBL	link	no	no
Baseline	link	simple	no
CAL	link	yes	no
CAF	user	yes	yes

the bottlenecks are in the middle or at the end of the paths, there will be some unnecessarily larger volume of the flow over the links before the bottleneck link. Such scheme is not good in networks whose bandwidth is very expensive, such as in an edge network, or in a wide area network (WAN).

In Table 1, some different designs are compared. Content-Blind Link dropping (CBL) is an approach that link acts as FIFO droptail queue and it is content-blind, not considering any frame information; Baseline is an approach that link acts as FIFO droptail queue and it is partially content-blind where I frames are transmitted if room permits; Content-Aware Link dropping (CAL) is an approach that link acts as a smart link where the frame dropping is content-aware, i.e., the link will find out the threshold of dropping frames based on the importance of every frame; and Content-Aware distortion-Fair (CAF) is an approach proposed in this work that user drops frames based on importance, and the links cooperate and signaling is used to pass the massage, where the entire network including users and links is intelligent.

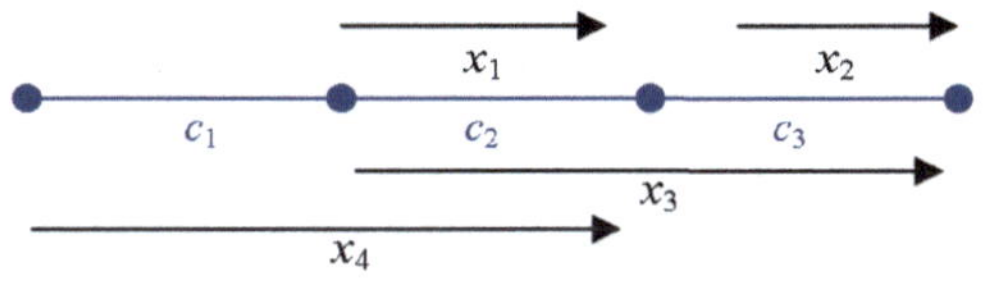

Fig. 5. A network with 3 links and 4 users.

To see the advantage of user dropping versus link dropping, we present an example here.

Consider a network as in Fig. 5.

Assume c_1, c_2, c_3 = 200, 540, 200 kbps. Each user has a constant bit rate video with 200 kbps. Assume each user has the same utility function.

Table 2 shows the resulting transmitting rate x of users and the load y of links. User dropping saves bandwidth of link 2 by 8%, and rate of user 1 and 4 increases by 11.1%.

User dropping frames not only provides better end-to-end quality (see user 1 and 4), but also reduce the load of the links (see link 2). This is true for general network topology.

In addition, user dropping frames matches Transmission Control Protocol (TCP) used in nowadays Internet, which requires end users to adapt rate.

Table 2. Comparison of user dropping over link dropping

	x_1	x_2	x_3	x_4	y_1	y_2	y_3
Link dropping	180	100	100	180	200	540	200
User dropping	*200*	100	100	*200*	200	*500*	200

Since network link capacity may not have noticeable change in the scale of round trip time (RTT), in this work, we assume link capacities are constant for a short time interval.

Algorithm for the Optimal Threshold of Dropping Frames

We propose the following Content-Aware distortion-Fair (CAF) algorithm to find the optimal threshold of dropping frames for each user.

CAF Algorithm

For a given interval W, the links and users do the following,

Initialization:

All links set 'Done=0' and $v_l = maxU$; All users set 'Done=0' and $z_s = maxU$, where $maxU$ is a given upper bound of the frame importance.

Iterations:

Do

1. Threshold computation at link l:
 If link l 'Done=0',
  Link l reduces its capacity c_l by the rate (determined by z_s) used by users marked 'Done=1',
  Link l calculates a common threshold v_l for the remaining users within the remaining capacity.
  If link current v_l= previous v_l
   Link l set itself 'Done=1' and the link sets all its users 'Done=1' and $z_s = v_l$
  End If
 End If
2. Threshold update at user s:
 If user s 'Done=0',
  User s set the frame dropping level $z_s = \max v_l$ over all the links on its path.
  Need a backward message passing.
 End If

until all users' 'Done'=1.

User source adaptation:

User s drops the frames whose importance is less than z_s.

In the algorithm, each link finds an equal distortion level to its users not marked 'Done'; each user sets its frame dropping level as the most stringent one over all the links on its path; afterwards, if the link is bottlenecked based on its users updated dropping level, the link marks itself and the users on it 'Done', and other links reduce its bandwidth by the flow of the users marked 'Done'; then the iteration goes to another round. The algorithm iterates among links and then flows, until the thresholds can not be reduced further. The criterion to decide whether a link is bottlenecked is that if the link is fully loaded or if the dropping level is zero at this link in the current iteration round, equivalently, if the current threshold would not change in the next iteration. Note that the concept of the bottleneck is extended to the bottleneck for the case that the flows are fully elastic, which means the flow can always increase if the link permits. In our case for the video clip which is already coded, the flow has a fixed maximum rate, hence if the link has plenty of bandwidth which can support the maximum transmission rate of video, the frame dropping level is zero, but if the flow is fully elastic (no upper bound), the link will be fully loaded.

Number of Iterations of the Algorithm

In each iteration in CAF algorithm, there will be at least a link whose current threshold v_l will not change in the next iteration, and such link is a link which would be fully utilized if the flows on it are fully elastic without upper bound. Hence we have the following Proposition.

Proposition 1. *The number of iterations of the algorithm is at most the number of bottlenecks, where the bottleneck is counted in the same network where all the flows are fully elastic without upper bound.*

2.5 Simulations

Here we show some simulation results. These results are only meant to demonstrate the effectiveness of the proposed scheme, not a full scale simulation with real network.

For the advantage of distortion-fairness over bandwidth-fairness, it can be readily viewed through Fig. 3 from the motion-rich video 'foreman' (as User 1) and video 'akiyo' (as User 2) which has less motion, as illustrated in Fig. 4. We omit the details here.

When the capacity at some links in the network is not sufficient to support the coming traffic, if we do not allow frame dropping, bottlenecks will appear at those links, with resulting delays creating playback "freeze ups" for certain flows. This is not desirable from end users' point of view. Instead, we allow certain frames to be dropped to clear up bottlenecks and to meet the stringent delay QoS requirements.

Obviously, most routers nowadays have packet drops in a content-blind fashion, for example, if a link is a first-in-first-out (FIFO) queue with a fixed maximum queue length limit, and if the queue meets the maximum length, the incoming frames are dropped. When there is congestion, if the packet drops are

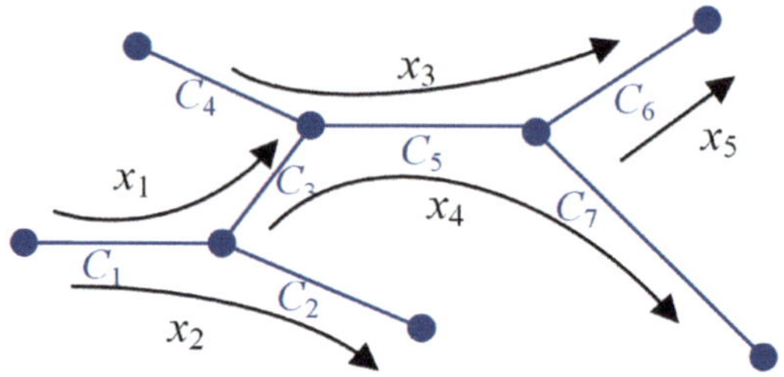

Fig. 6. A 7-link 5-user network. The link capacities are 150, 200, 250, 150, 200, 150 and 200 Kbps for links 1-7 respectively.

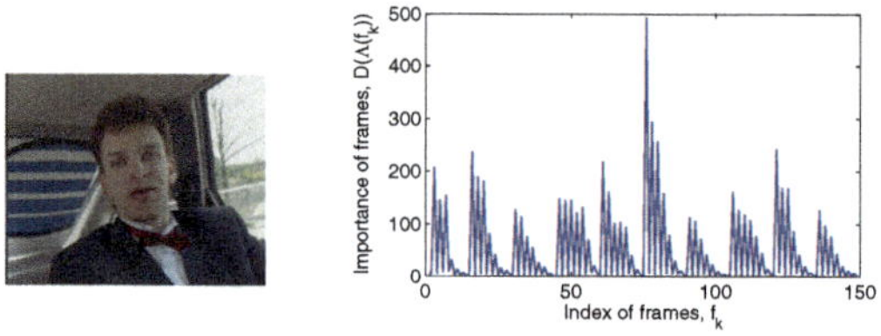

(a) Carphone (b) Frame utility

Fig. 7. Video clip Carphone, utility of each frame taking into account of frame dependency.

content blind, an I frame could be dropped with resulting catastrophic loss of quality for the whole GoP.

The above approaches are obviously not good when the links are congested. Due to the space limitation, we omit the comparison of CAF with these approaches. Instead, we show the comparison with the following baseline approach.

A *baseline* approach for comparison is that the frames from users are served at each link in a FIFO fashion, and assume for time interval W (seconds) of one GoP, link l can only serve frames with total length of Wc_l (kbits) and the remaining frames in the GoP of the users on this link are lost, for the purpose of stabilizing the queueing. This also guarantees that the most important frame, I frame, in each GoP will be served if link capacity permits.

We tested CAF approach in networks and our experiments demonstrate effectiveness of CAF. Here we show a random example where the network has 7 links and 5 users, as shown in Fig. 6. Users 1, 3 send video Foreman, users 2, 4 send Akiyo, user 5 sends Carphone (vbr, 144 Kbps). The utility of each frame taking into account of frame dependency for "Carphone" is shown in Fig. 7.

There are 3 bottlenecks in the network, links 1, 5, and 6. We verified that after 3 iterations of our algorithm, as indicated in the Proposition. Our CAF approach yields the distortion after frame dropping: for example, for user 3 and 5, the distortion is shown in Fig. 8; and for user 4, it only has one B frame dropped and the distortion is almost zero. Note that user 3 and 5 have almost

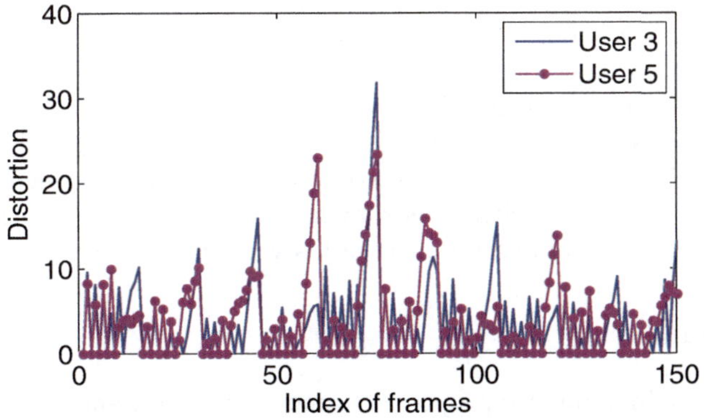

Fig. 8. Distortion after frame dropping.

Table 3. Max distortion level.

	User1		User2		User3		User4		User5	
	BL	CA	BL	CA	BL	CA	BL	CA	BL	CA
1	190	7.93	3.56	7.93	190	19.5	1.07	0	155	19.5
2	77.1	14.1	22.9	14.1	77.1	29.6	2.03	0	82.7	29.6
3	121	6.45	6.45	6.45	121	28.6	0.85	0	56.6	28.6
4	62.3	4.70	3.62	4.70	62.3	24.4	1.15	0	133	24.4
5	296	9.74	4.84	9.74	296	41.9	0.55	0	95.5	41.9
6	135	7.07	5.99	7.07	135	31.3	0.67	0	79.8	31.3
7	65.5	7.22	4.19	7.22	65.5	22.9	0.67	0	24.6	22.9
8	162	5.15	2.68	5.15	162	19.1	0.37	0	79.1	19.1
9	95.6	6.75	10.3	6.75	95.6	21.0	1.87	0	88.0	21.0
10	136	8.85	9.03	8.85	136	29.3	1.03	0.2	21.6	29.3

Table 4. Total distortion

	User1	User2	User3	User4	User5
Baseline	1802	537.4	1802	83.03	1094
CAF	307.5	136.6	596.6	0.17	630.4

the same distortion. Since user 3 is bottlenecked at link 6, its dropping at link 6 makes room for user 4 at link 5. We also examined the resulting video at the receiving ends, the subject perception matches the distortion metric.

Table 3 illustrates the performance in terms of the maximum distortion levels between baseline approach (BL) and CAF approach (CA). It can be seen that CAF achieves lower distortion level. Table 4 shows the total perceptual distortion of each user after frame dropping. In both cases CAF outperforms baseline solution by a large margin.

Note that CAF has limited computation and communication overhead. For stored video, the side information of frame length and importance is calculated in advance, so there is no much computational overhead for this. If CAF is used for real-time encoded video, it adds approximately 5% computation complexity to the encoder, as we found from our simulation, but note that our work here is mainly for stored video because most of the video over Internet is stored. For communication overhead, it mainly comes from notifying the side information of frame importance and length to links. If we use 20 bits to send such information, the overhead is 0.6kbps for each video stream of 30 frames per second. Within a given interval for CAF algorithm, as shown in the algorithm in Section 2.4, the computation and communication overhead is very limited and negligible.

2.6 CAF Summary

For the IPTV engineering, we propose a framework of Content-Aware distortion-Fair (CAF) video networking for joint video source adaptation and network resource allocation, by combining the intelligence of the temporal scalability of the video source coding at the application layer with the congestion control at the transport layer, such that the min-max end-to-end video distortion fairness is achieved among users in a general multi-hop multi-user video delivery network.

In the framework, we adopt the approach of users dropping less important frames when the network is congested. Architecturally we illustrate that the end users drop less important frames is in general more bandwidth-efficient than links drop less important frames when the network is congested.

Based on the framework, we propose an algorithm with a provable small number of iterations for users and links to cooperate, to achieve the min-max distortion fairness. With the algorithm, the application layer video streaming pruning and distortion metrics are utilized to guide transport layer resource allocation and coordination.

The advantages of our CAF approach include: 1) The drops of frames when links are congested are content-aware and distortion-fair among all the users. 2) It provides the possibility that video clips help each other to get a fair quality of service (in the sense of perceptual distortion) when they share the resources (link bandwidth), i.e., lazy sequence helps out busy sequence. 3) Users drop frames based on computed thresholds, where the thresholds are obtained by the cooperations among users and links. Users' dropping frames is more bandwidth-efficient in general compared with links' dropping frames, which is very important for the networks whose bandwidth is expensive. 4) The threshold of dropping frame can be computed on-the-fly, based on the time-varying video content. There is no need to estimate a utility function which is hard to track the time-varying characteristics of video, but instead, we only need to signaling the side information of the number of bits of the frames and the pre-computed frame importance to the links. 5) It can naturally handle random events such as user joining, leaving, etc.

The solution involves simple signaling between links and sources. Simulations demonstrated the effectiveness of this solution. Our approach has significant performance improvement compared with the one without content-awareness.

3 Intelligent Video Scheduling in Peer-to-Peer Networks

3.1 Introduction

Peer-to-Peer (P2P) live streaming has become a viable solution for IPTV [33] services with medium quality video for a large number of concurrent users [30]. With the popularity of video on demand applications over Internet, the traditional client-server and content server at edge solutions are not adequate in handling dynamic viewer behaviors and do not scale well with a large audience. On the other hand, the P2P based solutions utilizing application layer overlay are becoming popular, because it is easy to implement and cheaper than duplicating content servers at edges. The core benefit of P2P based solution is that it utilizes the buffering and uploading capacities of the participating peers, and provides a more scalable and robust content delivery solution.

With the great success of many P2P streaming systems, e.g., PPLive [2], PPStream [3], CoolStreaming [84], there comes some work analyzing the operation and tradeoff of such systems (see [69] [40] [61] [25] and references therein). In [69] the authors investigate the scaling law by quantitatively studying the asymptotic effects and tradeoffs. Recently in [40] the authors analyze the pull-based P2P streaming systems based on fluid model, and construct a simple 2-hop scheduling algorithm for a buffer-less system. Systems with buffer are also studied using extensive simulations. In [61] the authors consider the resilience utilizing path diversity for multicast tree based P2P streaming systems. In [25] the authors present a survey of the media distribution methods, overlay structures and error control solutions proposed for P2P live streaming. However, none of these works gives an analytical model of the P2P streaming system under heterogeneous buffer occupancies.

P2P downloading system, like Bit Torrent [63] has been extensively studied. However, P2P downloading system does not consider real time playback constraint and thus does not have the playback buffer underflow issues as is addressed in this work.

In a P2P streaming system, it is desirable to minimize the probability of "freeze-ups" during the live video playback. In other words, the extra buffered content (i.e., content reserve level) of each peer should be kept positive. Paper [50] uses the well-known α-fair utility function to model users' satisfactory level of their current content reserve levels, and allocates network resources (i.e., transmission rates) to all users to maximize the total utility. It shows that the optimal solution can balance the buffer occupancy level among users and it proposes a greedy heuristic solution that achieves the desirable result.

The heuristic in [50] is a centralized algorithm, which needs a coordinator to do all the computing for scheduling. In a P2P network, a distributed algorithm is favorable, where the computing load is distributed over peers. In this work, we propose successive water-filling (SWaF) scheduling algorithm. In general, SWaF can be implemented in a centralized way, moreover, for the case where the propagation delay of the message passing is small compared with the time interval for uploading capacity to vary, SWaF can be implemented in a distributed way, where every peer can solve its local problem, with some message passing to other peers. Theoretically we prove the optimality of proposed SWaF algorithm, and we show that our algorithm is even simpler than the heuristic in [50].

In the following, we first provide the system model and problem formulation, then we propose the successive water-filling algorithm, after which we present numerical examples, followed by our results and proposed future work.

3.2 System Model

As in [50], we consider a P2P system with a video source and N peers, with uploading bandwidth $C_0, C_1, \cdots, C_N$ respectively. Let the content be streamed over all peers with a constant bit rate (CBR) R_0. The playback buffer of each peer k is characterized by the buffer state tuple, $\{x_k, t_k, y_k\}$, where x_k and y_k are the buffer content starting point and ending point, respectively, and t_k is the current playback time. The content reserve level is $y_k - t_k$, which is the video segment remained in the buffer to be played.

For a scheduling interval T seconds, within its uploading capacity C_j, each peer j provides a content of playback time $z_{j,k}$ to peer k if the content is available ($0 \le z_{j,k} \le y_j - y_k$), and within its upload capacity C_0 the source provides a content of playback time $z_{0,k}$ to peer k. Then peer j transmits in a total bit rate of $\sum_k z_{j,k}R_0/T$ which should be no greater than C_j. The content reserve level of each peer k becomes $\sum_{j=0}^{N} z_{j,k} + (y_k - t_k)$.

An important consideration for P2P live video streaming system is to prevent peer buffers from underflow which may cause unpleasant "freezes" in the playback. Since each peer gets the content via the uploading link of other peers, and each peer wants high content reserve level, the peers are competing for the uploading bandwidth. The limited uploading bandwidth needs to be allocated reasonably to make each peer have enough content reserve level. To characterize the system requirement, as in [50], we adopt a model similar to the network utility maximization framework in [36], to maximize the total happiness about playback smoothness of all the peers, where the happiness of peer k is measured by a utility function, $U_k(l_k)$, for content reserve level l_k.

In general, $U_k(l_k)$ is a concave function in l_k, which means the increase of the happiness slows down as the reserve level l_k gets larger. An example of such concave function is, as adopted in [50], the well-known α-fair utility function [59], $U_k(l_k) = w_k \frac{l_k^{1-\alpha}}{1-\alpha}$, where w_k (a positive number) is the weight indicating the relative importance of peer k in the system and $\alpha \in (0, 1)$ is a parameter.

Hence, the problem of uploading bandwidth allocation is formulated as follows,

$$\begin{aligned} \text{maximize } & \textstyle\sum_k U_k(\sum_{j=0}^{N} z_{j,k} + (y_k - t_k)) \\ \text{subject to } & \textstyle\sum_k z_{j,k} \le \beta C_j, \\ & 0 \le \textstyle\sum_{j \in J} z_{j,k} \le \max\{0, \max_{j \in J} y_j - y_k\}, \\ & j \in \{0, 1, \cdots, N\}, k \in \{1, \cdots, N\} \\ \text{variables } & z_{j,k}, \end{aligned} \tag{11}$$

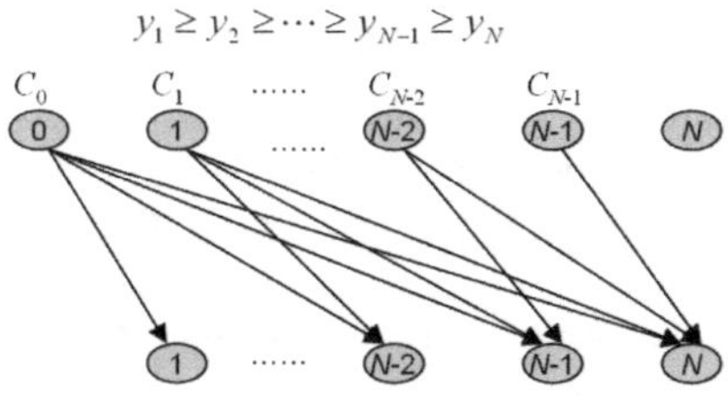

Fig. 9. Content provision pattern based on availability.

where $\beta = T/R_0$, J is any subset of $\{0, 1, \cdots, N\}\backslash\{k\}$. In this problem, the goal is to maximize the total utility over the playback buffer content reserve level, the first constraint states that the serving rate of the source or the peer should be no grater than the uploading capacity, and the second constraint states the content availability and no transmission repetition of content.

In [50], a centralized solution based on a greedy heuristic is proposed for a discrete version of problem (11) where the unit for the playback buffer is Group of Pictures (GoP), instead of the continuous time as in problem (11). In the following, we propose a distributed **successive water-filling (SWaF)** algorithm, where every peer solves its local problem, with some message passing to other peers. Theoretically we show that SWaF yields an optimal solution of (11). Computationally, SWaF is simpler than the algorithm in [50]. Note that in general, SWaF can be implemented as a centralized solution if a peer with good computing capacity is selected as a coordinator.

3.3 Successive Water-Filling (SWaF) Algorithm

For a practical solution, we assume a small group consisting of a source and N peers from all the peers in the system forms a cooperative group. The SWaF algorithm is within each cooperative group.

Denote θ_k as the content reserve level. Initially (at the beginning of current scheduling interval), $\theta_k = y_k - t_k$. Suppose $y_1 \geq y_2 \geq \cdots \geq y_N$, which means each C_j $(j = 0, 1, 2, \cdots, N-1)$ can serve peers $j+1, j+2, \cdots, N$. Note that $z_{j,k} = 0, j \geq k$ because peer j cannot serve peer $k \leq j$ due to the content availability. We assume $z_{j,k} \leq y_j - y_k$ for $j < k$, which means peer j has enough content to serve peer $k > j$. Figure 9 shows the content provision pattern.

We propose the following distributed SWaF algorithm.

Distributed SWaF Algorithm

---

1. Initialization:

Every peer k, $k = 1, 2, \cdots, N$, reports its y_k to peer 0 (this can be any other peer as well), then peer 0 sorts all the y_k's and then reports every peer k its neighbors $k-1$ and $k+1$ according to the sort result $y_1 \geq y_2 \geq \cdots \geq y_N$.

Peer k asks peer $k+1$ for its initial θ_{k+1}.

Every peer has one bit Cur_flag with initial value zero. Cur_flag=1 means current peer is running water-filling algorithm.

Peer 0 asks peer $N-1$ to set its Cur_flag=1.

2. Each peer** j ($j = N-1, N-2, \cdots, 1, 0$) **does the following:

IF peer j's Cur_flag=1

Peer j calculates $z_{j,k}$ for $k = j+1, \cdots, N$, using water-filling approach:

$$z_{j,k} = \max\{U_k'^{-1}(\lambda_j) - \theta_k, 0\}, \quad k = j+1, \cdots, N, \tag{12}$$

where λ_j is a positive number which is chosen such that $\sum_{k=j+1}^{N} z_{j,k} = \beta C_j$. λ_j is found by bisectional search.

Peer j sets its Cur_flag=0.

Peer j ($j \neq 0$) updates $\theta_k = \theta_k + z_{j,k}$, for $k = j+1, \cdots, N$, passes them to peer $j-1$.

Peer j ($j \neq 0$) asks peer $j-1$ to set its Cur_flag=1; Peer j ($j = 0$) reports 'The End'.

END IF

Note that $z_{j,k}$, as calculated in (12), is a water-filling solution of the following optimization problem,

$$\begin{array}{ll} \text{maximize} & \sum_{k=j+1}^{N} U_k(z_{j,k} + \theta_k) \\ \text{subject to} & \sum_{k=j+1}^{N} z_{j,k} \leq \beta C_j \\ \text{variables} & z_{j,k}, \quad k \in \{j+1, \cdots, N\}. \end{array} \tag{13}$$

By Lagrange dual approach and KKT condition [15], it is readily verified that the optimal solution of the problem (13) is the water-filling solution as calculated in (12), where λ_j is the Lagrange multiplier associated with the constraint in (13).

Optimality

Proposition 2. *SWaF algorithm yields an optimal solution for problem (11) assuming the second constraint is satisfied.*

Proof

First, we look at C_{N-1}. Since peer $N-1$ can only serve peer N, it is easy to check SWaF algorithm gives $z_{N-1,N} = \beta C_{N-1}$ which is the optimal solution.

Then, we look at C_{N-2}. Since peer $N-2$ can only serve peer $N-1$ and peer N, at the optimum we have $z_{N-2,N-1} + z_{N-2,N} = \beta C_{N-2}$. Note that in problem (11) the first inequality constraint becomes equality at the optimal. We have the following,

$$
\begin{aligned}
&\max_{\substack{\sum_{k=j+1}^{N} z_{j,k} = \beta C_j \\ j=0,1,\cdots,N-1}} \sum_{k=1}^{N} U_k(\sum_{j=0}^{k-1} z_{j,k} + \theta_k) \\
&= \max_{\substack{\sum_{k=j+1}^{N-2} z_{j,k} \le \beta C_j \\ j=0,1,\cdots,N-3}} \sum_{k=1}^{N-2} U_k(\sum_{j=0}^{k-1} z_{j,k} + \theta_k) \\
&+ \max_{z_{j,N-1}+z_{j,N}=\beta C_j - \sum_{k=j+1}^{N-2} z_{j,k}, j=0,1,\cdots,N-2} \\
&[U_{N-1}(\sum_{j=0}^{N-2} z_{j,N-1} + \theta_{N-1}) \\
&+U_N(\sum_{j=0}^{N-2} z_{j,N} + \beta C_{N-1} + \theta_N)] \\
&= \max_{\substack{\sum_{k=j+1}^{N-2} z_{j,k} \le \beta C_j \\ j=0,1,\cdots,N-3}} \sum_{k=1}^{N-2} U_k(\sum_{j=0}^{k-1} z_{j,k} + \theta_k) \\
&+ \max_{z_{j,N-1}+z_{j,N}=\beta C_j - \sum_{k=j+1}^{N-2} z_{j,k}, j=0,1,\cdots,N-3} \\
&[U_{N-1}(\sum_{j=0}^{N-2} z_{j,N-1} + \theta_{N-1}) \\
&+U_N(\sum_{j=0}^{N-2} z_{j,N} + \beta C_{N-1} + \theta_N)]|_{z_{N-2,N-1}, z_{N-2,N}:P} \\
&= \max_{\sum_{k=j+1}^{N} z_{j,k} = \beta C_j, j=0,1,\cdots,N-1} \\
&\sum_{k=1}^{N} U_k(\sum_{j=0}^{k-1} z_{j,k} + \theta_k)|_{z_{N-2,N-1}, z_{N-2,N}:P}
\end{aligned}
\tag{14}
$$

where P is the problem of max $[U_{N-1}(z_{N-2,N-1} + \theta_{N-1}) + U_N(z_{N-2,N} + \theta_N)]$ over variables $z_{N-2,N-1}, z_{N-2,N}$ with constraint $z_{N-2,N-1} + z_{N-2,N} = \beta C_{N-2}$. Equation (14) shows that an optimal $z_{N-2,N-1}$ and $z_{N-2,N}$ can be derived by solving P, where P can be solved by a water-filling algorithm.

Similarly, given the optimal $z_{N-1,N}$, $z_{N-2,N-1}$ and $z_{N-2,N}$ derived in previous steps, we further assign C_{N-3} by a corresponding water-filling algorithm getting an optimal $z_{N-3,k}$ for $k = N-2, N-1, N$ where $\sum_{k=N-2}^{N} z_{N-3,k} = \beta C_{N-3}$.

Successively, when SWaF algorithm finishes assigning C_0, we get an optimal solution. ∎

Complexity

The complexity of SWaF algorithm is $O(N \log \Lambda)$, where N is the number of peers and Λ is the number of equal-partitions of the searching interval of Lagrangian multiplier of the water-filling algorithm where the partition is based on the accuracy tolerance.

The complexity of the algorithm in [50] is $O(NM)$ where M is the number of GoPs transmitted by one peer in a scheduling interval. As our numerical results show, SWaF is simpler than the algorithm in [50], and this advantage is more obvious when M is large. Note that if the accuracy tolerance for the scheduling by SWaF algorithm is about one GoP, SWaF has a complexity $O(N \log M)$. This explains that when M is large, the simplicity of SWaF is more obvious.

Remarks

We have the following comments on SWaF.

Remark 1. We assume the second constraint in problem (11) is satisfied above, which means the solution of z happens to satisfy the constraint. If we do not have such assumption, the water-filling in (12) becomes

$$z_{j,k} = \min\{y_j - (y_k + \theta_k), \max\{U_k'^{-1}(\lambda_j) - \theta_k, 0\}\}, \tag{15}$$

and correspondingly, in the initialization of SWaF, peer j needs peer k, $k = j+1, \cdots, N$ to pass its y_k to peer j.

Remark 2. If all the peers use the same concave utility function, the shape of the utility function does not affect the algorithm. This is because, in such case, $U_k'^{-1}(\lambda_j)$ in the water-filling algorithm (12) will become a common $U'^{-1}(\lambda_j)$, which can be interpreted as 'water level' [15], and a direct bisectional search for the 'water level' can be done instead of the bisectional search for λ_j. The 'water level' is of our interest for the optima, not the dual variable λ_j. The illustration for this case is shown in Fig. 10. It is easy to see that the 'water level' is in the interval of $[\min \theta_k + \beta C_j/(N-j), \max \theta_k + \beta C_j/(N-j)]$ and the bisectional search is easy and fast.

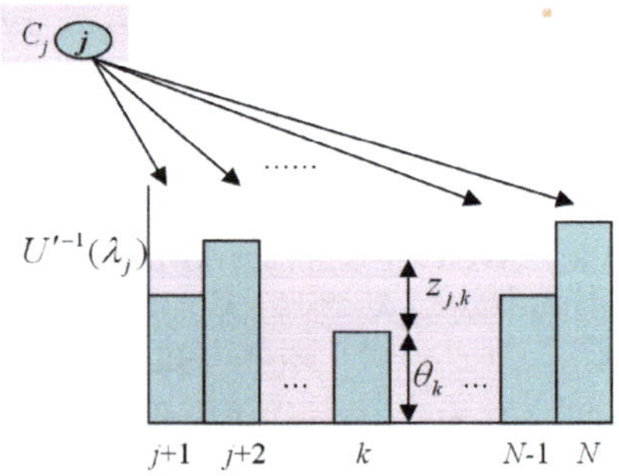

Fig. 10. Illustration of water-filling.

Remark 3. SWaF only gives one optimal solution of problem (11). Note that the optimal solution of problem (11) may not be unique.

Remark 4. SWaF here is in a distributed fashion, but in general, it can be implemented centrally, if a peer with good computing capacity is selected as a coordinator and it does all the computing.

Remark 5. When SWaF is in a distributed fashion, the dynamics of bisectional search for λ is local, which means each peer k searches its own λ_k, and would not affect the upload-link dynamics of others.

Remark 6. Distributed SWaF requires the scheduling interval greater than the total computing time and message passing time. Note that since SWaF is a successive algorithm, the propagation delay of the message passing accumulates. Hence for the case where uploading capacity varies slowly w.r.t. time, distributed SWaF is a good choice, while for the case where uploading capacity varies fast, SWaF in a centralized fashion may fit well.

3.4 Numerical Results

To verify the performance of the proposed algorithm, we set up P2P sessions with the following parameters similar to [50], video playback rate $R_0 = 300$ kbps,

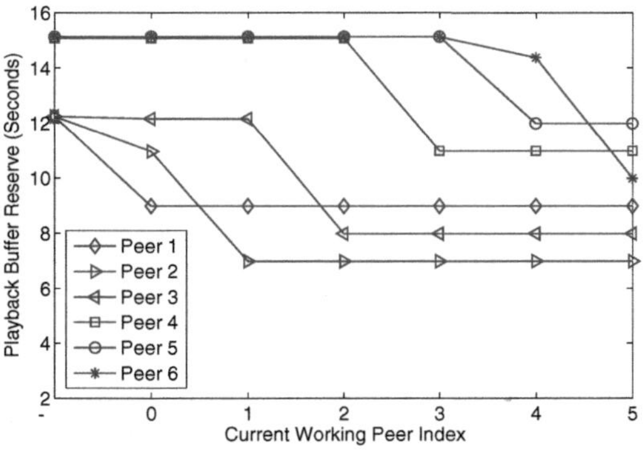

Fig. 11. Peer buffer content reserve states.

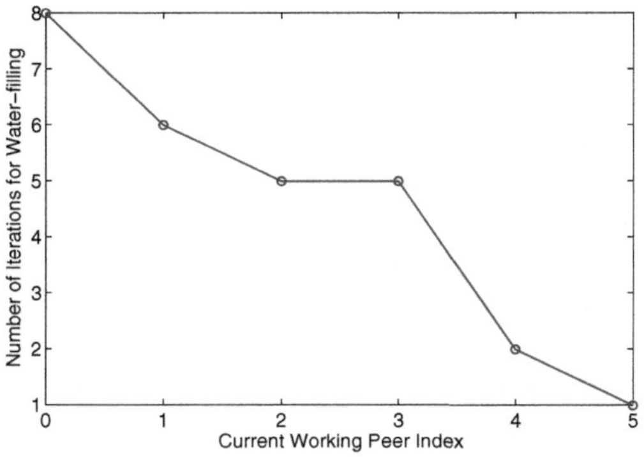

Fig. 12. Number of iterations for water-filling at each serving peer.

uploading bandwidth standard variation 40 kbps. The average uploading bandwidth is $C = \sum_{j=0}^{N} C_j/N$. We assume each peer has the same utility function. In [50], the advantage of utility-based scheduling compared with utility-blind scheduling has been illustrated and it is not our focus in this work. Here we focus on how the distributed SWaF algorithm works.

Consider a system with a source (peer 0) and 6 peers (peer $1, \cdots, 6$). Assume the scheduling interval $T = 4$ seconds. Assume the initial playback buffer reserves for peer $1, \cdots, 6$ are (9,7,8,11,12,10) seconds, respectively. Assume the initial buffer ending times are $(y_1, \cdots, y_6) = (30, 24, 20, 16, 11, 6)$. The average uploading bandwidth is $C = 324.6$ kbps$=1.28R_0$.

Figure 11 shows the peer buffer content reserve states for a successive water-filling carried out by a sequence of working (serving) peers (peer 5,4,3,2,1,0). It can be seen that SWaF has the tendency to make all the peers have similar (fair) content reserve level. It is readily verified that the greedy heuristic in [50] actually gives a solution similar to SWaF algorithm in a discrete version with segment of GoP.

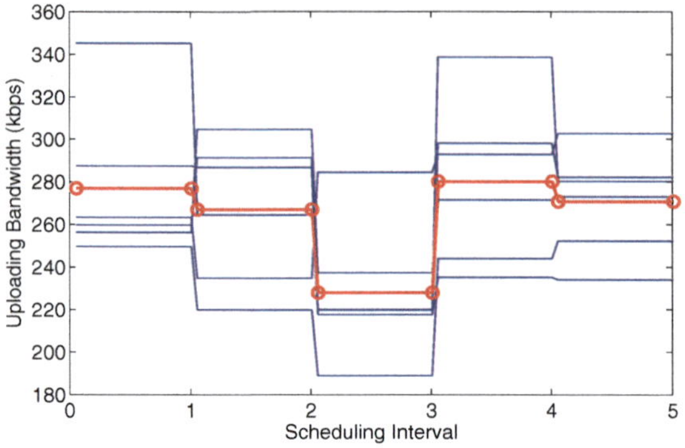

Fig. 13. Changes of uploading bandwidth.

Figure 12 shows the number of iterations for water-filling at each serving peer (peer 5,4,3,2,1,0), with an accuracy tolerance 2%, i.e., the bisectional search ending condition being $|\sum_{k=j+1}^{N} z_{j,k} - \beta C_j| \leq 0.02$ for all j. The average iterations over 6 peers is 4.5 in this figure which is just a random case. We run 10^5 cases for different uploading capacities, and we find an average iterations 4.58. For the same setting, the average iterations over 6 peers for the algorithm in [50] is 8, where for each iteration it has the similar computing complexity as in SWaF algorithm. Hence we can see SWaF algorithm has the advantage of low complexity.

The advantage of low complexity of SWaF can be more obvious when it is compared with the algorithm in [50] for a scheduling interval of more number of GoPs. For a scheduling interval $T = 8$ seconds, 10^5 cases of different uploading capacities are tested. An average number of iterations over 6 peers is about 5.54 with an accuracy tolerance 2%, while the heuristic in [50] needs 16 iterations.

Assume the uploading bandwidth varies from one scheduling interval to another, but fixed for each scheduling interval. The SWaF algorithm is effective for such system. For the same 6 peers set up, we let uploading bandwidth vary as plotted in Fig. 13, where the average bandwidth is marked in circle. Each scheduling interval T is 4 seconds. Figure 14 shows the content reserve level for 6 peers. The figure doesn't show the detailed reserve level for every water-filling, but it shows the reserve level after every 6 water-fillings in one scheduling interval. Here we assume the total time of computing and message passing in one scheduling interval is much less than T. As discussed in Remark 6, if the total time of computing and message passing in one scheduling interval is more than T, SWaF is better to be implemented centrally where a coordinator does all the computing.

3.5 SWaF Summary

We develop a successive water-filling algorithm (SWaF) to solve the video segment scheduling for P2P live video streaming system. We prove the optimality

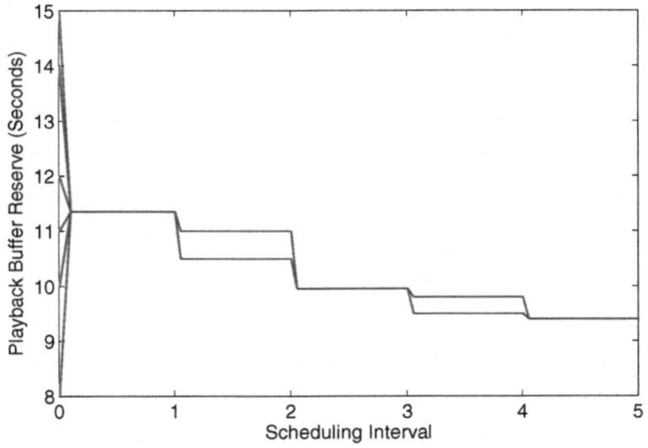

Fig. 14. Performance with time-varying uploading bandwidth.

of SWaF and show it has lower complexity compared with the heuristic algorithm in [50]. In general, SWaF can be carried out in a centralized fashion, (note that the algorithm in [50] is centralized as well), if a peer with good computing capacity is selected as a coordinator and it does all the computing. Moreover, for the case where the propagation delay of the message passing is small compared with the time interval for uploading capacity to vary, SWaF can be carried out in a distributed fashion where the optima can be found by each peer solving its own problem with some message passing over peers. The numerical results show the effectiveness of SWaF algorithm.

4 Conclusion and Future Work

For the purpose of providing end users with satisfying quality of service experience with multimedia applications, while keeping the system operations efficient in resource, robust, and scalable, in networks, we propose the utility based optimal resource allocation frameworks. The design degrees of freedom we consider include the network operations including the video source adaptation at the application layer, congestion control at the transport layer, video scheduling at the link layer, and so on. These design knobs can be jointly adjusted to maximize the network utility.

Our design philosophy makes the utility, which is the measure of user's happiness about the QoS, to drive the resource allocation and the design knobs (the network operations) to the optimal operating point, where the optimality is to make all the users happy, meanwhile the system is stable and efficient. We utilize the optimization theory (especially the decomposition) to come up with the distributed solution which networks favor of.

In this chapter, we have demonstrated examples on these design principles and developed solutions for the IPTV and P2P video applications.

In the work of new design on content-aware collaborative multimedia communication, we deliver better end-to-end video quality and achieve more efficient communication resource sharing, by distributed optimization in video content adaptation and resource allocation. Content awareness is derived from video sequences to define richer video quality metrics and utility, which helps intelligent video coding and communication decisions across application, networking and link layers. By exploiting multi-user diversity in channel states and content utility-resource tradeoffs, we develop distributed and collaborative solutions for resource allocation, as well as video adaptation, which is illustrated by an example of our design, max-min QoS fair video delivery over wired networks.

In the work of new design on how to utilize the playback buffer and upload bandwidth of peers to minimize the freeze-ups in playback, we propose a successive water-filling (SWaF) algorithm for the video transmission scheduling in P2P live streaming system. SWaF algorithm only needs each peer to optimally transmit (within its uploading bandwidth) part of its available video segments in the buffer to other peers requiring the content and pass small amount message to some other peers. Moreover, SWaF has low complexity and provable optimality.

Our new design yields stable and efficient networks with diverse traffic, including data traffic, multimedia traffic, etc., and users with more happiness in terms of the larger utility achieved. We provide frameworks for systematic networking with multimedia traffic, where utility shapes the resource allocation and drives how to adjust the design knob of the network operations over layers. Our work contributes to quantitative understanding to guide a systematic, rather than an ad hoc, process of designing layered protocol stack for wired and wireless networks.

In the future, we will explore new frameworks for content aware multimedia network optimization schemes for multi-access applications with mixed traffic, video broadcasting problems mapping video layers to embedded channels with rate and diversity tradeoff via space time coded MIMO channels [24], as well as incentives and game theoretical schemes for P2P video networks.

We may utilize more advanced coding features such as joint temporal-SNR scalability and metrics as well as multiple description coding schemes into the framework and develop new solutions with better efficiency and flexibility in content delivery network engineering, and better end-to-end QoS for users.

For P2P network, we will investigate alternative distributed algorithms. We have investigated a price based distributed algorithm similar to the algorithm in [36], but it seems a slow convergence. We will study more and compare it with SWaF. In addition, we will include the end-to-end transmission delays in the system modelling and simulations. We will also investigate richer video adaptation schemes and quality metrics in conjunction with underlying P2P distribution scheme, and work towards a graceful degradation in playback when system resources are limited.

References

1. MPEG-4 Visual. ISO/IEC 14496-2
2. Pplive system, `http://www.pplive.com`
3. Ppstream, `www.ppstream.com`

4. Alpcan, T., Basar, T.: Distributed algorithms for Nash equilibria of flow control games. In: Nowak, A. (ed.) Annals of dynamic games. Birkhauser, Cambridge (2003)
5. Alpcan, T., Basar, T.: A utility-based congestion control scheme for Internet-style networks with delay. In: Proc. IEEE INFOCOM (2003)
6. Altman, E., Basar, T., Srikant, R.: Nash equilibria for combined flow control and routing in networks: asymptotic behavior for a large number of users. IEEE Trans. on Autom. Contr. 47(6), 917–930 (2002)
7. Apostolopoulos, J.G., Wong, T., Tan, W., Wee, S.J.: On multiple description streaming with content delivery networks. In: Proc. IEEE INFOCOM (June 2002)
8. Baccichet, P., Noh, J., Setton, E., Girod, B.: Content-aware P2P video streaming with low latency. In: Proc. IEEE ICME (July 2007)
9. Bajaj, S., Breslau, L., Shenker, S.: Uniform versus priority dropping for layered video. In: Proc. ACM SIGCOMM (October 1998)
10. Bersekas, D., Tsitsiklis, J.: Parallel and Distributed Computation: Numerical Methods, 2nd edn. Athena Scientific, Belmont (1997)
11. Bertsekas, D.: Nonlinear Programming. Athena Scientific, Belmont (1999)
12. Blake, S., Black, D., Carlson, M., Davies, E., Wang, Z., Weiss, W.: An architecture for differentiated services. RFC 2475 (October 1998)
13. Bonald, T., Massoulié, L.: Impact of fairness on Internet performance. In: SIGMETRICS 2001: Proceedings of the 2001 ACM SIGMETRICS international conference on Measurement and modeling of computer systems, pp. 82–91. ACM, New York (2001)
14. Bonald, T., Massoulié, L., Proutière, A., Virtamo, J.: A queueing analysis of max-min fairness, proportional fairness and balanced fairness. Queueing Syst. Theory Appl. 53(1-2), 65–84 (2006)
15. Boyd, S., Vandenberghe, L.: Convex Optimization. Cambridge University Press, Cambridge (2004)
16. Braden, R., Clark, D., Shenker, S.: Integrated services in the Internet architecture: An overview. RFC 1633 (June 1994)
17. Cao, Z., Zegura, E.W.: Utility max-min: an application-oriented bandwidth allocation scheme. In: Proc. IEEE INFOCOM (March 1999)
18. Chakareski, J., Chou, P.A., Girod, B.: Rate-distortion optimized streaming from the edge of the network. In: Proc. Workshop on Multimedia Signal Process (December 2002)
19. Chakareski, J., Frossard, P.: Distributed collaboration for enhanced sender-driven video streaming. IEEE Trans. on Multimedia 10(5), 858–870 (2008)
20. Chen, J.C., Chan, S.H., Li, V.: Multipath routing for video delivery over bandwidth-limited networks. IEEE J. on Selected Areas in Communications 22(10), 1920–1932 (2004)
21. Chiang, M., Low, S.H., Calderbank, A.R., Doyle, J.C.: Layering as optimization decomposition: A mathematical theory of network architectures. Proceedings of the IEEE 95(1), 255–312 (2007)
22. Chou, P.A., Miao, Z.: Rate-distortion optimized streaming of packetized media. IEEE Trans. on Multimedia 8(2), 390–404 (2006)
23. Cover, T., Thomas, J.: Elements of Information Theory. Wiley, Chichester (2006)
24. Diggavi, S.N., Calderbank, A.R., Dusad, S., Al-Dhahir, N.: Diversity embedded space-time codes. IEEE Trans. on Information Theory 54(1), 33–50 (2008)
25. Fodor, V., Dan, G.: Resilience in live peer-to-peer streaming. IEEE Communications Magazine 45(6), 116–123 (2007)

26. Girod, B., Kalman, M., Liang, Y., Zhang, R.: Advances in channel-adaptive video streaming. IEEE Wireless Commun. Mobile Comput. 2(6), 549–552 (2002)
27. Yaiche, H., Mazumber, R.R., Rosenberg, C.: A game theoretic framework for bandwidth allocation and pricing in broadband networks. IEEE/ACM Trans. on Networking 8(5), 667–678 (2000)
28. Haskell, B.G.: Digital Video: An Overview of MPEG-2 (1997)
29. He, J., Bresler, M., Chiang, M., Rexford, J.: Towards robust multi-layer traffic engineering: optimization of congestion control and routing. IEEE J. on Selected Areas in Communications 25(5), 868–880 (2007)
30. Hei, X., Liang, C., Liang, J., Liu, Y., Ross, K.: Insight into pplive: a measurement study of a large-scale p2p IPTV system. In: Proc. WWW 2006 Workshop of IPTV Services over World Wide Web (2006)
31. Hormis, R., Linzer, E., Wang, X.: Joint diversity- and rate-control for video transmission on multi-antenna channels. In: Proc. IEEE Globecom (November 2007)
32. Huang, J., Li, Z., Chiang, M., Katsaggelos, A.K.: Joint source adaptation and resource pricing for multi-user wireless video streaming. IEEE Trans. on Circuits and Systems for Video Technology 18(5), 582–595 (2008)
33. Jain, R.: I want my IPTV. IEEE Multimedia 12(3) (2005)
34. Kalman, M., Ramanathan, P., Girod, B.: Rate-distortion optimized video streaming with multiple deadlines. In: Proc. IEEE ICIP (September 2003)
35. Kanakia, H., Mishra, P.P., Reibman, A.R.: An adaptive congestion control scheme for real time packet video transport. IEEE/ACM Trans. on Networking 3(6), 671–682 (1995)
36. Kelly, F., Maulloo, A., Tan, D.: Rate control for communication networks: Shadow prices, proportional fairness and stability. J. of Operational Research Society 49(3), 237–252 (1998)
37. Kelly, F.P.: Charging and rate control for elastic traffic. Eur. Trans. on Telecommun. 8, 33–37 (1997)
38. Kelly, F.P.: Fairness and stability of end-to-end congestion control. European Journal of Control 9, 159–176 (2003)
39. Kelly, F.P., Voice, T.: Stability of end-to-end algorithms for joint routing and rate control. ACM Comput. Commun. Rev. 35(2), 5–12 (2005)
40. Kumar, R., liu, Y., Ross, K.: Stochastic fluid theory for p2p streaming systems. In: Proc. IEEE INFOCOM (May 2007)
41. Lee, J.W., Chiang, M., Calderbank, R.A.: Price-based distributed algorithm for optimal rate-reliability tradeoff in network utility maximization. IEEE J. on Selected Areas in Communications 24(5), 962–976 (2006)
42. Li, Y., Chiang, M., Calderbank, R.A., Diggavi, S.N.: Optimal rate-reliability-delay tradeoff in networks with composite links. IEEE Trans. on Communications 57(5), 1390–1401 (2009)
43. Li, Y., Li, Z., Chiang, M., Calderbank, A.R.: Content-aware video-quality-fair (CAF) streaming. In: Proc. IEEE Globecom (November 2008)
44. Li, Y., Li, Z., Chiang, M., Calderbank, A.R.: Video transmission scheduling for peer-to-peer live streaming systems. In: Proc. IEEE ICME (June 2008)
45. Li, Y., Li, Z., Chiang, M., Calderbank, A.R.: Energy-efficient video transmission scheduling for wireless peer-to-peer live streaming. In: Proc. IEEE CCNC (January 2009)
46. Li, Y., Li, Z., Chiang, M., Calderbank, R.A.: Content-aware distortion-fair video streaming in congested networks. IEEE Trans. on Multimedia 11(6), 1–12 (2009)

47. Li, Y., Markopoulou, A., Apostolopoulos, J., Bambos, N.: Content-aware playout and packet scheduling for video streaming over wireless links. IEEE Trans. on Multimedia 10(5), 885–895 (2008)
48. Li, Y., Papachristodoulou, A., Chiang, M.: Stability of congestion control schemes with delay sensitive traffic. In: Proc. American Control Conf (June 2008)
49. Li, Y., Tian, C., Diggavi, S.N., Chiang, M., Calderbank, A.R.: Optimal network resource allocation for competing multiple description transmissions. In: Proc. IEEE Globecom (November 2008)
50. Li, Z., Huang, J., Katsaggelos, A.K.: Content reserve utility based video segment transmission scheduling for peer-to-peer live video streaming system. In: Proc. 2007 Allerton Conference on communication, control and computing (October 2007)
51. Li, Z., Katsaggelos, A.K., Schuster, G., Gandhi, B.: Rate-distortion optimal video summary generation. IEEE Trans. on Image Processing 14(10), 1550–1560 (2005)
52. Li, Z., Li, Y., Chiang, M., Calderbank, A.R., Chen, Y.C.: Optimal transmission scheduling for scalable wireless video broadcast with rateless erasure correction code. In: Proc. IEEE CCNC (January 2009)
53. Li, Z., Schuster, G., Katsaggelos, A.K.: MINMAX optimal video summarization and coding. Special issue on Analysis and Understanding for Media Adaptation, IEEE Trans. on Circuits and System for Video Technology 15(10), 1245–1256 (2005)
54. Low, S.: A duality model of TCP and queue management algorithms. IEEE/ACM Trans. on Networking 11(4), 525–536 (2003)
55. Low, S.H., Lapsley, D.E.: Optimal flow control, I: basic algorithm and convergence. IEEE/ACM Trans. on Networking. 7(6), 861–874 (1999)
56. Low, S.H., Paganini, F., Doyle, J.: Internet congestion control. IEEE Control Syst. Mag. 21, 28–43 (2002)
57. Marbach, P.: Priority service and max-min fairness. IEEE/ACM Trans. on Networking 11(5), 733–746 (2003)
58. Mastronarde, N., van der Schaar, M.: A queuing-theoretic approach to task scheduling and processor selection for video decoding applications. IEEE Trans. Multimedia 9(7), 1493–1507 (2007)
59. Mo, J., Walrand, J.: Fair end-to-end window-based congestion control. IEEE/ACM Trans. on Networking 8(5), 556–567 (2000)
60. Ohm, J.R.: Advances in scalable video coding. Special Issue on Video Coding, Proceedings of the IEEE 93(1), 42–56 (2005)
61. Padmanabhan, V.N., Wang, H.J., Chou, P.A.: Resilient peer-to-peer streaming. In: Proc. IEEE International Conference on Network Protocols (2003)
62. Palomar, D., Chiang, M.: A tutorial on decomposition methods and distributed network resource allocation. IEEE J. on Selected Areas in Communications 24(8), 1439–1451 (2006)
63. Qiu, D., Srikant, R.: Modeling and performance analysis of bittorrent-like peer-to-peer networks. In: Proc. ACM SIGCOMM (September 2004)
64. Quaglia, D., de Martin, J.C.: Delivery of MPEG video streams with constant perceptual quality of service. In: Proc. IEEE ICME (August 2002)
65. Schwarz, H., Marpe, D., Wiegand, T.: Overview of the scalable video coding extension of the H.264/AVC standard. IEEE Trans. on Circuits and Systems for Video Technology 17(9), 1103–1120 (2007)
66. Seferoglu, H., Markopoulou, A.: Opportunistic network coding for video streaming over wireless. In: Proc. Packet Video (2007)
67. Shenker, S.: Fundamental design issues for the future Internet. IEEE J. on Selected Areas in Communications 13(7), 1176–1188 (1995)

68. Shiang, H.P., van der Schaar, M.: Multi-user video streaming over multi-hop wireless networks: a distributed, cross-layer approach based on priority queuing. IEEE J. on Selected Areas in Communications 25(4), 770–785 (2007)
69. Small, T., Liang, B., Li, B.: Scaling laws and tradeoffs in peer-to-peer live multimedia streaming. In: Proc. ACM MULTIMEDIA (2006)
70. Srikant, R.: The Mathematics of Internet Congestion Control. Birkhäuser, Boston (2004)
71. Sullivan, G.J., Wiegand, T.: Video compression - from concepts to the H.264/AVC standard. Proceedings of the IEEE 93(1), 18–31 (2005)
72. Tang, A., Andrew, L., Chiang, M., Low, S.: Transport layer. Wiley Encyclopedia of Computer Science and Engineering (2008)
73. Tang, A., Wang, J., Low, S.H., Chiang, M.: Equilibrium of heterogeneous congestion control: existence and uniqueness. IEEE/ACM Trans. on Networking 15(4), 824–837 (2007)
74. Wang, W., Palaniswami, M., Low, S.H.: Application-oriented flow control: fundamentals, algorithms and fairness. IEEE/ACM Trans. on Networking 14(6), 1282–1291 (2006)
75. Wu, D., Ci, S., Wang, H.: Cross-layer optimization for video summary transmission over wireless networks. IEEE J. on Selected Areas in Communications 25(4), 841–850 (2007)
76. Wu, D., Hou, Y.T., Zhang, Y.Q.: Transporting real-time video over the Internet: challenges and approaches. Proceedings of the IEEE 88(12), 1855–1875 (2000)
77. Wu, F., Li, S., Zhang, Y.Q.: A framework for efficient progressive fine granular scalable video coding. IEEE Trans. on Circuits and Systems for Video Technology 11(3), 332–344 (2001)
78. Xin, J., Lin, C.W., Sun, M.T.: Digital video transcoding. Proceedings of the IEEE 93(1), 84–97 (2005)
79. Xu, D., Li, Y., Chiang, M., Calderbank, A.R.: Provisioning of elastic service availability. In: Proc. IEEE INFOCOM (May 2007)
80. Xu, D., Li, Y., Chiang, M., Calderbank, A.R.: Provisioning of elastic service availability. IEEE J. on Selected Areas in Communications (2008) (to appear)
81. Yi, Y., Shakkottai, S.: Hop-by-hop congestion control over a wireless multi-hop network. IEEE/ACM Trans. on Networking 15(1), 133–144 (2007)
82. Zhai, F., Berry, R., Pappas, T.N., Katsaggelos, A.K.: A rate-distortion optimized error control scheme for scalable video streaming over the internet. In: Proc. IEEE ICME (July 2003)
83. Zhang, Q., Zhang, Y.Q.: Cross-layer design for QoS support in multihop wireless networks. Proceedings of the IEEE 96(1), 67–76 (2008)
84. Zhang, X., Liu, J., Li, B., Yum, T.S.P.: Coolstreaming/donet: A data-droven overlay network for peer-to-peer live media streaming. In: Proc. IEEE INFOCOM (2005)
85. Zhang, Z., Sun, Q., Wong, W.C., Apostolopoulos, J., Wee, S.: An optimized content-aware authentication scheme for streaming JPEG-2000 images over lossy networks. IEEE Trans. on Multimedia 9(2), 320–331 (2007)

Media Coding for Streaming in Networks with Source and Path Diversity*

Nikolaos Thomos and Pascal Frossard

Signal Processing Laboratory (LTS4), Ecole Polytechnique Fédérale de Lausanne (EPFL), Lausanne, Switzerland

Summary. Novel network architectures such as overlay networks offer significant diversity that can compensate for the lack of strict quality of service in today's communication infrastructures. In order to take advantage of this diversity for delay-sensitive media streaming applications, the network systems can employ efficient mechanisms based on source, channel and even network coding. In particular, fountain codes offer interesting benefits for streaming with server diversity. When they are used independently at each server, they permit to avoid explicit coordination between the senders that only have to provide the receivers with enough innovative packets. In addition, network coding allows for improved throughput and error robustness in multipath transmission where the network nodes participate to increase the symbol diversity in the system. We review in this chapter the most popular rateless codes that enable the deployment of low-cost decentralized communication protocols in self-organized dynamic networks. We then describe their application in distributed multimedia streaming solutions. We further discuss the most popular network coding algorithms in practical media streaming schemes. Finally, we show that hybrid systems based on both rateless coding and network coding can deliver high quality media streams with low computational complexity, as they permit to benefit from both server and path diversity in overlay architectures.

1 Introduction

Recent advances on overlay architectures have motivated many research efforts for the development of efficient multimedia streaming systems that are able to exploit the structure of such networks. These overlay networks that can be built on peer-to-peer or wireless mesh networks for example, are characterized by a high diversity in terms of source nodes and transmission paths. In particular, a receiver peer can be connected simultaneously to several senders, which can further use multiple communication paths for media packet delivery. It is therefore important to design efficient streaming solutions that are able to exploit the network diversity for improved media quality at receivers.

* This work has been supported by the Swiss National Science Foundation, under grant PZ00P2-121906.

C.W. Chen et al. (Eds.): Intel. Multimedia Communication: Tech. and Appli., SCI 280, pp. 291–316.
springerlink.com

The traditional multimedia communication tools such as error correcting codes, retransmission mechanisms, or packet scheduling are challenged in overlay networks as they usually necessitate coordination between the nodes participating in the transmission process, or good estimation of the network status. Although there exist efficient solutions to improve the video quality in distributed settings [1], these cannot be deployed and maintained easily in real networks suffering from unexpected nodes departures and arrivals as well as non-ergodic link failures. Complex systems combining joint source and channel coding with adaptive network protocols and appropriate routing algorithms provide effective solutions for dealing with the delay-sensitive nature of highly heterogeneous media packet streams in traditional architectures [2]. However, they are usually difficult to be implemented in dynamic ad hoc architectures that provide sources and path diversity. They moreover do not scale well with the number of sources and the size of the network. They require a good knowledge of the end-to-end network characteristics for optimal performance. It is also usually difficult to build distributed implementations of such systems, without high communication costs for coordination or decision about the best streaming strategy.

Appropriate coding of the media streams however permits to flatten the differences in importance of the packets and thus relaxes the need for important coordination. In particular, fountain codes [3] have emerged as a solution that allows to take benefit from the network diversity with decentralized algorithms. The main characteristic of these codes relies on the fact that the receiver only needs to gather a sufficient number of innovative packets for decoding. The actual identity of the packets do no play a role with such codes, only the number of packets matters. Strict coordination is not needed anymore in distributed streaming since the sources only have to generate different packets, which happens with high probability even when the encoding is done independently. Another interesting feature of fountain codes relies on their high adaptivity due to their rateless property. As a potentially limitless set of packets can be generated from a given set of source packets, the sender can easily adapt the coding rate to the status of the communication channel. Another appealing advantage of fountain codes is their relatively low encoding and decoding complexity that enables their use in real-time applications. Fountain codes have been successfully applied in multimedia transmission [4, 5] and distributed storage systems [6] for example.

While fountain codes appear to be an ideal solution for coding at the sources, they are difficult to be implemented in the network nodes within the overlay. Successive decoding and encoding processes in the network nodes [7] might induce delays that are too important for delay-sensitive applications. In dynamic and large scale networks, it is however beneficial to request help from the network nodes as this provides higher adaptivity to exploit fully the network diversity. Such in-network processing is typically performed by network coding algorithms. Network coding [8] provides an interesting solution to exploit the path diversity with gains in terms of achievable throughput and delay time. It typically permits to approach the max-flow min-cut limit of the underlying network graph when the network nodes combine the received packets instead of forwarding them unprocessed. Practical network coding solutions are generally based on random packets combination in the network nodes. Random linear network coding (RLNC) does not require coordination between network nodes so that complex routing and scheduling algorithms can be avoided. Similarly to fountain codes, network coding increases the packet diversity in the network, with a larger overhead however. When the receiver gathers enough innovative packets on multiple transmission paths, it can recover the original information. Network coding is used in various fields like content distribution [9, 10], distributed storage [11] or data dissemination [12]. Network coding can even be combined with fountain codes for efficient systems that take benefit from both source and path diversity [13].

The aim of this book chapter is to present coding techniques that permit to exploit the source and path diversity in overlay networks for efficient media streaming algorithms. First, we provide a brief overview of fountain codes with an emphasis on the encoding and decoding procedures in Section 2. We then show how these codes can be used in distributed streaming solutions with multiple sources. In Section 3, we provide an overview of network coding and its application in streaming systems. We show how network coding permits to exploit path diversity in overlay networks. We also show how network coding can be adapted to support different classes of packets in media streaming. Finally, we describe in Section 4 some hybrid schemes that exploit the advantages from both fountain codes and network codes and appear to be quite appropriate for low-cost real-time multimedia streaming.

2 Fountain Codes in Distributed Streaming

2.1 Fountain Codes

Fountain codes [3] have interesting properties for adaptive streaming, possibly from multiple sources. They provide rate adaptivity and permit to avoid the need for coordination between the senders. In this section, we first describe the principles of fountain codes and present the main families of fountain codes. We later describe their application to streaming application with a special emphasis on distributed streaming scenarios and unequal error protection algorithms.

The fountain codes are beneficial for multimedia applications that undergo strict deadlines (short playback delays). Some implementations even offer linear encoding and decoding times in contrast to other channel codes such as the Reed-Solomon codes, which have quadratic decoding complexity that grows with the block size. Fountain codes have only probabilistic guarantees for successful decoding and incur a small performance penalty ϕ which means that a client should receive $(1 + \phi) \cdot k$ packets to decode successfully the k source packets. Fountain systems eliminate the need for retransmission (ARQ) mechanisms because of their rateless property that permits to easily adapt the coding rate since a large number of encoded symbols can be generated from the same subset of source packets. Obviously, they utilize the available resources more efficiently as acknowledgment messages (ACK) often worsen congestion problems.

The fountain codes have an implicit structure that depends on the degree distribution function

$$\Omega(x) = \sum_{i=1}^{n} \Omega_i \cdot x^i,$$

which determines the number of symbols (degree) that should be combined for generating an output symbol. The parameter n is the maximal symbol degree. The rateless encoded symbols are generated by randomly combining (XOR-ing) the source packets. The number of combined symbols is determined by randomly sampling the degree distribution function $\Omega(x)$.

Let's x_i denote the ith source symbol and y_j the jth transmitted symbol. Then, it holds

$$y_j = \sum_{l=1}^{d_j} \bigoplus x_l$$

where d_j is the degree of the jth symbol and $\sum \bigoplus$ is the bitwise XOR operation. A small header called ESI is appended to each packet. The ESI is usually the seed of the pseudorandom generator used for generating the encoded symbol. Interestingly, Fountain codes are universal and thus near-optimal for any packet erasure channel. Their universal property is attributed to their rateless property that permits to generate large sets of encoded packets with only a small overhead penalty. Note that Fountain codes often need a termination message for signifying the reception of a full rank set of symbols. This mechanism saves significant amounts of bandwidth as otherwise the senders should assume a predetermined transmission rate which may be inaccurate as network conditions generally vary in time. A survey of fountain codes is given in [14].

LT codes

Among the most popular fountain codes, we find the codes based on the Luby Transform (LT) [15]. LT codes are sparse random linear codes with nearly linear decoding time. Their encoding consists in random combinations of the source symbols, while their decoding can be done either by belief propagation or Gaussian elimination. The encoding and decoding procedures are summarized as follows:

Procedure 1. LT encoding

1: Choose randomly the degree d of the LT encoded symbol by sampling $\Omega(x)$.
2: Choose uniformly d distinct symbols.
3: Combine the symbols by XOR-ing the selected symbols.

Procedure 2. LT decoding

```
while S ≠ ∅ do
    Select a symbol x from S.
    if x ∉ L then
        Include x in L.
        XOR x with its neighbouring symbols ∉ L.
    end if
end while
```

where $\mathcal{S}$ in Procedures 1 and 2 denotes the set of symbols in the ripple and $\mathcal{L}$ the set of the recovered symbols. The ripple is the number of covered input symbols (i.e., symbols that have a unique neighbor in their Tanner graph representation after branches elimination in decoding) that have not been yet processed.

The LT codes employ a degree distribution function known as ideal soliton distribution (ISD) which is given by

$$\rho(d) = \begin{cases} \frac{1}{k}, & \text{for } d = 1 \\ \frac{1}{d \cdot (d-1)}, & \text{for } d > 1 \end{cases}$$

Although ISD performs well in expectation, in practice it is fragile to transmission errors because the ripple often stays empty. Ideally, the ISD ripple contains only one symbol at each decoding step. To enhance ISD's robustness to errors, it is slightly modified. The modified distribution is called robust soliton distribution (RSD) $\mu(d)$ and it is given by

$$\mu(d) = \frac{\rho(d) + \tau(d)}{\sum\limits_{d} \rho(d) + \tau(d)} \qquad (1)$$

where

$$\tau(d) = \begin{cases} \frac{S}{d \cdot k}, & d = 1, 2, \ldots, \frac{k}{S} - 1 \\ \frac{S}{k} \ln\left(\frac{S}{\delta}\right), & d = \frac{k}{S} \\ 0, & d = \frac{k}{S} + 1, \ldots, k \end{cases} \qquad (2)$$

The parameter δ controls the size of ripple and S is the average number of symbols of degree one, *i.e.* the number of recovered symbols that have not been processed yet. S is finally defined as

$$S = c \cdot \sqrt{k} \cdot \ln\left({}^{S}\!/_{\delta}\right)$$

Raptor codes

Raptor codes [16] are the most successful rateless codes because they have an overhead that asymptotically tends to zero for very large codeblocks. For typical multimedia communication scenarios, the Raptor codes (short codeblocks are

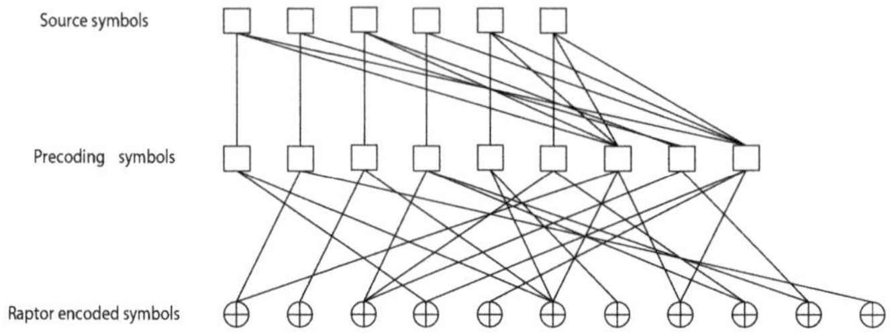

Fig. 1. Encoding of non-systematic Raptor codes. The source symbols are first pre-coded and successively LT encoded.

assumed) can recover the data with high probability if they receive a set of symbols slightly larger than the set of source symbols *i.e.* one or two symbols are enough. Their encoding consists of two steps: first, the data are encoded by codes like LDPC, LDGM or Tornado and then the encoded data are fed in the LT encoder. The first step is known as pre-coding and it allows the employment of weakened LT codes, *i.e.* LT codes with very sparse parity-check matrices. Due to the sparse LT codes generator matrices, linear encoding and decoding times are achieved. The pre-coding step makes the decoding procedure more robust to erasures as the errors remaining after LT decoding can be corrected by the pre-coder. Raptor encoding is schematically depicted in Fig. 1. Their degree distribution is a modified ISD distribution

$$\Omega_R(x) = \frac{1}{\mu+1} \cdot \left(\mu \cdot x + \sum_{i=2}^{D} \frac{x^i}{i \cdot (i-1)} + \frac{x^{D+1}}{D} \right) \tag{3}$$

where $\mu = (\epsilon/2) + (\epsilon/2)^2$, $D = \lceil 4 \cdot (1+\epsilon)/\epsilon \rceil$, and ϵ is a parameter that drives the decoding failure probability.

Raptor decoding is usually performed by Gaussian elimination of an equation system $\mathbf{A}$ that is constructed from the received symbols. An efficient implementation of Raptor codes has been presented in 3GPP standard [17]. The pre-coding of 3GPP codes consists of LDPC codes followed by Half codes, which have dense parity check matrices. The Raptor codes equations system is of the form

$$\mathbf{A} = \begin{bmatrix} \mathbf{G}_{LDPC} \; I_S \; O_{S\times H} \\ \mathbf{G}_{Half} \; I_H \\ \mathbf{G}'_{LT} \end{bmatrix} \tag{4}$$

where $\mathbf{G}_{LDPC}$ and $\mathbf{G}_{Half}$ are respectively the generator matrices of LDPC and Half codes, while $\mathbf{G}'_{LT}$ corresponds to the generator matrix of the received symbols. I_S and I_H are unitary matrices with size $S \times S$ and $H \times H$ where S and H denote the number of LDPC and Half codes constraints. Finally, $O_{S\times H}$ is a zero matrix of size $S \times H$. It is worth noting that LDPC and Half codes constraints should not be transmitted as they can be reproduced if the clients are aware of the number of source symbols k. 3GPP codes provide a fast algorithm for solving these equations systems, which consists in a variant of classical Gaussian elimination process.

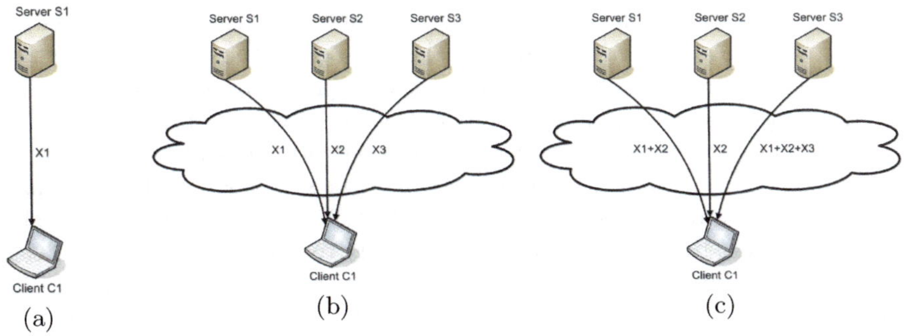

Fig. 2. (a) Direct server-client communication, (b) multiple servers-single client communication with scheduling and (c) multiple servers-single client communication with coding at the servers.

Other Fountain codes

A few other Fountain codes have been recently presented for specific communication scenarios. Specifically, Growth codes [18] improve the poor performance of Fountain codes when less symbols than the number of source symbols are received by a client. For such cases, most Fountain codes like common channel coding schemes cannot retrieve any data segment. Systematic channel codes can recover more data, but their performance is driven by the transmission of uncoded data. However, the systematic codes are quite restrictive for many applications due to their low symbol diversity. On the other hand, Growth codes offer high symbol diversity while they achieve intermediate performance by means of the recovery of more data symbols compared to systematic codes. This performance comes with an increased overhead, which can still be affordable in catastrophic scenarios and in emergency situations.

Shifted codes [19] are variants of LT codes that outperform original LT codes when servers are aware about the number of packets that have been already decoded. They are useful for data streaming over wireless sensor networks where a feedback about the number of decoded symbols can be provided. Due to this knowledge, the LT codes distribution can be shifted towards higher degrees. Symbols of higher degrees are typically sent in order to be useful for more clients. Similarly, reconfigurable codes [20] have been proposed for wireless channels. They can be considered as LT codes with time varying degree distribution. Reconfigurable codes adapt their distribution based on a one bit ACK message signaled from clients to servers.

2.2 Streaming Systems with Fountain Codes

The traditional server-client paradigm (point-to-point communication) depicted in Fig. 2(a) typically represents a scenario where one client requests data from a single server that dedicates part of its resources to serve this receiver. The paradigm is replaced by a new model in overlay networks, where a client can request media streams from several servers simultaneously, as illustrated in Fig. 2(b). In this way, the throughput and the error robustness of the system are increased. However, this often necessitates the employment of complex routing and scheduling protocols to minimize the probability of the reception of several identical packets, which represents a waste of bandwidth resources. Fountain codes permit to solve this limitation due to their rateless property where the multiple packet reception probability becomes very low. Fountain codes are applied at the multiple servers that send data simultaneously to the same client, as illustrated in Fig. 2(c). The receiver only needs to gather a sufficient number of different packets in order to decode the media stream. Strict coordination between servers is not required anymore in this case. The importance of the encoded packets becomes quite homogeneous, hence the media delivery is facilitated.

The interesting properties of the Fountain codes have not escaped from the attention of the media streaming community. Many researchers have adopted them in recent multimedia streaming systems for broadcast, one-to-one and many-to-one transmission scenarios. Multimedia broadcast and multicast service (MBMS) delivery over UMTS systems has been investigated in [21]. This system uses Turbo codes at physical layer to cope with bit errors and Raptor codes [16] at the application layer to deal with packet erasures. The Raptor codes permit the system to alleviate the need for coordination and to benefit from inherent network diversity while keeping the complexity low. It becomes clear that the careful balancing of available resources between application and physical layer boosts system's performance. This technique further lowers the power consumption as the error protection can be lighter. Higher packet loss rates can be afforded at the application layer, which can be successively corrected by Raptor codes. MBMS systems based on 3GPP Raptor codes [4] benefit from the partial recovery of packets corrupted by errors at the body of a packet as well as at the overhead. Their advanced performance comes with a slight increment of the computational cost. When a client receives an insufficient number of symbols for successful decoding, two post repair algorithms are used: one that finds the minimum set of source packets to be requested and another that determines the minimum number of consecutive packets to be requested. Transmission of Raptor encoded scalable videos over Mobile Ad Hoc Networks (MANET) has been proposed in [22]. Raptor codes are used to improve the obtained video quality and allow better adaptation to the rapid changing network conditions and network dynamics. A heuristic distributed rate allocation algorithm is also presented, which fairly allocates the available resources to various clients requesting various videos.

Unicast video streaming over packet erasures channel is further examined in [23]. The video is Raptor encoded to combat packet losses. A reliable feedback channel is used for sending a signal that informs the server that enough packets have reached the client. After the reception of such a message the server stops the transmission process. However, in the meanwhile the server continues to stream packets to the client, which wastes valuable resources. To reduce the wasted bandwidth the packet loss probability distribution is taken into account and the servers adaptively change the transmission rate starting from an optimistic guess for the channel conditions. The server waits for a given amount of time to receive a message for terminating the transmission process, if the message does not arrive upon the deadline; the transmission rate is repeatedly increased to meet the real channel conditions. The optimal transmission strategies are determined beforehand by a greedy algorithm. However, [23] makes several unrealistic assumptions such as the employed codes have a fixed overhead, the pdf of the channel loss rate is known and the communication is performed through a reliable channel by a single symbol message. These shortcomings are addressed in [24] where the channel conditions and thus the pdf are derived from real time measurements. Raptor codes are replaced by LT codes for utilizing the low cost belief propagation decoder. Moreover, it is assumed that the communication over the feedback channel may face delays. The results show that the on-the-fly update of the channel statistics incurs only a small performance penalty compared to an ideal situation.

Fountain codes have also been used in many-to-one communication scenarios. In [5], Raptor codes are employed to avoid coordination among several servers streaming scalable video to a single client. Independent packet losses are assumed between any pair of server-client. Each video layer of each Group of Pictures (GOP) is independently encoded with Raptor codes. Optimal sending strategies as well as close to optimal heuristics that take into account the importance of each data layer are defined. Later, in [25], correlated packet losses between several pairs of servers-clients are considered and the optimal coding strategies are analyzed. Block-based Raptor codes [16] are also used for video communication between Bluetooth nodes in [26] where the channels are modeled as two-state Markov chains whose states represent the reception or loss of a packet. After a packet erasure some additional Raptor encoded blocks are incorporated into the payload of the next Raptor packets. This leads to low energy consumption compared to that of typical Bluetooth FEC schemes. Furthermore, the added overhead due to the block-based approach is shown to stay low. Finally, Raptor codes have been shown to be appropriate in the streaming of multiview videos in [27]. The transmitted streams consist of large number of packets. With good models for predicting Raptor codes performance, an efficient rate-distortion optimized encoding is shown to lead to a good exploitation of the available resources with good video quality.

2.3 UEP and Rateless Codes

Unequal data protection (UEP) of media data is critical in applications with strict decoding deadlines as well as large differences in the importance of the media packets for the reconstruction quality. Fountain codes can be modified to provide unequal error protection of media packets. For example, the LT codes can be re-designed so that they involve more important symbols in more encoding operations (more symbols) [28]. Alternatively, expanding window fountain codes (EWF) [29] have been proposed to protect unequally the data without modifying LT encoding. Specifically, the EWF codes encode successively several expanded

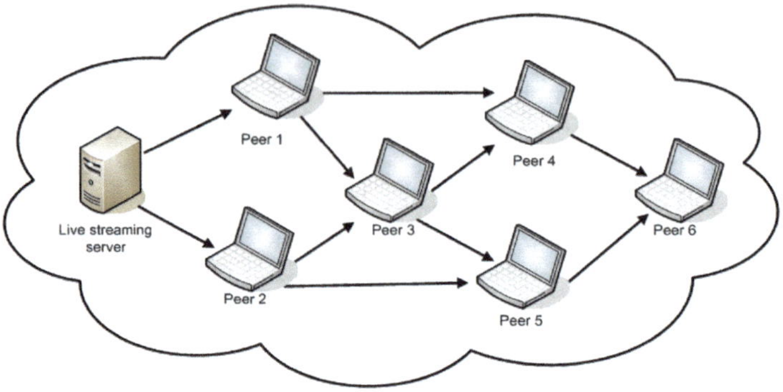

Fig. 3. Streaming with path diversity. The stream originated from the server is forwarded by the different network peers so that it can reach all the clients. The network nodes can implement network processing in order to increase the throughput and the robustness of the system.

windows of source symbols with LT codes. These EWF codes have been applied to layered multimedia in [30]. In [31], the LT encoding procedure is modified such as shorter coding paths are assigned to the more important symbols. The decoding probability of these symbols is thus increased. Finally, in [32], the LT codes distribution remains unaltered but the LT encoded symbols of degree one and two are not uniform combinations of symbols from all classes. Therefore, the most important classes have a higher probability to be sent as degree one or combined in order to generate symbols of degree two. This process does not harm the overall performance of LT codes, but rather increases the probability for the most important symbols to be correctly decoded at the receiver.

3 Network Coding in Multipath Streaming

3.1 In-Network Processing

Streaming in overlay networks introduces a new transmission model, where the network nodes can be used advantageously for improved performance, as illustrated in Fig. 3. Processing in the network nodes and not only at the ends of the transmission channels permits a better adaptation to network dynamics and a better exploitation of the diversity of the network. We describe in this section the benefits of network coding and its application to practical streaming applications, with an emphasis on unequal error protection strategies that are specific to media streaming problems.

One of the first attempts to realize some type of coding in nodes of an overlay network is presented in [33, 34]. The nodes are organized in multicast trees. Some of them implement Reed-Solomon (RS) channel coding operations to increase the robustness of the system. These are called network-embedded FEC (NEF) nodes and perform RS decoding on the packets they receive. They encode them again with RS codes, before passing them to the children peers. NEF nodes permit to increase the resiliency of the system, while avoiding waste of resources with a strong end-to-end protection. A greedy algorithm determines the number of NEF nodes and their location. Only a few well-positioned NEF nodes are sufficient to provide significant network throughput gains which results into a high video quality.

Similarly, decoding and encoding based on fountain codes is performed in the network nodes in [7]. The LT codes [15] are used in this work since they perform close to perfect codes and eliminate the need for reconciliation among network peers and packet scheduling. The intermediate network nodes wait for receiving a sufficient number of packets to recover the source content. Then the source packets are re-encoded into a new set of LT packets that differ from the packets produced independently in the other nodes. This is made possible by the rateless property of LT codes which allows for the generation of an infinite number of different packets. Decoding and encoding in the nodes however come at the price of increased complexity and delay. The network topology is however constructed such that minimal delays can be achieved. The streaming system is shown to be resilient to network dynamics with an increased throughput due to the rateless properties of the LT codes.

Decoding and recoding in the network nodes are difficult to implement in large networks, since they introduce quite important delays. However, when the network provides path diversity, the packet diversity in the system can be maintained by effective network coding algorithms. Network coding basically consists in combining packets in the network nodes. The encoded packets are then forwarded to the next nodes. When the receiver eventually gathers enough innovative packets, it can invert the full coding system and recover the original data. Network coding permits to increase the network throughout and the error resiliency in streaming systems with path diversity. The benefits of network coding methods are illustrated in a simple streaming scenario in the peer-to-peer network depicted in Fig. 4 where all links have capacity of one packet per time slot. When all intermediate nodes follow the store and forward approach, there is no guarantee that Peer 6 receives packets $X1$ and $X2$ because of the bottlenecks at Peers 4 and 5. If, however, Peers 3, 4 and 5 are network coding points, the probability of receiving an undecodable set of packets by Peer 6 is approximately zero when coding operations are in a large galois field.

We refer the interested readers to [35, 36, 37, 38, 39, 40] for detailed descriptions of network coding theory and principles. The theoretical works in network coding have made apparent that network coding can be beneficial for various applications like streaming or distributed storage. We focus, in the rest of this section, on the application of network coding to practical multipath streaming scenarios.

3.2 Practical Network Coding

Early network coding schemes [41, 42] necessitate the use of computationally complex algorithms for defining the coding coefficients. In addition, the network

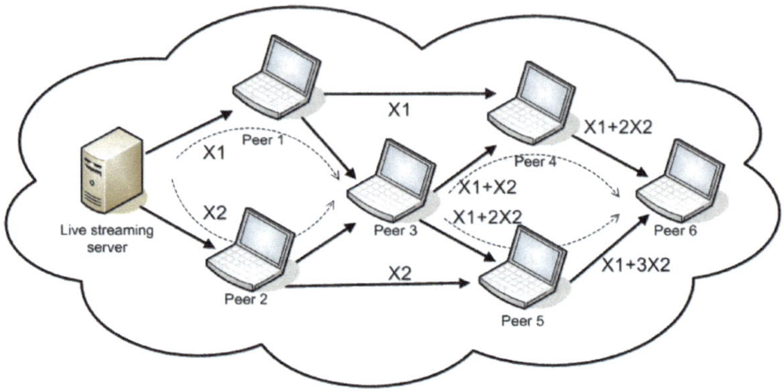

Fig. 4. Multimedia streaming in peer-to-peer networks with network coding techniques.

coding design methods generally assume that the servers have a full knowledge of the network topology, which is quite unrealistic in large scale networks. In order to simplify the design of network codes and improve the robustness in network dynamics, several works [43, 44, 45] have proposed to implement network coding with a random selection of the coefficients. If the coefficients are chosen in a sufficiently large Galois Field GF(q), random linear network coding can achieve the multicast capacity with a probability that asymptotically approaches one for long code-lengths (high number of source packets). The probability that all receivers are able to decode the source message is larger than

$$(1 - d/q)^r ,$$

where d is the number of receivers and q represents the size of the field [44]. The parameter r corresponds to the maximum number of links receiving signals with independent randomized coefficients in any set of links constituting a flow solution from all sources to any receiver. This lower bound on the decoding probability holds for independent or linearly correlated sources and for networks with or without delays. RLNC provides therefore a low complexity alternative towards the design of practical algorithms, since it permits to relax the requirements on the full knowledge of the network topology. It permits to implement distributed solutions with independent coding decisions in each node. Such a distributed algorithm is particularly interesting in ad hoc networks.

Motivated by the results of RLNC, the authors in [46] propose a practical network coding system for streaming applications. They define a proper format that can be used in random network graphs without the need for a hypernode that is aware of all coding coefficients and the overall topology. A header is assigned to each packet and contains the coding coefficients. As the packets travel through the network they are subject to successive coding operations, which modify both the message and header parts of every packet. All packets therefore contain encoded symbols along with the coefficients that have been used for their computation. The header part thus defines a global encoding vector that can be used at any decoder to recover the original message by Gaussian elimination, typically. It is shown that the equation system built by the successive network coding operations is with probability 99.6% decodable when the computations are performed on GF(2^{16}). Smaller fields like GF(2^8) are sufficient in practice [46].

Even if network coding a priori permits to combine any packet in the network nodes, the coding choices clearly have an influence on the end-to-end delay of the transmission system. In order to cope with the buffering delay problem, the authors in [46] introduce the concept of *generation*. A generation is a group of packets with similar decoding deadlines, which can be combined together by the network coding operations. The generation of every packet is identified by a small header of one or two bytes that is added to each packet. At the network nodes, the packets are stored into the buffer upon their reception. Whenever there is a transmission opportunity, the network coding node linearly combines the available packets and transmit the encoded packet. As coding becomes constrained to the packets of the same generation, the resulting delay is limited. However, the redundancy might become rapidly large since the number of packets in a generation stays limited. In order to maximize the robustness of the system, the packets that are not innovative with respect to the information that has been previously received, are simply discarded, and not forwarded nor encoded [46]. The clients finally implement a progressive decoding strategy by Gaussian elimination typically. The decoding becomes successful if the number of received packets is equal to the size of the generation. The delay in the system is mostly driven by the time that is needed for each client to collect enough packets.

3.3 Network Coding in Multipath Media Streaming

Due to its benefits in terms of increased throughput and robustness, practical network coding has been applied to diverse multipath streaming applications. It first finds a perfect application in peer-to-peer multicast applications. Such applications have become recently very popular, as they rely on the bandwidth contributions from peers in order to reduce the load on the main streaming server. Live multicast streaming can therefore be implemented by transmitting the media packets from the server to all the clients via other peers that are grouped in an overlay or ad hoc configuration. The packet distribution is mostly organized in two modes, which are the push or pull strategies. In the first case, the packets are simply pushed through the different peers in a way that is determined by the senders. In the pull scenario, the clients request specific packets or group of packets from the source peers. Network coding can be beneficial in both cases, as it helps to cope with the network dynamics. It permits to achieve a sustainable throughput with reduced transmission delays.

One of the first works that has studied the performance of network coding in peer-to-peer (p2p) streaming has been proposed in [47]. Randomized linear network coding is implemented in a system called "Lava" in order to evaluate the tradeoffs and benefits of network coding in live p2p streaming. The system offers network coding as an option in a pull-based p2p streaming solution that allows for multiple TCP connections for multiple upstream peers. Prior to transmission, the streams are divided into segments of specific duration, similar to the idea of generations proposed in [46]. These segments are further divided into blocks that undergo network coding operations in the different peers. The peers periodically exchange messages to announce the availability of segments in a pull-based manner. At any time, peers make concurrent requests for segments that are missing in their playback buffer by addressing randomly one of the peers that possess the segment of interest. The peers then decode the segments from their playback buffer in a progressive manner using Gauss-Jordan elimination. The evaluation shows that the network coding scheme is resilient to network dynamics, maintains stable buffering levels and reduces the number of playback freezes. Network coding is shown to be most instrumental when the bandwidth supply barely meets the streaming demand.

Based on the encouraging results of [47], the same authors redesign the peer-to-peer streaming algorithm and propose the R^2 architecture in [48]. In R^2, randomized linear network coding is combined with a randomized push algorithm to take full advantage of coding operations at peer nodes. The peers periodically exchange buffer maps that indicate the segments that have not been fully downloaded yet. The R^2 system sends the buffer maps together with the data packets whenever is possible, otherwise they are transmitted separately. The frequency of this information exchange has to be chosen high enough, in order to avoid the transmission of redundant segments. Whenever a coding opportunity is detected, a peer randomly chooses a video segment that the downstream peer has not completely received and generates a network coded block. The segment selection is inspired from [49]. The system also uses large segment sizes in order to avoid the transmission of too much overhead information by buffer map exchanges. The streams are progressively decoded by Gauss-Jordan elimination, similarly to the Lava system described above [47]. The R^2 system provides several advantages in terms of buffer level and delay, as well as resilience to network dynamics. The scalability of the system is also increased. Most of these advantages are due to the combination of push-based methods with randomized linear network coding.

The organization of the peers in the overlay network has a large influence on the performance of the streaming system. In particular, the delivery has to be organized in such a way that the bandwidth constraints can be respected, and such that the clients with the smallest bandwidth do not penalize the performance of the overall system. A method for constructing peer-to-peer overlay networks for data broadcasting is proposed in [50]. The overlay construction imposes that all the peers have the same number of parents nodes, which are the nodes that forward them the data packets. Such a constraint tends to distribute the load over the network. Network coding is then used in the peer nodes for increasing throughput and improving system robustness. In order to avoid limiting the performance of the system by the smallest capacity peers, one could organize the overlay into several layered meshes. Heterogeneous receivers can then subscribe to one of several meshes, depending on their capacities [51]. The data are similarly organized into layers, and network coding is performed on packets of the same layers. The practical network coding scheme of [46] is adopted in this work due to its low complexity. The construction of the layered meshes takes into consideration the overlapping paths in order to exploit the network coding benefits. Depending on the network state and the clients' requirements, every receiver determines the proper number of meshes it has to subscribe to. The network throughput is finally increased by network coding combined with appropriate peer organization. In the same spirit, the work in [52] proposes to split the bitstream into several sub-bitstreams for streaming over peer-to-peer networks. A neighbourhood management algorithm is then used to schedule appropriately the transmission of the different encoded sub-bitstreams. Finally, the problem posed by the heterogeneity of the receivers could also be solved by combining network coding with multiple description coding as proposed in [53].

3.4 Prioritized Network Coding

Media streams are generally characterized by packets with different importance with respect to their contribution to the quality at the decoder. Network coding can adapt to this property by handling the packets according to their priority. Network coding based on Prioritized Encoding Transmission (PET) [54] principles has been initially proposed in [46], where data of high importance receive a high level of error protection by a proper arrangement of the data blocks in the encoding matrix. Unequal error protection is also proposed in [55]. The PET algorithm is however replaced by a MD-FEC scheme [56], which seeks for the distortion-minimal source and channel rate allocation for the given channel conditions. Prioritized network coding is applied to scalable video streams in [57]. Data are segmented and interleaved in the coding matrix, in such a way that base layer typically receives more redundancy bits than the enhancement layers. Classical network coding is then performed on packets within the same generation. The proposed scheme is shown to outperform other solutions based on either routing or routing with replication policies.

The above methods face several problems that limit their application in real settings. Specifically, the MD-FEC based schemes often overprotect the information since they require the end-to-end network statistics and full knowledge of network topology to find the optimal protection. The application of MD-FEC based network coding methods in a distributed manner is possible only when the unequal amounts of protection are determined in hop-by-hop basis. Specifically in such a setting, each peer node first decodes the received network coded packets and then successively re-encodes the recovered video packets. Unfortunately, this procedure is computationally expensive and pronounces latency issues. Furthermore, it is questionable whether MD-FEC based NC methods are appropriate for overlay networks since ideally all received packets by a peer should be generated by the same MD-FEC matrix. Otherwise significant rate is lost as each node can decode only when it has received a decodable set of packets from the same MD-FEC matrix. Without such coordination the nodes receive packets from multiple MD-FEC codes that cannot be jointly decoded.

One could also achieve different levels of protection by changing the network coding scheme itself, where the coding operations are adapted to the importance of the packets. Priority random linear codes are proposed in [58] for data dissemination in peer-to-peer and sensor networks. Improved data persistence is achieved due to the fact that the most important video data represents a combination of fewer source packets. Prioritized coding can also be achieved by modifying the network coding operations in practical streaming applications. For example, the work in [59] addresses the problem of streaming H.264/AVC encoded video content in MBMS networks. Packets are grouped in different classes, and frame dependencies are further taken into account for determining the optimal network coding operations for each class. The coding choices are determined locally in each node by estimating the number of innovative packets received by each client. However, the coding decisions are still complex to compute due to the high number of dependencies between packets.

However, the distributed method of [59] imposes frequent packet exchanges to determine the optimal coding operations. Obviously, this increases significant the computational cost. Moreover, the application of this distributed algorithm for video streaming in overlay mesh networks is not straightforward as for MBMS broadcast networks the extended communication between peer nodes is reasonable, however, in mesh networks it may worsen congestion problems and lead to an explosion of erased packets. Techniques based on UEP rateless codes like EWF codes [29] are inefficient for scalable video multicasting due to network dynamicity which results in poor estimation of the expected EWF codes overhead when central estimation is considered. EWF codes can work efficiently for large codeblocks which is not typical for video streaming where strict timing constraints exist. Smaller codeblocks penalize significant EWF codes performance as they are associated with large overheads.

Differently from the works described thus far a receiver-driven network coding technique for prioritized video transmission in overlay networks is presented in [60]. The specific importance of media packets is taken into account to prioritize the delivery of the most important packets. The overlay nodes perform RLNC on incoming video packets in order to improve the robustness to failures or erasures in the delivery process without the need for any centralized control. As media packets are grouped into classes of different importance, prioritized transmission is achieved by varying the number of packets from each class that are used in the network coding operations at a node. The mixing operations are not uniform across all packets arriving at a node, but instead packets with higher importance are involved in more coding operations. Thus, each peer is interested in selecting the vector of coefficients $\mathbf{w} = [w_1, \ldots, w_L]$ such that the network coding strategy employed by its parent nodes will maximize the peer's video quality. L stands for the number of video layers and w_i is a weighted coefficient determining the number of packets from class i[1]. An optimization problem solved by each node tries to determine the number of packets it should request from each class,

[1] Class i is a set of video packets from first i video layers and the combinations of these packets.

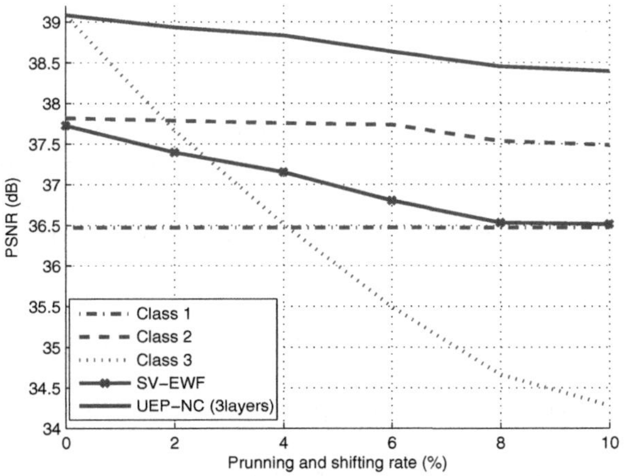

Fig. 5. Influence of the network regularity (pruning and shifting probabilities) of the proposed UEP scheme with baseline NC solutions and the *SV-EWF* scheme.

so that the total video distortion is minimized or alternatively the cumulative distortion reduction, as contributed by the requested packets, is maximized. This optimization problem is actually log-concave and can be solved by a low-cost greedy algorithm [60]. The algorithm starts from a pivotal distribution and then examines single packet exchanges between consecutive classes. This procedure is repeated till there is no beneficial exchange that further maximizes the distortion reduction.

The influence of the network topology on the performance of the above method termed *UEP-NC* is investigated in [60] and we illustrate here the performance of several UEP scheme based on network coding. The *UEP-NC* is evaluated for irregular networks generated following the process described in [61] where network link's are pruned and shifted according to two probabilities P_{pr} and P_{sf} respectively. The *UEP-NC* scheme of [60] is compared with basic network coding schemes called *Class-i* and a UEP scheme based on *EWF* codes (*SV-EWF*). *Class-i* scheme assumes video packets from first *i* video layers that are uniformly combined through RNC. Three video layers are assumed and the coding operations are restricted into GOPs to deal with timing issues and the frame rate is set to 30 *fps*. The pruning and the shifting probabilities are set to be equal, however their values vary in the range $[0, \ldots, 10]\,\%$. For each value of P_{pr} and P_{sf}, a seven-stage irregular network was generated. The packet loss rate on a link is fixed to 5% and the link capacity value equals to 360 kbps. The results of the evaluation are illustrated in Fig. 5. It can be seen that, when P_{pr} and P_{sf} are low then *Class-3* scheme performs equally well with *UEP-NC* as both are able to exploit the sufficient network resources. However, when P_{pr} and P_{sf} increase, the performance of *UEP-NC* degrades gracefully while *Class-3* scheme exhibits a significantly lower performance. The other two basic network coding schemes seems also to be robust to network variations, but are limited by the smaller number of video layers that they consider. The performance of *SV-EWF* for low pruning and shifting probabilities is comparable to that of *Class-2* scheme, however as network topologies become more random its performance approaches that of *Class-1* scheme. This is because bandwidth variations force *SV-EWF* to generate symbols from the first class only as the available bandwidth is insufficient to transmit data from other classes. Furthermore, the second class can often not be retrieved and the source symbols diversity in the network degrades quickly. For detailed comparisons interested readers are referred to [60].

4 Streaming with Rateless Coding and Network Coding

The practical network coding scheme presented in [46] has proposed efficient solutions to cope with inefficiencies of early network coding systems for deployment in streaming applications. However, there are still many issues that should be resolved before network coding can be applied to real time communication. One of the major drawbacks of today's network coding schemes is their relative high decoding computational cost as a dense equation system should be solved by Gaussian elimination. Another limiting factor is the significant coding overhead that grows linearly with the generation size. The above issues have been addressed in [13] where network coding is combined with Raptor codes [16] to take benefit from linear encoding and decoding times of the latter. Packets are encoded with non-systematic Raptor codes at the servers, as illustrated in Fig. 6. The network nodes then selectively combine packets when they have to compensate for packet losses and bandwidth variations. This scheme requires a smaller network coding header than that of typical network coding schemes. It is simply a concatenation of the combined Raptor coded packets headers. This saves significant amounts of rate which can be used for transmitting video streams of higher quality. Compared to implementations of fountain codes with in network processing [7], decoding operations in the network nodes are avoided

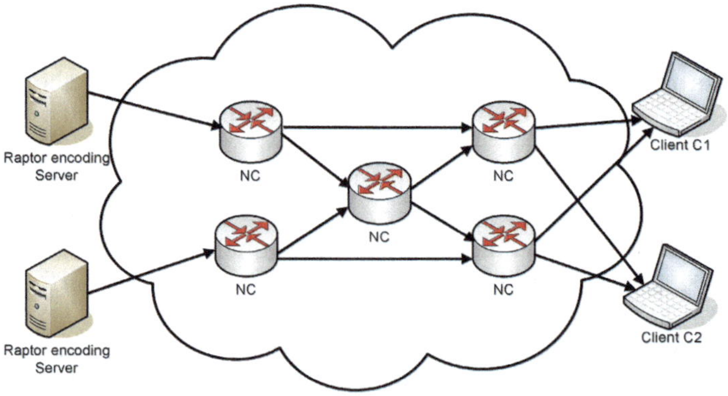

Fig. 6. Raptor network coding.

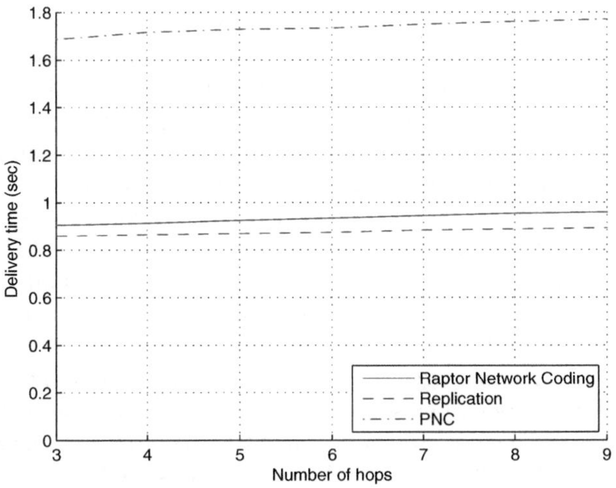

Fig. 7. Time delivery for regular networks with three nodes per hop with respect to the network size (number of hops).

in [13] and replaced by linear packets combinations. Such a system does not necessitate the use of large buffers in the nodes and the coding operations are kept very simple. This solution is advantageous in terms of both delay and complexity. The combination of Raptor coding and network coding permits to benefit from both source and path diversity in overlay networks, with controllable delay and complexity. We describe this algorithm in more details in the rest of this section, where we focus on the analysis of the delay and the coding strategies in Raptor network coding.

The Raptor based network coding system of [13] resembles RLNC based systems in that it performs simple linear operations with packets for generating network coded packets. The implicit coding structure of the Raptor network coding schemes is communicated as a header that is appended to each Raptor encoded packet. Raptor codes as RLNC systems are endowed with the rateless property that allows easy adaptation to bandwidth variations. The servers encode the video with non-systematic Raptor codes in order to provide initially high symbol diversity to the network, while the intermediate nodes perform encoding with the received packets. Specifically, the packets are XOR-ed before the peers forward them. Unlike other network coding systems, the peers do not replace the received packets with random combinations of them, but they perform selectively network coding to cope with bandwidth variations and packet erasures. This selective network coding keeps high the symbol diversity as all symbols are not involved in every coding operation. This approach also reduces the complexity since the equation system constructed at clients with the received packets is less dense.

The Raptor network coding scheme does not need large buffers. The packets are combined as soon as they arrive to the peer node, hence limiting end-to-end delays. Each node maintains a list with the transmitted packets in order to avoid transmission of a network coded packet over the same link with the packets combined to generate this packet. Then, these packets are removed from buffer to avoid future re-combinations of the same packets. This strategy assists the system to maintain a high symbol diversity in the network. The delivery times of the Raptor network coding scheme with respect to various network sizes are presented in Fig. 7. This scheme is compared with the practical network coding scheme of [46] and a scheme that applies Raptor codes in an end-to-end fashion. The intermediate nodes just replicate randomly some of the received packets to replace missing packets. Regular mesh networks, where the nodes are organized into different transmission stages that depend on the hop distance to the sources are examined. Three nodes per stage are considered. The link's capacity is set to 400 kbps. The packet erasure rate is fixed to 5% for each network link and the buffer size equal to 32 packets. As it can be seen from Fig. 7 results in larger delivery times as each node forwards packets when the buffer is full. It is worth noting that there are not significant performance variations as the network size grows. The advanced performance of the Raptor network coding scheme is attributed to fact that network coded packets are generated upon request. Since the packet loss rate is rather low and the buffer size sufficiently large, the probability of not finding orthogonal packets is also very low.

Although the computational cost of the Raptor network coding scheme is relative low, it increases with the successive coding operations. Specifically, due to losses and network coding in successive stages, the distribution of the degree of the symbols received at the decoder is altered and might differ significantly from the degree distribution used in the source Raptor encoders. In particular, the generator matrix becomes denser due to combinations of Raptor symbols by network coding, which leads to increased decoding complexity. Apparently, the received symbols by a client do not follow anymore the original Raptor codes degree distribution[2]. In order to guarantee linear decoding times, in [62] the re-design of the degree distribution of the Raptor codes $\Omega(x)$ used at the senders can be studied. The optimization goal is the design of a $\Omega^{'}(x)$ that grows as packets travel through the network towards clients and ends up to a distribution similar to that of well performing Raptor codes. The design optimization problem can be formulated as a geometric programming (GP) [63] problem. When the servers are aware of the network statistics (network losses and topology), the optimal source degree distribution can be determined so that linear decoding times are preserved. This method is more generic than [64] where the design of LT codes degree distribution for simple relay topologies has been investigated and the RSD [15] is decomposed into two component distributions prior to deconvolution of RSD. Although the algorithm in [64] ensures that clients receive symbols whose degree distribution is close to RSD, it imposes rather complicated encoding rules.

[2] The degree stands for the number of symbols combined for generating an Raptor coded packet.

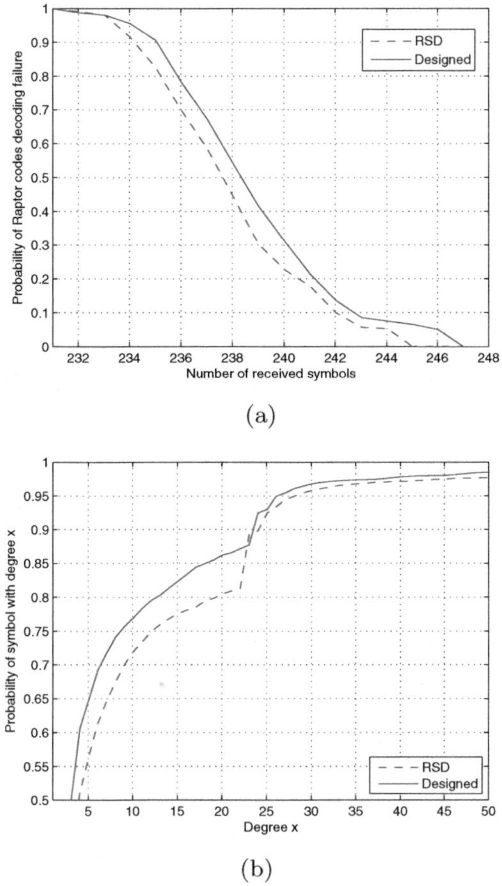

Fig. 8. Performance evaluation of (231, 249) Raptor codes for regular network topologies with three intermediate nodes per stage. The compared network coding schemes employ: a 3GPP variant with RSD distribution and the low-complexity codes. (a) Raptor decoding probabilities at clients with respect to the number of received symbols. (b) Cumulative distribution of symbols degree received by each client.

The extension of this method to complex network topologies is furthermore not trivial.

The Raptor decoding probabilities with respect to the number of received packets for nine-stage regular network topologies with three nodes per stage is shown in Fig. 8(a). It is obvious that the designed distribution performs slightly worse, but it remains close to the performance of a scheme employing 3GPP codes with RSD distribution. This performance degradation is due to the inefficiency of the design method to preserve the exact value of the spike which guarantees the convergence of the decoding process. However, when the methods are compared in terms of the cumulative degree distribution function (cdf) as

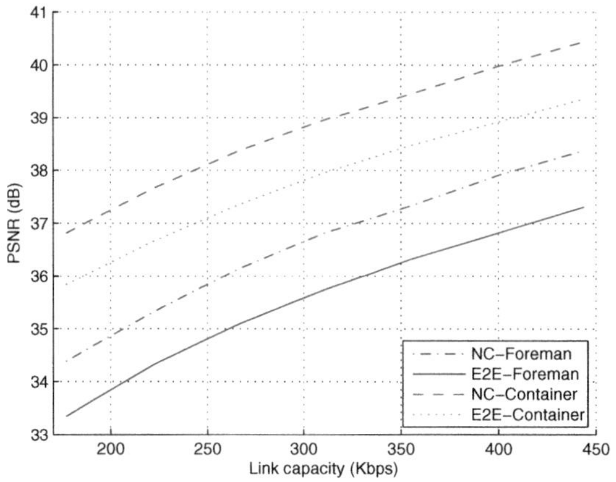

Fig. 9. PSNR performance evaluation of Raptor network coding (NC) and end-to-end Raptor coding (E2E), for regular topologies with six coding stages and 3 nodes per stage. The PSNR is measured as a function of the average link bandwidth in the network, the packet loss ratio is 5 %.

depicted in Fig. 8(b), it is clear that the proposed codes outperform the 3GPP variant with RSD as they correspond to sparser generator matrices and thus decoding is faster.

Finally, the optimal determination of source and channel rate allocation for the Raptor network coding system is investigated in [65]. The optimization algorithm is centralized and seeks for the optimal allocation based on the network statistics gathered backwards. The algorithm estimates the rank decrement of the equation system built on the received network coded packets. The centralized optimization algorithm is formulated as a *minmax* problem. Therefore, the algorithm seeks for the channel rate allocation which minimizes the maximal distortion among the clients. The optimization is performed under the constraints that the number of packets sent over a link can not surpass link capacity.

We illustrate the performance of the Raptor network coding scheme with efficient rate allocation in in Fig. 9. The average quality is given as a function of the average link bandwidth is for the transmission of "Foreman" and "Container" CIF sequences over a regular topology with six encoding stages between servers and decoders and three nodes per coding stage. The packet loss ratio is set to 5% while the link bandwidth varies between 170 kbps and 450 kbps. The Raptor network coding scheme is compared with a scheme that applies Raptor codes as a form of end-to-end error protection. The Raptor network coding algorithm performs better than the end-to-end solution due to its improved adaptivity that permits a better exploitation of the path diversity. We also observe that the performance gap remains unaltered as the link capacity increases. This is due to the fact that the schemes are not very sensitive to bandwidth variations, but rather to the overall packet loss rate. A more detailed analysis of Raptor network video coding system performance for a variety of transmission scenarios can be found in [65].

5 Conclusions

In this book chapter, we have shown that fountain codes and network coding algorithms can be interesting solutions for building efficient streaming systems in networks with diversity. These techniques are similar in some concepts as they are based on random operations, do not target at a predetermined rate and try to maintain a high packet diversity in the streaming system. Although both types of algorithms have appealing characteristics, their use is not trivial for multimedia applications with strict deadlines. For example, rateless codes may still require some coordination especially in large scale networks, while network codes use computationally expensive decoding algorithms such as gaussian elimination. Hybrid schemes combining network coding with rateless codes enable the use of low-cost decoding algorithms without imposing any requirements for coordination of the servers. Hence, they probably represent an interesting solutions for streaming in overlay networks with proper exploitation of source and path diversity.

References

1. Zhang, X., Liu, J., Li, B., Yum, Y.S.P.: CoolStreaming/DONet: a data-driven overlay network for peer-to-peer live media streaming. In: Proc. of the 24th Annual Joint Conf. of the IEEE Computer and Communications Societies, INFOCOM 2005, vol. 3, pp. 2102–2111 (2005)
2. Chakareski, J., Frossard, P.: Adaptive systems for improved media streaming experience. IEEE Communications Magazine (2007)
3. Byers, J., Luby, M., Mitzenmacher, M., Rege, A.: A Digital Fountain Approach to Reliable Distribution of Bulk Data. In: Proc. ACMSIGCOMM 1998, Vancouver, BC, Canada, pp. 56–67 (1998)
4. Gasiba, T., Xu, W., Stockhammer, T.: Enhanced System Design for Download and Streaming Services Using Raptor Codes. European Trans. on Telecommunications 20(2), 159–173 (2009)
5. Wagner, J.P., Chakareski, J., Frossard, P.: Streaming of Scalable Video from Multiple Servers using Rateless Codes. In: Proc. of IEEE Conf. on Multimedia and Expo, ICME 2006, Toronto, Ontario, Canada, pp. 1501–1504 (2006)
6. Aly, S.A., Kong, Z., Soljanin, E.: Raptor Codes Based Distributed Storage Algorithms for Wireless Sensor Networks. In: Proc. of IEEE Int. Symp. on Information Theory, ISIT 2008, Toronto, Ontario, Canada, pp. 2051–2055 (2008)
7. Wu, C., Li, B.: rStream: Resilient and Optimal Peer-to-Peer Streaming with Rateless Codes. IEEE Trans. Parallel and Distributed Systems 19(1), 77–92 (2008)
8. Ahlswede, R., Cai, N., Li, S.Y.R., Yeung, R.W.: Network Information Flow. IEEE Trans. Information Theory 46(4), 1204–1216 (2000)
9. Gkantsidis, C., Miller, J., Rodriguez, P.: Comprehensive View of a Live Network Coding P2P system. In: Proc. ACM SIGCOMM/USENIX IMC 2006, Brasil (2006)

10. Gkantsidis, C., Rodriguez, P.R.: Network Coding for Large Scale Content Distribution. In: Proc. of the 24th Annual Joint Conf. of the IEEE Computer and Communications Societies, INFOCOM 2005, vol. 4, pp. 2235–2245 (2005)
11. Dimakis, A.G., Godfrey, P.B., Wainwright, M.J., Ramchandran, K.: Network Coding for Distributed Storage Systems. In: 26th IEEE Int. Conf. on Computer Communications, ICCC 2007, Anchorage, Alaska, USA (2007)
12. Deb, S., Médard, M., Choute, C.: On Random Network Coding Based Information Dissemination. In: IEEE Int. Symp. on Information Theory 2005, ISIT 2005, Adelaide, Australia, pp. 278–282 (2005)
13. Thomos, N., Frossard, P.: Raptor Network Video Coding. In: Proc. of the 1st ACM Int. Workshop on Mobile video (in conjunction with ACMMM 2007), Augsburg, Germany (2007)
14. Mitzenmacher, M.: Digital Fountains: A Survey and Look Forward. In: Proc. of the Information Theory Workshop, ITW 2004, San Antonio, Texas, USA, pp. 271–276 (2004)
15. Luby, M.: LT codes. In: Proc. of the 43rd Annual IEEE Symposium on Foundations of Computer Science (FOCS 2002), Vancouver, Canada, pp. 271–280 (2002)
16. Shokrollahi, A.: Raptor codes. IEEE Trans. Information Theory 52(6), 2551–2567 (2006)
17. 3GPP TS 26.346 V7.1.0, Technical Specification Group Services and System Aspects; Multimedia Broadcast/Multicast Service; Protocols and Codecs (2005)
18. Kamra, A., Feldman, J., Misra, V., Rubenstein, D.: Growth Codes: Maximizing Sensor Network Data Persistence. In: Proc. of ACM SIGCOMM 2006, Pisa, Italy (2006)
19. Agarwal, S., Hagedorn, A., Trachtenberg, A.: Adaptive Rateless Coding Under Partial Information. In: Proc. Information Theory and Applications Workshop, ITA 2008, San Diego, CA, USA, pp. 5–11 (2008)
20. Bonello, N., Zhange, R., Chen, S., Hanzo, L.: Reconfigurable Rateless Codes. In: Proc. VTC 2009-Spring, Barcelona, Spain (2009)
21. Luby, M., Gasiba, T., Stockhammer, T., Watson, M.: Reliable Multimedia Download Delivery in Cellular Broadcast Networks. IEEE Trans. on Broadcasting 53(1), 235–246 (2007)
22. Schierl, T., Johansen, S., Perkis, A., Wiegand, T.: Rateless Scalable Video Coding for Overlay Multisource Streaming in MANETs. Journal on Visual Communication and Image Representation 19(8), 500–507 (2008)
23. Ahmad, S., Hamzaoui, R., Akaidi, M.A.: Robust Live Unicast Video Streaming with Rateless Codes. In: Proc. of 16th Int. Workshop on Packet Video, PV 2007, Lausanne, Switzerland, pp. 77–84 (2007)
24. Ahmad, S., Hamzaoui, R., Akaidi, M.A.: Practical Channel-adaptive Video Streaming with Fountain Codes. In: Proc. of IEEE Int. Conf. on Pervasive Computing and Applications, ICPCA 2008, Alexandria, Egypt (2008)
25. Wagner, J.P., Frossard, P.: Adaptive and Robust Media Streaming over Multiple Channels with Bursty Losses. In: Proc. European Signal Processing Conference, EUSIPCO 2007, Poznan, Poland (2007)
26. Razavi, R., Fleury, M., Ghanbari, M.: Block-based Rateless Coding for Energy-efficient Video Streaming over Bluetooth. In: Proc. of IEEE Symp. on Computers and Communications, ISCC 2008, Marrakech, Morocco (2008)
27. Tan, A.S., Aksay, A., Akar, G.B., Arikan, E.: Rate-distortion Optimization for Stereoscopic Video Streaming with Unequal Error Protection. EURASIP Journal on Advances in Signal Processing 2009(632545) (2009)

28. Dimakis, A.G., Wang, J., Ramchandran, K.: Unequal Growth Codes: Intermediate Performance and Unequal Error Protection for Video Streaming. In: Proc. of Multimedia Signal Processing, MMSP 2007, Pisa, Italy, pp. 107–110 (2007)
29. Sejdinovic, D., Vukobratovic, D., Doufexi, A., Piechocki, R., Senk, V.: Expanding Window Fountain Codes for Unequal Error Protection. In: Proc. of 41st Annual Asilomar 2007 Conf. on Signals, Systems and Computers, Pacific Grove, CA, USA (2007)
30. Vukobratovic, D., Stankovic, V., Sejdinovic, D., Stankovic, L., Xiong, Z.: Scalable Video Multicast using Expanding Window Fountain Codes. IEEE Trans. Multimedia 11(6), 1094–1104 (2009)
31. Chang, S.K., Yang, K.C., Wang, J.S.: Unequal-Protected LT Code for Layered Video Streaming. In: Proc. of IEEE Int. Conf. on Communications, ICC 2008, Beijing, China, pp. 500–504 (2008)
32. Woo, S.S., Cheng, M.K.: Prioritized LT codes. In: Proc. of 42nd Annual Conf. on Information Sciences and Systems, CISS 2008, Princeton, NJ, USA, pp. 568–573 (2008)
33. Karande, S., Wu, M., Radha, H.: Network Embedded FEC (NEF) for Video Multicast in Presence of Packet Loss Correlation. In: Proc. IEEE Int. Conference on Image Processing, Genoa, Italy, vol. 1, pp. 173–176 (2005)
34. Wu, M., Karande, S., Radha, H.: Network Embedded FEC for Optimum Throughput of Multicast Packet Video. EURASIP Journal on Applied Signal Processing 20(8), 728–742 (2005)
35. Fragouli, C., Soljanin, E.: Network Coding Fundamentals. Source. In: Foundations and Trends in Networking archive. Now Publisers Inc., Boston (2007)
36. Fragouli, C., Soljanin, E.: Network Coding Applications. Source. In: Foundations and Trends in Networking archive. Now Publisers Inc., Boston (2007)
37. Yeung, R.W., Li, S.Y.R., Cai, N., Zhang, Z.: Network Coding Theory. In: Foundations and Trends in Communications and Information Theory. Now Publisers Inc., Boston (2005)
38. Yeung, R.W.: Information Theory and Network Coding. In: Information Technology: Transmission, Processing and Storage. Springer, Heidelberg (2008)
39. Ho, T., Lun, D.S.: Network Coding: An Introduction. Cambridge University Press, Cambridge (2008)
40. Thomos, N., Frossard, P.: Network Coding for Theory to Media Streaming. Journal of Communications 4(9) (2009)
41. Koetter, R., Médard, M.: An Algebraic Approach to Network Coding. IEEE/ACM Trans. on Networking 11(5), 782–795 (2003)
42. Jaggi, S., Sanders, P., Chou, P.A., Effros, M., Egner, S., Jain, K., Tolhuizen, L.M.G.M.: Polynomial Time Algorithms for Multicast Network Code Construction. IEEE Trans. Information Theory 51(6), 1973–1982 (2005)
43. Ho, T., Medard, M., Shi, J., Effros, M., Karger, D.R.: On Randomized Network Coding. In: 41st Allerton Annual Conf. on Communication, Control, and Computing, Monticello, IL, USA (2003)
44. Ho, T., Koetter, R., Medard, M., Karger, D.R., Effros, M.: The Benefits of Coding Over Routing in a Randomized Setting. In: IEEE Int. Symp. on Information Theory, ISIT 2003, Kanagawa, Japan (2003)
45. Ho, T., Medard, M., Koetter, R., Karger, D.R., Effros, M., Jun, S., Leong, B.: A Random Linear Network Coding Approach to Multicast. IEEE Trans. Information Theory 52(10), 4413–4430 (2006)
46. Chou, P.A., Wu, Y., Jain, K.: Practical Network Coding. In: Proc. 41st Allerton Conf. on Communication Control and Computing, Monticell, IL, USA (2003)

47. Wang, M., Li, B.: Lava: A Reality Check of Network Coding in Peer-to-Peer Live Streaming. In: IEEE INFOCO, Anchorage, Alaska (2007)
48. Wang, M., Li, B.: R^2: Random Rush with Random Network Coding in Live Peer-to-Peer Streaming. IEEE Journal on Selected Areas in Communications 25(9), 1655–1666 (2007)
49. Maymoukov, P., Harvey, N.J.A., Lun, D.: Methods for Efficient Network Coding. In: Proc. 44th Allerton Conf. on Communication, Control, and Computing, Monticello, IL, USA (2006)
50. Jain, K., Lovász, L., Chou, P.A.: Building Scalable and Robust Peer-to-Peer Overlay Networks for Broadcasting Using Network Coding. Journal on Distributed Computing 19(4), 301–311 (2006)
51. Zhao, J., Yang, F., Zhang, Q., Zhang, Z., Zhang, F.: LION: Layered Overlay Multicast With Network Coding. IEEE Trans. Multimedia 8(5), 1021–1032 (2006)
52. Liu, Y., Peng, Y., Dou, W., Guo, B.: Network Coding for Peer-to-Peer Live Media Streaming. In: Proc of the Fifth Int. Conf. Grid and Cooperative Computing, GCC 2006, Monticello, IL, USA, pp. 149–155 (2006)
53. Shao, M., Wu, X., Sarshar, N.: Rainbow Network Flow with Network Coding. In: Proc. of the fourth Workshop on Network Coding, Theory and Applications, NetCod 2008, Hong Kong, China, pp. 1–6 (2008)
54. Albanese, A., Bloemer, J., Edmonds, J., Luby, M., Sudan, M.: Priority Encoding Transmission. IEEE Trans. Information Theory 42, 1737–1744 (1996)
55. Ramasubramonian, A.K., Woods, J.W.: Video Multicast Using Network Coding. In: SPIE VCIP, San Jose, CA, USA (2009)
56. Puri, R., Ramchandran, K., Bharghavan, K.W.L.V.: Forward Error Correction (FEC) Codes Based Multiple Description Coding for Internet Video Streaming and Multicast. Signal Processing: Image Communication 16(8) (2001)
57. Wang, H., Xiao, S., Kuo, C.C.J.: Robust and Flexible Wireless Video Multicast with Network Coding. In: IEEE Global Telecommunications Conference, GLOBECOM 2007, Adelaide, Australia, pp. 2129–2133 (2007)
58. Lin, Y., Li, B., Liang, B.: Differentiated Data Persistence with Priority Random Linear Codes. In: 27th Int. Conf. on Distributed Computing Systems of Contents, ICDCS 2007, Toronto, ON, Canada, pp. 47–55 (2007)
59. Liu, X., Cheung, G., Chuah, C.N.: Structured Network Coding and Cooperative Local Peer-to-peer Repair for MBMS Video Streaming. In: IEEE Int. Workshop on Multimedia Signal Processing, MMSP 2008, Cairns, Queensland, Australia (2008)
60. Thomos, N., Chakareski, J., Frossard, P.: Randomized network coding for UEP video delivery in overlay networks. In: International Conference on Multimedia and Expo, ICME 2009, Cancun, Mexico (2009)
61. Thomos, N., Frossard, P.: Collaborative Video Streaming with Raptor Network Coding. In: Inter. Conf. on Multimedia and Expo, ICME 2008, Hannover, Germany, pp. 497–500 (2008)
62. Thomos, N., Frossard, P.: Degree Distribution Optimization in Raptor Network Coding. EPFL Technical report (2010)
63. Chiang, M.: Geometric Programming for Communications Systems. Now Publisers Inc., Boston (2005)
64. Puducheri, S., Kliewer, J., Fuja, T.E.: The Design and Performance of Distributed LT codes. IEEE Trans. Information Theory 53(10), 3740–3754 (2007)
65. Thomos, N., Frossard, P.: Network Coding of Rateless Video in Streaming overlays EPFL Technical report LTS-REPORT-2009-011 (2009)

Peer-Assisted Media Streaming: A Holistic Review

Yuan Feng and Baochun Li

Department of Electrical and Computer Engineering
University of Toronto
10 King's College Road
Toronto, ON, M5S 3G4
Canada
{yfeng,bli}@eecg.toronto.edu

Summary. This chapter presents a holistic review of recent research advances in peer-assisted streaming systems, including both live and on-demand streaming. We approach this task by first presenting design objectives of streaming systems in general, and then discuss differences between live and on-demand streaming. These common and different design objectives motivate the protocol design space in streaming systems, in categories of peer selection, segment scheduling, and distributed caching protocols. We present main results from the existing literature in each of these dimensions, with a particular focus on the pivotal role of network coding within such a protocol design space. We conclude the chapter with an outlook towards future research directions, especially in the application of network coding in peer-assisted streaming systems.

1 Introduction

To meet the demand of explosively growing multimedia applications, media streaming over the Internet has been a research topic attracting substantial interests in the literature over the past two decades. The "holy grail" of Internet media streaming is to satisfy the media streaming needs of as many end users as possible, with sustainable server bandwidth costs. The traditional client/server architecture advocates the use of large data centers to sustain streaming to end users at a large scale.

To maintain a satisfactory user experience, the bandwidth costs on servers grow rapidly as the user population scales up, and may not be bearable in corporations with limited resources. New architectural designs in the past two decades, such as IP multicast and content delivery networks (CDNs), attempted to address the problem by conserving resources in the edge or core routers, or by load balancing across a large number of edge servers that are geographically distributed. However, the problem of scalability to a large user population in media streaming systems is only mitigated to a certain degree, not solved.

C.W. Chen et al. (Eds.): Intel. Multimedia Communication: Tech. and Appli., SCI 280, pp. 317–340.
springerlink.com

Peer-to-peer (P2P) networks propose a different architectural design perspective: it offloads part of the bandwidth burden from dedicated streaming servers hosted by the content providers, and shifts them to end hosts themselves when they serve content to each other. The total volume of bandwidth consumed is not reduced, and in most cases, is even increased due to overhead of protocol-induced messaging and redundancy. Peer-to-peer networks are not panacea: the architecture does not *replace* streaming servers — they simply *complement* them. Such an architectural paradigm is referred to as *peer-assisted media streaming* in this chapter, to highlight the complementary nature of peer-to-peer networks.

Existing peer-assisted media streaming systems can be further divided into two categories, *live* and *on-demand* media streaming, with the latter often referred to as *video-on-demand* (VoD) in the literature. At a high level, a peer-assisted media streaming system typically contains three elements: (1) a number of dedicated streaming servers that serve media content, and can be treated in an aggregate fashion as a single dedicated media source; (2) one or a small number of index servers that keep track of state information of the system, such as existing end users (or *peers*); and (3) a web portal that provides information about media channels. An example of such a system is shown in Fig. 1.

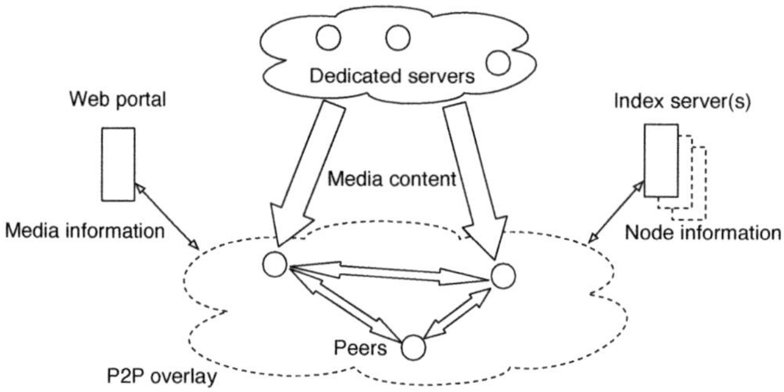

Fig. 1. The architecture of peer-assisted streaming systems.

Fundamentally, architectural and protocol designs in peer-assisted media streaming systems need to successfully address three characteristics observed in real-world streaming systems:

- *Scale:* A real-world peer-assisted streaming system needs to support at least hundreds of thousands of users and thousands of media channels simultaneously, with a reasonable playback quality to each of the end users. The CBS broadcast of NCAA tournament in March 2006 [26] has peaked at 268,000 simultaneous viewers. UUSee Inc. experienced a flash crowd of 871,000 peers on the Chinese New Year Eve in 2007 [35].
- *Dynamics:* Real-world peer-assisted streaming systems have shown a high level of dynamics. End users, as peers who serve media content to one another, may join or leave a media channel or the system at any time. Network bandwidth between a pair of peers vary over time, sometimes quite drastically

due to the shifting of bandwidth bottlenecks. Such dynamics challenge the design of any system architecture or protocol. Recent measurement studies have clearly shown that peer dynamics represent one of the most influential factors that affect the overall performance of peer-assisted streaming systems [22].

▷ *Heterogeneity:* According to a large set of live streaming traces obtained from the Coolstreaming system, it has been discovered that there exists a highly skewed distribution with respect to upload bandwidth contributions from peers. Such heterogeneity may have significant implications on resource allocation in peer-assisted streaming systems.

This chapter provides a taxonomy of recent research achievements in peer-assisted media streaming systems, with a special focus on on-demand streaming systems (VoD). We start our discussion with an in-depth coverage of common challenges in both live and on-demand media streaming systems. We then shift our focus to specific challenges and their corresponding solutions in on-demand streaming systems. In particular, we point out that *network coding* may be a feasible and practical solution to improve streaming performance in peer-assisted streaming systems. Finally, we present an outlook to future research directions in peer-assisted streaming systems.

2 Fundamental Challenges in Peer-Assisted Streaming Systems

In peer-assisted streaming systems that include both live and on-demand streaming, dedicated streaming servers and end users (peers) are organized into a topology at any given time, so that media content can be broadcast from the servers to all the participating peers in a streaming channel. In such a topology, a peer is directly connected with a subset of other peers in the same channel, referred to as its *neighbors*. The topology that peers and servers are organized into is sometimes referred to as a streaming *overlay*. In this section, we present two fundamental challenges and their corresponding solutions in such a streaming overlay for a media channel, common to both live and on-demand streaming systems.

In order to allow peers to serve one another and to maximize its efficiency, the following challenges naturally arise. First, which subset of peers should become neighbors of a participating peer? Second, when being served by neighbors, which segment of the media stream should be served first, and thus be given priority? Solutions to the first question are referred to as *neighbor selection* (or *overlay construction*) algorithms, and those to the second are referred to as *segment scheduling* algorithms.

2.1 Neighbor Selection Algorithms

One of the important design goals in peer-assisted media streaming systems is load balancing, *i.e.,* the redistribution of content should be balanced on the participating peers so that the costs for the system are shared, much of the existing research literature focused on designing suitable neighbor selection algorithms. We present several examples of this category of algorithms.

With an inter-overlay optimization scheme, Anysee, proposed by Liao *et al.* [25], uses an *inter-overlay manager* to find appropriate streaming paths with low delays in the peer-assisted streaming system. The main task of the inter-overlay manager is to maintain a backup streaming path set and an active streaming path set, using a reverse tracking algorithm. When the size of the backup set is less than a threshold, the peer will send out a message to some of its neighbors, which will be relayed till the receiver finds the delay from the initial peer to the current peer is greater than the minimum of all source-to-end delays from the current peer to the streaming source on different paths. Using such an algorithm, a peer is able to construct the best overlay path accordingly. When the total bit rates from active streaming paths are lower than a certain threshold, the manager will check whether a better path should be activated to replace the current one, selected from the backup streaming path set.

An alternative approach is to use a score-based neighbor selection algorithm, where a score is assigned as the criteria to decide which set of peers are selected as neighbors. OCTOPUS [24], for example, adopts a *score* that represents the degree of data overlapping between any two neighbors, *i.e.*, the amount of time slots it can provide to its neighbor and the amount of time slots the neighbor can provide to the peer. Under the assumption that underlying media streaming system provides the functionality of a Distributed Hash Table (DHT), Graffi *et al.* [17] propose another scoring function that calculates the costs for choosing a specific peer by taking into account of peer characteristics such as the uptime, network condition, and the number of tasks already performed for the system. As identified peers for a specific block may randomly leave the system, peer churn may become a critical challenge for this partner selection algorithm.

All of the aforementioned algorithms are under the assumption that users can and are willing to collaborate, which is not always the case realistically. Measurement studies have shown that, in some peer-to-peer streaming systems, a small set of nodes are requested to contribute 10 to 35 times more uploading bandwidth than downloading bandwidth [2]. As a result, peers are not willing to voluntarily contribute their own upload bandwidth, which may seriously affect the overall performance of peer-assisted streaming systems. Therefore, an appropriate incentive mechanism is critical to the performance of a peer-assisted streaming system, and the design of neighbor selection algorithms offers an opportunity to tackle this challenge.

Silverston *et al.* [30] discover that the incentive mechanism in BitTorrent file sharing systems, called *tit-for-tat*, is not well suited to peer-assisted streaming systems. Media flows impose temporal constraints that do not exist in bulk content distribution. Sometimes peers cannot serve data in return, not because they are non-cooperative, but because the temporal constraints make it pointless. As such, the temporal nature of the content makes the incentive mechanisms of BitTorrent obsolete. Existing solutions are mainly score-based, in which each peer is a credit entity, and peers will allocate resources according to the credits of requesting peers to maximize their own credits. However, these mechanisms suffer from a high computational complexity that prevents their real-world implementation. The design of a scalable, light-weighted incentive mechanism that can be incorporated into peer-assisted streaming systems remains to be an open problem.

Finally, recent measurement studies have unveiled that there is a highly unbalanced distribution with respect to uploading contributions from peers, which has significant implications on the resource allocation in such a system [22]. Further, by investigating the distribution of inter-peer bandwidth in various peer ISP categories, Wu *et al.* [35] shows that the ISPs that peers belong to are more correlated to inter-peer bandwidth than their geographic locations. In addition, there exist excellent linear correlations between peer last-mile bandwidth availability and inter-peer bandwidth within the same ISP, or between a subset of ISPs. The joint consideration of highly skewed resource distribution, incentive awareness and inter-peer bandwidth interference may lead to more appropriate neighbor selection algorithms in future research.

2.2 Segment Scheduling Algorithms

Different from other P2P applications such as file sharing, peer-assisted streaming systems need to consider timeliness requirements of streaming media. If media segments do not arrive in a timely fashion, they have to be skipped at playback, which degrades the playback quality, which is one of the most important performance metrics in streaming systems. Experiments show that peer-assisted streaming systems do not perform congestion control correctly, in scenarios where peer access links get congested [1]. Their response is to increase the level of redundancy, which further increases the download rate, and further exacerbate the congestion situation. These observations call for a judicious design of segment scheduling algorithms.

The first strategy for segment scheduling in peer-assisted streaming systems is a *tree-based per-slice push* strategy, in which a media stream is first divided into substreams as its slices, and then each of these slices is pushed to downstream peers in a tree. Though tree-based strategies offer minimal source-to-peer delays (the delay between the occurrence of a live event in a live stream and its playback at a peer), it suffers from the complexity in maintaining such tree structures when peers arrive and depart frequently. In contrast, a *mesh-based per-segment pull* strategy, first proposed by Zhang *et al.* [39], has been implemented in most real-world streaming systems. In such a strategy, the media stream is divided into segments in the time domain, and peers make explicit requests for the desired media segment from its neighboring peers. Neighboring peers exchange buffer availability bitmaps periodically. This strategy is simple to implement, but increases the source-to-peer delay in live streaming systems, due to the need for periodic buffer availability exchanges. The source-to-peer delay, however, is not an issue for on-demand streaming systems.

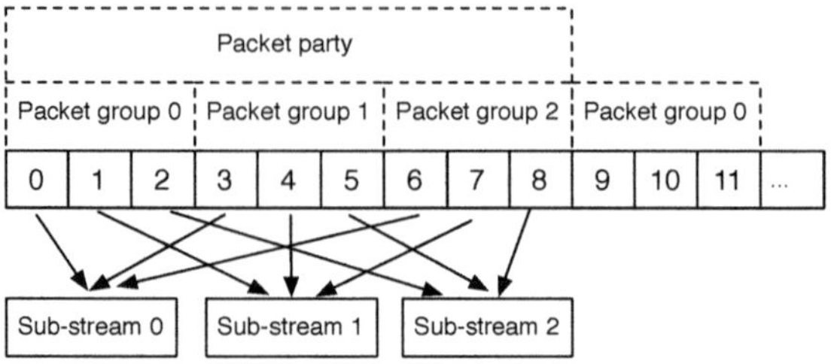

Fig. 2. Sub-streams in a push-pull hybrid strategy.

Zhang *et al.* [38] has proposed that push and pull strategies be combined, so that a hybrid strategy can be designed to be both robust to peer departures, and to support smaller source-to-peer delays. In the proposed push-pull hybrid strategy, segments are pulled in the first time slot when a peer first joins the system, and then pushed directly by the neighbors afterwards. To achieve this, the stream is evenly partitioned into n sub-streams, and each sub-stream is composed of the packets whose sequence numbers are congruent to the same value modulo n. Every continuous n packets belong to different sub-streams separately are grouped into a *packet group* and every continuous g packet groups are further clustered into a *packet party*. Each packet group in a packet party is numbered from 0 to $g-1$ and hence the packet group number can be computed by $\lfloor s/n \rfloor \bmod g$, where s is the packet sequence number. An example is shown in Fig. 2, where the stream is partitioned into 3 sub-streams and $n=3, g=3$.

Once a packet in packet group 0 in one packet party is requested successfully from a neighbor, the peer will send this neighbor a *sub stream subscription* to let it directly push the remaining packets in the same sub-stream. There are two types of streaming packets — the pulled packets and the pushed packets. When over 95% packets are pushed directly from the neighbors, the node will stop requesting buffer maps. And once the delivery ratio drops below 95% or a neighbor who provides over 5% streaming data quits, the peer will start to request buffer maps and pull streaming packets from its neighbors. Thus, most of the packets received will be pushed packets from the second time interval.

Zhou *et al.* [40] form a stochastic model to analyze some widely used segment selection strategies, including the *greedy*, *rarest first*, and *mixed* strategies. With M peers in the system, each peer maintains a buffer B that can cache up to n segments received from the system, as shown in Fig. 3.

The *greedy* strategy intends to fill the empty buffer locations closest to the playback point first. The segment selection function for the greedy strategy, $s(i)$, which is the probability of selecting $B(i)$, can be expressed as follows:

$$s(i) = 1 - (p(n) - p(i+1)) - p(1), i = 1, ..., n-1, \tag{1}$$

where the term $p(n) - p(i+1)$ is the probability that any particular segment is downloaded into buffer positions between $B(n)$ to $B(i+1)$; and the term $p(1)$ is the probability that any particular segment is downloaded directly from the server. The *rarest first* strategy will select a segment that has the fewest number

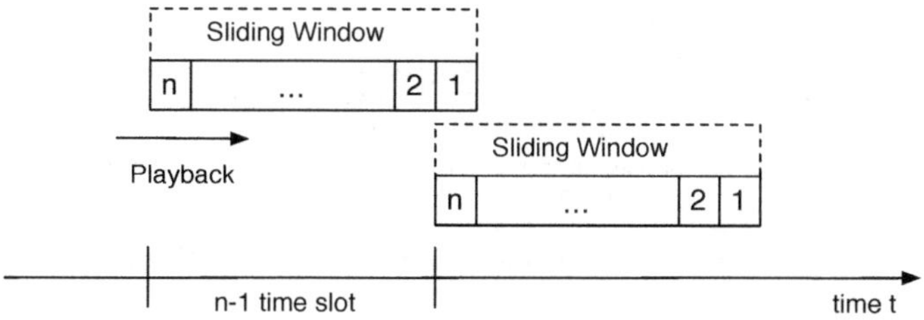

Fig. 3. The sliding window of buffer B.

of replicas in the system, for which the segment selection function is:

$$s(i) = 1 - p(i), \tag{2}$$

where $p(i)$ represents the probability that any particular segment is downloaded into buffer positions between $B(1)$ to $B(i-1)$. A *mixed* strategy implies that the the rarest first strategy is used for a portion of the buffer $B(1), ..., B(m), 1 \leq m \leq n$. If no segment can be downloaded using the rarest first strategy, the greedy strategy is then used for the remainder of the buffer $B(m+1), ..., B(n)$. In this case, the segment selection function for the mixed strategy can be expressed as:

$$p(1) = 1/M \tag{3}$$

$$p(i+1) = p(i) + p(i)(1 - p(i))^2, i = 1, ..., m - 1. \tag{4}$$

Simulation results in [40] have shown that the rarest first strategy is much better with respect to scalability, whereas the greedy strategy is able to produce better playback performance (continuity) in small scale systems.

While analytical results have indicated that a sequential selection strategy is well-suited for use in peer-assisted media streaming systems, many choose to implement a hybrid segment selection algorithm, which divides the cache into a high-priority set and the remaining set according to their time to the playback deadline, as indicated in Fig. 4. Intuitively, a peer is likely to choose segments in the high-priority set with higher probability for downloading [28, 14]. Much effort has been devoted to design the appropriate distribution in these two sets. For example, the piece picking policy in Toast performs in-order selection in high-priority set and Beta random selection in remaining set, which is implemented using Python's generator for a beta distribution in remaining set. It adds a given-up area between the current streaming position and high-priority region, which corresponds to pieces that are too close to the current stream position to download on time. Simulation results show that the performance beats rarest-first and in-order; however, the simulation is only conducted in static case in which peers all are assumed to stay until the end of a video, which is less likely to happen in the real world. Then in R^2, a uniform distribution is used in high-priority set and a randomized choice subjected to Weibull distribution is adopted in the remaining set.

Combined with network coding, Annapureddy *et al.* propose a worst-seeded-policy, which is to upload a block from a lesser represented segment whenever

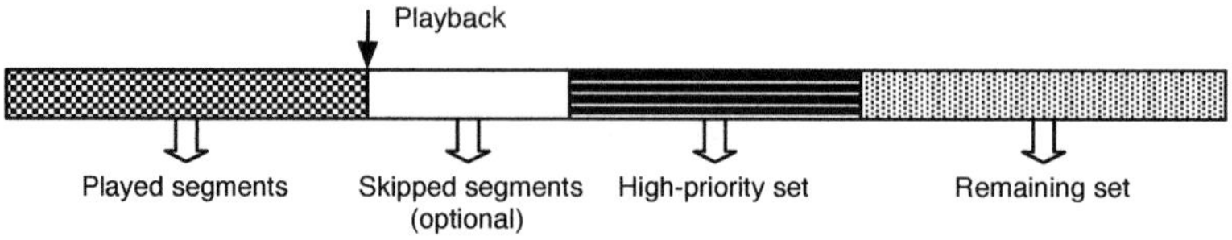

Fig. 4. Hybrid segment scheduling algorithms

possible [5, 4]. Compared with greedy approach that peers greedily request blocks from their earliest incomplete segments, system throughput increases significantly. However, the assumption of source node has knowledge of the rarity of segments aggregated across the other nodes in the network is not realistic for the real world. They also examine the effect of pre-fetching, i.e. fetching a block that is needed later than the first segment of interest. By evaluating probabilistic-first-range-random-second-range-rarest policy, which performs consistently well in pre-fetching policies, it is shown that pre-fetching provides easy access to blocks when they are needed by creating additional sources for the blocks.

However, most papers discuss the situation of only one or two channels exist, which is definitely an over-simplified case. The peer-assisted streaming systems in nowadays like PPStream, PPLive and so on have hundreds of channels available. From the server's point of view, we have to consider how to allocate bandwidth in multi-channel streaming systems. In a server capacity provisioning algorithm—Ration, which is proposed by Wu *et al.* [36], server capacity allocated to each channel in next time slot is proactively computed by an optimization problem *Provision*$(t+1)$, as follows ($\forall t = 1, 2, ...$):

$$\begin{aligned} \text{maximize} \quad & \sum_{c \in C} p^c n^c_{t+1} q^c_{t+1} \\ \text{subject to} \quad & \sum_{c \in C} s^c_{t+1} \leq U, \\ & q^c_{t+1} = F^c_{t+1}(s^c_{t+1}, n^c_{t+1}), \forall c \in C, \\ & 0 \leq q^c_{t+1} \leq 1, s^c_{t+1} \geq 0, \forall c \in C \end{aligned} \tag{5}$$

where p^c denotes the assigned priority level, n_c represents the estimated population and q^c the streaming quality for each channel c. At each time t, Ration proactively computes the amount of server capacity s^c_{t+1} to be allocated to each channel c for time $t+1$. With full ISP awareness, Ration is carried out on a per-ISP base, and is able to guide the deployment of server capacities and channels in each ISP to maximally constrain P2P traffic inside ISP boundaries. Cheng *et al.* also evaluate the improvement of multi-video caching (MVC) [12]. They show that for any given number of peers, MVC further reduces the aggregate load on the sources and MVC has lower chunk cost for the same concurrency compared with single-video caching. Further, it is presented that MVC increases chunk replica counts across the system which improves the continuity by decreasing jitter and seek latencies. In general, designing peer-assisted streaming systems with multiple channels is more complicated and challenging, which may raise the interests of researchers in the future.

3 Special Challenges in P2P VoD Streaming Systems

While peer-assisted live streaming appears to be evolving towards a mature stage, VoD services based on P2P overlays have appeared as a more attractive area to most researchers. A peer-assisted VoD service is more challenging to design than a live streaming system because the peers in a live streaming scenario have a shared temporal focus, while in VoD case they have greater temporal diversity of requests. The peer dynamics resemble those of file downloading, while still requiring low startup delays for the sequential playback of large media objects. Besides, equipped with larger storage space—cache—than the limited buffer in live streaming scenario, peers in VoD case can store and further upload more content. Intuitively, questions will arise as in case of larger cache, what content should be chosen to store in each peer. The fact that different users may be watching different parts of the video at any time can greatly impact the efficiency of a swarming protocol. The lack of synchronization among users reduces the block sharing opportunities and increases the complexity of the block transmission algorithms. This naturally bring the problem of how to search for the desired content efficiently, which results in numerous searching algorithms.

3.1 Cache Management

With larger storage capacities, peers can store redundant copies of a file near the end user and it has been proven to be extremely effective in many P2P VoD applications. Recent measurement results show that a simple static cache that stores the top long-term popular files uses 84% less space than a dynamic infinite cache solution which stores all videos, and still manages to save about 75% of the load in the server [9]. As a result, a number of algorithms that intend to choose the most "popular" content wisely to store in a finite storage space are proposed, a summary of which is shown in Table 1.

Popularity Prediction

In order to minimize VoD request rejection rates for a very large content library, one of the key question is how to translate the popularity distribution into content availability in the network. The common way of doing so is to use number of requests for a specific video content to represent its popularity and the number of replicas in the system to reflect its availability. Through numerical simulations and empirical validations, the popularity distribution of relatively popular video categories exhibits power-law behavior (a strait line in a log-log plot) with a tail truncation, which is affected by both the average requests per user and the number of videos in a category. While the popularity distribution of the unpopular content behaves more like a Zipf's law, later research shows that despite

Table 1. Cache management in peer-assisted VoD systems

Topics	[31]	[3]	[11]	[27]	[21]	[18]	[37]
Popularity Prediction	×	×	×	√	√	√	√
Peer Availability Detection	×	×	×	√	×	×	×
Replication Mechanism	√	√	√	×	×	√	√
Bandwidth Allocation	×	×	×	×	√	×	×
Dynamic Repair	×	×	√	×	×	×	×

the traffic variation among different days, the popularity distributions are quite similar and the distribution is more skewed than a Zipf distribution [9], which may due to the fact that on any given day, there are several highly popular news and business clips, with each of these clips having roughly the same popularity. Using these distributions, we can use the video's past behavior to predict their future more precisely.

Since the popularity of a video changes during a typical day and the change between two continuous short intervals is smooth, the popularity in the most recent interval is useful to predict the future requests. In addition, chunks requested recently are more likely to be requested later. Based on these observations, a chunk request predictor is constructed weighting the contribution of recent requests to decay with time. Then by discovering the fact that for the same popularity level and content availability, the VoD rejection rate for short content will be much lower than that for long content, a popularity correction method is proposed for that based on Zipf distribution, it further takes into consideration of the expected average number of simultaneous sessions (ESS) per content, which depends on both content duration and average number of requests received for this content. Thus the corrected popularity can be expressed as $P_i' = P_i \cdot (1 + ESS_i \cdot \alpha)$, where α is introduced to study how much weight the ESS parameter should be given as a large α means popular titles would be more available in the network with more copies spread around the network [27]. Guo *et al.* define the popularity of a segment as

$$p = \frac{\frac{S_{sum}}{S_0}}{T_r - T_0} \times \min\left(1, \frac{\frac{T_r - T_0}{n}}{t - T_r}\right), \qquad (6)$$

where t is the current time instant, $\frac{\frac{S_{sum}}{S_0}}{T_r - T_0}$ represents the average access rate of the segment in the past, normalized by the segment size, and $\min(1, \frac{\frac{T_r - T_0}{n}}{t - T_r})$ reflects the probability of future access: $\frac{T_r - T_0}{n}$ is the average time interval of accesses in the past; if $t - T_r > \frac{T_r - T_0}{n}$, the possibility that a new request arrives is small; otherwise, it is highly possible a request is coming soon [18].

Peer Availability Detection

To ensure the maintenance of desired number of replicas of a video in the system, we should detect the availability of the peers. Due to the extensive influence of user interaction, the detection faces two requirements. First, the system must empirically measure the short-term availability distribution of its host population on an ongoing basis. Second is the non-transient failures that take stored data out of the system for infinite periods [27]. The availability monitor (AM) tracks host availability, maintains host availability metrics which are short-term host availability and long-term decay rate, and notifies the redundancy engine when the system reaches availability thresholds that trigger repair. Through observation of real systems, it is shown that online time is a predictor of peer departure, which can be used for a simple peer departure predictor. A peer is predicted to leave if it has been online for fewer than a certain period of time. Otherwise, it is predicted to stay in the system.

Replication Mechanism

The Replication mechanisms consists of redundancy computation and replace algorithm, which we will discuss below consequently.

▷ *Redundancy computation*: With sufficient redundancy across many peers, at any moment enough peers will be in the system to make a given data item available with high probability. However, it is not trivially clear how much redundancy is necessary for a given level of availability or what redundancy mechanism is most appropriate for a given context. The simplest form of redundancy is pure replication. Given a target level of availability A and a mean host availability of μ_H, the number of required replicas c can be calculated directly.

$$A = 1 - (1 - \mu_H)^c \tag{7}$$

$$c = \frac{\log(1 - A)}{\log(1 - \mu_H)} \tag{8}$$

However, it can be highly inefficient in low-availability environments caused by peer churn since many storage replicas are required to tolerate potential transient failures. An alternative way is to use erasure coding. Given the number of blocks in a file b, and the stretch factor c specifying the erasure code's redundancy, the delivered availability can be expressed as:

$$A = \sum_{j=b}^{cb} \binom{cb}{j} \mu_{H^j} (1 - \mu_H)^{(cb-j)} \tag{9}$$

$$c = \left(\frac{k\sqrt{\frac{\mu_H(1-\mu_H)}{b}} + \sqrt{\frac{k_2\mu_H(1-\mu_H)}{b} + 4\mu_H}}{2\mu_H} \right)^2 \tag{10}$$

However, the price for this efficiency is a quadratic coding time and a requirement that reads and writes require an operation on the entire object [7].

▷ *Replace algorithm*: We then look at the different cache strategies to manage the cached content. In general, instead of depending on peers to manually decide how much cache space is shared and which video to be kept to share or not, peer-assisted VoD systems require peers provide cache space and let the cache management algorithm to manage cache. Peers may be asked to cache media content they are not watching at the moment or even not planning to watch in the future but will be beneficial from the global point of view. The simplest one is *Least Recently Used* (LRU) strategy [3], which maintains a queue of each file sorted by when it was last accessed. When a file is accessed, it is located in the queue, updated, and moved to the front. If it is not in the cache already, it is added immediately. When the cache is full the program at the end of the queue is discarded. Another one is *Least Frequently Used* (LFU) strategy. To compute the cache contents, the index server keeps a history of all events that occur within the last N hours (where N is a parameter to the algorithm). It calculates the number of accesses for each program in this history. Items that are accessed the most frequently are stored in the cache, with ties being resolved using an LRU strategy.

Under the belief of the greater the number of peers with video cache, the higher the probability that the video is available, and the higher the effective bandwidth, Ying *et al.* propose a cache replacement strategy by introducing *cache_needed_rank* [37] by taking account both the video popularity and the number of peers that currently are caching that video, which can be expressed as:

$$cache\_needed\_rank = \frac{\text{the number of peers demanding}}{\text{the number of online peers with cache}}. \tag{11}$$

This cache replacement strategy is based on the this evaluation metric, where stream with a lower *cache_needed_rank* is replaced by one with a higher *cache_needed_rank*.

To replace both those media segments with diminishing popularities as they rarely get accessed and those popular media segments with too many copies being cached, Guo *et al.* [18] introduce another cache replacement policy. To achieve the optimal distribution, they define the *utility function* of a segment as

$$u = \frac{(f(p) - f(p_{\min})) \times (f(p_{\max}) - f(p))}{r^{\alpha+\beta}}, \tag{12}$$

where p represents the popularity of the segment, $p_{\min}$ and $p_{\max}$ estimate the minimum and maximum of segment popularities in the system respectively. r is the number of replicas of this segment and $f(p)$ is a monotonic nondecreasing function, which is called the *adjustment factor* of the utility function. The term $\frac{(f(p)-f(p_{\min})}{r^{\alpha}}$ captures the segments with small popularities and large number of replicas while $\frac{f(p_{\max})-f(p)}{r^{\beta}}$ captures the segments with large popularities and large number of replicas. These two kinds of segment replicas should be replaced by the replicas of segments with moderate popularities but a small number of copies. As a result, segments with the smallest utility value are chosen to be replaced when a peer's cache is full.

Vratonjić *et al.* present another cache management algorithm, which is used in their newly proposed system—BulletMedia [31]. This algorithm examines the problem to ensure all content blocks are replicated at least k times system-wide, where a typical value of k is 4. To achieve this, a peer examines its local content cache and determines the set of chunkIds for chunks it is currently not replicating. It then selects chunkIds at random from this set, and performs a lookup in the DHT. The DHT simply includes a count of the number of peers replicating the chunk in the response. If the number of replicas is below a per-defined level (e.g. 4), then the peer begins to retrieve the blocks associated with the chunk. If not, the peer chooses another chunkId at random and queries it. This process proceeds until a chunk is found that requires more replications.

Bandwidth Allocation Policy for Prefetching

When peers have surplus upload capacity, it can be used to prefetch content for the future use, which has been proved to leverage the server load dramatically [21]. Then how to allocate the instantaneous surplus upload capacity among the peers in the system becomes a critical question. One of the allocation scheme is *water-leveling policy*, which aims to equalize the reservoir levels of perfected content across all the peers. Once the reservoir level of all the peers reach the same level, an attempt is made to equally distribute the surplus capacity among all the peers. Another one being considered is *greedy policy*, where each user simply dedicates its remaining upload bandwidth to the next user right after itself. Through simulation, it is observed that the greedy policy does slightly better than the water-leveling policy and both of them are near optimal.

Dynamic Repair Policy

Over longer periods, peers leave the system permanently, which requests the system to "repair" this lost redundancy by continuously writing additional redundant data onto new peers [7]. The two key parameters in repairing file data are the degree of redundancy used to tolerate availability transients and how quickly the system reacts to peer departures. In general, the more redundancy used to store file data, the longer the system can delay before reacting to peer departures. The repair policy spectrum can be defined in terms of two categories: eager and lazy.

Eager repair means the system immediately repairs the loss of redundant data when a peer fails. Using this policy, data only becomes unavailable when peers fail more quickly than they can be detected and repaired. The primary advantage of eager repair is its simplicity. Every time a peer departs, the system only needs to place redundant data on another peer in reaction. Moreover, detecting peer failure can be implemented in a completely distributed fashion since it isn't necessary to coordinate information about which peers have failed. However, the eager policy makes no distinction between permanent departures that require repair and transient disconnections that do not. Consequently, in public peer-to-peer environments, many repair actions may be redundant and wasteful.

Lazy repair tries to defer immediate repair and use additional redundancy to mask and tolerate peer departures for an extended period. The key advantage of lazy repair is that, by delaying action, it can eliminate the overhead of redundant repairs and only introduce new redundant blocks when availability is threatened. However, lazy repair also has disadvantages. In particular, it must explicitly track the availability of individual peers and what data they carry. This is necessary to determine when an object's availability is threatened and a repair should be initiated. Consequently, the system must maintain explicit metadata about which peers hold what data.

Cheng *et al.* propose and evaluate a framework for lazy replication in GridCast, a P2P VoD system [11]. Their algorithm uses a peer departure predictor to choose when to replicate (just before a peer leaves) and to whom (peers that are predicted to be most unlikely to leave) and uses a chunk request predictor to choose what to replicate (the chunks that will be most popular in the coming sessions). They use a lazy factor α to control the upload used for replication. Then through simulation evaluation, though eager replication reduces more departure missed than lazy replication, the total number of replicated chunks in lazy case is 45% of that in eager algorithm and the cost is much lower. Further, the efficiency of lazy algorithm more than three times that of eager replication, which suggests that lazy replication can replicate more efficiently than eager replication.

Interestingly, almost all the existing articles on peer-assisted VoD systems are under the assumption of small cache or even no cache. Each peer in the network is only equipped with a larger buffer that is able to store a part of the media the user is watching currently, which requires a wise cache management mechanism to make better use of the limited storage space. With the development of hard devices, every peer can have a large storage space that is able to store not only the current media, but also the other media it has watched a certain time ago, at a reasonable price. With the consideration of large cache, the desired content can be found either on the peers who are watching the same media stream at this time or some "offline" peers who are watching something else at the moment.

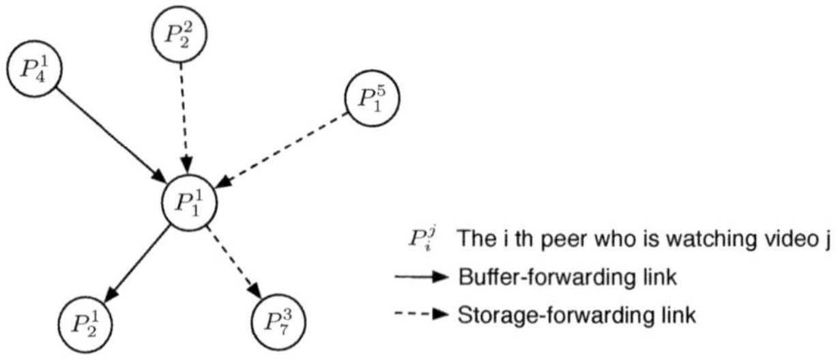

Fig. 5. A hybrid-forwarding P2P VoD system

Then the challenging task of designing more efficient content lookup algorithms that are capable of detecting offline peers with the interested content is desired. What is more, from peers' point of view, how to allocate the limited upload bandwidth to contribute to both the currently watching media and the stored content might be an interesting research topic in the future. Till now, to the best of our knowledge, only [20] considers the hybrid-forwarding architecture which integrates both buffer-forwarding approach and storage-forwarding approach, as shown in Fig. 5. And they try to maximize the throughput by solving an optimization problem.

3.2 Content Searching Algorithm

For peer-assisted VoD systems, one salient additional characteristic is the allowance of user interaction with video streams. Examples include pause, fast-forward, back-forward, random seek operations and so on, which further lead to more complicated system design. To address the issue of the unpredictability of these user interactions, different kinds of searching algorithms aiming at finding the desired content efficiently with shorter initial buffering delay have been proposed.

Used to get the optimal match between clients and servers, Shaking is a searching algorithm proposed to find the server that can provide the fastest service [8]. Since the server schedules the pending requests in a FIFO manner, a shorter waiting time can be achieved by trying to move the requests that arrived earlier at the server to other servers. That is, given a set of server candidates, a client can contact them for their pending requests and try to find each of them a new server. The key idea can be found in Fig. 6.

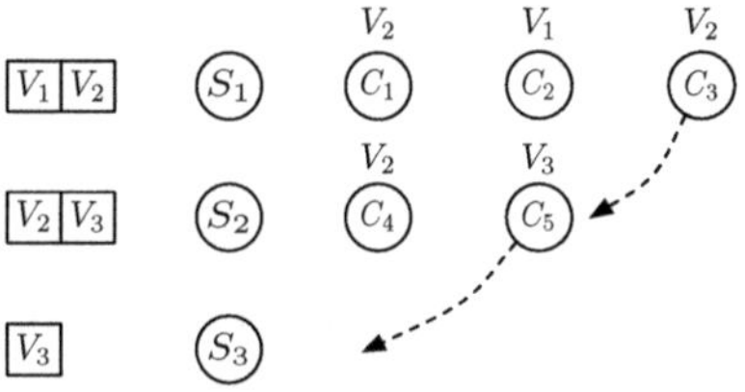

Fig. 6. Shaking algorithm

A client C trying to shake request $[C_3, V_2]$ out of server S_1 may find server, S_2 has V_2. Although this server cannot serve $[C_3, V_2]$ earlier than S_1, the client can try to see if it can shake any pending requests out of S_2. For example, it may launch a search for V_3 and find server S_3, which is free of workload at this moment. It shows that in order to successfully move out a request, a client may need to shake a chain of servers. To address the challenge of chaining effect, Shaking uses a 3-step solution, building closure set, shaking closure set and executing shaking plan. Further, Shaking makes it possible to for a client to be served by a server that is beyond the client's search scope.

RINDY, which is proposed by Cheng *et al.*, is a ring-assisted overlay topology in P2P VoD systems for efficient content lookup [10]. Under this scheme, a peer only needs $O(\log(T/w))$ hops (where T is the total time length of the video stream and w is the buffer window size of peers) to identify and connect to a new group of peers close to the destination playing positions when a random seek occurs. In RINDY, each peer organizes all its neighbors into a series of concentric and non-overlapping logical rings according to their relative distances calculated from the current playing positions. The rings are separated into two types, *gossip-ring* and *skip-ring*. A gossip-ring is a peer's innermost ring, which is mainly responsible for collecting some near neighbors with close playing positions and overlapped buffer windows. And all the outer rings, which are mainly used to improve the speed of lookup operations and reduce the load of the tracker server are called skip-ring. When a peer seeks to a target position d, it first checks whether d is within its gossip-ring. If so, it performs a local search to find enough near-neighbors (neighbors in the gossip-ring) and far-neighbors (neighbors in the skip-ring). Otherwise, it finds the closest neighbor to d as its next hop, pushes its address into the *routing_path* and *result_set* fields of the query message, and finally forwards this query to that nest hop neighbor. Upon receipt of this query, the next-hop neighbor executes the same procedure, which will be iterated until this request arrives at some peer whose gossip-ring covers d. The final peer adds all of its near-neighbors into the *result_set* field of the query message and returns the message to the source peer along the routing path. Then the source peer adds all the members of *result_set* into its mCache and then change its current playing position to d.

Inspired by the fact that if a peer's buffer range is fully covered by other peers, removing it from the index structure will not cause much trouble as the peers who use this removed peer as a partner can still find other possible partners, Chi *et al.* introduce Buffer Assisted Search (BAS) to select as few index peers as possible without reducing search effectiveness, i.e., the total buffer coverage [13]. The problem can be formulated as the Minimum Buffer Cover (MBC) problem that can be solved using a distributed algorithm. The basic flow is to divide the existing index peers into two groups according to whether they overlap with the newcomer, and then apply the dynamic programming algorithm to the newcomer plus the group overlapping with the newcomer. When a first-join-peer tries to find the index peers with buffer overlapping by tracing backward and forward along the closest index peer's predecessor and successor, the expected number of hops to be traced is a constant. Then the dynamic programming algorithm is applied to the found peers plus the new peer to figure out which peers should be pruned from the index overlay. The expected number of nodes a new client should contact is also a constant.

Wang *et al.* introduce the Dynamic Skip List (DSL) to inherently accommodates dynamic and asynchronous peers [32]. A skip list is an ordered list of *keys* with additional parallel links. The key here is chosen to be the playback offset of each node. Assume that there are N keys in the list, indexed from 1 to N. The $(i \times 2^l)th$ key, $i = 1, 2, ..., l = 0, 1, ...$, will have links to the $((i-1) \times 2^l)th$

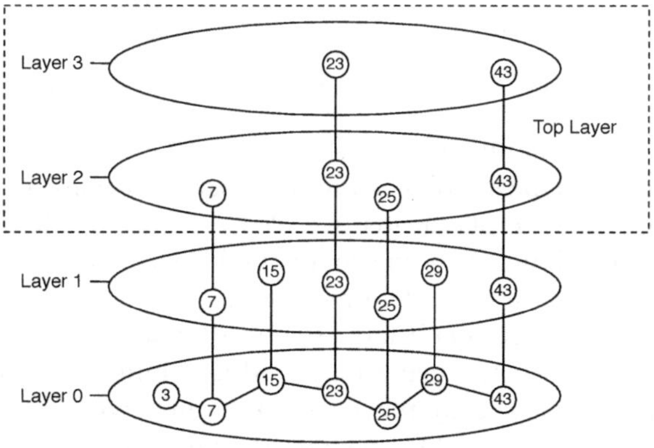

Fig. 7. A DSL of 7 nodes. The number in each node represents the key of this node.

keys respectively. Besides, DSL is built with no MaxLayer limit and a newly inserted key will stop promoting itself only when it fails. As unnecessary neighbor information has to be maintained, which reduces the search efficiency, DSL compresses all the layers above L into a single *top layer*, where $L = \log(\frac{N}{\log N})$. Its structure is shown in Fig. 7.

In this layered representation, a single key in the list is mapped into multiple logical nodes along the same column. Since the parallel links in higher layers skip geometrically more than those in lower layers, a key search can be started from the highest layer so as to quickly skip unnecessary parts and then progressively move to lower layers until it hits a logical node having the key. The complexity of this top-down search is $O(\log N)$ [29]. The key is updated over time according to the playback progress. The server periodically checks the size of a randomly selected layer, estimates the overlay size N, and then accordingly releases or merges layers to ensure that there are $O(\log N)$ logical nodes in the top layer. It is shown that each overlay node in the system, including the server, maintains at most $O(\log N)$ extra information.

In a new distributed cache management model (DCMM) proposed by Liang *et al.*, peers are assigned fixed segments in turn according to the order peer join the system and all the peers whose cache content compose an integrated program into some chord-like rings [23]. The size of each ring equals the number of segments of a media file, i.e., if the media file is divided into M segments, then the ring consists of M peers. All segments have a number from 1 to M according to the playback time. Each peer in ring caches the corresponding media segment according to the time when peers join the system in turn. Segments of a program can be quickly located within these rings. When searching a specific segment, peer tries to find peers caching this segment in its own ring. If it fails, the requests will be transfered to the lower ring. This process is repeated until the peer is found or the message arrives at the ring 0.

GridCast uses two content proximity metrics *playhead* and *playcache* to estimate the sharing potential between pairs of peers [12], which can be described as follows:

$$playhead_{(B \to A)} = |offset_B - offset_A| \quad (13)$$

$$playcache_{(B \to A)} = \frac{\sum_{i=\min}^{\max} chunkmap_{(B)}[i]}{\max - \min + 1}. \quad (14)$$

Playhead proximity roughly estimates the potential for sharing using only the current locations of the peers' playheads. Two peers are considered to have better sharing potential if their playheads are closer. On the other hand, *playcache* measures how many chunks can be provided by the other peer by comparing the contents of two caches directly. The more chunks B caches within the prefetch window of A, the better sharing it has. Neighbors who share the metadata but do not share chunks are selected using playhead proximity while partners used for chunk sharing are promoted by playcache proximity when the number of them fails below a certain degree, 50 for neighbors and 25 for partners in this case.

While most of content searching algorithms in the peer-assisted VoD systems support random seek, few of them consider the problem of enabling advanced VCR functions such as fast forward or fast rewind. Offering more freely user interactivity that is lacking in the traditional video broadcasting services, advanced VCR operations are highly desirable, yet, challenging to implement in online streaming environment as they change the playback speed of the media and require the extra bandwidth at the server and the underlying network.

To the best of our knowledge, only two papers discuss this problem. In [19], Guo *et al.* try to solve the fast forward and fast rewind problem by adjusting segment deadlines to reflect the change of playback speed, so that the new deadlines can be used for the proposed real-time scheduling based peer selection algorithm to recompute the uploading urgency metrics. Inspired by the idea of realizing these two functions by playing one segment out of v segments, where v is the speed of fast forward or rewind, Wang *et al.* present the number of logical nodes to be skipped ahead is $\frac{n(v-1)}{d}$ for each jump operation for fast forward, where d is the average difference between two consecutive nodes, and it needs $O(1)$ time [32]. Further, they obtain the upper bound of the speed a system can provide. Under simplified assumptions and through roughly evaluation, the ability of solving the fast forward or rewind problem of these mechanisms are not that convincing. Thus, designing peer-assisted VoD systems that can better support advanced VCR functions is still in its infancy.

4 Network Coding in Peer-Assisted Streaming Systems

In this section, we illustrate the role of network coding in peer-assisted streaming systems. Due to the ability of making optimal use of bandwidth resources, network coding is proposed to be used in segment scheduling algorithms in peer-assisted streaming systems. By using random network coding, peers are allowed to retrieve fractionalized segments in the parallel machine scheduling context and any received segment is useful with high probability. An excellent property of network-coding based segment scheduling algorithm is that the missing segment on a downstream peer can be served by *multiple* seeds, as each uses its

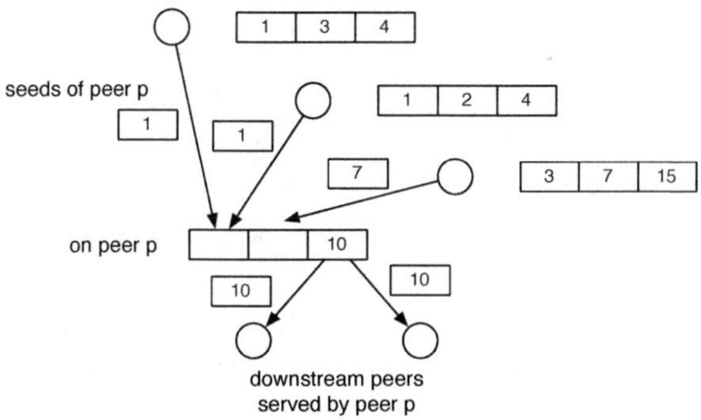

Fig. 8. An illustration of network-coding based segment scheduling algorithm

randomized selection algorithm to select a segment to send coded blocks, which is illustrated in Fig. 8.

However, the limit lies in that peers have to wait to download the complete file before they can start decoding. This problem is solved by restricting network coding into one segment, which reduces the start-up delay into one segment size. What is more, coding prevents the occurrence of rare blocks within a segment, and ensures that the bandwidth is not wasted in distributing the same block multiple times. In essence, coding minimizes the risk of making the wrong uploading decision.

4.1 Core Idea

Here we introduce the core idea of network-coding based segment scheduling algorithm. In it, each segment of the media streaming is further divided into n blocks $[b_1, b_2, ..., b_n]$, with each b_i a fixed number of bytes k (refereed to as the block size). When a segment is selected by a seed p to code for its downstream peer, the seed independently and randomly chooses a set of coding coefficients $[c_1^p, c_2^p, ..., c_m^p](m \leq n)$ in a finite field F_q for each coded block. It then randomly chooses m blocks —$[b_1^p, b_2^p, ..., b_m^p]$ —out of all the blocks in the segment is has received so far, and produces one coded block x of k bytes, as shown in Fig. 9

$$x = \sum_{i=1}^{m} c_i^p \cdot b_i^p \tag{15}$$

As soon as a peer receives the first coded block of this segment, it starts to progressively decode using Gauss-Jordan elimination. As a total of n coded blocks $\mathbf{x} = [x_1, x_2, ..., x_n]$ have been received, the original blocks can be immediately recovered as Gauss-Jordan elimination computes:

$$\mathbf{b} = \mathbf{A}^{-1}\mathbf{x}^T, \tag{16}$$

where $\mathbf{A}$ is the matrix formed by coding coefficients of $\mathbf{x}$.

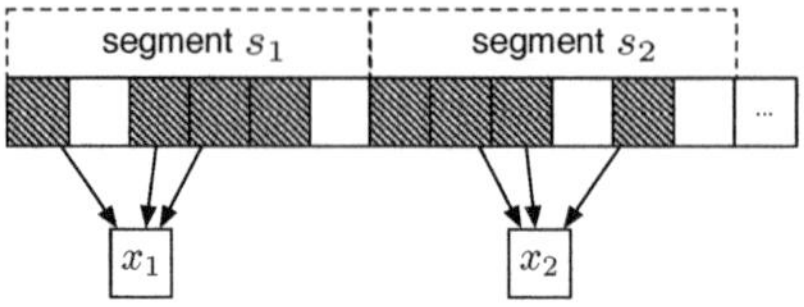

Fig. 9. An example of coding operation on peer p, where m=3, n=6

▷ *Coding window optimization*: The key difference of network-coding based segment scheduling algorithms lies in the way of choosing m, which is also called the coding window. In Wang *et al.*'s algorithm [32], m is simply chosen to be the length of the segment, which means the coding range covers all the expected segment. Though a larger coding window makes better utilization of resource, it causes a higher probability of segments being missing upon the playback deadline due to the longer waiting time before decoding. R^2 [34], which is proposed by Wang *et al.*, refers the ratio m/n to as density, and a low ratio leads to a sparse decoding matrices. Their previous work [33] has experimentally shown that the density can be as low as 6% without leading to linear dependency among coded blocks. Then in deadline-aware network coding scheduling introduced by Chi *et al.* [13], it tries to find the as large an m as possible for better coding efficiency while controlling m so that no segment will miss its playback deadline by transforming the formulated coding window assignment problem in to a max-flow problem.

4.2 Analytical Study of the Basic Model

Other than protocol designs, Feng *et al.* seek to mathematically analyze peer-assisted streaming systems with network coding, with a focus on playback quality, initial buffering delays, server bandwidth costs, as well as extreme peer dynamics [15]. They consider the coding performance in flash crowd scenarios when most peers join the system approximately at the same time and highly dynamic scenarios in which peer join and leave the system in a highly volatile fashion.

In particular, the server strength δ is defined as $\delta = \frac{U_s}{NU_p}$, where U_p is the average upload capacity of participating peers, and U_s is the server upload capacity. It is proved that the sufficient conditions on smooth playback at a streaming rate R during any flash crowd with scale N is as follows:

$$U_s + NU_p = (1+\epsilon)NR, \qquad (17)$$
$$\epsilon = \alpha + \frac{\ln(1+\delta) - \ln\delta}{m}.$$

where m is the number of coded blocks in each segment and α denotes the fraction of linearly dependent coded blocks induced by network coding. This demonstrates that using network coding can achieve near-optimal performance in term of sustainable streaming rate during a flash crowd. Also, they show that in terms of initial buffering delay, algorithms using coding are within a factor of $2(1+\epsilon)$ of the optimal streaming scheme, which reflects the ability of guarantee very short initial buffering delays during a flash crowd of network coding. Besides, an upper bound of the additional server capacity to handle peer dynamics in current time slot is given to be $W_1U_1 - (1+\epsilon)W_1R$, where W_1 denotes the number of departures of high-bandwidth peers in current time slot.

In general, network-coding based segment scheduling algorithms are shown sufficient to achieve provably good overall performance in realistic settings.

4.3 Implementation Results

Annapureddy *et al.* evaluate the efficiency of network-coding based segment scheduling protocols through simulation [5, 6]. Considering a flash crowd where 20 peers join the network, by comparing with an optimal scheme without network coding—global-rarest first policy, where peer requests the globally rarest block in the target segment of its interest, it is shown that with network coding, the segment scheduling protocol provides a greater throughput than the global-rarest policy (about 14% times better) and results in significantly less variance and more predictable download times.

5 Conclusion

The study of peer-assisted media streaming systems has boomed for recent years as videos has become the dominant type of traffic over the internet, dwarfing other types of traffic. On one hand, peer-to-peer solutions have shown great promise in supporting media broadcast, which is witnessed by their increasingly widespread deployments. On the other hand, there still exists some technical challenges that might be the hurdles need to be overcome, or otherwise they may obstacle the development of peer-assisted streaming systems.

In the analytical works that try to model and characterize the behavior of such peer-assisted systems, the majority of them are under the assumption that all the peers are homogeneously, which means most of them model the peers with same upload and download capacities. The problem is that heterogeneity exists among peers in the real world. For example, peers behind Ethernet can have an upload and download bandwidth of up to $1Mbps$ to $10Mbps$, while in contrast, peers behind ADSL only have an upload bandwidth of $256Kbps$ to $512Kbps$ and their download capacity is about $1Mbps$ to $2Mbps$. Treating them all equally is obviously not a good idea.

The conventional method of proving the accuracy of analysis is by simulation. While most of the existing empirical studies only have a few hundred peers involved typically, with the only exception of [15, 16], which have a scale of more than $200,000$ peers at times throughout their simulations. By stating characterizing the large scale property of peer-assisted streaming systems, only a few hundred of peers can definitely affect the accuracy of their analysis.

By implementing real peer-assisted media streaming systems over the Internet, NATs and firewalls impose fundamental restrictions on pair-wise connectivity of nodes on an overlay, and may prohibit direct communication with one another. Whether communication is possible between two nodes depends on several factors such as the transport protocol(UDP or TCP), the particular kind of NAT/firewall, and whether the nodes are located behind the same private network. For example, under UDP protocol, the connectivity form NAT to firewall is only possible for some cases. When under TCP protocol, the connectivity from NAT to firewall is not possible [26].

In this chapter, we reviewed the state-of-art of peer-assisted media streaming systems, paying special attention to peer-assisted VoD systems. Bedsides, we also present the implementation of network coding in peer-assisted media streaming systems and the benefits it brought. While networking researchers have achieved substantial progress in this area over the past years, there are still several open problems, such as incentive-based mechanism, multichannel consideration, lager storage space in VoD systems, advanced VCR functions supporting and so on, which may present some areas of future research.

References

[1] Alessandria, E., Gallo, M., Leonardi, E., Mellia, M., Meo, M.: P2P-TV Systems under Adverse Network Conditions: A Measurement Study. In: Proc. IEEE INFOCOM, pp. 100–108 (2009), doi:10.1109/INFCOM.2009.5061911

[2] Ali, S., Mathur, A., Zhang, H.: Measurement of Commercial Peer-To-Peer Live Video Streaming. In: Proc. of ICST Workshop on Recent Advances in Peer-to-Peer Streaming (2006)

[3] Allen, M.S., Zhao, B.Y., Wolski, R.: Deploying Video-on-Demand Services on Cable Networks. In: Proc. 27th International Conference on Distributed Computing Systems (ICDCS), p. 63 (2007), `http://dx.doi.org/10.1109/ICDCS.2007.98`

[4] Annapureddy, S., Gkantsidis, C., Rodriguez, P., Massoulie, L.: Providing Video-on-Demand using Peer-to-Peer Networks. In: Proc. Internet Protocol TeleVision Workshop (IPTV), vol. 6 (2006)

[5] Annapureddy, S., Guha, S., Gkantsidis, C., Gunawardena, D., Rodriguez, P.: Exploring VoD in P2P Swarming Systems. In: Proc. IEEE INFOCOM, pp. 2571–2575 (2007), doi:10.1109/INFCOM.2007.323

[6] Annapureddy, S., Guha, S., Gkantsidis, C., Gunawardena, D., Rodriguez, P.R.: Is High-quality VoD Feasible using P2P Swarming? In: Proc. 16th International Conference on World Wide Web (WWW), pp. 903–912 (2007), doi:10.1145/1242572.1242694

[7] Bhagwan, R., Tati, K., Cheng, Y.C., Savage, S., Voelker, G.M.: Total Recall: System Support for Automated Availability Management. In: Proc. 2nd Symposium on Networked Systems Design and Implementation (NSDI), pp. 337–350 (2004)

[8] Cai, Y., Natarajan, A., Wong, J.: On Scheduling of Peer-to-Peer Video Services. IEEE Journal on Selected Areas in Communications 25(1), 140–145 (2007)

[9] Cha, M., Kwak, H., Rodriguez, P., Ahn, Y.Y., Moon, S.: I Tube, You Tube, Everybody Tubes: Analyzing the World's Largest User Generated Content Video System. In: Proc. 7th SIGCOMM Conference on Internet Measurement (IMC), pp. 1–14 (2007), doi:10.1145/1298306.1298309

[10] Cheng, B., Jin, H., Liao, X.: Supporting VCR Functions in P2P VoD Services Using Ring-Assisted Overlays. In: Proc. 16th IEEE International Conference on Communications (ICC), pp. 1698–1703 (2007), doi:10.1109/ICC.2007.284
[11] Cheng, B., Stein, L., Jin, H., Zhang, Z.: A Framework for Lazy Replication in P2P VoD. In: Proc. 18th International Workshop on Network and Operating Systems Support for Digital Audio and Video (NOSSDAV), pp. 93–98 (2008), doi:10.1145/1496046.1496068
[12] Cheng, B., Stein, L., Jin, H., Zhang, Z.: Towards Cinematic Internet Video-on-Demand. In: Proc. of the 3rd European Conference on Computer Systems (EuroSys), pp. 109–122 (2008), doi:10.1145/1352592.1352605
[13] Chi, H., Zhang, Q., Jia, J., Shen, X.: Efficient Search and Scheduling in P2P-based Media-on-Demand Streaming Service. IEEE Journal on Selected Areas in Communications 25(1), 119–130 (2007)
[14] Choe, Y.R., Schuff, D.L., Dyaberi, J.M., Pai, V.S.: Improving VoD Server Efficiency with BitTorrent. In: Proc. 15th ACM International Conference on Multimedia, pp. 117–126 (2007), doi:10.1145/1291233.1291258
[15] Feng, C., Li, B.: On Large-Scale Peer-to-Peer Streaming Systems with Network Coding. In: Proc. 16th ACM International Conference on Multimedia, pp. 269–278 (2008), doi:10.1145/1459359.1459396
[16] Feng, C., Li, B., Li, B.: Understanding the Performance Gap between Pull-based Mesh Streaming Protocols and Fundamental Limits. In: Proc. IEEE INFOCOM (2009)
[17] Graffi, K., Kaune, S., Pussep, K., Kovacevic, A., Steinmetz, R.: Load Balancing for Multimedia Streaming in Heterogeneous Peer-to-Peer Systems. In: Proc. 18th International Workshop on Network and Operating Systems Support for Digital Audio and Video (NOSSDAV), pp. 99–104 (2008), doi:10.1145/1496046.1496069
[18] Guo, L., Chen, S., Zhang, X.: Design and Evaluation of a Scalable and Reliable P2P Assisted Proxy for On-Demand Streaming Media Delivery. IEEE Transanctions on Knowledge and Data Engineering 18(5), 669–682 (2006), `http://dx.doi.org/10.1109/TKDE.2006.79`
[19] Guo, Y., Yu, S., Liu, H., Mathur, S., Ramaswamy, K.: Supporting VCR Operation in a Mesh-Based P2P VoD System. In: Proc. 5th Consumer Communications and Networking Conference (CCNC), pp. 452–457 (2008), doi:10.1109/ccnc08.2007.107
[20] He, Y., Lee, I., Guan, L.: Distributed Throughput Maximization in Hybrid-Forwarding P2P VoD Applications. In: Proc. International Conference on Acoustics, Speech and Signal Processing (ICASSP), pp. 2165–2168 (2008), doi:10.1109/ICASSP.2008.4518072
[21] Huang, C., Li, J., Ross, K.W.: Can Internet Video-on-Demand be Profitable? In: Proc. ACM SIGCOMM, vol. 37, pp. 133–144 (2007), doi:10.1145/1282427.1282396
[22] Li, B., Xie, S., Keung, G., Liu, J., Stoica, I., Zhang, H., Zhang, X.: An Empirical Study of the Coolstreaming+ System. IEEE Journal on Selected Areas in Communications 25(9), 1627–1639 (2007), doi:10.1109/JSAC.2007.071203
[23] Liang, W., Huang, J., Huang, J.: A Distributed Cache Management Model for P2P VoD System. In: Proc. International Conference on Computer Science and Software Engineering (ICCSSE), vol. 3, pp. 5–8 (2008), doi:10.1109/CSSE.2008.1059
[24] Liao, X., Jin, H.: OCTOPUS: A Hybrid Scheduling Strategy for P2P VoD Services. In: Proc. 6th International Conference on Grid and Cooperative Computing (GCC), pp. 26–33 (2007), doi:10.1109/GCC.2007.89

[25] Liao, X., Jin, H., Liu, Y., Ni, L.M., Deng, D.: AnySee: Peer-to-Peer Live Streaming. In: Proc. IEEE INFOCOM, pp. 1–10 (2006), doi:10.1109/INFOCOM.2006.288
[26] Liu, J., Rao, S.G., Li, B., Zhang, H.: Opportunities and Challenges of Peer-to-Peer Internet Video Broadcast. Proceedings of the IEEE 96, 11–24 (2008)
[27] Nafaa, A., Murphy, S., Murphy, L.: Analysis of a Large-Scale VoD Architecture for Broadband Operators: A P2P-Based Solution. IEEE Communications Magazine 46(12), 47–55 (2008)
[28] Parvez, N., Williamson, C., Mahanti, A., Carlsson, N.: Analysis of Bittorrent-like Protocols for On-Demand Stored Media Streaming. SIGMETRICS Performance Evaluation Review 36(1), 301–312 (2008)
[29] Pugh, W.: Skip Lists: A Probabilistic Alternative to Balanced Trees. Communications of the ACM 33, 668–676 (1990)
[30] Silverston, T., Fourmaux, O., Crowcroft, J.: Towards an Incentive Mechanism for Peer-to-Peer Multimedia Live Streaming Systems. In: Proc. 8th International Conference on Peer-to-Peer Computing (P2P), pp. 125–128 (2008), `http://dx.doi.org/10.1109/P2P.2008.25`
[31] Vratonjić, N., Gupta, P., Knežević, N., Kostić, D., Rowstron, A.: Enabling DVD-like Features in P2P Video-on-Demand Systems. In: Proc. Workshop on Peer-to-Peer Streaming and IP-TV (P2P-TV), pp. 329–334 (2007), doi:10.1145/1326320.1326326
[32] Wang, D., Liu, J.: A Dynamic Skip List-Based Overlay for On-Demand Media Streaming with VCR Interactions. IEEE Transactions on Parallel and Distributed Systems 19(4), 503–514 (2008), `http://dx.doi.org/10.1109/TPDS.2007.70748`
[33] Wang, M., Li, B.: How Practical is Network Coding?. In: Proc. 14th International Workshop on Quality of Service (IWQoS), pp. 274–278 (2006), doi:10.1109/IWQOS.2006.250480
[34] Wang, M., Li, B.: R^2: Random Push with Random Network Coding in Live Peer-to-Peer Streaming. IEEE Journal on Selected Areas in Communications 25(9), 1655–1666 (2007)
[35] Wu, C., Li, B., Zhao, S.: Characterizing Peer-to-Peer Streaming Flows. IEEE Journal on Selected Areas in Communications 25(9), 1612–1626 (2007)
[36] Wu, C., Li, B., Zhao, S.: Multi-Channel Live P2P Streaming: Refocusing on Servers. In: Proc. of IEEE INFOCOM, pp. 1355–1363 (2008)
[37] Ying, L., Basu, A.: pcVOD: Internet Peer-to-Peer Video-On-Demand with Storage Caching on Peers. In: Proc. 11th International Coference on Distributed Multimedia Systems (DMS), pp. 218–223 (2005)
[38] Zhang, M., Luo, J.G., Zhao, L., Yang, S.Q.: A Peer-to-Peer Network for Live Media Streaming Using a Push-Pull Approach. In: Proc. 13th ACM International Conference on Multimedia, pp. 287–290 (2005), doi:10.1145/1101149.1101206
[39] Zhang, X., Liu, J., Li, B., Yum, Y.S.: CoolStreaming/DONet: A Data-Driven Overlay Network for Peer-to-Peer Live Media Streaming. In: Proc. IEEE INFOCOM, vol. 3, pp. 2102–2111 (2005), doi:10.1109/INFCOM.2005.1498486
[40] Zhou, Y., Chiu, D.M., Lui, J.: A Simple Model for Analyzing P2P Streaming Protocols. In: Proc. 15th International Conference on Network Protocols (ICNP), pp. 226–235 (2007), doi:10.1109/ICNP.2007.4375853

FTV (Free-Viewpoint TV)

Masayuki Tanimoto

Graduate School of Engineering, Nagoya University, Chikusa-ku, Furo-cho, Nagoya, Japan

Abstract. We have developed a new type of television named FTV (Free-viewpoint TV). FTV is an innovative visual media that enables us to view a 3D scene by freely changing our viewpoints. We proposed the concept of FTV and constructed the world's first real-time system including the complete chain of operation from image capture to display. At present, FTV is available on a single PC and FTV with free listening-point audio is also realized. FTV is based on the ray-space method that represents one ray in real space with one point in the ray-space. We have developed new type of ray capture and display technologies such as a 360-degree mirror-scan ray capturing system and a 360 degree ray-reproducing display. MPEG regarded FTV as the most challenging 3D media and started the international standardization activities of FTV. The first phase of FTV was MVC (Multi-view Video Coding) and the second phase is 3DV (3D Video). MVC enables the efficient coding of multiple camera views. 3DV is a standard that targets serving a variety of 3D displays.

1 Introduction

We have developed a new type of television named FTV (Free-viewpoint TV) [34, 36, 37, 38, 39, 40, 41]. FTV is an innovative visual media that enables us to view a 3D scene by freely changing our viewpoints as if we were there. It is easy to realize the free viewpoint for virtual scenes made by computer graphics. However, it is very difficult and has not yet been realized for real scenes. Therefore, FTV that has achieved such a function for real scenes will bring an epochal change in the history of visual media.

We proposed the concept of FTV and verified its feasibility with the world's first real-time system including the complete chain of operation from image capture to display. The viewpoint was controlled with a mouse, and free-viewpoint images of real moving scenes were displayed on a 2D display in real time [30, 31].

We have been developing technologies for FTV and realized FTV on a single PC and a mobile player. We also realized FTV with free listening-point audio [28].

FTV is based on the ray-space method [6, 8, 10, 43]. Fig. 1 shows the evolution of visual systems. In the past, visual systems such as photography, film and TV were individual systems. At present, they are digitized and can be treated on the same platform as pixel-based systems. These pixel-based systems are developing toward using more pixels. This trend is exemplified by Super High-Definition TV [16]. However, the demand for more pixels will saturate, and more views will be needed. This is the way to 3DTV and FTV. This will result in the evolution from a pixel-based system to a ray-based system. We have been developing ray capture, processing, and display technologies for FTV.

C.W. Chen et al. (Eds.): Intel. Multimedia Communication: Tech. and Appli., SCI 280, pp. 341–365.
springerlink.com

FTV is closely related to 3DTV. The relation between FTV and 3DTV is described in the reference [17]. 2DTV generates a single view and 3DTV generates 2 or more views for display. On the other hand, FTV generates infinite number of views since the viewpoint can be placed anywhere. Therefore, we can regard FTV as an ultimate 3DTV. FTV captures and transmits the information of all rays in a 3D scene. We can also regard FTV as a natural interface between human and environment and an innovative tool to create new types of content and art.

We proposed FTV to MPEG (Moving Picture Experts Group) [35]. MPEG regarded FTV as the most challenging 3D media and started the international standardization activities of FTV. The first phase of FTV was MVC (Multi-view Video Coding) [57], which enables the efficient coding of multiple camera views. MVC was completed in March 2009. The second phase of FTV is 3DV (3D Video) [58]. 3DV is a standard that targets serving a variety of 3D displays.

FTV technologies and its international standardization are overviewed in the following.

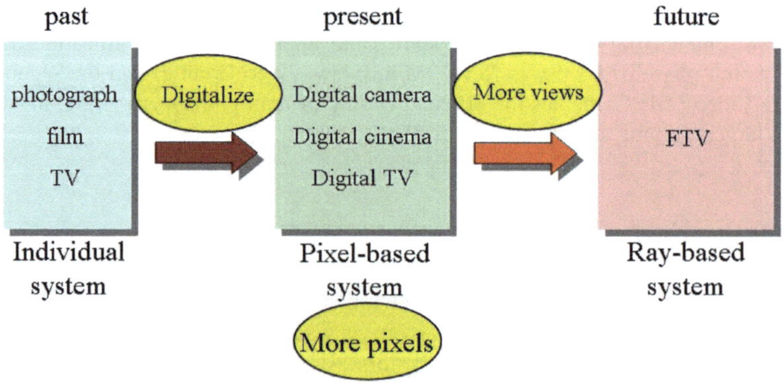

Fig. 1. Evolution of visual systems.

2 Ray-Space Representation for 3d Visual Communication

We developed FTV based on the ray-space representation [6, 10, 35, 43]. The ray-space was proposed to describe rays in 3D space. It can be a common data format for 3D visual communication as shown in Fig. 2.

In ray-space representation, one ray in the 3D real space is represented by one point in the ray space. The ray space is a virtual space. However, it is directly connected to the real space. The ray space is generated easily by collecting multi-view images while giving consideration to the camera parameters.

Let (x, y, z) be three space coordinates and θ,ϕ be the parameters of direction. A ray going through space can be uniquely parameterized by its location (x, y, z) and its direction (θ,ϕ); in other words, a ray can be mapped to a point in this 5D, ray-parameter space. In this ray-parameter space, we introduce the function *f*, whose value

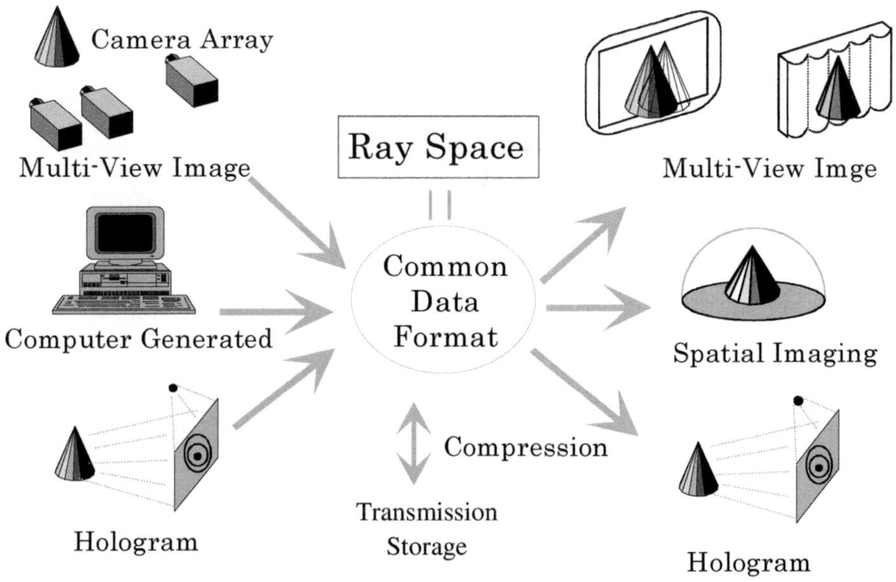

Fig. 2. Ray-space as a common data format for 3D visual communication.

corresponds to the intensity of a specified ray. Thus, all the intensity data of rays can be expressed by

$$f(x, y, z; \theta, \phi).$$
$$-\pi \leq \theta < \pi, -\pi/2 \leq \phi < \pi/2. \tag{1}$$

We call this ray-parameter space the "ray-space."

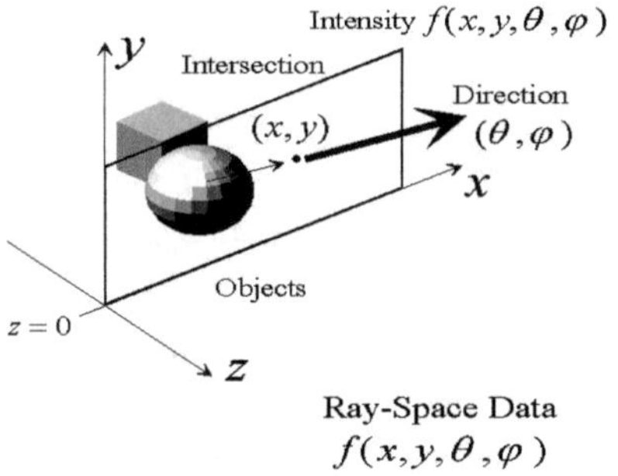

Fig. 3. Definition of orthogonal ray-space

Although the 5D ray-space mentioned above includes all information viewed from any viewpoint, it is highly redundant due to the straight traveling paths of the rays. Thus, when we treat rays that arrive at a reference plane, we can reduce the dimension of the parameter space to 4D.

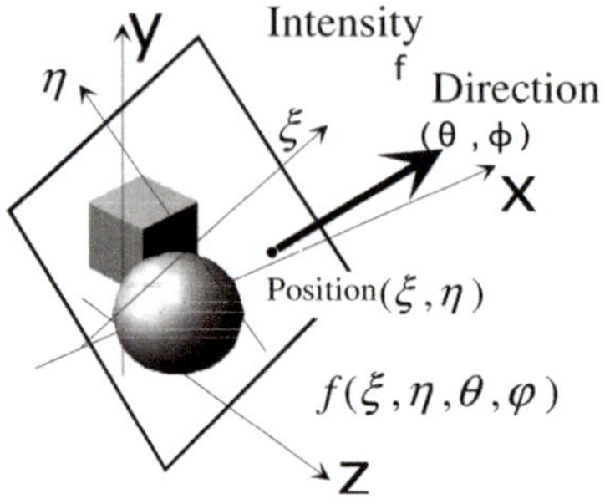

Fig. 4. Definition of spherical ray-space

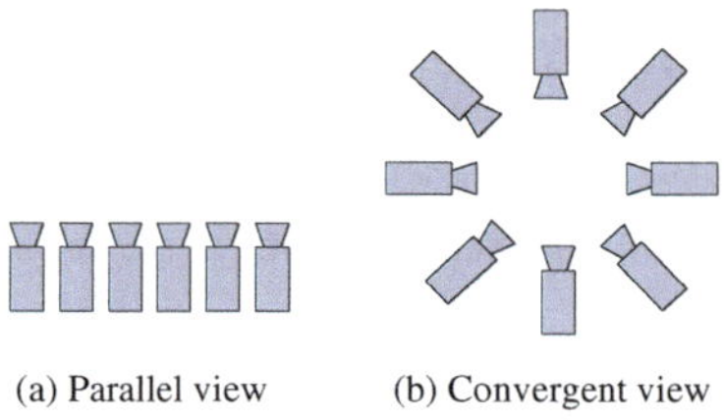

Fig. 5. Camera Arrangements for FTV

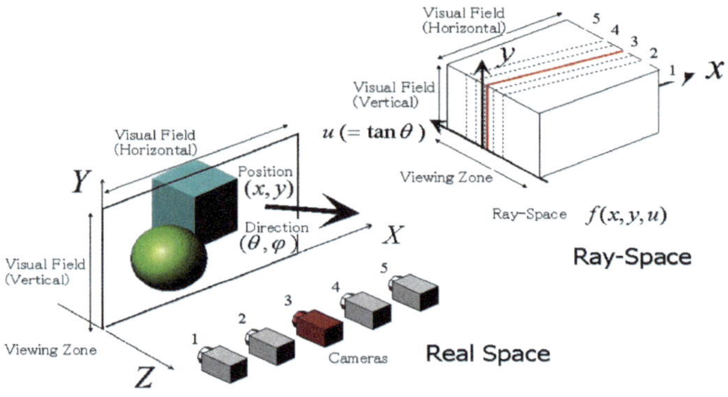

Fig. 6. Acquisition of orthogonal ray-space

We use two types of ray-space for FTV. One is the orthogonal ray-space, where a ray is expressed by the intersection of the ray and the reference plane and the ray's direction as shown in Fig. 3. Another is the spherical ray-space, where the reference plane is set to be normal to the ray as shown in Fig.4. The orthogonal ray-space is used for FTV with a linear camera arrangement, whereas the spherical ray-space is used for FTV with a circular camera arrangement. The linear camera arrangement is used for parallel view and the circular camera arrangement is used for convergent view as shown in Fig. 5.

Both the orthogonal ray-space and the spherical ray-space are 4D and 5D, including time. If we place cameras within a limited region, the obtained rays are limited,

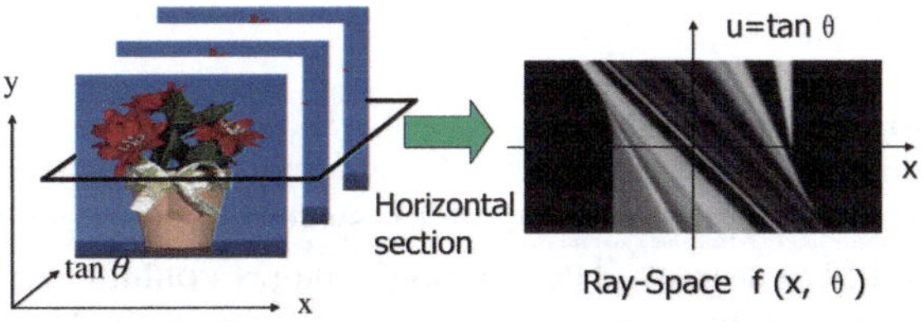

Fig. 7. Example of orthogonal ray-space

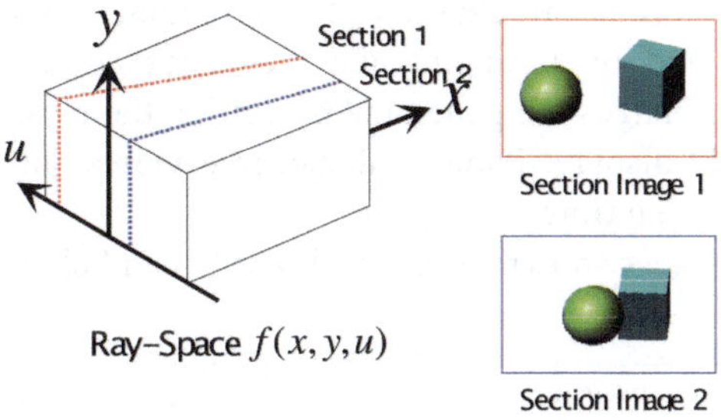

Fig. 8. Generation of view images

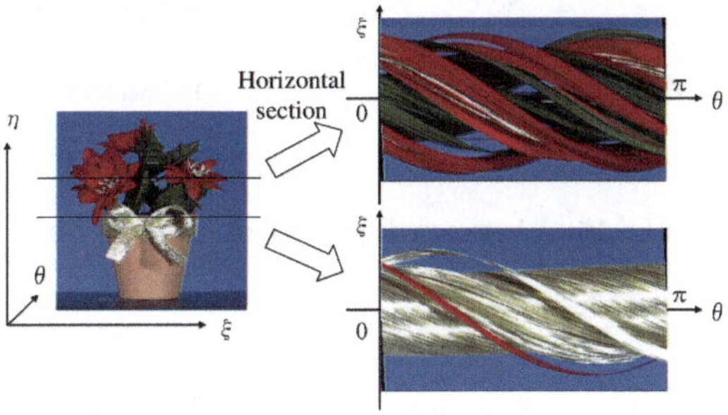

Fig. 9. Example of spherical ray-space

and the ray-space constructed from these rays is a subspace of the ray-space. For example, if we place cameras in a line or in a circle, we have only one part of the data of the whole ray-space. In such cases, we define a smaller ray-space.

For the linear camera arrangement, the ray-space is constructed by placing many camera images upright and parallel, as shown in Fig. 6, forming the FTV signal in this case. The FTV signal consists of many camera images, and the horizontal cross-section has a line structure as shown in Fig. 7. The line structure of the ray-space is used for ray-space interpolation and compression. Vertical cross-sections of the ray-space give view images at the corresponding viewpoints as shown in Fig. 8.

For the circular camera arrangement, the spherical ray-space is constructed from many camera images, and its horizontal cross-section has a sinusoidal structure as shown in Fig. 9. The sinusoidal structure of the ray-space is also used for ray-space interpolation and compression.

3 FTV System

3.1 Configuration of FTV System

Fig. 10 shows a basic configuration of the FTV system. At the sender side, a 3D scene is captured by multiple cameras. The captured images contain the misalignment and luminance differences of the cameras. They must be corrected to construct the ray-space. The corrected images are compressed for transmission and storage by the MVC (Multi-view Video Coding) encoder.

At the receiver side, reconstructed images are obtained by the MVC decoder. The ray-space is constructed by arranging the reconstructed images and interpolating them. Free-viewpoint images are generated by cutting the ray-space vertically and are displayed on a 2D/3D display. Photo-realistic, free-viewpoint images of the moving scene are generated in real time.

Each part of the process shown in Fig. 10 is explained below.

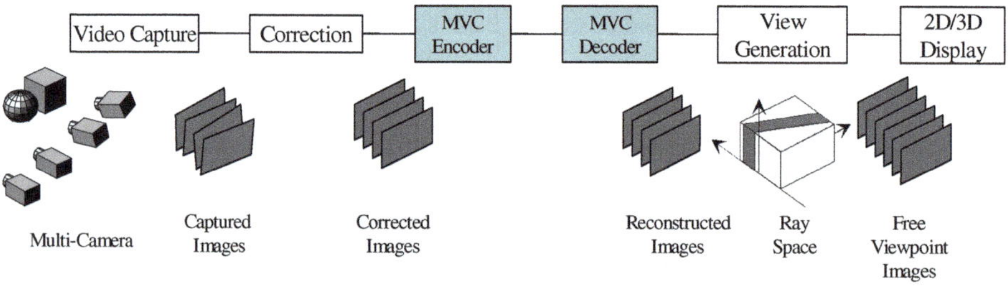

Fig. 10. A basic configuration of FTV system.

3.2 Video Capture

We constructed a 1D-arc capturing system shown in Fig. 11 for a real-time FTV system [22, 23]. It consists of 16 cameras, 16 clients and 1 server. Each client has one camera and all clients are connected to the server with Gigabit Ethernet.

Fig. 11. 1D-arc capturing system.

Table 1. Specification of 100-camera system.

Image resolution	1392(H) x 1040(V)
Frame rate	29.4118 [fps]
Color	Bayer matrix
Synchronization	Less than 1 [us]
Sampling rate of A/D	96 [kS/s] maximum
Maximum number of nodes	No limit. (128 max for one sync output)

(a) linear arrangement (b) circular arrangement (c) 2D-array arrangement

Fig. 12. 100-camera system.

Pantomime
500 frames

Champagne_tower
500 frames

Dog
300 frames

Fig. 13. MPEG test sequences.

A "100-camera system" has been developed to capture larger space by Nagoya University (Intelligent Media Integration COE and Tanimoto Laboratory) [7]. The system consists of one host-server PC and 100 client PCs (called 'nodes') that are equipped with JAI PULNiX TM-1400CL cameras. The interface between camera and PC is Camera-Link. The host PC generates a synchronization signal and distributes it to all of the nodes. This system is capable of capturing not only high-resolution video with 30 fps but also analog signals of up to 96 kHz. The specification of the 100-camera system is listed in table 1.

The camera setting is flexible as shown in Fig. 12. MPEG test sequences "Pantomime", "Champagne_tower" and "Dog" shown in Fig. 13 were taken in linear arrangement Fig. 12 (a).

3.3 Correction

The geometric correction [15, 19] and color correction [49] of multi-camera images are performed by measuring the correspondence points of images. This measurement is made once the cameras are set.

Examples of the geometric correction and color correction are shown in Fig. 14 and Fig. 15, respectively.

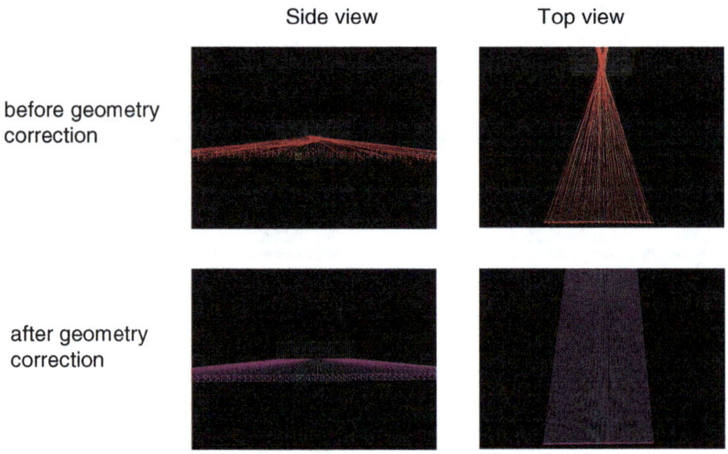

Fig. 14. Camera positions and directions before and after geometric correction.

3.4 MVC Encoding and Decoding

An example of time and view variations of multi-view images is shown in Fig. 16. They have high temporal and interview correlations. MVC (Multi-view Video Coding) reduces these correlations [50, 52, 57]. The standardization of MVC is described in Section 5.

3.5 View Generation

Ray-space is formed by placing the reconstructed images vertically and interpolating them. Free-viewpoint images are generated by making a cross-section of the ray-space.

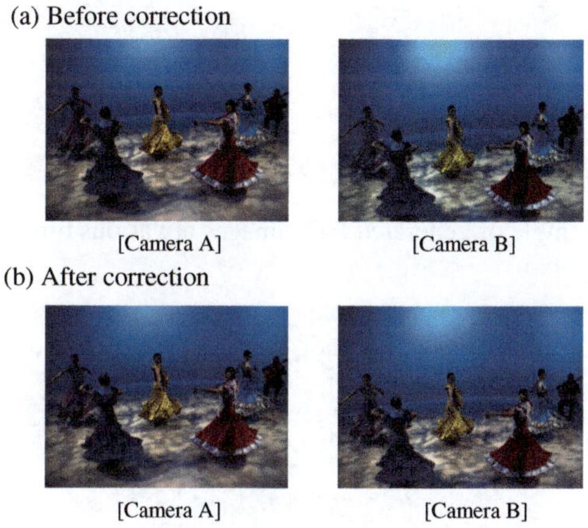

Fig. 15. An example of color correction.

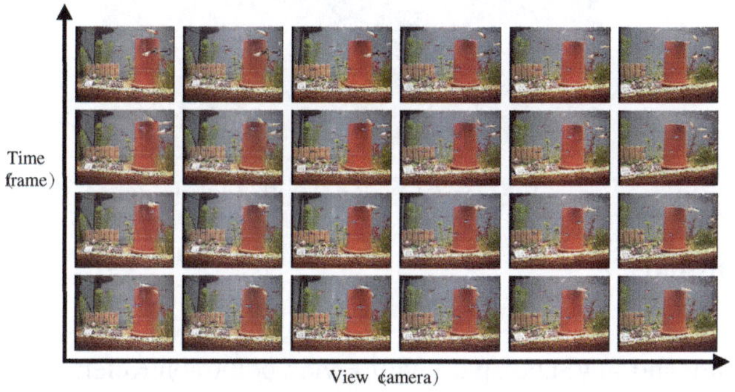

Fig. 16. Time and view variations of multi-view images.

Examples of the generated free-viewpoint images are shown in Fig. 17. Complicated natural scenes, including sophisticated objects such as small moving fish, bubbles and reflections of light from aquarium glass, are reproduced very well.

The quality of the generated view images depends on the ray-space interpolation. The ray-space interpolation is achieved by detecting depth information pixel by pixel from the multi-view video. We proposed several interpolation schemes of the ray-space [4, 5, 12, 13, 14, 21, 24].

Previously, free-viewpoint images were generated by a PC cluster. Now, they can be generated by a single PC, and FTV on a PC can be accomplished in real time [13]. FTV is also implemented on a mobile player as shown in Fig.18.

Fig. 17. An example of generated FTV images at various times and viewpoints.

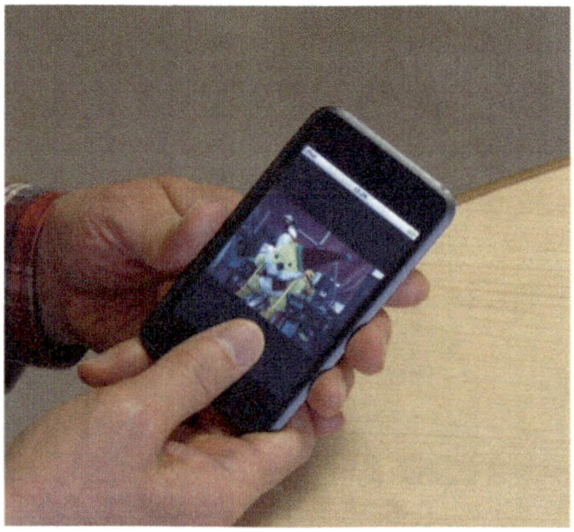

Fig. 18. FTV on a mobile player.

3.6 2D/3D Display

FTV needs a new user interface to display free-viewpoint images. Two types of display, 3D display and 2D/3D display with a viewpoint controller, are used for FTV as shown in Fig. 19.

Viewpoint control by head-tracking is shown here. Many head-tracking systems have been proposed using magnetic sensors, various optical markers, infrared cameras, retroreflective light from retinas, etc. Our head-tracking system uses only a conventional 2D camera and detects the position of a user's head by image processing. The user doesn't need to attach any markers or sensors.

In the user interface using a 2D display, the location of the user's head is detected with the head-tracking system and the corresponding view image is generated. Then, it is displayed on the 2D display as shown in Fig. 20.

Automultiscopic displays enable a user to see stereoscopic images without special glasses. However, there are two problems: a limited viewing zone and discreteness of motion parallax. Because the width of the viewing zone for each view approximates

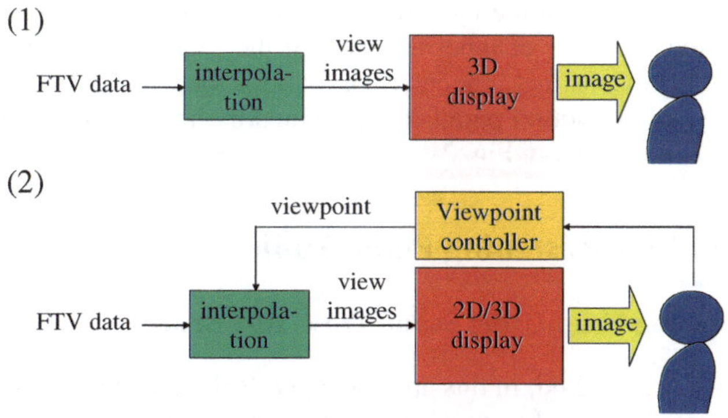

Fig. 19. Display of FTV.

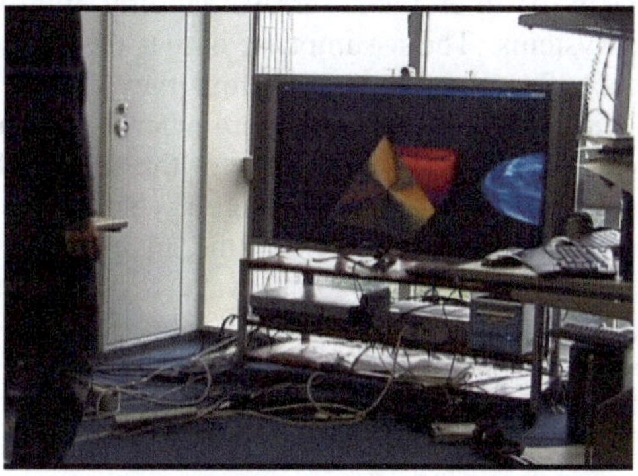

Fig. 20. 2D display with eye tracking.

(a) without head tracking (b) with head tracking

Fig. 21. 3D display with and without head tracking.

the interpupillary distance, the view image does not change with the viewer's movement within the zone. On the other hand, when the viewer moves over the zone, the view image changes suddenly.

In the user interface using the automultiscopic display, the function of providing motion parallax is extended by using the head-tracking system. The images fed to the system change according to the movement of the head position to provide small motion parallax, and the view channel for feeding the images is switched for handling large motion. This means that binocular parallax for the eyes is provided by automultiscopic display, while motion parallax is provided by head tracking and changing the image adaptively as shown in Fig. 21.

4 FTV with Free-Listening Point Audio

4.1 Free Listening-Point Using Sound Wave Ray-Space

Similar to ray-space method, in this approach [27, 28] we record the sound data in the format of an image, namely sound-image (simage), where the light equivalent of sound is captured using array of microphone with beamforming for every pixel or pixel-group in several frequency layers. The simage has certain duration of time, e.g. a video frame. An array of microphone arrays captures multiview simages as it is done in multicamera systems. The assumption is that the sound waves can be travelled directionally and reflected as light rays, thus they can be represented based on ray-space method. Similarly, sound ray parameters construct a 5D ray-space $f(x,y,z,\theta,\varphi)$ of simages, where f is the magnitude of the specified sound ray in a given time (t). In the 2D subspace $f(x,\theta)$ of 5D simage ray-space the sound rays that pass through a point (X,Z) in the real-space form a line, given by ($X = x + uZ$, $u(t) = \tan \theta$), in the ray-space. The capturing of sound wave ray-space, and its representation are shown in Fig 22(a) and (b).

Fourthly, free simage is generated using dense sound wave ray-space by cutting it along the locus and extract the section of simage as shown in Fig. 22(c) and (d). Dense sound-wave ray-space is generated using stereo matching of multiview

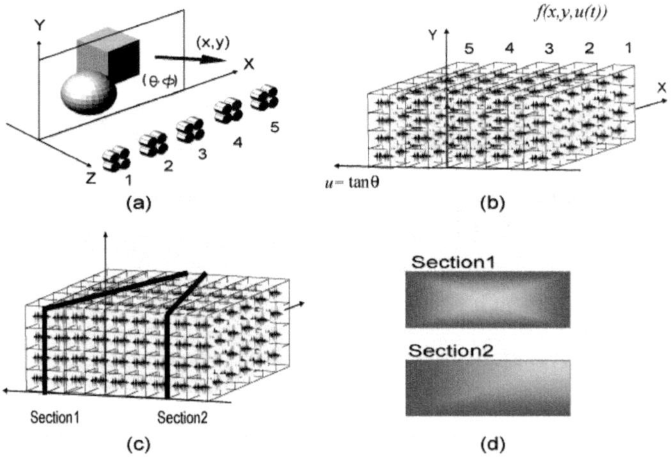

(a) Capturing rays in real space, (b) sampled ray-space, (c) dense ray-space, and (d) generated simages.

Fig. 22. Sound wave ray-space.

simages, as it is done for multiview images. The corresponded sound of a simage is generated by averaging the sound wave in each pixel or group of pixels.

In this case, we represent 3D image/simage data in ray-space format. Hence, we are able to integrate 3DAV data together in ray-space format [28]. Each ray data in "integrated 3DAV ray-space" shows the corresponded sound wave and light ray. Note that, stereo matching of light rays can be used for geometry compensation of simages in ray-space, and vice versa. Fig 23 depicts the integration approach.

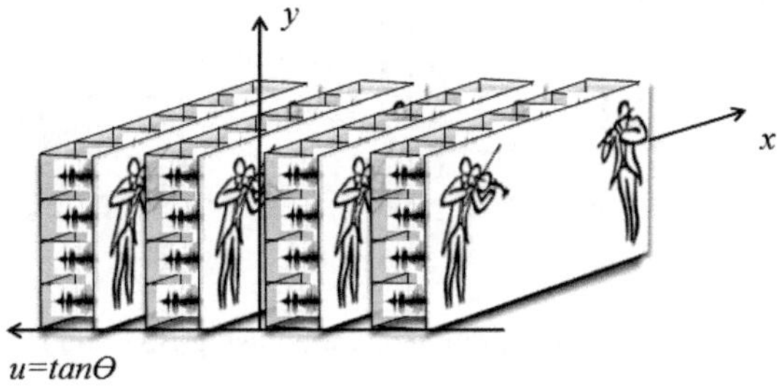

Fig. 23. Integration of audio and video in ray-space.

4.2 Free Listening-Point Using Acoustic Transfer Function

The acoustic transfer functions from sound sources to all far field microphones are estimated using a Time-Stretched Pulse (TSP) by dividing the frequency responses at both sides. ATF-pool [28] is generated between pairs or on the plane among microphones by applying a weighted linear interpolation of ATFs at far-field in frequency domain.

We use BSS (Blind Signal Separation) method based on a frequency domain independent component analysis (FD-ICA), and separate recorded signals into each part (e.g. each musical instrument) [26]. Separated signals are obtained by multiplying observed signals recorded by microphone and separation filter in the frequency domain. The separation filter is estimated from the observed signals using a fast fixed point algorithm [1] and Kullback-Leibler information minimization [29] for every frequency bands. However, since the FD-ICA cannot decide the magnitude and the order of separated signals, we have to solve the scaling problem and the permutation problem. For solving the scaling problem, we used the minimal distortion principal method [20]. This method extracts a diagonal of separation filter matrix. We modified the sorting method based on estimating sound source direction using the separation filter [26] for solving the permutation problem. Since the BSS method was applied to the instrumental signal, we conducted the solution method as following order; (i) sound source direction, (ii) correlation of a harmonic structure and (iii) correlation between neighbor frequency bands.

For free listening-point generation, ATFs from all sound sources to the desired listening-point are synthesized by convolving the source sound data obtain by BSS, with the ATF of each sound source in desired location in head related format, and then combining them [25,48], i.e. Fig. 24.

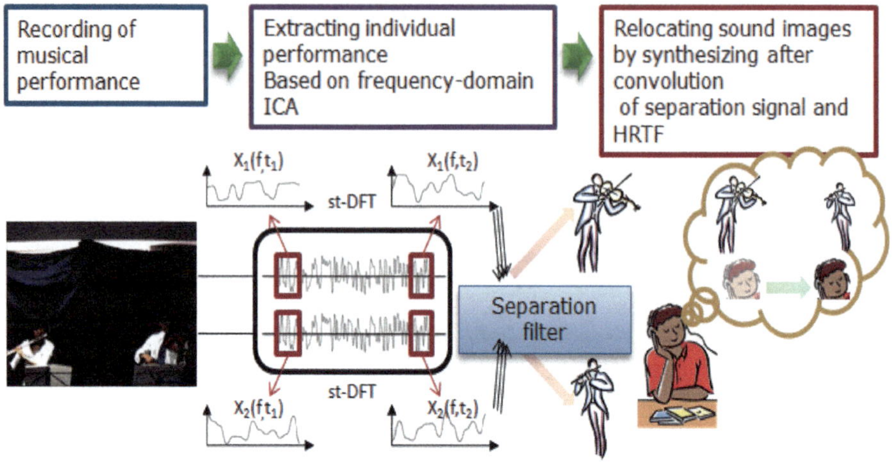

Fig. 24. Free listening-point generation using BSS and HRTF.

The system captures all viewpoint images and corresponded sound signal with two microphones (i.e. stereo case) located close to each camera to estimate the ATFs. To integrate the audio signal corresponds to each frame of video signal; each cameras video signal and its ATFs (e.g. two ATFs for stereo case) are represented as integrated 3DAV data [28]. Note that switching from previously requested audio frame to the next audio frame causes switching noise. It is eliminated using fade in/out.

4.3 Integrated FTV System

We constructed a multipoint cameras and microphone system [28]. For the integrated 3DAV data, we developed a viewer that can smoothly generate free-viewpoint video and free-listening point audio among the multipoint observations. The viewer is able to display video frames and play the audio frame in realtime.

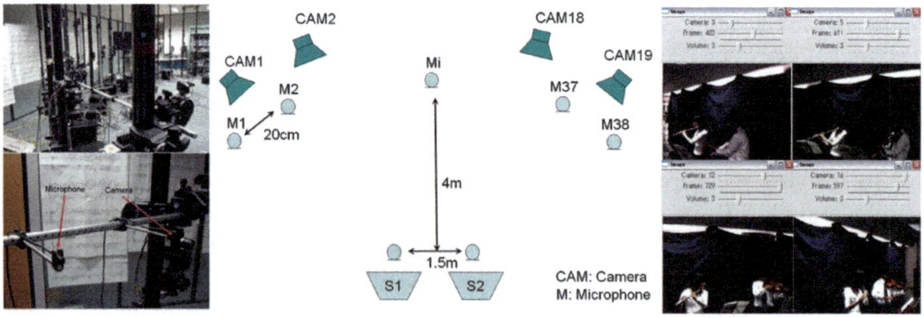

Fig. 25. Experimental setup and integrated viewer. (a) cameras and microphones array, (b) configuration of the cameras and microphones array, and (c) Integrated viewer.

The experimental setup is as shown in Fig. 25 and Table 2. The viewer performance on a general PC (Xeon 3.60GHz, 3GB RAM) for color image size (320x240) is 15-frame per second, with a good sound quality due to subjective evaluation. The larger image size makes the slower frame rate, down to 10-frame per second for 640x480 color image. The developed 3DAV viewer has the layout shown in Fig. 25 (c). Demonstration results are available in the following URL: "http://www.sp.m.is.nagoya-u.ac.jp/~niwa/fvlp-j.html".

Table 2. Hardware description of multipoint camera and microphone array used for the integrated FTV

Part	Description
Camera	PULNiX (TMC-1400CL) -1392x1040x1 - Bayer Matrix - 29.411fps
Microphone	Sony ECM-77B, - 16 Bits - 96~8 KS/sec
Node General purpose PC	- Celeron 2GHz, 256 RAM, OS: Linux Inputs: one camera, two microphones
Sever General purpose PC	- Xeon 3.60GHz Dual, OS: Windows
Network	Gigabit Ethernet
Synchronizer	Trigger all Nodes to record synchronized using the GPS signal - Wired - Wireless

5 Ray-Space Technology

5.1 Ray Capture

We are developing ray-space technology such as ray capture [9, 11, 18], ray processing [2, 3, 32, 33] and ray display [51].

We have developed a 360-degree mirror-scan ray capturing system [11, 18] as shown in Fig. 26. This system uses two parabolic mirrors. Incident rays that are parallel to the axis of a parabolic mirror gather at the focus of the parabolic mirror. Hence, rays that come out of an object placed at the focus of the lower parabolic mirror gather at the focus of the upper parabolic mirror. Then, the real image of the object is generated at the focus of the upper parabolic mirror and a rotating aslope mirror scans rays at the focus of the upper parabolic mirror. Finally, the image from the aslope mirror is captured by a high-speed camera. By using this system, we can capture all-around convergent views of an object as shown in Fig. 27.

5.2 Ray Display

Figure 28 shows the SeeLINDER [51], a 360-degree, ray-producing display that allows multiple viewers to see 3D FTV images. It consists of a cylindrical parallax

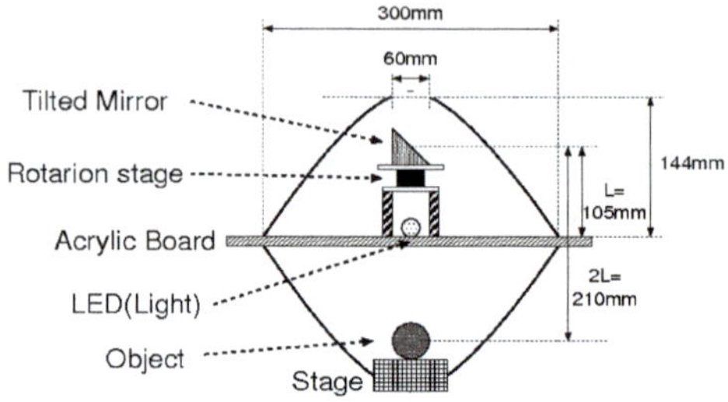

(c) Configuration of proposed system

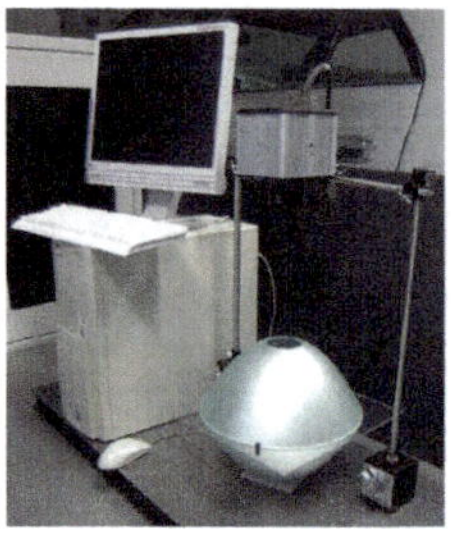

(a) Overview of ray-space acquisition system

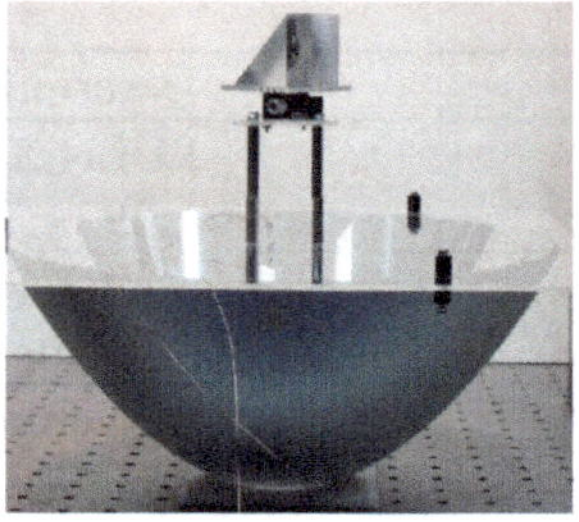

(b) Ray-space acquisition system without upper parabolic mirror

Fig. 26. 360-degree mirror-scan ray capturing system.

(a) Object
(about 10mm size)

(b) Angle of rotating stage : 0 °

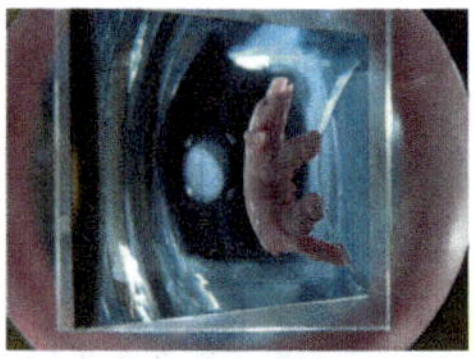

(c) 90 °

(d) 180 °

(e) 315 °

Fig. 27. Object and captured images of 360-degree ray capturing system.

barrier and one-dimensional light-source arrays. LEDs are aligned vertically for the one-dimensional light-source arrays. The cylindrical parallax barrier rotates quickly, and the light-source arrays rotate slowly in the opposite direction. If the aperture width of the parallax barrier is sufficiently small, the light going through the aperture becomes a thin flux, and its direction is scanned by the movement of the parallax barrier and the light-source arrays. By synchronously changing the intensity of the light sources with the scanning, pixels whose luminance differs for each viewing direction can be displayed. We can see the 3D image naturally, and the images have the strong depth cues of natural binocular disparity. When we move around the display, the image changes corresponding to our viewing position. Therefore, we perceive the objects just as if they were floating in the cylinder.

We are going to connect these two systems directly in real time.

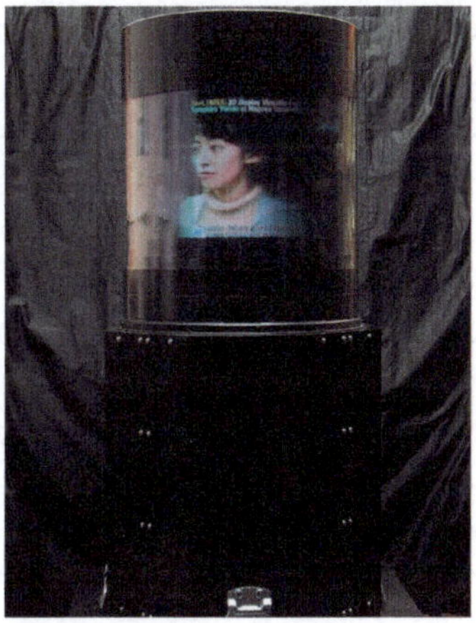

Fig. 28. The SeeLINDER, a 360 degree ray-reproducing display.

6 International Standardization

We proposed FTV to MPEG in December 2001. Fig. 29 shows the history of FTV standardization in MPEG.

In the 3DAV (3D Audio Visual) group of MPEG, many 3D topics such as omni-directional video, FTV [35], stereoscopic video and 3DTV with depth disparity information were discussed. The discussion was converged on FTV in January 2004 because FTV got strong support from industry in response to the "Call for Comments on 3DAV" [54].

Succeeding in "Call for Evidence on Multi-view Video Coding" [55] led the start of standardization of MVC in January 2005. "Call for Proposals on Multi-view Video Coding" [56] was issued in July 2005. The background of MVC is described in [57]. The proposals were evaluated in January 2006 [60]. The MVC activity moved to the Joint Video Team (JVT) of MPEG and ITU-T (International Telecommunication Union Telecommunication Standardization Sector) for further standardization processes in July

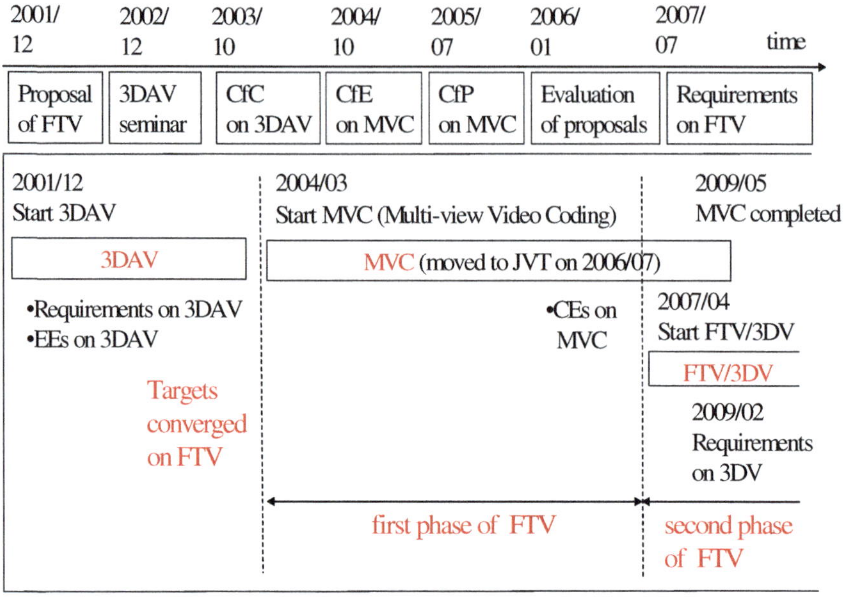

Fig. 29. History of FTV Standardization jn MPEG.

2006. The standardization of MVC is based on H.264/MPEG4-AVC and was completed in May 2009. MVC was the first phase of FTV

FTV cannot be constructed by coding part alone. We proposed to standardize the entire FTV [45] and MPEG started a new standardization activity of FTV in April 2007 [53].

The function of view generation in Fig. 10 is divided into depth search and interpolation. As shown in Fig. 30, an FTV system can be constructed in various ways, depending on the location of depth search and interpolation. In case A, both depth search and interpolation are performed at the receiver side as in Figure 10. In case B, depth search is performed at the sender side and interpolation is performed at the receiver side. In case C, both depth search and interpolation are performed at the sender side. Case B is suitable for the download/package and broadcast services since processing at the sender side is heavy and processing at the receiver side is light.

Case B was adopted by the FTV reference model [59] shown in Figure 31. Possible standardization items are FTV data format, decoder and interpolation. Interpolation might be informative.

At the sender side of the FTV reference model, multi-view images are captured by multiple cameras. The captured images contain the misalignment and color differences of the cameras. They are corrected and the depth of each camera image is obtained. The corrected multi-view images and multi-view depth are compressed for transmission and storage by the encoder. At the receiver side, the multi-view images and multi-view depth are reconstructed by the decoder. Free-viewpoint images are generated by interpolating the reconstructed images using multi-view depth information and displayed on a 2D/3D display.

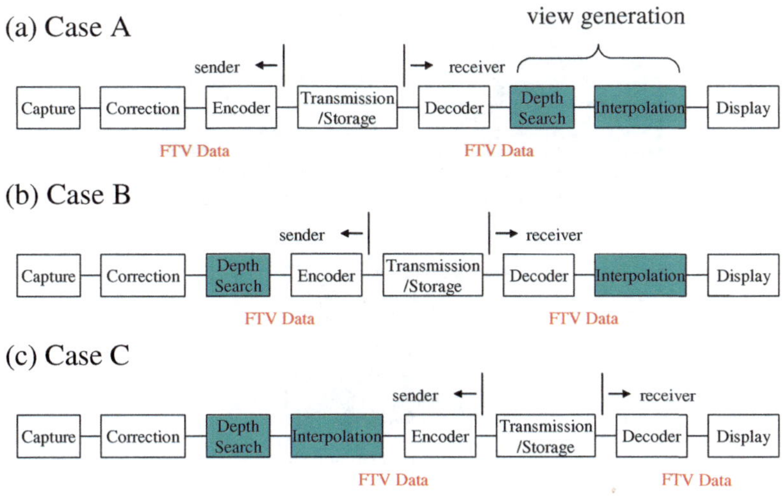

Fig. 30. 3 cases of FTV configuration based on the positions of depth search and interpolation.

Thus, FTV is a new framework that includes a coded representation for multi-view video and depth information to support the generation of high-quality intermediate views at the receiver. This enables free viewpoint functionality and view generation for 2D/3D displays.

Intermediate view is generated by interpolating multi-view images using multi-view depth as shown in Fig. 32 [46]. Compared to the view synthesis by using one view and one depth, the number of pixels which are not filled is greatly reduced. Especially, the occluded area is successfully interpolated by the reliable pixel data.

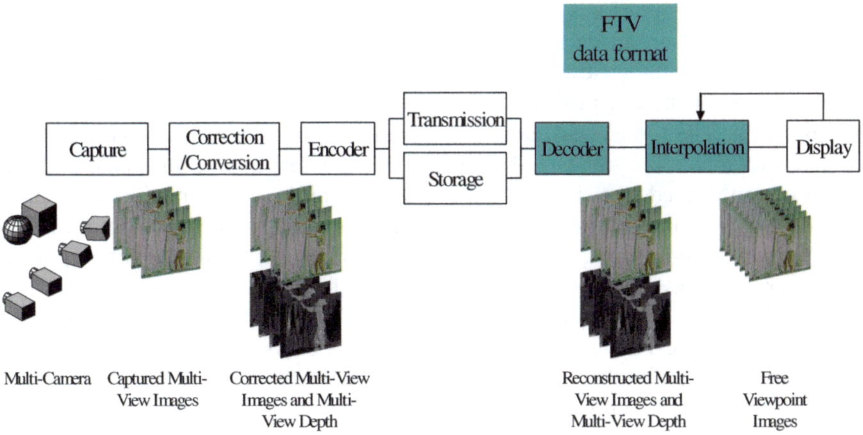

Fig. 31. FTV reference model and standardization items.

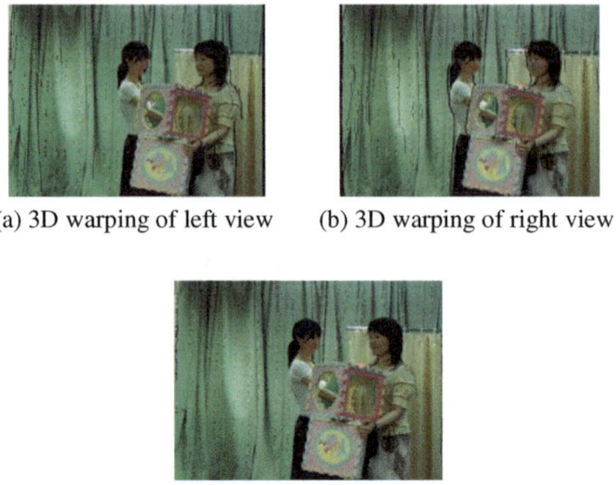

(a) 3D warping of left view (b) 3D warping of right view

(c) Integration of (a) and (b)

Fig. 32. An example of view generation by FTV reference model

We have proposed an FTV data unit (FDU) so that we can express FTV data of Case B by the combination of FDUs [42, 47]. Here, we assume linear camera arrangement. However, the same concept can be applied to other camera arrangements.

An FDU consists of view, depth, additional information for right view synthesis and additional information for left view synthesis as shown in Fig. 33. The view and depth of the FDU are used to generate free viewpoint images in the coverage of this FDU. However, occluded parts can't be generated by using one view and one depth. The additional information for right view synthesis and additional information for left view synthesis are used to generate the occluded parts of free viewpoint images. The FDU can synthesize views at the viewpoints in the coverage shown in Fig. 33. The coverage for view synthesis is extended by using two or more FDUs.

The FTV data unit is derived from the multiview and depth representation as shown in Fig. 34. Three views and depths at the center, right and left positions are used. An image at the left position is synthesized by using the view and depth at the center position. Then, make difference between the original left view and the synthesized left view. This is synthesized error of left view. Similarly, an image at the right position is synthesized by using the view and depth at the center position. Then, make difference between the original right view and the synthesized right view. This is synthesized error of right view. The synthesized error of left view and the depth of the left view are used as additional information for left view synthesis. The synthesized error of right view and the depth of the right view are used as additional information for right view synthesis. Thus, an FDU that consists of the view and depth at the center position, the additional information for right view synthesis and the additional information for left view synthesis is obtained.

The FDU is capable of synthesizing high quality views. It is clear that the FDU can generate original views at center, right and left positions. Furthermore, the FDU can reduce synthesis error in the coverage [42, 47].

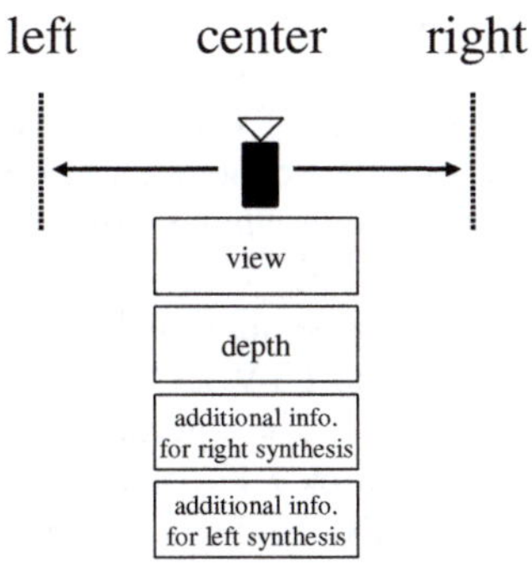

Fig. 33. FDU (FTV data unit) and its coverage for view synthesis.

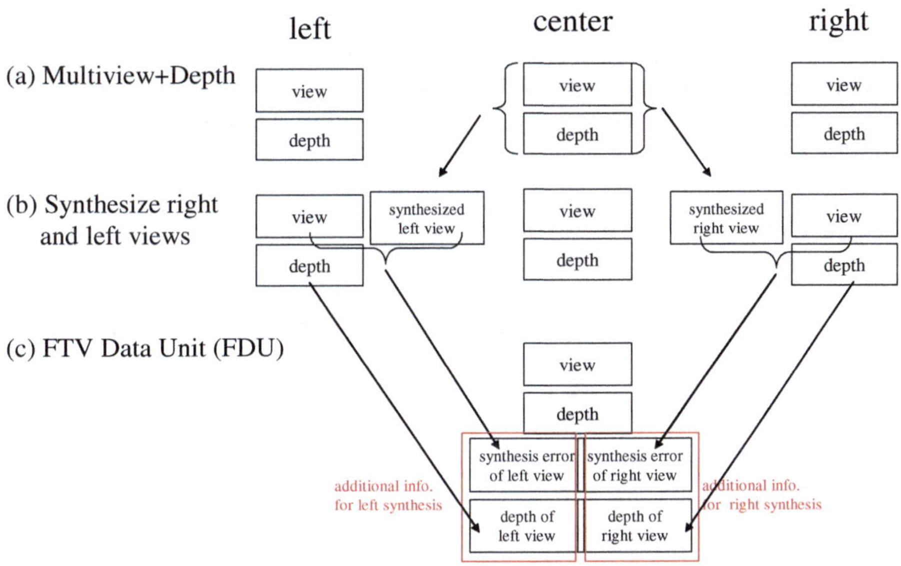

Fig. 34. Derivation of FDU from Multiview+Depth.

In January 2008, MPEG-FTV targeted the standardization of 3DV (3D Video). 3DV is a standard that targets serving for a variety of 3D displays. 3DV is the second phase of FTV. The introduction to 3DV is described in the reference [58].

For further information on MPEG-FTV activity, please follow the link and subscribe MPEG-FTV.

https://mailman.rwth-aachen.de/mailman/listinfo/mpeg-ftv

7 Conclusion

FTV is an innovative media that allows one to view a 3D world by freely changing the viewpoint and will bring an epochal change in the history of television.

FTV will find many applications in the fields of broadcast, communication, amusement, entertainment, advertising, exhibition, education, medicine, art, archives, security, surveillance, and so on since FTV is an ultimate 3DTV that captures and transmits all visual information of a 3D scene, a natural interface between human and environment, and an innovative tool to create new types of content and art.

The most essential element of visual systems is ray. FTV is not a conventional pixel-based system but a ray-based system. We have been developing ray capture, processing and display technologies for FTV.

The introduction of FTV is not far. The international standardization of FTV has been conducted in MPEG. We can enjoy FTV on a PC or a mobile player if the standards of FTV are available and FTV contents are delivered over internet or with packaged media.

Acknowledgments

This research is partially supported by Strategic Information and Communications R&D Promotion Programme (SCOPE) 093106002 of the Ministry of Internal Affairs and Communications, and National Institute of Information and Communications Technology, Japan (NICT).

References

[1] Bingham, E., Hyvarinen, A.: A fast fixed-point algorithm for independent component analysis of complex valued signals. International Journal of Neural System 10(1), 1–8 (2000)

[2] Chimura, N., Yendo, T., Fujii, T., Tanimoto, M.: New Visual Arts by Processing Ray-Space. In: Proc. of Electronic Imaging & the Visual Arts (EVA), Florence, pp. 170–175 (2007)

[3] Chimura, N., Yendo, T., Fujii, T., Tanimoto, M.: Image Generation with Special Effects by Deforming Ray-Space. In: Proc. of NICOGRAPH, S1-4 (2007)

[4] Droese, M., Fujii, T., Tanimoto, M.: Ray-Space Interpolation based on Filtering in Disparity Domain. In: Proc. 3D Conference 2004, Tokyo, Japan, pp. 213–216 (2004)

[5] Droese, M., Fujii, T., Tanimoto, M.: Ray-Space Interpolation Constraining Smooth Disparities Based On Loopy Belief Propagation. In: Proc. of IWSSIP 2004, Poznan, Poland, pp. 247–250 (2004)

[6] Fujii, T.: A Basic Study on Integrated 3-D Visual Communication. Ph.D dissertation in engineering. The University of Tokyo (1994) (in Japanese)

[7] Fujii, T., Mori, T., Takeda, K., Mase, K., Tanimoto, M., Suenaga, Y.: Multipoint Measuring System for Video and Sound: 100-camera and microphone system. In: IEEE 2006 International Conference on Multimedia & Expo (ICME), pp. 437–440 (2006)

[8] Fujii, T., Tanimoto, M.: Free-viewpoint Television based on the Ray-Space representation. In: Proc. SPIE ITCom 2002, pp. 175–189 (2002)

[9] Fujii, T., Tanimoto, M.: Real-Time Ray-Space Acquisition System. In: SPIE Electronic Imaging, vol. 5291, pp. 179–187 (2004)

[10] Fujii, T., Kimoto, T., Tanimoto, M.: Ray Space Coding for 3D Visual Communication. In: Picture Coding Symposium 1996, pp. 447–451 (1996)

[11] Fujii, T., Yendo, T., Tanimoto, M.: Ray-Space Transmission System with Real-Time Acquisition and Display. In: Proc. of IEEE Lasers and Electro-optics Society Annual Meeting, 2007, pp. 78–79 (2007)

[12] Fukushima, N., Yendo, T., Fujii, T., Tanimoto, M.: Real-time arbitrary view interpolation and rendering system using Ray-Space. In: Proc. SPIE Three-Dimensional TV. Video. and Display IV, vol. 6016, pp. 250–261 (2005)

[13] Fukushima, N., Yendo, T., Fujii, T., Tanimoto, M.: An Effective Partial Interpolation Method for Ray-Space. In: Proc. of 3D Image Conference 2006, pp. 85–88 (2006)

[14] Fukushima, N., Yendo, T., Fujii, T., Tanimoto, M.: Free Viewpoint Image Generation Using Multi-Pass Dynamic Programming. In: Proc. SPIE Stereoscopic Displays and Virtual Reality Systems XIV, vol. 6490, pp. 460–470 (2007)

[15] Fukushima, N., Yendo, T., Fujii, T., Tanimoto, M.: A Novel Rectification Method for Two-Dimensional Camera Array by Parallelizing Locus of Feature Points. In: Proc. of IWAIT 2008, B5-1 (2008)

[16] Kawakita, M., Iizuka, K., Nakamura, H., Mizuno, I., Kurita, T., Aida, T., Yamanouchi, Y., Mitsumine, H., Fukaya, T., Kikuchi, H., Sato, F.: High-definition real-time depth-mapping TV camera: HDTV Axi-Vision Camera. Optics Express 12(12), 2781–2794 (2004)

[17] Kubota, A., Aljoscha, S., Marcus, M., Tanimoto, M., Tsuhan, C., Zhang, C.: Multiview Imaging and 3TDTV. IEEE Signal Processing Magazine 24(6), 10–21 (2007)

[18] Manoh, K., Yendo, T., Fujii, T., Tanimoto, M.: Ray-Space Acquisition System of All-Around Convergent Views using a Rotation Mirror. In: Proc. of SPIE, vol. 6778, 67780C-1–8 (2007)

[19] Matsumoto, K., Yendo, T., Fujii, T., Tanimoto, M.: Multiple-Image Rectification for FTV. In: Proc. of 3D Image Conference 2006. P-19, pp. 171–174 (2006)

[20] Matsuoka, K., Nakashima, S.: Minimal, distortion principle for blind source separation. In: Intl Symp. ICA, pp. 722–727 (2001)

[21] Mori, Y., Fukushima, N., Yendo, T., Fujii, T., Tanimoto, M.: View Generation with 3D Warping Using Depth Information for FTV. Signal Processing Image Communication 24(1), 265–272 (2009)

[22] Na Bangchang, P., Fujii, T., Tanimoto, M.: Experimental System of Free Viewpoint TeleVision. In: Proc. IST/SPIE Symposium on Electronic Imaging, vol. 5006-66, pp. 554–563 (2003)

[23] Na Bangchang, P., Panahpour Tehrani, M., Fujii, T., Tanimoto, M.: Realtime System of Free Viewpoint Television. The journal of the institute of Image information and Television Engineers (ITE) 59(8), 63–70 (2005)

[24] Nakanishi, A., Fujii, T., Kimoto, T., Tanimoto, M.: Ray-Space Data Interpolation by Adaptive Filtering using Locus of Corresponding Points on Epipolar Plane Image. The Journal of the Institute of Image Information and Television Engineers (ITE) 56(8), 1321–1327 (2002)

[25] Nishino, T., Kajita, S., Takeda, K., Itakura, F.: Interpolation of the head related transfer function on the horizontal plane. Journal of Acoust. Society of Japan 55(2), 1–99 (1999)

[26] Niwa, K., Nishino, T., Miyajima, C., Takeda, K.: Blind source separation of musical signals applied to selectable-listening-point audio reconstruction. Jr. of ASA, 3pSP20, Hawaii (2006)

[27] Panahpour Tehrani, M., Hirano, Y., Fujii, T., Kajita, S., Takeda, K., Mase, K.: The Sub-band Sound Wave Ray-Space Representation. IEEE ICASSP MMSP-p3.11, v541–v544 (2006)

[28] Panahpour Tehrani, M., Niwa, K., Fukushima, N., Hirano, Y., Fujii, T., Tanimoto, M., Kazuya, M., Takeda, K., Mase, K., Ishikawa, A., Sakazawa, S., Koike, A.: 3DAV Integrated System Featuring Arbitrary Listening-point and Viewpoint Generation. In: Proc. of IEEE Multimedia Signal Processing. MMSP 2008, vol. PID-213, pp. 855–860 (2008) (Best paper award)

[29] Sawada, H., Mukai, M., Araki, S., Makino, S.: A robust and precise method for solving the permutation problem of frequency-domain blind source separation. In: Intl Symp. ICA, pp. 505–510 (2003)

[30] Sekitoh, M., Toyota, K., Fujii, T., Kimoto, T., Tanimoto, M.: Virtual Bird's-Eye View System based on Real Image. In: EVA 2000, Gifu, 8, 8-1– 8-7 (2000)

[31] Sekitoh, M., Fujii, T., Kimoto, T., Tanimoto, M.: Bird's Eye View System for ITS. In: IEEE Intelligent Vehicle Symposium, pp. 119–123 (2001)

[32] Takano, R., Yendo, T., Fujii, T., Tanimoto, M.: Scene Separation in Ray-Space. In: Proc. of IMPS 2005, pp. 31–32 (2005)

[33] Takano, R., Yendo, T., Fujii, T., Tanimoto, M.: ene Separation and Synthesis Processing in Ray-Space. In: Proc. of IWAIT 2007, vol. P6-23, pp. 878–883 (2007)

[34] Tanimoto, M.: Fee Viewpoint Television. The Journal of Three Dimensional Images 15, 17–22 (2001) (in Japanese)

[35] Tanimoto, M., Fujii, T.: FTV - Free Viewpoint Television. ISO/IEC JTC1/SC29/WG11. M8595 (2002)

[36] Tanimoto, M.: Free Viewpoint Television – FTV. In: Picture Coding Symposium 2004. Special Session 5 (2004)

[37] Tanimoto, M.: FTV (Free Viewpoint Television) Creating Ray-Based Image Engineering. In: Proc. of ICIP 2005, pp. II-25–II-28 (September 2005)

[38] Tanimoto, M.: Overview of Free Viewpoint Television. Signal Processing: Image Communication 21(6), 454–461 (2006)

[39] Tanimoto, M.: Free Viewpoint Television, OSA Topical Meeting on Digital Holography and Three-Dimensional Imaging, DWD2(2007) (invited paper)

[40] Tanimoto, M.: FTV (Free viewpoint TV) and Creation of Ray-Based Image Engineering. ECTI Transaction on Electrical Engineering, Electronics and Communications 6(1), 3–14 (2008) (invited paper)

[41] Tanimoto, M.: FTV (Free viewpoint TV) and Ray-Space Technology. In: IBC 2008. The cutting edge. part two (2008)

[42] Tanimoto, M., Wildeboer, Menno: Frameworks for FTV Coding. In: Picture Coding Symposium 2009. Special Session - 2: Multiview and 3D Video Coding (2009)

[43] Tanimoto, M., Nakanishi, A., Fujii, T., Kimoto, T.: The Hierarchical Ray-Space for Scalable 3-D Image Coding. In: Picture Coding Symposium 2001, pp. 81–84 (2001)

[44] Tanimoto, M., Fujii, T., Sakazawa, S., Kimata, H.: Proposal on Standardization of Free Viewpoint TV (FTV). ISO/IEC JTC1/SC29/WG11. M13612, JVT-T140 (2006)

[45] Tanimoto, M., Fujii, T., Kimata, H., Sakazawa, S.: Proposal on Requirements for FTV. ISO/IEC JTC1/SC29/WG11, M14417 (2007)

[46] Tanimoto, M., Fujii, T., Suzuki, K.: Experiment of view synthesis using multi-view depth, ISO/IEC JTC1/SC29/WG11, M14889 (2007)

[47] Tanimoto, M., Fujii, T., Suzuki, K.: Data Format for FTV. ISO/IEC JTC1/SC29/WG11. M16093 (2009)

[48] Wightman, F.L., Kistler, D.J.: A Model of HRTFs Based on Principal Component Analysis and Minimum-phase Reconstruction. Jr. of the ASA 91(3), 1637–1647 (1992)

[49] Yamamoto, K., Yendo, T., Fujii, T., Tanimoto, M.: Colour Correction for Multiple-camera System by using Correspondences. The journal of the institute of Image Information and Television Engineers 61(2), 213–222 (2007)
[50] Yamamoto, K., Kitahara, M., Yendo, T., Fujii, T., Tanimoto, M., Shimizu, S., Kamikura, K., Yashima, Y.: Multiview Video Coding Using View Interpolation and Color Correction. IEEE Transactions on Circuits and Systems for Video Technology 17(11), 1436–1449 (2007) (invited paper)
[51] Yendo, T., Kajiki, Y., Honda, T., Sato, M.: Cylindrical 3-D Video Display Observable from All Directions. In: Proc. of Pacific Graphics 2000, pp. 300–306 (2000)
[52] Yun, H., Joern, O., Tanimoto, M., Aljoscha, S.: Introduction to the Special Section on Multiview Video Coding. IEEE Transactions on Circuits and Systems for Video Technology 17(11), 1433–1435 (2007)
[53] AHG on FTV (Free Viewpoint Television), ISO/IEC JTC1/SC29/WG11 N8947 (2007)
[54] Call for Comments on 3DAV, ISO/IEC JTC1/SC29/WG11 N6051 (October 2003)
[55] Call for Evidence on Multi-view Video Coding. ISO/IEC JTC1/SC29/WG11 N6720 (2004)
[56] Call for Proposals on Multi-view Video Coding. ISO/IEC JTC1/SC29/WG11 N7327 (2005)
[57] Introduction to Multi-view Video Coding, ISO/IEC JTC 1/SC 29/WG11, N7328 (2005), http://www.chiariglione.org/mpeg/technologies/mp-mv/index.htm
[58] Introduction to 3D Video, ISO/IEC JTC1/SC29/WG11 N9784 (2008), http://www.chiariglione.org/mpeg/technologies/mp3d/index.htm
[59] Preliminary FTV Model and Requirements, ISO/IEC JTC1/SC29/WG11, N9168 (2007)
[60] Subjective test results for the CfP on Multi-view Video Coding. ISO/IEC JTC1/SC29/WG11 N7779 (2006)

UGC Video Sharing: Measurement and Analysis

Xu Cheng[1], Kunfeng Lai[2], Dan Wang[2], and Jiangchuan Liu[1]

[1] School of Computing Science, Simon Fraser University, British Columbia, Canada
xuc,jcliu@cs.sfu.ca

[2] Department of Computing, The Hong Kong Polytechnic University
cskflai,csdwang@comp.polyu.edu.hk

Abstract. User-generated content (UGC) site has become a new killer Internet application in the recent four years. Among those popular sites, YouTube is the most representative and successful one providing a new generation of short video sharing service. Today, YouTube is a dominant provider of online video in the Internet, and is still growing fast. Understanding the features of YouTube and similar video sharing sites is thus crucial to their sustainable development and to network traffic engineering.

We investigate the YouTube site from two perspectives, internal and external. Using traces crawled in a 1.5-year span, we systematic measure the characteristics of YouTube videos. We find that YouTube videos have noticeably different statistics compared to traditional streaming videos, ranging from length, access pattern, to their active life span. The series of datasets also allows us to identify the growth trend of this fast evolving Internet site in various aspects, which has seldom been explored before. We also look closely at the social networking aspect of YouTube, as this is a key driving force toward its success. In particular, we find that the links to related videos generated by uploaders' choices form a small-world network. This suggests that the videos have strong correlations with each other, and creates opportunities for developing novel caching or peer-to-peer distribution schemes to efficiently deliver videos to end users.

We also provide an in-depth study into the effects of the external links of YouTube. We collected nearly one million videos' external link information, and traced different types of videos for more than two months. Our study shows interesting characteristics of external links of YouTube. In particular, we find that views from external links are independent from total views in each category. Also, videos benefit more from external links in the early stage. Our work can serve as a initial step for the study of the external environment.

1 Introduction

Online videos existed long before UGC sites entered the scene. However, uploading videos, managing, sharing and watching them were very cumbersome due to a lack of an easy-to-use integrated platform. More importantly, the videos distributed by traditional streaming media servers or peer-to-peer file downloads like BitTorrent were standalone units of content, as each single video was not connected to other related video clips. The new generation of video sharing sites

C.W. Chen et al. (Eds.): Intel. Multimedia Communication: Tech. and Appli., SCI 280, pp. 367–402.
springerlink.com

however offer integrated Web 2.0 [39] platforms to overcome these problems. The systems allow content suppliers to upload videos effortlessly, and to tag uploaded videos with keywords and links to other videos. Users can easily share videos by mailing links to them, or embedding them in blogs. The videos are no longer independent from each other with the clients browsing them following the links. Consequently, popular videos can rise to the top in a very organic fashion.

The most successful UGC site, YouTube [8], now enjoys more than 6 billion videos viewed every month [16]. The success of similar sites like Yahoo! Video [6] and Youku [7] (the most popular video sharing site in China), and the expensive acquisition of YouTube by Google [12], further confirm the mass market interest. Their great achievement lies in the combination of the content-rich videos, and more importantly, the establishment of a social network. These sites have created a video village on the web, where anyone can be a star, from lip-synching teenage girls to skateboarding dogs. With no doubt, they are changing the content distribution landscape and even the popular culture [13].

Established in 2005, YouTube is one of the fastest-growing UGC sites, and has become the fourth most accessed site in the Internet by the survey of Alexa [1]. On the other hand, an April 2008 report estimated that YouTube consumed as much bandwidth as did the entire Internet in year 2000 [14], and industry insiders estimate that YouTube spends roughly $1 million a day to pay for its server bandwidth [15]. Moreover, a survey revealed that the delay of YouTube is much longer than many other measured sites [42]. Therefore, understanding the features of YouTube and similar video sharing sites is crucial to network traffic engineering and to the sustainable development of this new generation of service.

On the other hand, the overall popularity of these UGC sites are attributed by both internal environment and the external links. The exterior environment are the external links that reference the video object outside the UGC sites. These external links embed the videos in the web pages, and people are also able to view the videos as well as follow the related video links. In other word, the external links provide a way of advertising the videos outside the UGC sites. We believe the impacts from external links are substantial. According to Alexa [1], YouTube has more than 560 thousand cites linking in, and the even large number of pages linking in will closely related the visits of these sites with the popularity of YouTube. Since the Internet are interconnected and the number of sites linking to UGC sites are huge, the study of the impacts from external links to online social networks are of directive significance, and the results of our study will help the promotion of UGC sites from the external circumstance.

In this chapter, we present an in-depth and systematic measurement study on the characteristics of YouTube. We measure the YouTube site from internal and external. We conducted two rounds of crawling of the YouTube site, from February 2007 to July 2008. We have collected more than three million distinct videos' information in the first round, and obtained 59 datasets totaling $5,043,082$ distinct videos' information in the second round, which is, to our knowledge, the largest dataset crawled so far. From this large collection of datasets, we find that

YouTube videos have noticeably different statistics from traditional streaming videos, in aspects from video length to access pattern. The long span of the measurement also enables us to examine new features that have not been addressed in previous measurement studies, for example, the growth trend and active life span.

We also look closely at the social networking aspect of YouTube, as this is a key driving force toward the success of YouTube and similar sites. In particular, we find that the links to related videos generated by uploader's choices form a small-world network. This suggests that the videos have strong correlations with each other, and creates opportunities for developing novel caching or peer-to-peer distribution schemes to efficiently deliver videos to end users.

We also present a progressive measurement study on how the external links affect the popularity of the videos in YouTube. Our measurement is based on the data of nearly one million videos gathered from video crawler for more than two month. The study furthers the understanding of connection of social networks by stressing on analyzing the impacts on the popularity from external links. In particular, we find that views from external links are independent from total views in each category. Also, videos benefit more from external links in the early stage. Our work can serve as a initial step for the study of the external environment.

In the remaining part of the chapter, we will present background, related work, and our methodology, and then analyze the YouTube site from internal and external. We will also characterize the YouTube video popularity and its evolution. The unique characteristic of social network is also analyzed. Finally, we will discuss the implications of the results, and suggest ways that the YouTube service could be improved.

2 Background and Related Work

2.1 YouTube Background

YouTube is a UGC video sharing site established in February 2005. Since then, YouTube becomes a dominant provider of online video in the Internet, with a market share of around 43 percent, and had 6 billion videos viewed in January 2009 [16]. In November 2006, YouTube was acquired by Google for $1.65 billion dollars [12].

YouTube's video playback technology is based on Adobe's Flash Player. This technology allows YouTube to display videos with quality comparable to well established video playback technologies (such as Windows Media Player, QuickTime and Realplayer). YouTube accepts uploaded videos in most formats, including WMV, AVI, MOV, MPEG, MKV, MP4, SWF, FLV, 3GP formats, which are converted into FLV (Adobe Flash Video) format after uploading [8]. It is well recognized that the use of a uniform easily-playable format is critical in the success of YouTube.

Table 1. Comparison of YouTube video formats

	Standard	HD	Mobile
Container	FLV	MP4	3GP
Video encoding	H.263	H.264/MPEG-4 AVC	H.263/AMR
Video resolution	320×240	1280×720	176×144
Audio encoding	MP3	AAC	AAC

Originally, YouTube used the Sorenson Spark H.263 video codec with pixel dimensions of 320×240, with mono MP3 audio. Later YouTube introduced "mobile" format for viewing on mobile phones and "high quality" format for better viewing quality. The high quality videos use the H.264 codec and stereo AAC audio. The YouTube player was also changed from a 4 : 3 aspect ratio to widescreen 16 : 9. Table 1 shows the comparison of YouTube video formats [11].

2.2 Workload Measurement of Traditional Media Servers

There have been significant research efforts into understanding the workloads of traditional media servers, looking at, for example, the video popularity and access locality [18, 19, 28, 21, 45, 27, 47, 32]. We have carefully compared their measurement results with ours, and have found that, while sharing similar features, many of the video statistics of these traditional media servers are quite different from YouTube-like sites, e.g., the video length distribution, user access pattern and active life span. More importantly, these traditional studies lack a social network among the videos, and links from the external either.

2.3 Workload Measurement of New Generation Media Servers

The reason why the online social networks are widely believed capable is that users participate in the resource management and the resource popularization. As a whole, online social networks comprises of a swarm of nodes, which may be users or web pages, and links among the nodes. With this structure, online social networks enable communities or groups in which the nodes share common characteristics, and the concentration of nodes with common features greatly alleviates the cumbersome of managing the resources in the system.

To understand this advantage and to further dig-out of the potential inside this structure, the study of the online social networks has been widely carried out in the past four years. According to the currently existing UGC sites study, we summarize the characteristics of the online social networks as follows:

- The popularity of the nodes: Cha *et al.* studied YouTube and Daum UCC, the most popular UGC service in Korea, studying the nature of the user behavior and identified the key elements that shape the popularity distribution, and also proposed some improvement for UGC design, such as cache design [24]. Gill *et al.* tracked YouTube transactions in a campus network, focusing on deriving video access patterns from the network edge perspective, and also

discussed the improvement approaches such as caching and CDNs [29]. Zink *et al.* obtained the trace of YouTube traffic in a campus network, investigating the properties of local popularity, and they also discussed some implications such as P2P and caching [49].
- The links in the social networks: Halvey *et al.* were the first to study the social network aspect in YouTube, focusing more on users [30]. Mislove *et al.* studied four online social networking sites (Flickr, YouTube, LiveJournal and Orkut), and confirmed the power-law, small-world and scale-free properties of online social networks [38]. Paolillo found the social core appearing in YouTube, also from the user perspective [40].
- User behavior: Singla *et al.* found that people who usually contact each other in online social networks are more likely to share more interests, and they also have the inclination that share the same characteristics, such as their age, their interests and their locations [44].

Our study complements these existing works, including our previous work [25, 33], by the long-term large-scale measurement that spanning 1.5 years, and an external link analysis of YouTube. It facilitates our understanding of the evolution and the latest development of this rapidly evolving service.

3 Methodology of Measurement

In this chapter, we study the YouTube site from two perspectives, internal and external. We have built a YouTube crawler to collect the YouTube videos' information through a combination of the YouTube API and scrapes of YouTube video web pages. We also use the crawler to collect external link information of YouTube through a universal JavaScript engine. In this section, we will first describe our YouTube crawler, and then present the crawled datasets for internal measurement and traces for external measurement.

3.1 YouTube Crawler for Internal Measurement

YouTube assigns each video a distinct 11-digit ID composed of 0-9, a-z, A-Z, -, and _. Each video contains the following intuitive meta-data: video ID, user who uploaded it, date when it was added, category, length, number of views, ratings and comments, and a list of "related videos". The related videos are links to other videos that have a similar title, description, or tags, all of which are chosen by the uploader. A video can have hundreds of related videos, but the webpage only shows at most 20 at once, so we limit our scrape to these top 20 related videos. A typical example of the meta-data is shown in Table 2.

We consider all the YouTube videos to form a directed graph, where each video is a node in the graph. If video b is in the related video list of video a, then there is a directed edge from a to b. Our crawler uses a breadth-first search to find videos in the graph. We define the initial set of a list of IDs, which the crawler

Table 2. Meta-data of A YouTube Video

ID	YiQu4gpoa6k
Uploader	NewAgeEnlightenment
Date Added	August 08, 2008
Category	Sports
Video Length	270 seconds
Number of Views	924, 691
Number of Ratings	1, 039
Number of Comments	212
Related Videos	ri1h2_jrVjU, 0JdQlaQpOuU, ...

reads into a queue at the beginning of the crawl. When processing each video, it checks the list of related videos and adds any new ones to the queue. Given a video ID, the crawler first extracts information from the YouTube API [9], which contains all the meta-data except date added, category and related videos. The crawler then scrapes the video's webpage to obtain the remaining information.

We started our crawl on February 22nd, 2007. The first round ended on May 18th, 2007, collecting 3, 269, 030 distinct videos. We started second round of crawling on March 27th, 2008, and ran the crawler every two days. On average, the crawl finds 81 thousand distinct videos each time. The crawl ended on July 27th, 2008, collecting 5, 043, 082 distinct videos, in which only 8.3% of the videos were also crawled in the first round, suggesting that YouTube is rapidly growing.

To study the growth trend of the video popularity, we also use the crawler to update the statistics of some previously found videos. For this crawl we only retrieve the number of views for relatively new videos. In 2007, we obtained 7 datasets, in two-month period. In 2008, we re-collect this information, crawling once a week from April 16th to September 3rd 2008 (six-month period), which results in 21 datasets. We will be focusing on the 2008 data, which represents the latest development of YouTube, and we will point out noticeable and interesting differences between the 2008 and 2007 data.

We also separately crawled the file size and bitrate information. To get the file size, the crawler retrieves the response information from the server when requesting to download the video file and extracts the information on the size of the download. Some videos also have the bitrate embedded in the FLV video meta-data, which the crawler extracts after downloading the meta-data of the video file.

Finally, we have collected the information about YouTube users. The crawler retrieves information on the number of uploaded videos and friends of each user from the YouTube API, for a total of more than 2 million users.

All the crawled data for internal measurement are available online. A more detailed description and the dataset can be found at `http://netsg.cs.sfu.ca/youtubedata.html`.

3.2 YouTube Crawler for External Measurement

To widely distribute the content videos and attract more users, the UGC websites provide a key mechanism, the *external links* for the videos. The users can easily obtain an embedded link of a video and paste the link to any other websites, such as a forum, or their blogs. Note that such functionality is usually not provided by other commercial websites. We define the *internal links* as those maintaining a relationship within the websites. These links include the user-video, user-user, video-video relationship, etc. We define the *external links* as the links to the videos that are embedded in other websites. In this work, we would like to evaluate the following two key aspects of the external links:

- What is the impact of the external links on the videos, e.g., how many views of the video are contributed by external links;
- How long is the active life cycle of the external links.

Our experiment follows the following steps. To analyze the cumulative impacts of external links, we use the crawler to collect information of 983, 453 videos. In detail, we started our sampling from the most popular videos on March 24th, 2009. We recursively crawled all the related videos from December 1st to 4th, 2008. In order to study how much the external links affect the daily popularity of the videos. We collect the recent videos as well as the videos uploaded before December 31st, 2007. In this stage, we kept tracking 131, 633 videos every day and their daily information changes. The traces are summarized in Table 3.

Table 3. Video Traces Collected for External Measurement

Trace	Objective	Duration	Number of categories	Number of videos
1	Study on category and cumulative impact	Dec 1st, 2008 to Dec 4th, 2008	15	9.8×10^5
2	Study on active life cycle of external links	Oct 22nd, 2008 to Dec 1st, 2008	All categories	1.3×10^5 per day

To collect the external links of YouTube, we use a universal JavaScript engine provided by Google [5]. This engine can track the external link information maintained by YouTube internally. With this engine, we can get the URL of the external links as well as the number of views of each external link.

Unfortunately, YouTube only maintains the external links of each video which are the top-five external links in terms of the number of external views. Thus, we can only use the top five external links in our study. We have yet to find another method that can collect the information of all the external links of the videos. We admit that collecting information only from top five external links affects the study of the total views from external links. Our argument is, our study shows that the investigation of external links meaningful, because only these five external links has already substantially boost the videos' popularity in

Table 4. List of YouTube Video Categories

Rank	Category	Count	Pct.	Pct.(2007)
1	Entertainment	1,304,724	25.4	25.2
2	Music	1,274,825	24.8	22.9
3	Comedy	449,652	8.7	12.1
4	People & Blogs	447,581	8.7	7.5
5	Film & Animation	442,109	8.6	8.3
6	Sports	390,619	7.6	9.5
7	News & Politics	186,753	3.6	4.4
8	Autos & Vehicles	169,883	3.3	2.6
9	Howto & Style	124,885	2.4	2.0
10	Pets & Animals	86,444	1.7	1.9
11	Travel & Events	82,068	1.6	2.2
12	Education	54,133	1.1	–
13	Science & Technology	50,925	1.0	–
14	Unavailable	42,928	0.8	0.9
15	Nonprofits & Activism	16,925	0.3	–
16	Gaming	10,182	0.2	–
17	Removed	9,131	0.2	0.5

their early stage (as verified in Section 6.3). To be more significant, most of our studies in external links are referred to the trend of views from external links, but not the total external views. In this case, top five external links from a huge number of different videos are already representative enough to demonstrate the trend of external views.

4 Characteristics of YouTube Video

Our crawled videos for internal measurement constitute a good portion of the entire YouTube video repository (as there are estimated nearly 120 million videos as of September, 2008). Also, because most of these videos can be accessed from the YouTube homepage in less than 10 clicks, they are generally active and thus representative for measuring characteristics of the repository.

4.1 Video Category

One of 15 categories is selected by the user when uploading the video. Table 4 lists the number and percentage of all the categories, which are also shown graphically in Figure 1. In our entire dataset, we can see that the distribution is highly skewed: the most popular category is "Entertainment", at about 25.4%, and the second is "Music", at about 24.8%. These two categories of videos constitute half of the entire YouTube videos.

In the table, we also list two other categories. "Unavailable" are videos set to private, or videos that have been flagged as inappropriate video, which the crawler can only get information for from the YouTube API. "Removed" are

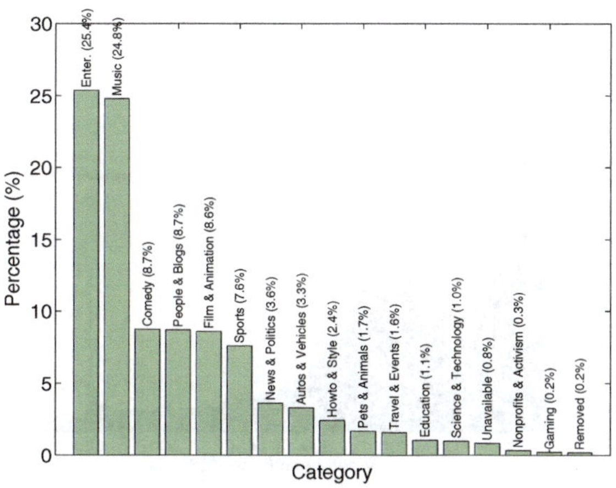

Fig. 1. Distribution of YouTube video categories

videos that have been deleted by the uploader, or by a YouTube moderator (due to the violation of the terms of use), but still are linked to by other videos.

4.2 Video Length

The length of YouTube videos is the most distinguished difference from traditional media content servers. Whereas most traditional servers contain a significant portion of long videos, typically 1-2 hour movies (e.g., HPLabs Media Server [45]), YouTube is mostly comprised of videos that are short clips.

In our entire dataset, 98.0% of the videos' lengths are within 600 seconds, and 99.6% are within 700 seconds. This is mainly due to the limit of 10 minutes imposed by YouTube on regular users uploads. We do find videos longer than this limit though, as the limit was only established in March, 2006, and also the YouTube Director Program allows a small group of authorized users to upload videos longer than 10 minutes [10].

Figure 2 shows the distribution of YouTube videos' lengths within 700 seconds, which exhibits three peaks. The first peak is within one minute, and contains 20.0% of the videos, which shows that YouTube is primarily a site for very short videos. The second peak is between 3 and 4 minutes, and contains about 17.4% of the videos. This peak is mainly caused by the large number of videos in the "Music" category, which is the second most popular category for YouTube, and the typical length of a "Music" video is often within this range, as shown in Figure 3. The third peak is near the maximum of 10 minutes, and is caused by the limit on the length of uploaded videos. This encourages some users to circumvent the length restriction by dividing long videos into several parts, each being near the limit of 10 minutes. We can also observe that there are peaks at around every exact minute.

Figure 3 shows the video length distributions for the top six most popular categories. "Entertainment" videos have a similar distribution as the entire videos',

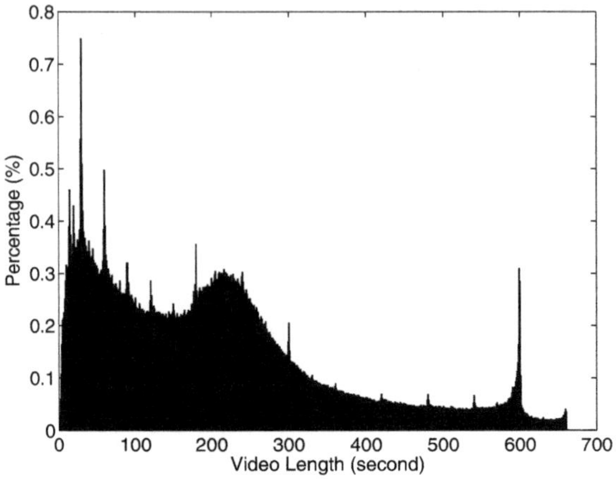

Fig. 2. Distribution of YouTube video length

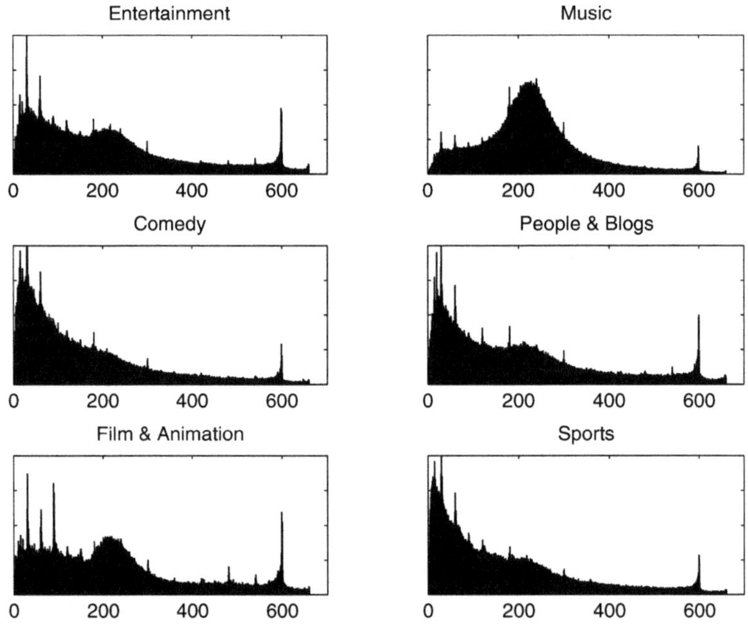

Fig. 3. Distribution of YouTube video length for the six most popular categories

and have the greatest peak at around 10 minutes, probably because a great portion of these videos are talk shows, which are typically a half hour to several hours in length, but have been cut into several parts near 10 minutes. "Music" videos have a very large peak between three and four minutes (29.1%), and "Film & Animation" videos have a similar (though smaller) peak. "Comedy",

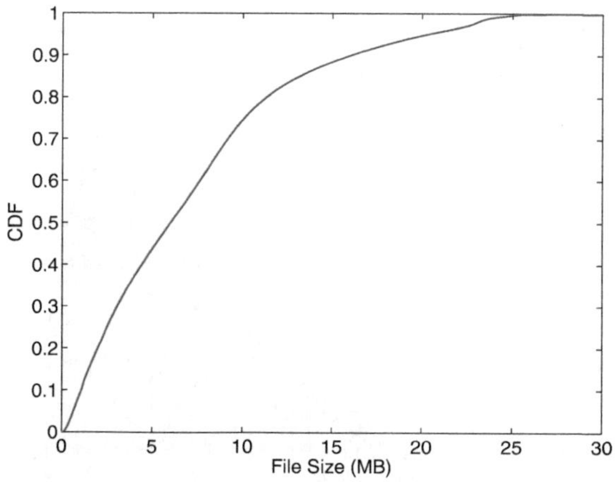

Fig. 4. Distribution of YouTube video file size

"People & Blogs" and "Sports" videos have more videos within two minutes (53.1%, 41.7% and 49.3% respectively), corresponding to "highlight" type clips.

4.3 Video File Size and Video Bitrate

Using video IDs from a normal crawl, we retrieved the file size of more than 130 thousand videos. Not surprisingly, we find that the distribution of video sizes is very similar to the distribution of video lengths. This is because the H.263 video codec used in YouTube videos is constant bitrate (CBR). We plot the cumulative distribution function (CDF) of YouTube video file size in Figure 4. In our crawled data, 99.1% of the videos are less than 25 MB. We calculate an average video

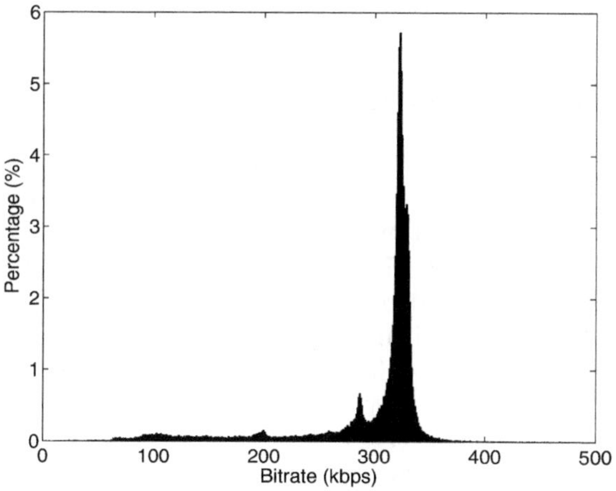

Fig. 5. Distribution of YouTube video bitrate

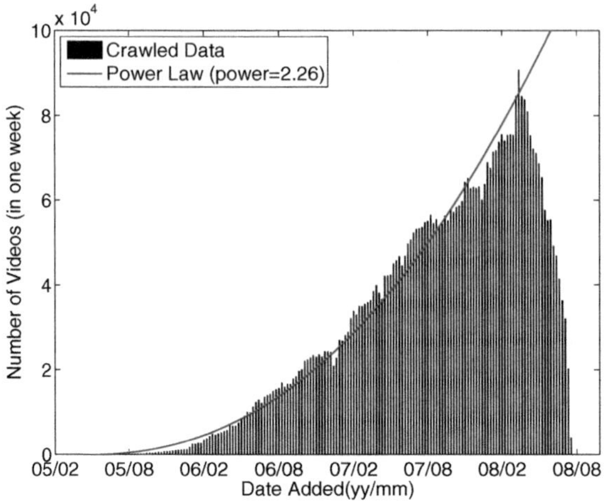

Fig. 6. Number of YouTube videos added among crawled data

file size to be about 7.6 MB. This size is less than the 2007 dataset, which is about 8.4 MB, so there are more and more short videos uploaded. However, considering there are nearly 120 million YouTube videos, the total disk space required to store all the videos is nearly 900 TB! Smart storage management is thus quite demanding for such an ultra-huge and still growing site, which we will discuss in more detail in Section 7.

We found that 99.6% of the videos we crawled contained FLV meta-data specifying the video's bitrate in the beginning of the file. For the rest of the videos that do not contain this meta-data, we calculate an average bitrate from the file size and its length. In Figure 5, the videos' bitrate has a clear peak at around 320 kbps, with two other peaks at around 285 kbps and 200 kbps. This implies that YouTube videos have a moderate bitrate that balances quality and bandwidth.

4.4 Date Added – Uploading Trend

During our crawl we record the date that each video uploaded, so we can study the YouTube's uploading trend. Figure 6 shows the number of new videos added every two weeks in our entire crawled dataset of 2008.

Our first crawl was on March 27th, 2008, thus we can get the early videos only if they are still very popular videos or are linked to by other videos we crawled. YouTube was established on February 15th, 2005, and we can see there is a slow start, as the earliest video we crawled was uploaded 8 days after that day. After 6 months from YouTube's establishment, the number of uploaded videos increases steeply. We use a power law curve to fit this trend.

In the dataset we collected, the number of uploaded videos decreases steeply starting in March, 2008. However, this does not imply that the uploading rate of YouTube videos has suddenly decreased. The reason is that many recently

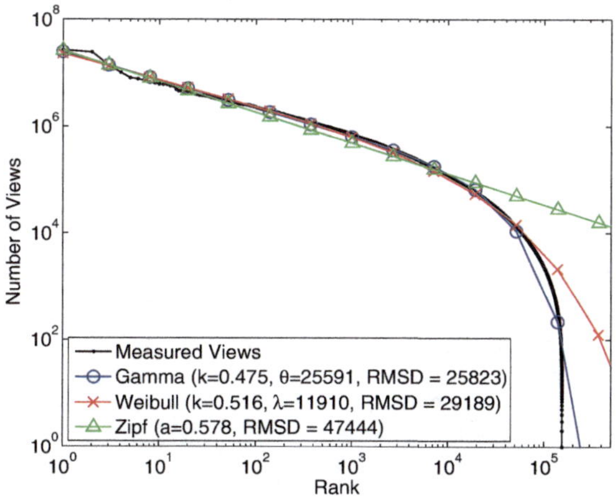

Fig. 7. Number of views against rank (in popularity) of YouTube videos

uploaded videos are probably not in other videos related videos' list. Since few videos link to those new videos, they are not likely to be found by our crawler. We have found that the 2007 data also shows this feature, and the 2008 date indeed confirms that the uploading trend does not decrease.

4.5 Views - User Access Pattern

The number of views a video has had is one of the most important characteristics we measured, as it reflects the popularity and access patterns of the videos. Because a video's view count is changing over time, we cannot use the entire dataset that combines all the data together since the crawl spans 1.5 years. Therefore, we use a single dataset from April 2nd, 2008, containing more than 150 thousand videos, which is considered to be relatively static. It is worth noting that similar results are observed with all other datasets.

Figure 7 shows the number of views as a function of the rank of the video by its popularity. Though the plot has a long tail on the linear scale, it does NOT follow the Zipf distribution, which should be a straight line on a log-log scale. This is consistent with some previous observations [21, 45, 32] that also found that video accesses on a media server does not follow Zipf's law. We can see in the figure, the beginning of the curve is linear on a log-log scale, but the tail (after 10^4 videos) decreases tremendously, indicating there are not so many less popular videos as Zipf's law predicts. We believe that this is because the links among videos enable each of them be browsed by interested views through various paths, as will be illustrated later. Another reason might be a user will access to his/her own video several times after uploading it to check if it is successfully uploaded, thus there are few videos have never been accessed or only once. The result differs from previous measurements on traditional media servers, in which the curve either is skewed from beginning to end or does not

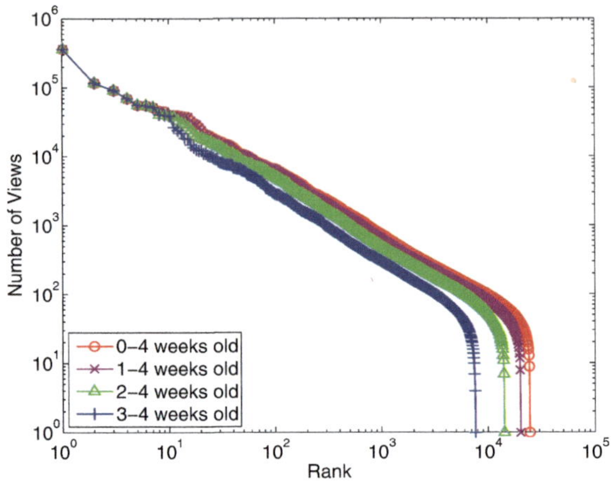

Fig. 8. Recently added YouTube videos views against rank

have the heavy tail. Their results indicate that the popular videos are also not as popular as Zipf's law predicts, which is not the case in YouTube.

To fit the skewed curve, previous studies have used a concatenation of two Zipf distributions [21] or a generalized Zipf distribution [45]. Because our curve is different, we attempted to use three different distributions: Zipf, Weibull, and Gamma. We find that Gamma and Weibull distributions both fit better than Zipf, due to the heavy tail (in log-log scale) that they have, also as the root mean square deviations (RMSD) show.

We were initially concerned that the crawled data might be biased, as popular videos may appear in our BFS more likely than non-popular ones. Since the entire video name space is too large (2^{66}), random sampling directly can be quite difficult. Therefore, we have been saving the recently added videos from the YouTube RSS feed for four weeks, as sampling from these is close to random. We update the views counts of these videos, and plot in Figure 8. The leftmost (blue) plot is only the videos added during the first week (i.e., all the 3-4 weeks old videos), while the rightmost (red) plot contains all the videos (i.e., all the 0-4 weeks old videos). There is a clear heavy tail in all the plots, verifying that our BFS crawl does find non-popular videos just as well as it finds popular ones.

Using the 21 datasets of updated views, we calculate the incremental views of the videos for different span of time, as shown in Figure 9. The downmost (red) plot is the incremental views for one week, and the upmost (blue) plot is the incremental views for 20 weeks. We can see as the time passes, the curve has a more and more heavy tail. This indicates that the links from other videos make the unpopular video more possible to be accessed, than the traditional media server, in which videos are independent with each other. This also demonstrates that Zipf distribution in local popularity will gradually become heavy tail like.

Next, we investigate the correlation between video's length and number of views. We divide the dataset into five groups and calculate the statistics of

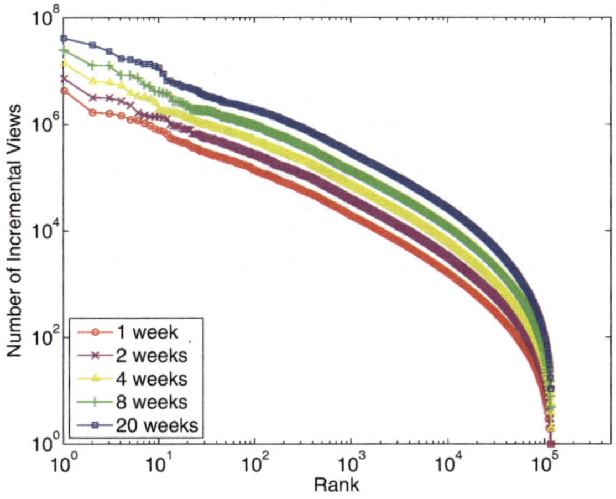

Fig. 9. YouTube videos incremental views against rank

Table 5. Correlation between video length and number of views

Length	Count	Number of views				
		Min	Max	Mean	Median	Std. Dev.
≤ 60*s*	32, 296	0	5, 696, 432	27, 066	3, 675	108, 881
61s–150s	34, 876	0	7, 179, 286	38, 286	4, 562	127, 795
151s–240s	36, 379	0	26, 601, 428	46, 118	5, 973	237, 230
241s–420s	34, 096	0	24, 192, 720	46, 224	7, 101	212, 712
> 420s	18, 180	0	4, 386, 483	19, 095	6, 399	107, 909

views. Table 5 lists the fact of these statistics. We can see that medium length videos are slightly more popular than very short videos and long videos, and the deviations in all of the five groups are very large. We calculate the correlation coefficient of video length and number of views as 0.007. Therefore, we can claim that the correlation is not strong.

Finally, we examine the correlation between video age and number of views, which is shown in Table 6. We also calculate the correlation coefficient of video age and number of views as 0.166. Not surprisingly, the video age affects the number of views, because older videos have more opportunity to be accessed. However, we can see in the younger video groups there are very popular videos, and in the older video group there are very unpopular videos. In fact, the deviations in all the groups are quite large. This indicates that different videos have different *growth trends*, making videos's popularity increasing in various speed.

4.6 Growth Trend and Active Life Span

Comparing the popularity of YouTube videos, we find that some are very popular (their number of views grows very fast), while others are not. Also, after a certain

Table 6. Correlation between video age and number of views

		Number of views				
Age	**Count**	**Min**	**Max**	**Mean**	**Median**	**Std. Dev.**
≤ 1 month	10,910	0	768,422	5,169	449	28,134
1–3 month	20,019	3	7,179,286	8,677	1,198	66,324
3–6 month	26,394	7	3,398,017	14,549	2,829	62,925
6–12 month	49,613	3	6,624,546	31,852	6,514	122,968
> 12 month	46,040	4	26,601,428	74,045	17,829	275,270

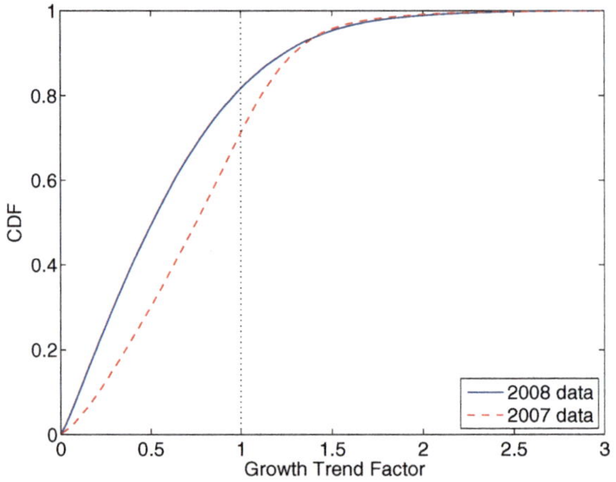

Fig. 10. Distribution of YouTube video growth trend factor

period of time, some videos are almost never accessed. Starting from April 16th, 2008, we updated the number of views of relatively new videos (uploaded between February 15th, 2008 and April 6th, 2008) every week for 21 weeks. To be sure the growth trend will be properly modeled, we eliminate the videos that have been removed and thus do not have the full 21 data points, resulting in 120 thousand videos.

We have found that the growth trend can be modeled better by a power law than a linear fit. Therefore, a video can have a increasing growth (if the power is greater than 1), a constant growth trend (power near 1), or a slowing growth (power less than 1). The trend depends on the exponent factor, which we call the growth trend factor p. We model the number of views after x weeks as

$$v(x) = v_0 \times \frac{(x+\mu)^p}{\mu^p} \tag{1}$$

where μ is the number of weeks before April 16th that the video has been uploaded, x indicates the week of the crawled data (from 0 to 20), and v_0 is the number of views the video had in the first dataset.

We modeled the 120 thousand videos using Equation 1 to get the distribution of growth trend factors p, which is shown in Figure 10. We also plot the growth

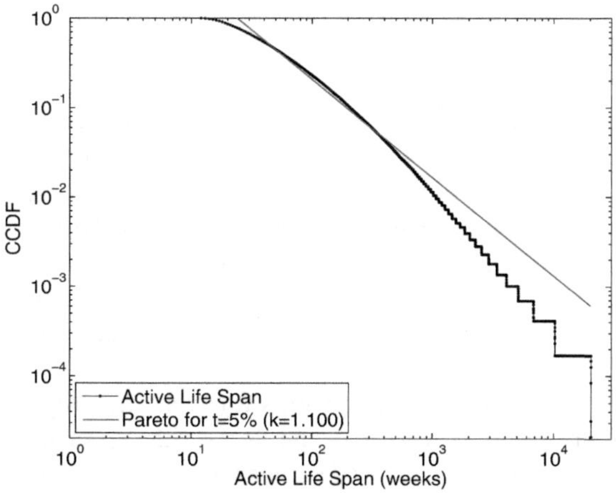

Fig. 11. Distribution of estimated active life span

trend factor of 2007 data. Over 80% of the videos have a growth trend factor that is less than 1, indicating that most videos grow more and more slowly as time passes.

Since YouTube has no policy on removing videos after a period of time or when their popularity declines, the life span of a YouTube video is almost infinite. However, when the video's popularity grows more and more slowly, and will almost stop growing after some time, which we define as the video's *active life span*. From this active life span, we can extract the feature of a video's temporal locality.

If a video's number of views increases by a factor less than t from the previous week, we define the video's active life span to be over. We prefer this relative comparison to an absolute comparison, since we are only concerned with the shape of the curve instead of the scale. For each video that has a growth trend factor p less than 1, we can compute its active life span l from

$$\frac{v(l)}{v(l-1)} - 1 = t \tag{2}$$

which can be solved for the active life span

$$l = \frac{1}{\sqrt[p]{1+t} - 1} + 1 - \mu \tag{3}$$

Thus we see that the active life span depends on the growth trend factor p and the number of weeks the video has been on YouTube.

Figure 11 shows the complementary cumulative distribution function (CCDF) for the active life span of the approximately 95 thousand videos (with p less than 1), for a life span factor of $t = 5\%$. The solid line is the Pareto CCDF $(\frac{x_m}{x})^k$ fit to the data, which can be roughly fitted, and results in a parameter k of 1.100. From looking at multiple fits with various values of t, we find that they all result in the similar parameter k, the only difference is the location (x_m) of the line. Although Pareto distribution cannot perfectly fit the active life span, the figure still shows that a great portion of videos have very short active life spans.

Since the server logs of YouTube are not publicly available, we cannot accurately measure the characteristic of global temporal locality, which would show whether recently accessed videos are likely to be accessed in the near future. However, the active life span gives us another way to estimate the temporal locality of YouTube videos. Figure 11 implies that most videos have a short active life span, which means the videos have been watched frequently in a short span of time, and after the video's active life span is complete, fewer and fewer people will access them.

This characteristic has good implications for web caching and server storage. We can design a predictor to forecast the active life span using our active life span model, which can help a proxy or server to make more intelligent decisions, such as when to drop a video from the cache. We will discuss this in more detail in Section 7.

5 The Social Network in YouTube

YouTube is a prominent social media application: there are communities and groups in YouTube, there are statistics and awards for videos and personal channels, and so videos are no longer independent from each other. It is therefore important to understand the social network characteristics of YouTube. We next examine the social networks among YouTube users and videos, which is a very unique and interesting aspect of this new kind of video sharing sites, as compared to traditional media services.

5.1 User Ratings and Comments, Friends and Upload

We first study the statistics of number of ratings and comments from the same dataset as we did number of views. Whenever a video webpage has been accessed, the number of views increases, whereas users need to log in to rate and comment so that the number of ratings and comments will increase. Therefore, the number of ratings and comments reflect the user behavior. Figure 12 plots the number of ratings against the rank, and similarly for the comments. The two have a similar distribution, and we note that the tails do not drop so quickly as that of the number of views.

We also list the statistics of ratings and comments in Table 7, along with views, and statistics of 2007 data for comparison. We can see that comments are fewer than ratings, and both are much fewer than views. Moreover, many videos do not have a single rating or comment. Not surprisingly, compared with the 2007 data, the number of views, ratings and comments becomes greater, likely because the new dataset we crawled is one year older than the previous one. Interestingly, the number of videos that have no rating or comment decreases, indicating that users are more willing to rate and make comments than before.

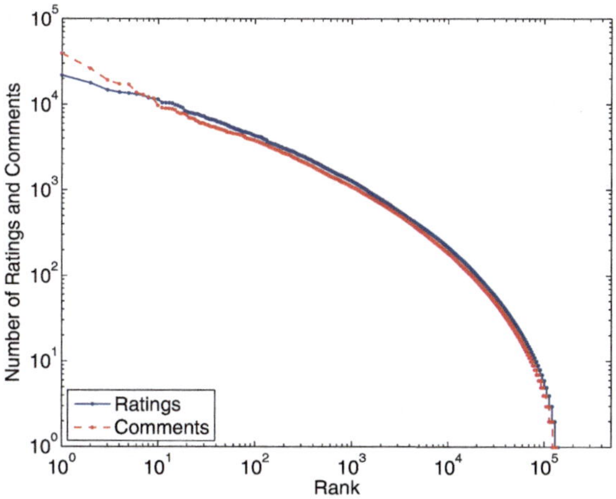

Fig. 12. YouTube videos ratings and comments ranks by the numbers of ratings and comments

Table 7. Facts of views, ratings and comments

	min	max	mean	median	std.	zeros
views	0	26,601,428	37,595	5,357	174,780	0.1%
ratings	0	22,137	68	12	280	8.5%
comments	0	39,742	59	10	283	12.2%
views (2007)	0	2,839,893	4,771	741	24,892	0.1%
ratings (2007)	0	34,401	15	3	128	21.6%
comments (2007)	0	3,899	9	2	39	33.6%

YouTube currently has more than 11 million registered users.[1] Users need to login to upload video or watch some limited videos. A user can add another user to their friend list so that it is convenient to watch their friends' videos. From the crawl of user information, we can extract two statistics of YouTube users: the number of uploaded videos and the number of friends. We did this for about 2.1 million users found by our crawler in all the crawls.

YouTube is a typical user generated content website, in which all the videos are uploaded by the users. These active users are a key in the success of YouTube. In Figure 13, the blue line plots the rank of user uploaded videos' count. The distribution is similar to views, ratings and comments. We have found that many users have uploaded a few videos, and a small number of users have uploaded many videos. Since we collected the user IDs from the previous crawled data, all of these users have uploaded at least one video. However, the information of users that do not upload videos cannot been directly crawled, though we believe that there are a huge number of such users.

[1] A channel search for "*" as a wildcard character in YouTube returns more than 11 million channels, of which a registered user has one.

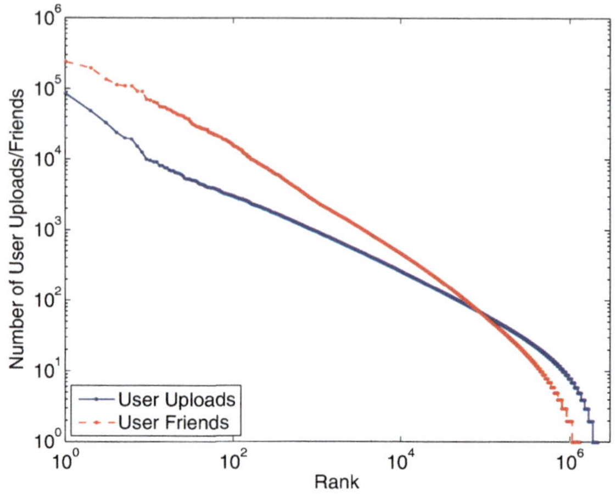

Fig. 13. User uploads and friends ranks by the number of uploads and friends

We also plot the rank of number of friends each user has in Figure 13 as the red line. We also calculate the average and median number of friends each user has to be 18.7 and 1, respectively. Interestingly, in all these users data, we have found that 40.3% of the users have no friends. Therefore, we can see that having friends does not affect the access pattern, so that the correlation between users is not very strong. This suggests that the social network existing among users has less impact than that among videos. We will thus focus on the social networks among videos, which are indirectly formed with user interactions.

5.2 Small-World Phenomenon

Small-world network phenomenon is probably the most interesting characteristic for social networks. It has been found in various real-world situations, e.g., URL links in the Web [20], Gnutella's search overlay topology [34], and Freenet's file distribution network [31].

The concept of a small-world was first introduced by Milgram [37] to refer to the principle that people are linked to all others by short chains of acquaintances (AKA. six degrees of separation). This formulation was used by Watts and Strogatz to describe networks that are neither completely random, nor completely regular, but possess characteristics of both [46]. They introduce a measure of one of these characteristics, the cliquishness of a typical neighborhood, as the *clustering coefficient* of the graph. They define a small-world graph as one in which the clustering coefficient is still large, as in regular graphs, but the measure of the average distance between nodes (the *characteristic path length*) is small, as in random graphs.

Given the network as a graph $G = (V, E)$, the clustering coefficient C_i of a node $i \in V$ is the proportion of all the possible edges between neighbors of the node that actually exist in the graph. The clustering coefficient of the

(a) 500 nodes

(b) 1000 nodes

(c) 4000 nodes

Fig. 14. Three sample graphs of YouTube videos and their links (directions are omitted)

graph $C(G)$ is then the average of the clustering coefficients of all nodes in the graph. The characteristic path length d_i of a node $i \in V$ is the average of the minimum number of hops it takes to reach all other nodes in V from node i.

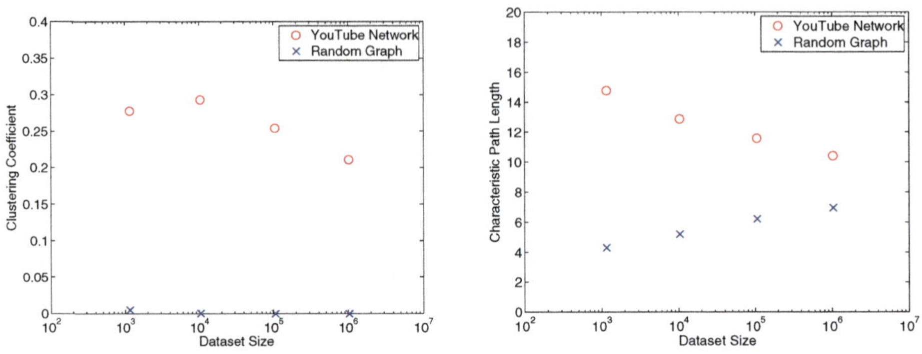

(a) Clustering coefficient (b) Characteristic path length

Fig. 15. Small-world characteristic of YouTube videos

The characteristic path length of the graph $D(G)$ is then the average of the characteristic path lengths of all nodes in the graph.

5.3 The Small-World in YouTube Videos

We measured the graph topology for the YouTube videos, by using the related links in YouTube pages to form directed edges in a video graph. Videos that have no outgoing or no incoming links are removed from the analysis. Some visual illustrations for parts of the network (about 500, 1000 and 4000 nodes) are shown in Figure 14.

From the entire crawled data, we obtain four datasets for measurement, each consisting different order of magnitude number of videos. Since not all of YouTube is crawled, the resulting graphs are not strongly connected, making it difficult to calculate the characteristic path length. Therefore, we use the largest strongly connected component (LCC) of each graph for the measurements. For comparison, we also generate random graphs that are strongly connected. Each of the random graphs has the same number of nodes and average node degree of the LCC of the datasets.

Figure 15a shows the clustering coefficient for the graph, as a function of the size of the dataset. The clustering coefficient is quite high (between 0.2 and 0.3), especially in comparison to the random graphs (nearly 0). There is a slow decreasing trend in the clustering coefficient, showing that there is some inverse dependence on the size of the graph, which is common for some small-world networks [41].

Figure 15b shows the characteristic path length for the graphs. We can see that the average diameter (between 10 and 15) is only slightly larger than the diameter of a random graph (between 4 and 8), which is quite good considering the still large clustering coefficient of these datasets. Moreover, as the size of graph increases, the characteristic path length decreases for YouTube network, but increases for random graph. This phenomena better verifies that YouTube network is a small-world network.

The network formed by YouTube's related videos list has definite small-world characteristics, which can also be verified from the visual illustrations of the graphs. The clustering coefficient is very large compared to the same sized random graph, while the characteristic path lengths are approaching the short path lengths measured in the random graphs. This finding is expected, due to the user-generated nature of the tags, title and description of the videos that is used by YouTube to find related ones.

These results are similar to other real-world user-generated graphs that exist, yet their parameters can be quite different. For example, the graph formed by URL links in the world wide web exhibits a much longer characteristic path length of 18.59 [20]. This could possibly be due to the larger number of nodes (8×10^8 in the web), but it may also indicate that the YouTube network of videos is a much closer group.

6 Analysis of External Links

In this section, we characterize the external link properties on the videos in YouTube. To understand these characteristics, we study the features of the external links from these aspects: the category inclination of the external links, the percentage of the videos with external links among different age groups, the percentage of the external links contribute to the popularity of different age group videos, and the external links impacts on the daily video views in different stages.

6.1 Overview

Based on the information of videos retrieved, we find that the top five external links popularize the videos in YouTube with an average reasonable high level. Still, we also find that the percentage of views from external links will be different according to different categories and different video ages. In general, we summarize the characteristics of YouTube external links as follows:

- The number of views from external links does not have significant relationship with the number of total views in each category. "Sports" (6.7%), "Science & Technology" (6.6%), "News & Politics" (4.5%) and "Gaming" (4.3%) enjoy the largest percentage of views from their top five external links;
- The young videos are more likely to be promoted by external links. Throughout the time, the videos which are less than one month old, have approximately 17% of views from external links. Comparatively, the videos over one years old only have around 4.5% views from external links;
- Most of the external links take substantial part (varying from 10% to 30%) of YouTube videos views only within 4 days they published, this is because the external links are still within their life cycle;

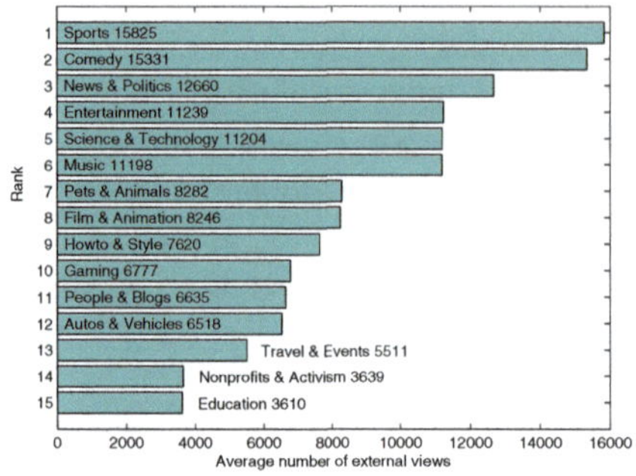

(a) The external views vs. categories

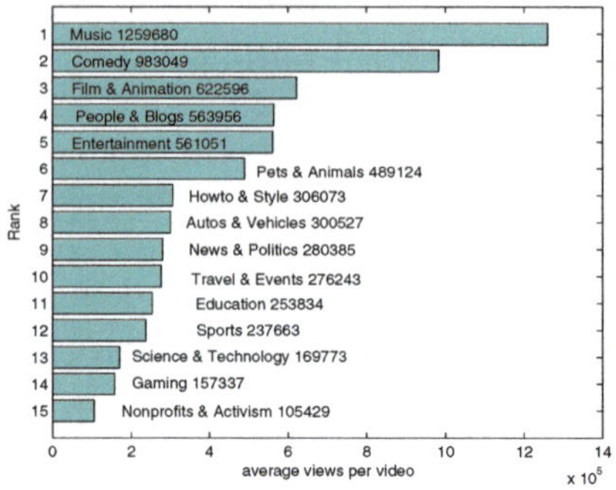

(b) The total views vs. categories

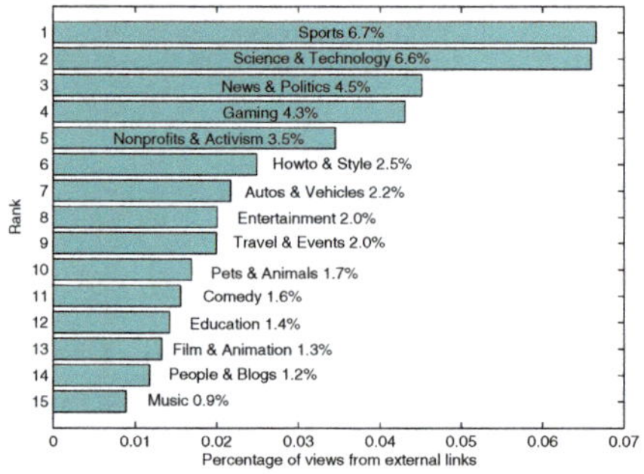

(c) External view percentage vs. category

Fig. 16. External views vs. total views

- The videos which are more than one year old can generally enjoy about 40 views from external links per day.

Overall, one important conclusion can be drawn as videos benefit more from external links in the early stage. Even when the videos are out of date within a certain user group, a new external link can act as a bridge among the websites and keep these videos popular.

6.2 The Category of External Links

The category inclination feature of external links can help us understand which categories of videos are more likely to be linked and how much they are affected by these links. Obviously, different categories of videos are welcomed in varying degrees and this discrepancy also strikingly reflects on the external links. In particular, we focus on the different impacts of external links on the different categories of videos. Therefore, we analyze the data retrieved from Trace 1 shown in Table 3. Figure 16a and Figure 16b show the average views and the total views per video received from the external links, and Figure 16c outlines the percentage of views from external links of all categories.

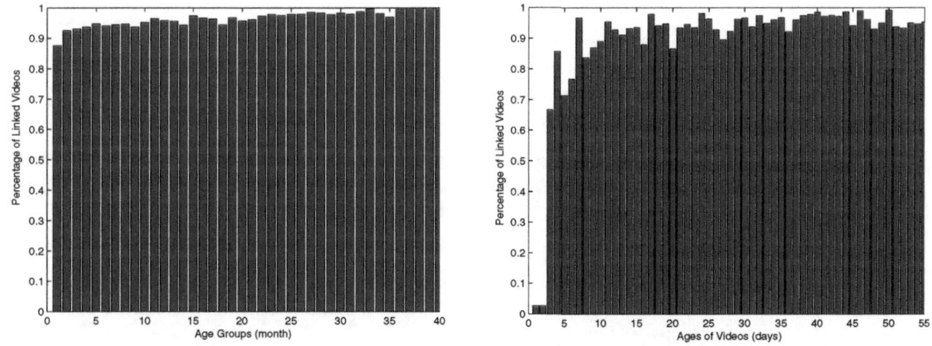

(a) Percentage of the linked video throughout the time (monthly basis)

(b) Percentage of the linked video throughout the time (daily basis)

Fig. 17. The percentage of videos linked vs. video age

In Figure 16a, it is evident that "Sports" (15825), "Entertainment" (15331) and "New & Politics" (12660) videos have highest external view per video. It is not difficult to get the result that "Sports" (6.7%), "Science & Technology" (6.6%), "News & Politics" (4.5%) and "Gaming" (4.3%) videos enjoy the largest percentage of views from external links, as shown in Figure 16c.

However, one of the interesting results we can find is that the views from external links are quite independent from the total views in each category. Although we find "Comedy" videos have very high visit rate from both external links and YouTube website, it is also evident that "Sports" videos, which have the fourth least total views, gain the largest visits from external links. On the other hand, the result is also interesting that the "Music" videos, which have the largest number of total views, has only 0.9% of the views from external links. This dramatic comparison indicates that there is not significant relationship between the external views and the total views.

6.3 The Time the Videos Get Linked

In this section, we are going to study how the videos get their first external links. Two groups with two sampling intervals are set up for this study, the Trace 1 videos (shown in Table 3) are divided into two groups by different sampling interval, on daily basis and on monthly basis. The information of these two groups is depicted in Figure 17.

In detail, Figure 17a shows the percentage of the videos linked in all age groups. Specifically, it is only in the first month that less than 90% videos are linked. The percentage grows slowly but steadily with videos aging. It is also quite meaningful that nearly all the videos aged more than 3 years are linked.

From Figure 17a, it is obvious that most of the videos get their first external links in the first month. To study this detail, Figure 17b provides us the thorough information. From this figure, we can see that there are no more than 10% videos with external links in the first two days. However, the most dramatic rise

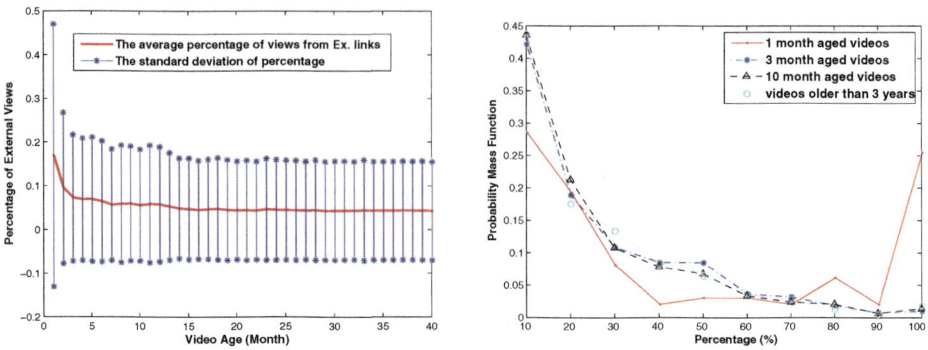

(a) External view percentage over the time

(b) Distribution of each age group

Fig. 18. Cumulative contribution of external links

happens in the third day, and the percentage soars to over 60%, before it remains an average level of 90% after the twelfth day.

In general, this delivers us a message that, an overwhelming majority of the videos are linked and most of the videos get their first links within 12 days. That is, external links can be regarded as a very common phenomenon in YouTube.

6.4 External Views over the Time

In this section, we carry out the study to find out how much the external links affect the popularity of videos throughout the time. To study the time function of external links, we look at the average views from external links as well as their standard deviations through the time. We also study the distribution percentage of the external link contribution in four age groups. The detail information is described in Figure 18.

Figure 18a presents the percentage of views from external links throughout the three years' time on monthly basis. The solid horizontal curve shows the average percentage of the views from external links throughout the time, while the vertical solid line indicates the standard deviation. The videos which are not older than one month can enjoy a high proportion (up to 17%) of views from external links. However, this number drops steeply from the second day to the fifth day, after which the percentage almost remains at 4.5%. On the other hand, the standard deviation of the percentage also follows a similar trend. When the videos are within one month old, the standard deviations are also high up to 29.7%, while the videos older than five months have a relatively stable standard deviations at nearly 14.1%.

Another striking feature is that the absolute value of the standard deviations are larger than the average values. To examine the cause, we investigate the distribution of the external views percentage in detail. Figure 18b presents the distribution of the external views percentage in four age groups. From the figure, the videos of the one-month-aged group are distinctly different from the

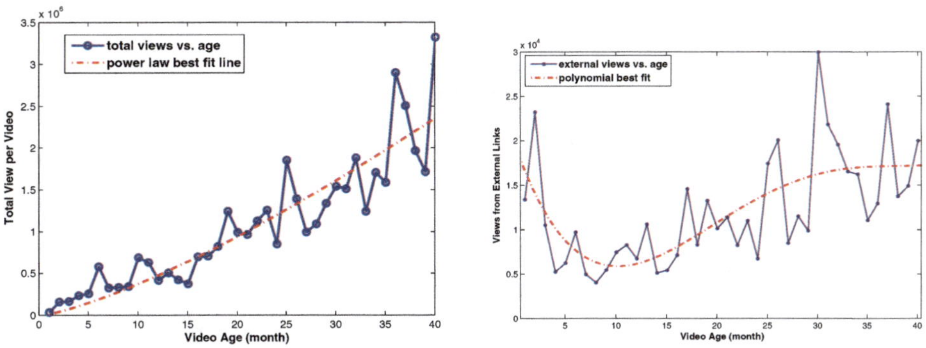

(a) The trend of total views

(b) The trend of external views

Fig. 19. The comparison of total views and external views

other groups. Generally, the percentage of the views from external links drops dramatically. However, one exception is that the videos that are not more than 1 month old also have considerably high proportion of one quarter between 90% and 100%. This result may be accounted by some videos uploaded for being viewed outside YouTube.

To get an insight into how external links affect the videos, we retrieve the information of how views from external links and the total views grow over the time. Figure 19 provides us a complete view of both trend of external views and total views. Figure 19a informs us that the average number of total views fluctuates but grows steadily, while Figure 19b shows that compared with the total views, the average number of views from external links fluctuates but the increase is not so dramatic. Moreover, the red dotted line in Figure 19a, which is the best fit curve for the average total views, shows that the increasing rate of the total views slightly grows during these three year time.

As shown in Figure 18, the trend of external view percentage clearly reduces with a power law distribution over the time. As shown in Figure 20, which is depicted in log-log scale, the starred line follows the trend of the dotted line (the power best fit line) in most part, except that the waist drops more quickly than the power law fit line, and the tail maintains a level above the power law fit line.

Figure 19b offers the explanation of the first exception. The waist below the power law line is caused by the views from external links reducing steeply from the third month to the 15th month. In total 2.1 million external links, we find that although only 9.82% of the external links are the main pages, the others come from secondary folders of the websites, such as news, articles from blogs, and simply posts in the forums. Compared with the powerful promote mechanism inside YouTube, most of the external pages have a very short period of active time, during which pages are popular and have high view rate every day. Therefore, external links are quite different from videos in YouTube, which may be added into the lists of "most viewed videos", "most discussed videos" and "most recent videos", etc, and enjoy a long popular period. Nevertheless, it is

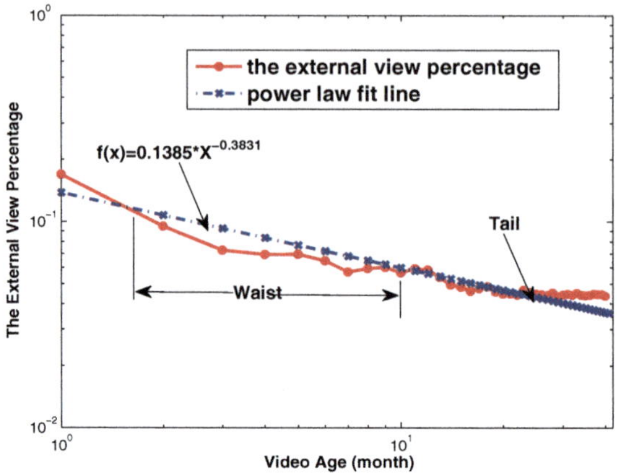

Fig. 20. The power law fit curve for external view percentage

common that people only focus on the recent articles in blogs or forums for a relatively much shorter period. After several days, the posts or the articles will be neglected by people or even be deleted by websites' administrators because they are out of date. If the old external links are no longer active and the new ones do not come, the proportion of views from external links will reduce steeply.

The reason of the tail that is above the power law fit line can be explained by the increasing line after the 15th month in Figure 19b. We believe that this is caused by frequent renew of external links. Anderson demonstrated that there exist huge opportunities in the unlimited number of non-popular items to form the long tail below the power law level [22]. Different from this but still similar in the principle, the tail above the power law line indicates the huge opportunities in the unlimited number of still popular items. In our circumstance, this means that there are still a large number of external links active in the old videos. Note that we can only get top five external links for each video in YouTube. As the time goes by, the old top five external links can be replaced by new and more popular ones. While these new external links may be still in their active time, they keep the percentage of the external view above power law line for the old videos. To more specifically study these two points, we believe that we should still study the daily views from external links.

6.5 Daily Views from External Links

To further understand the impact from external links, we study the daily external views of the "recent videos" and the videos older than one year. We conduct the experiment as following. Every day we collected 2000 to 3000 videos from "recent videos" in YouTube and saved them as a group. As the same time, we also collected $51,124$ videos which are uploaded before December 1st, 2007 at one time. Both the stored "recent videos" and the old videos were tracked every day, so that we obtain the daily increase of the external views and total views.

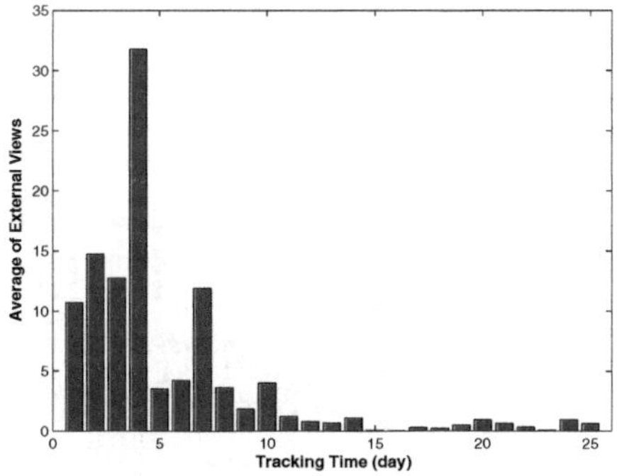

(a) Daily external views

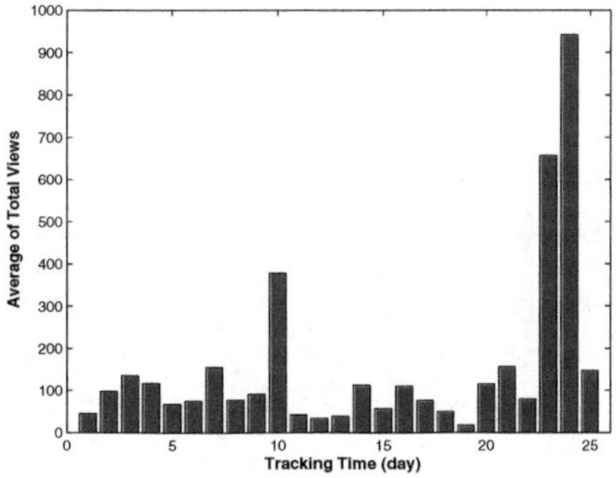

(b) Daily total views

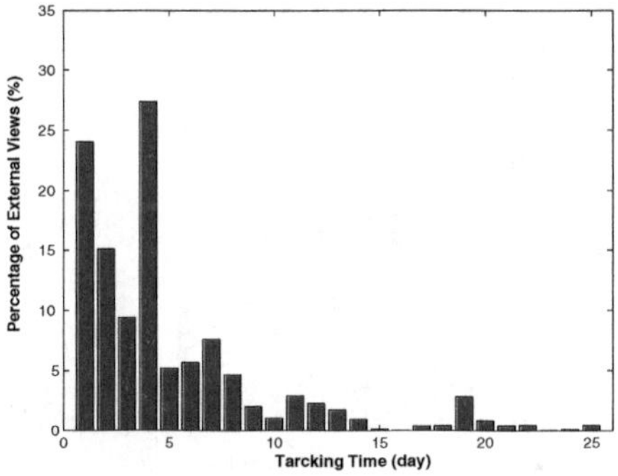

(c) External view percentage over the time

Fig. 21. The daily external links of the most recent videos

However, during this period, some videos were deleted, and also in some days we cannot get all of the videos tracked, thus we delete these days and videos.

Figure 21 provides the views of how external links affect videos' popularity in their early stages. From Figure 21a and Figure 21b, it is clear that the average of daily total views and the views from external links fluctuates. In detail, although the number of total views changes frequently, it generally has stable income every day and we can also see an outburst of the views in the end of the sampling days. However, the number of external views maintains a high level only in the first four days but it dramatically drops in the next seven days. After that, the number stays on a low level near zero. In the first four days, the proportion of the external views is higher than 10% and even reaches to 30%. But the continuous drop in the daily external view number is responsible for the quick reduction in the percentage.

Compared with the new videos, the situation of the old videos is relatively simple. After tracking videos which is published earlier than December 31st 2007 for 41 days, we plot the results in Figure 22. Although both of the total daily view number and the daily external view number rebound but they maintain a smooth average level. We believe that this accounts for the long tail in Figure 20.

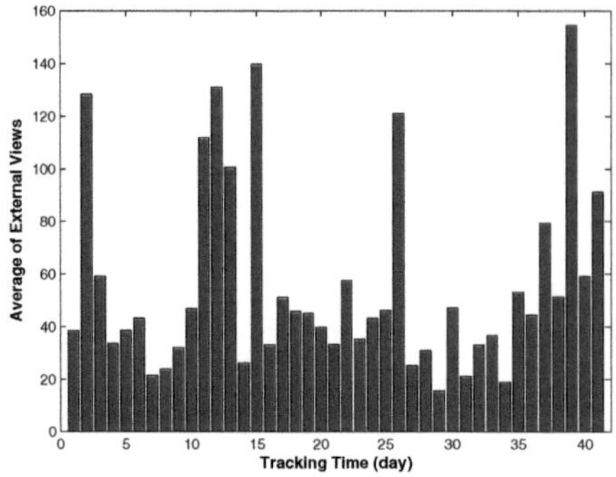

(a) Daily external views

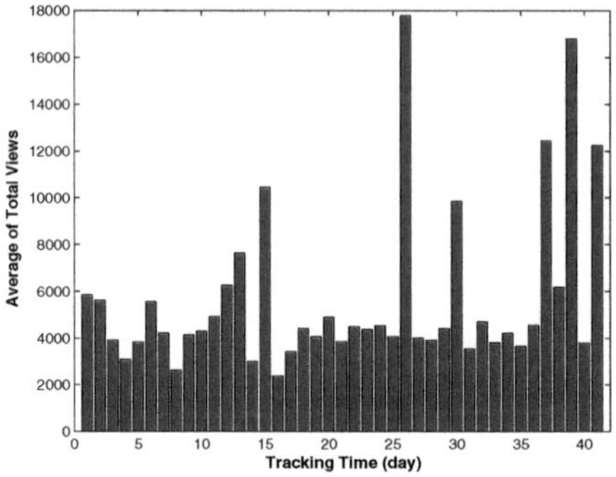

(b) Daily total views

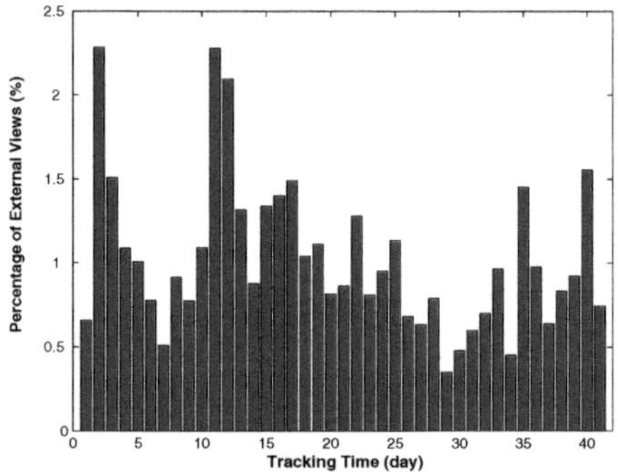

(c) External view percentage over the time

Fig. 22. The daily external links of the old videos

To summarize, it fully demonstrates that videos gains more proportion of views from external links in their early stages. However, the videos which are even more than one years old still have a stable part of views from external links.

7 Further Discussions

An April 2008 report estimated that YouTube consumed as much bandwidth as did the entire Internet in year 2000 [14], and industry insiders estimate that YouTube spends roughly $1 million a day to pay for its server bandwidth [15]. This high and rising expense for network traffic was one of the major reasons YouTube was sold to Google [12].

In a recent study, the authors revealed that the delay of YouTube is much longer than the other measured sites [42]. Scalability is no doubt the biggest challenge that YouTube faces, particularly considering that websites such as YouTube survive by attracting more users. In this section, we briefly discuss the implications of our measurement results toward improving the scalability of YouTube.

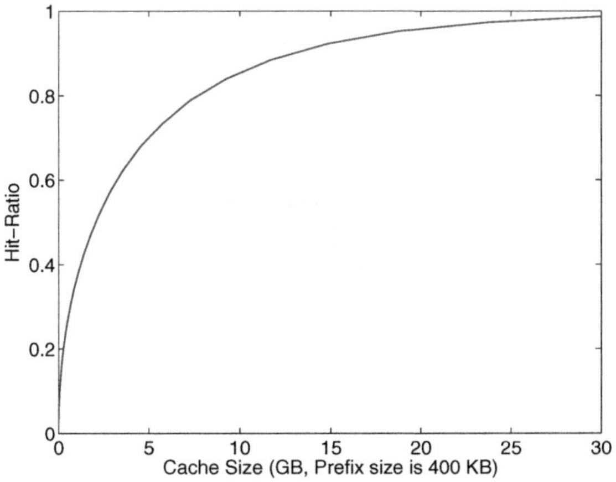

Fig. 23. Prefix caching hit-ratio as a function of cache size

7.1 Implications on Proxy Caching and Storage Management

Caching frequently used data at proxies close to clients is an effective way to save backbone bandwidth and prevent users from experiencing excessive access delays. Numerous algorithms have been developed for caching web objects or streaming videos [36]. While we believe that YouTube will benefit from proxy caching, three distinct features call for novel cache designs. First, the number of YouTube videos (120 million) is orders of magnitude higher than that of traditional video streams (e.g., 2999 for HPC and 412 for HPL [45]). Second, the size of YouTube videos is also much smaller than a traditional video (99.1% being less than 25 MB in YouTube versus a typical MPEG-1 movie of 700 MB). Finally, the view frequencies of YouTube videos do not well fit a Zipf distribution, which has important implications on web caching [23].

Considering these factors, full-object caching for web or segment caching for streaming video are not practical solutions for YouTube. Prefix caching [43] is probably the best choice. Assume for each video, the proxy will cache a 10 second initial prefix, i.e., about 400 KB of the video. Given the Gamma distribution of view frequency suggested by our measurements, we calculate and plot the hit-ratio as a function of the cache size in Figure 23, assuming that the cache space is devoted only to the most popular videos. To achieve a 80% hit-ratio, the proxy would require less than 8 GB of disk space for the current YouTube video repository, which is acceptable for today's proxy servers.

The cache efficiency can be further improved by exploring the small-world characteristic of the related video links. That is, if a group of videos have a tight relation, then a user is likely to watch another video in the group after finishing the first one. This expectation is confirmed by Figure 24, which shows a clear correlation (correlation coefficient being 0.749) between the number of views and the median of the neighbors' number of views. Once a video is played

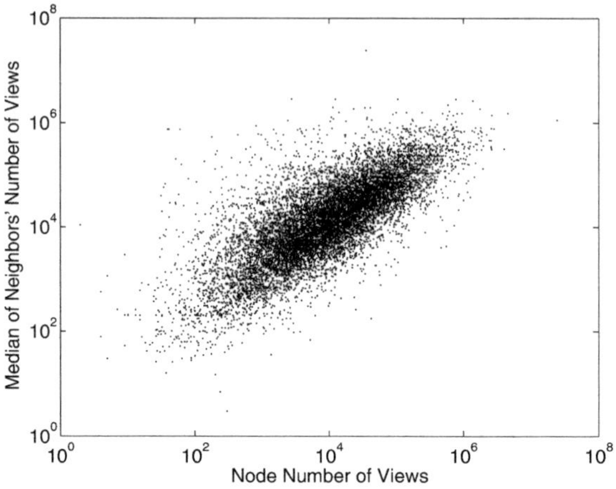

Fig. 24. Correlation of views and median of neighbors' views

and cached, the prefixes of its directly related videos can also be pre-fetched and cached, if the cache space allows.

A remaining critical issue is when to release the space for a cached prefix, given the constant evolution of YouTube's video repository. We have found in Section 4.6 that the active life span of YouTube videos roughly follows a Pareto distribution, implying that most videos are popular during a relatively short span of time. Therefore, a predictor can be developed to forecast the active life span of a video. With the predictor, the proxy can decide which videos have already passed their life span, and replace it if the cache space is insufficient. The life span predictor can also facilitate disk space management on the YouTube server. Currently, videos on a YouTube server will not be removed by the operator unless they violate the terms of service. With around 15 hours of video are uploaded every minute [17], the server storage will soon become a problem. A hierarchical storage structure can be built with videos passing their active life span being moved to slower and cheaper storage media.

7.2 Can Peer-to-Peer Save YouTube?

In the mean time of the booming of YouTube-like sites, peer-to-peer has evolved into a promising communication paradigm for large-scale content sharing. With each peer contributing its bandwidth to serve others, a peer-to-peer overlay scales extremely well with larger user bases. Besides file sharing, it has been quite successful in supporting large-scale live streaming (e.g., CoolStreaming [48] and PPLive [4]) and on-demand video streaming (e.g., GridCast [2]) [35], thus naturally being believed as an accelerator of great potentials for YouTube-like video sharing services.

Unfortunately, using peer-to-peer delivery for short video clips can be quite challenging, evidently from our measurement on YouTube. In particular, the length of a YouTube video is short (many are even shorter than the typical startup time in a peer-to-peer overlay), making the long startup delay unacceptable. And a user often quickly loads another video when finishing the previous one, so the overlay will suffer from an extremely high churn rate. Moreover, there are a huge number of videos with highly skewed popularity, thus many of the peer-to-peer overlays will be too small to function well. They together make existing per-file based overlay design suboptimal, if not entirely inapplicable. The study on MSN Video has suggested a peer-assisted VoD (PA-VoD) [32]. However, we notice that the statistics for MSN Video are quite different from YouTube, particularly in terms of lengths, and consequently PA-VoD may not perform well. We believe that, to guarantee acceptable QoS, a hybrid implementation with complementary servers, proxy caches, and client peers will be an viable solution, and the social networks can again be explored in this context.

We have successfully implemented a social network assisted peer-to-peer system customized for short video sharing [26]. In our system, all the peers that have previously downloaded a video can serve as potential suppliers, forming an overlay for this video. The neighbors in this overlay also serve as the starting point for a peer to locate suppliers for its next video to watch. The social network plays two important roles in this context: first, it enables efficient searching of suppliers having the video that a peer to watch next; second, it improves the success ratio for a peer to pre-fetch its video to watch next. Both of them leverage the strong clustering among the videos, i.e., the next video is likely to be available in the neighbors of a peer or their neighbors. We have performed both simulations and PlanetLab [3] prototype experiments to verify the effectiveness of our proposed design. The detailed designs and the performance evaluation results can be found in the paper [26].

8 Conclusion

This chapter has presented a detailed investigation of the characteristics of YouTube, the most popular Internet short video sharing site to date. Through examining millions of long-term YouTube video data, we have demonstrated that, while sharing certain similar features with traditional video repositories, YouTube exhibits many unique characteristics, especially in length distribution, access pattern and growth trend. These characteristics introduce novel challenges and opportunities for optimizing the performance of short video sharing services.

We have also investigated the social network of YouTube videos, which is probably the most unique and interesting aspect, and has substantially contributed to the success of this new generation of service. We have found that the networks of related videos, which are chosen based on user-generated content, have both small-world characteristics of a large clustering coefficient indicating the grouping of videos, and a short characteristic path length linking any two videos. We have suggested that these features can be exploited to facilitate the design of novel caching and peer-to-peer strategies for short video sharing.

Other than the internal network of YouTube, we have also presented a measurement study of the impacts on the YouTube videos from external links. With the information gathered, our study shows interesting characteristics of external links of YouTube. In particular, we find that views from external links are independent from total views in each category. Also, videos benefit more from external links in the early stage. The objective of this study is to unravel the impacts on online social networks from exterior environment.

References

1. Alexa, http://www.alexa.com
2. GridCast, http://www.gridcast.cn
3. PlanetLab, http://www.planet-lab.org
4. PPLive, http://www.pplive.com
5. V8 JavaScript Engine, http://code.google.com/p/v8
6. Yahoo! Video, http://video.yahoo.com
7. Youku, http://www.youku.com
8. YouTube, http://www.youtube.com
9. YouTube API, http://youtube.com/dev_docs
10. YouTube Blog, http://youtube.com/blog
11. YouTube (Wikipedia), http://en.wikipedia.org/wiki/YouTube#Video_quality
12. Google to buy YouTube for $1.65 billion. CNN (2006), http://money.cnn.com/2006/10/09/technology/googleyoutube_deal/index.htm
13. YouTube video-sharing site is changing popular culture. KCRW (2006), http://www.kcrw.com/news/programs/ww/ww061122youtube_video-sharin
14. Web could collapse as video demand soars. Telegraph (2008), http://www.telegraph.co.uk/news/uknews/1584230/Web-could-collapse-as-video-demand-soars.html
15. YouTube looks for the money clip. CNN (2008), http://techland.blogs.fortune.cnn.com/2008/03/25/youtube-looks-for-the-money-clip
16. YouTube Surpasses 100 Million U.S. Viewers for the First Time. comScore (2009), http://www.comscore.com/press/release.asp?press=2741
17. YouTube's Chad Hurley: We Have The Largest Library of HD Video On The Internet. TechCrunch (2009), http://www.techcrunch.com/2009/01/30/youtubes-chad-hurley-we-have-the-largest-library-of-hd-video-on-the-internet
18. Acharya, S., Smith, B.: An Experiment to Characterize Videos Stored on the Web. In: Proceedings of the ACM/SPIE Multimedia Computing and Networking, MMCN (1998)
19. Acharya, S., Smith, B., Parnes, P.: Characterizing User Access To Videos On The World Wide Web. In: Proceedings of ACM/SPIE Multimedia Computing and Networking, MMCN (2000)
20. Albert, R., Jeong, H., Barabási, A.: The Diameter of the World Wide Web. Nature 401, 130–131 (1999)
21. Almeida, J., Krueger, J., Eager, D., Vernon, M.: Analysis of Educational Media Server Workloads. In: Proceedings of ACM International Workshop on Network and Operating Systems Support for Digital Audio and Video, NOSSDAV (2001)

22. Anderson, C.: A Problem with the Long Tail (2006), http://www.longtail.com/scifoo.ppt
23. Breslau, L., Cao, P., Fan, L., Phillips, G., Shenker, S.: Web Caching and Zipf-like Distributions: Evidence and Implications. In: Proceedings of IEEE INFOCOM (1999)
24. Cha, M., Kwak, H., Rodriguez, P., Ahn, Y., Moon, S.: I Tube, You Tube, Everybody Tubes: Analyzing the World's Largest User Generated Content Video System. In: Proceedings of ACM SIGCOMM conference on Internet measurement, IMC (2007)
25. Cheng, X., Dale, C., Liu, J.: Statistics and Social Network of YouTube Videos. In: Proceedings of IEEE International Workshop on Quality of Service, IWQoS (2008)
26. Cheng, X., Liu, J.: NetTube: Exploring Social Networks for Peer-to-Peer Short Video Sharing. In: Proceedings of IEEE INFOCOM (2009)
27. Cherkasova, L., Gupta, M.: Analysis of Enterprise Media Server Workloads: Access Patterns, Locality, Content Evolution, and Rate of Change. IEEE/ACM Transactions on Networking 12(5), 781–794 (2004)
28. Chesire, M., Wolman, A., Voelker, G., Levy, H.: Measurement and Analysis of a Streaming Media Workload. In: Proceedings of Conference on USENIX Symposium on Internet Technologies and Systems, USITS (2001)
29. Gill, P., Arlitt, M., Li, Z., Mahanti, A.: YouTube Traffic Characterization: A View From the Edge. In: Proceedings of ACM SIGCOMM conference on Internet measurement, IMC (2007)
30. Halvey, M., Keane, M.: Exploring Social Dynamics in Online Media Sharing. In: Proceedings of ACM International Conference on World Wide Web, WWW (2007)
31. Hong, T.: Performance. In: Oram, A. (ed.) Peer-to-Peer: Harnessing the Power of Disruptive Technologies. O'Reilly & Associates, Inc., Sebastopol (2001)
32. Huang, C., Li, J., Ross, K.: Can Internet Video-on-Demand be Profitable? In: Proceedings of ACM SIGCOMM (2007)
33. Lai, K., Wang, D.: A Measurement Study of External Links of YouTube. In: Proceedings of IEEE Global Communications Conference, GLOBECOM (2009)
34. Liu, G., Hu, M., Fang, B., Zhang, H.: Measurement and Modeling of Large-Scale Peer-to-Peer Storage System. In: Proceedings of Grid and Cooperative Computing Workshops, GCC (2004)
35. Liu, J., Rao, S., Li, B., Zhang, H.: Opportunities and Challenges of Peer-to-Peer Internet Video Broadcast. Proceedings of the IEEE 96(1), 11–24 (2008)
36. Liu, J., Xu, J.: Proxy Caching for Media Streaming over the Internet. IEEE Communications Magazine 42(8), 88–94 (2004)
37. Milgram, S.: The Small World Problem. Psychology Today 2(1), 60–67 (1967)
38. Mislove, A., Marcon, M., Gummadi, K., Dreschel, P., Bhattacharjee, B.: Measurement and Analysis of Online Social Networks. In: Proceedings of ACM SIGCOMM conference on Internet measurement, IMC (2007)
39. O'Reilly, T.: What Is Web 2.0: Design Patterns and Business Models for the Next Generation of Software (2005), http://www.oreilly.com/pub/a/oreilly/tim/news/2005/09/30/what-is-web-20.html
40. Paolillo, J.: Structure and Network in the YouTube Core. In: Proceedings of Hawaii International Conference on System Sciences, HICSS (2008)
41. Ravasz, E., Barabási, A.: Hierarchical Organization in Complex Networks. Physical Review E 67(2), 26112 (2003)
42. Saxena, M., Sharan, U., Fahmy, S.: Analyzing Video Services in Web 2.0: A Global Perspective. In: Proceedings of International Workshop on Network and Operating Systems Support for Digital Audio and Video, NOSSDAV (2008)

43. Sen, S., Rexford, J., Towsley, D.: Proxy Prefix Caching for Multimedia Streams In: Proceedings of IEEE INFOCOM (1999)
44. Singla, P., Richardson, M.: Yes, There Is a Correlation - From Social Networks to Personal Behavior on the Web. In: Proceedings of ACM International Conference on World Wide Web, WWW (2008)
45. Tang, W., Fu, Y., Cherkasova, L., Vahdat, A.: Long-term Streaming Media Server Workload Analysis and Modeling Technical report, HP Labs (2003)
46. Watts, D., Strogatz, S.: Collective Dynamics of "Small-World". Networks Nature 393(6684), 440–442 (1998)
47. Yu, H., Zheng, D., Zhao, B., Zheng, W.: Understanding User Behavior in Large-Scale Video-on-Demand Systems. ACM Operating Systems Review (SIGOPS) 40(4), 333–344 (2006)
48. Zhang, X., Liu, J., Li, B., Yum, T.: CoolStreaming/DONet: A Data-Driven Overlay Network for Peer-to-Peer Live Media Streaming. In: Proceedings of IEEE INFOCOM (2005)
49. Zink, M., Suh, K., Gu, Y., Kurose, J.: Watch Global, Cache Local: YouTube Network Traffic at a Campus Network - Measurements and Implications. In: Proceeding of SPIE/ACM Multimedia Computing and Networking, MMCN (2008)

Terrestrial Television Broadcasting in China: Technologies and Applications

Wenjun Zhang, Yunfeng Guan, Xiaokang Yang, and Weiqiang Liang

Shanghai Jiao Tong University

Abstract. This chapter will discuss the two main terrestrial television broadcasting systems widely used in China. One is the China Terrestrial Television Broadcasting (CTTB) standard which is named as "Frame structure, channel coding and modulation for digital television terrestrial broadcasting system". It was issued in August 2006 as a mandatory standard for traditional terrestrial broadcasting and had been put into execution from August 2007. The other is China Multimedia Mobile Broadcasting (CMMB) which is a mobile television and multimedia standard developed and specified by the State Administration of Radio, Film and Television (SARFT). Based on the statellite and terrestrial interactive multiservice infrastructure (STiMi), it is a mixed satellite and terrestrial wireless broadcasting system designed to provide audio, video and data service for handheld receivers with less than 7 inch wide LCD display.

The first part focuses on CTTB standard. The comparisons with ATSC and DVB-T standard and the considerations on primary working modes selection are discussed. The receiver design for both single carrier and multi-carrier options is also addressed. Then the applications of CTTB, including fixed, mobile and portable reception is presented. Especially, the signal coverage scheme for railway lines is introduced.

As for the CMMB standard, this chapter firstly introduces the system infrastructure. Then its key techniques, combined with the comparison with another popular mobile TV standard DVB-H are presented. At last, the development of CMMB services in China is briefly depicted.

1 CTTB Standard

Years of development in digital television technologies leads to three main digital TV terrestrial transmission standards around the world. The first is the 8-level Vestigial Side-Band (8-VSB) modulation system, developed by ATSC (Advanced Television System Committee) in the USA [1]. The second one is Digital Video Broadcasting-Terrestrial, which uses traditional Coded Orthogonal Frequency Division Multiplexing (COFDM) technology and was widely used in Europe, Australia, Singapore and other areas [2]. The third, the Band Segmented Transmission (BST)-OFDM system is the basis for Japan's Terrestrial Integrated Service Digital Broadcasting standard (ISDB-T) [3,4].

China has the world's largest consumer base and there are more than 400 million TV sets nationwide, which is around 30% of the 1.4 billion TV sets worldwide [5]. Particularly, nearly 65% of television audiences use the terrestrial method [5]. At the same time, China has its specific attributes and needs for DTTB. Therefore, it is very

C.W. Chen et al. (Eds.): Intel. Multimedia Communication: Tech. and Appli., SCI 280, pp. 403–423.
springerlink.com

important for China to develop a terrestrial broadcasting standard to provide high quality multimedia service.

In this chapter, we first introduce the China Terrestrial Television Broadcasting standard from a technical point of view, including development history block diagram of transmitters, receiver design considerations, and also the comparisons with ATSC and DVB-T systems. Then, we will discuss the applications including the traditional fixed reception and high speed mobile reception.

1.1 Technical Review of CTTB

1.1.1 History and Roadmap

China started to develop its own digital television terrestrial broadcast standard in 1994. In 1995, the central government founded the HDTV Technical Executive Experts Group (TEEG) whose members come from various domestic universities and research institutes to develop the HDTV prototype. After three years' effort, the first HDTV prototype--including high-definition encoder, multiplexer, modulator, demodulator, de-multiplexer and high-definition decoder--was developed. All the key parts, from transmitter to receiver, were successfully applied to the live broadcast of the 50th National Day Parade in 1999 [5].

In 2001, China called for proposals for the Digital Terrestrial Television Broadcasting (DTTB) standard. Five proposals were submitted including Advanced Digital Television Broadcasting-Terrestrial (ADTB-T), the single-carrier approach from HDTV TEEG and Digital Multimedia Broadcasting-Terrestrial (DMB-T), the multi-carrier approach from Tsinghua University. In 2003, another multi-carrier approach named as Terrestrial Interactive Multiservice Infrastructure (TiMi) was proposed by Academy of Broadcasting Science. After lab tests, field trials and intellectual property analyses, a united working group led by Chinese Academy of Engineering was founded in 2004 to merge the three proposals together [5].

Finally, in August 2006, the merged standard, called "Frame structure, channel coding and modulation for digital television terrestrial broadcasting system" was issued and has been mandatory since August 1st, 2007 [6]. It contains both single carrier and multi-carrier options and supports various multi-program SDTV/HDTV terrestrial broadcasting services. The announcement of this standard is indeed a milestone of Chinese DTV industry.

1.1.2 Brief Description of CTTB

DTTB system is used to convert the input data stream to the output RF signal. The following baseband processing will be applied to input data stream [6]:

- Scrambler
- Forward Error Correction (FEC)
- Constellation mapping
- Interleaving
- Multiplex of Basic data block and system information
- Combine the Frame body and Frame header to the signal frame

After these baseband processing, input data stream will be up-converted to RF signal within 8MHz bandwidth in UHF or VHF band.

The block diagram of CTTB transmit system is shown in Fig.1. The functional descriptions of main blocks are listed as follows.

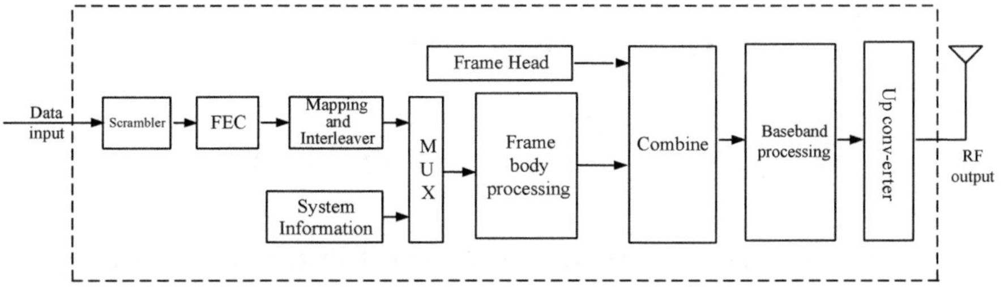

Fig. 1. Block diagram of CTTB transmission system

1. Scrambler

In order to ensure adequate binary transitions, the multiplexed input data shall be randomized in accordance with the configurations depicted in Fig.2.

The polynomial for the Pseudo Random Binary Sequence (PRBS) generator shall be:

$$G(x) = 1 + x^{14} + x^{15} \tag{1}$$

The initial state of this LFSR is: "100101010000000".

The first bit at the output of the PRBS generator shall be applied to the first bit (i.e. MSB) of the first input stream byte by XOR operation to randomize the data. Scrambler will be reset to its initial state at the beginning of each signal frame [6].

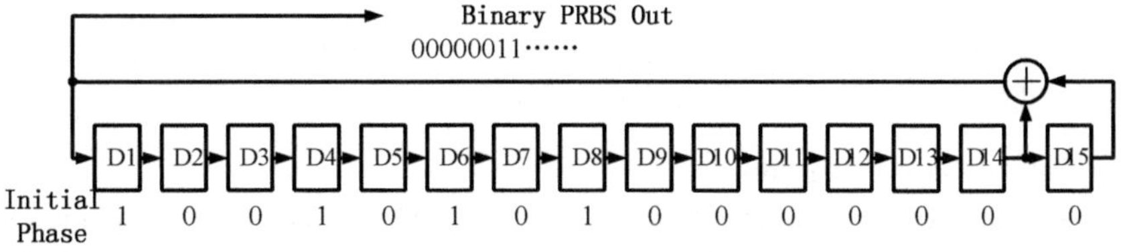

Fig. 2. Scrambler schematic diagram

2. FEC

FEC will be done after the input data stream is scrambled. It is concatenation of outer BCH (762, 752) and inner Low density parity check (LDPC) codes [6].

BCH(762,752) code is derived from BCH(1023,1013) by adding 261 zeros in front of each of the 752 information bits before BCH encoding and these bits will then be removed from the output of the BCH encoder. BCH (762, 752) can only correct one bit error. Without an interleaver between the BCH code and the LDPC code, BCH cannot contribute much for error correction but be ensured that a frame always includes integral TS packets [7].

The generation matrix of LDPC is given by [6]

$$\mathbf{G}_{qc} = \begin{bmatrix} \mathbf{G}_{0,0} & \mathbf{G}_{0,1} & \cdots & \mathbf{G}_{0,c-1} & \mathbf{I} & \mathbf{O} & \cdots & \mathbf{O} \\ \mathbf{G}_{1,0} & \mathbf{G}_{1,1} & \cdots & \mathbf{G}_{1,c-1} & \mathbf{O} & \mathbf{I} & \cdots & \mathbf{O} \\ \vdots & \vdots & \mathbf{G}_{i,j} & \vdots & \vdots & \vdots & \ddots & \vdots \\ \mathbf{G}_{k-1,0} & \mathbf{G}_{k-1,1} & \cdots & \mathbf{G}_{k-1,c-1} & \mathbf{O} & \mathbf{O} & \cdots & \mathbf{I} \end{bmatrix} \quad (2)$$

Where I is b×b identity matrix with b≡127, O is b×b zero matrix, and $G_{i,j}$ is b×b circulant matrix with 0≤i≤k-1 and 0≤j≤c-1. Here k and c are parameters used to define the LDPC code.

There are three different FEC coding rates with parameters and performances (TOV: BER<3e-6) shown in Table 1. [7,8]

Table 1. Parameters for FEC

Code Rate	Block length (bits)	Information bits	*K*	*c*	E_b/N_0 (dB)	Performance away from Shannon Limit(dB)
0.4	7488	3008	24	35	2.1	2.3
0.6	7488	4512	36	23	2.3	1.6
0.8	7488	6016	48	11	3.3	1.2

The encoded block length from Eq. (1.2) is 7493 and the first 5 check bits are then punctured to reduce the block length to 7488.

3. Constellation and mapping

The output of FEC will be evenly converted to nQAM (n: constellation points) symbol stream with the first FEC coding input bit of each symbol being LSB. This standard supports the following constellations: 64QAM、32QAM、16QAM、4QAM、4QAM-NR (Nordstrom Robinson). Power normalization is considered when mapping the symbol, which helps the average power of different mappings roughly remain at the same level.

The so-called 4QAM-NR mapping actually means that NR coding is added before 4QAM mapping. Interleaving described in next section is done bit-wise on the FEC output, then the NR coding called Quasi-orthogonal pre-mapping is applied [9].

4. Time Interleaving

Time domain Interleaving is applied across many signal frames. Convolutional interleaver is used for the time interleaving, shown in Fig.3 where B is the number of interleaving branches and M is the interleaving depth. The branch 0 synchronizes to the first symbol of basic data block. The total delay of interleaving and de-interleaving is B * (B-1) * M symbols.

There are two modes for the time interleaving:

a) Mode1: B=52, M=240 symbols. The total delay of interleaving and de-interleaving is 170 signal frame;

b) Mode2: B=52, M=720 symbols. The total delay of interleaving and de-interleaving is 510 signal frame.

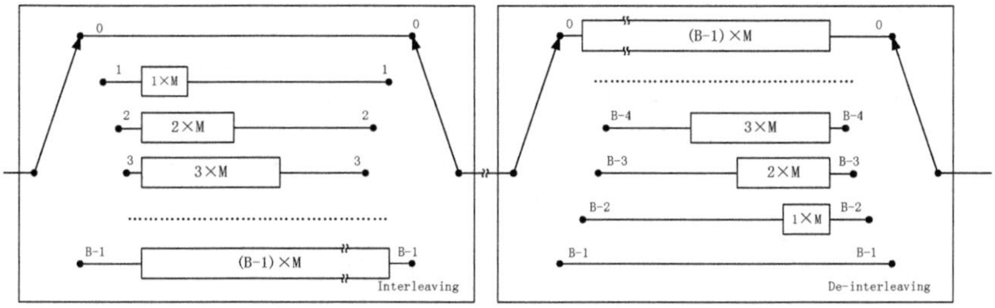

Fig. 3. The structure of convolutional interleaver

5. Frame structure

The physical channel frame structure is hierarchical with 4 layers, which is shown in Fig.4. The bottom layer, which is called signal frame, is the basic element of system

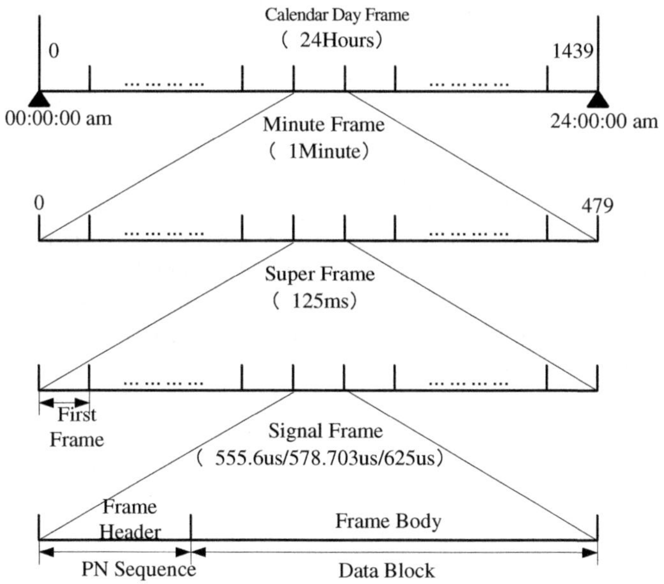

Fig. 4. The frame structure of CTTB

frame structure. The super frame and minute frame are defined as a group of signal frames and super frames, respectively. The top layer is called Calendar Day Frame (CDF). The physical channel is periodical and synchronized to the absolute time.

Each signal frame consists of two parts, namely, Frame Header (FH) and Frame Body (FB), while FB consists of system information and coded data. Both FH and FB have the same baseband symbol rate that is 7.56 Msps. There are totally three kinds of signal frame structure, with three length options on the FH shown as Fig.5a,b,c. While the frame body and super frame always have fixed time periods, one super frame will have 225, 216, and 200 signal frames corresponding to the frame header length options shown in Fig.5a to 5c, respectively [7].

FH(420 symbols) (55.6 μs)	FB(3780 symbols) (500 μs)

a) frame structure 1

FH(595 symbols) (78.7 μs)	FB(3780 symbols) (500 μs)

b) frame structure 2

FH(945 symbols) (125 μs)	FB(3780 symbols) (500 μs)

c) frame structure 3

Fig. 5. Signal frame structure

The FHs are filled with three different lengths of Pseudo-Noise (PN) sequences which are abbreviated to PN420, PN595 and PN945 respectively. PN420 and PN945 are made up of a complete M-sequence of length 255 and length 511, respectively and with their respective cyclical extensions as pre-ambles and post-ambles, while PN595 is made up of the first 595 symbols of an M-sequence of length 1023 and it has no cyclical characteristic. The initial phase of the generation polynomial is fixed for PN595, but for PN420 and PN945, it can be optionally rotated in each super-frame. The power of FH is also different between PN595 and the other two: PN595 has the same average power with the FB while PN420 and PN945 have an average power that is twice as much as that of the FB [7].

From the above description, it is shown that FHs are of great importance for receiver designs. FH can be not only the training sequence for equalizers but also the guard interval for the FB. It should be pointed out in particular that regardless of the length, cyclical characteristics, power, FHs are always modulated in time domain (single carrier) with BPSK mapping. This fundamental characteristic of the CTTB standard makes it possible that all the receiver signal processing blocks including AGC, frame synchronization, carrier recovery, symbol clock recovery, channel estimation, channel equalization are all based on time domain signal processing technologies so that the single carrier modes (C1) and multi-carrier modes (C3780) can share the same architecture for modulators and demodulators. As a result, while in achieving the great flexibilities to satisfy various requirements for single carrier and multi-carrier modes, the hardware complexity increase is minimized [7].

6. Frame body (FB) processing
As for the FB processing block, there are two options on the parameter C which denotes the number of carriers. One is C1 mode which indicating single carrier modulation scheme, the other is C3780 mode indicating multi-carrier modulation scheme with the sub-carrier number of 3780. In both options, the input data is the symbol after LDPC coding and constellation mapping [7].

Please be noted that in C3780 modes, a block interleaver is further applied within each FB.

7. System information
Each FB contains 36 system information symbols that are BPSK mapped. The first 4 symbols are the indicator of C1 or C3780 modes, and the last 32 symbols are spread spectrum protected Walsh codes to indicate necessary system information including constellation mapping, LDPC code rate and interleaving mode etc [7].

8. SRRC
In the post base band processing procedure, a square root raised cosine (SRRC) filter with a roll-off factor of 5% is defined as shaping filter to limit the bandwidth of the transmitted signal to 8 MHz. Additionally, in C1 mode, dual-pilots are optionally inserted at ±0.5 symbol rate onto the transmitted data whose power is -16dB lower than the total average power. The base band frequency spectrum is shown in Fig.6.

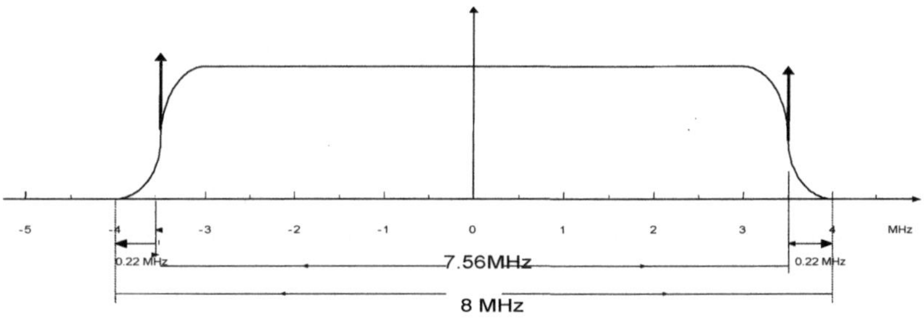

Fig. 6. The baseband frequency spectrum of CTTB with dual pilots

1.1.3 Comparisons of CTTB with Other Existing DTTB Standards

Compared with ATSC, DVB-T and ISDB-T, the most significant feature of CTTB is that it contains both single carrier and multi-carrier options. Next, we will briefly compare the single carrier and multi-carrier option of CTTB with ATSC and DVB-T systems respectively.

Compared with ATSC, which adopts 8-VSB single carrier modulation, the single carrier option of CTTB has the following primary differences.

1. The frame structure of ATSC is base on segment, field and frame. Each segment consists of 4 synchronization symbols, which are usually used for timing recovery, and 828 data symbols [1]. Each field consists of one field synchronization segment and 312 data segments. The field synchronization segment is composed of a PN sequence with 511-symbol length and three PN sequences with 63-symbol length, which can be used as training sequence of equalizer. The time interval of each field synchronization segment is about 24ms [1]. While in CTTB, the signal is structured in frames with a long PN sequence acting as frame header. Just take the PN595 mode as an example, which is commonly used in single carrier mode. The time interval of two consecutive frame headers is about 0.58ms, which is much shorter than that of ATSC. The purpose of much more frequently insertion of PN sequence is to make it easier for tracking the time variety of channel. However, its penalty is reducing the bandwidth efficiency.
2. In ATSC system, the Reed-Solomon is adopted as outer code and trellis code is the inner code [1]. In the receive side, the decision feedback equalizer commonly uses the output of trellis decoder as the slicer output to reduce the phenomenon of error propagation [10]. In CTTB, LDPC is adopted as inner code, which can provide the superior error correction capability for a better sensitivity. However, to use the output of LDPC decoder as the slicer output is too hard to implement, because we need to re-mapping and interleave the LDPC decoder out, of which the overall delay and resource are not acceptable. Even in the 4QAM-NR mode, the NR code can be combined with the equalizer to track fast channel changes in a mobile reception environment [11], but the algorithm used in NR decoding which is a block code is still quite different with that of traditional trellis decoding.
3. The ATSC system uses 8-VSB as its constellation scheme. The useful information is all concentrated on the in-phase component and the quadrature component is the Hilbert transformation of the in-phase component. In CTTB, QAM constellation is adopted, in which both the in-phase and quadrature components carry useful information. The difference of constellation will lead to the difference of implementation structure of equalizer.

As for the comparison between the multi-carrier option of CTTB and DVB-T, besides the frame structure and FEC mentioned above, the main difference lies in the insertion scheme of known information in aid of channel estimation. DVB-T adds cyclic prefix before an OFDM symbol to mitigate the inter-symbol interference (ISI). Continual and scattered pilots are added in each OFDM symbol, aiding the synchronization and channel estimation [4]. In multi-carrier option of CTTB, by choosing the known PN sequence as the FH to mitigate ISI, it brings the benefits of fast channel acquisition since this can be done directly in time domain as well as the high spectrum efficiency as PN can also be used for the channel estimation, avoiding both continual and scattered pilot insertion into the FB by COFDM approach [8].

1.1.4 Considerations on Mode Selection

From the above introduction, the reader may find that the CTTB standard contains many combinations such as single carrier modulation (C1) and multi-carrier modulation

(C3780), 3 frame header (FH) options, 3 FEC coding rates, 5 constellation mappings, 2 interleaver depths, fixed or rotated PN in frame header and so on, which results in hundreds of operation modes [8]. The demodulator chip required to support all the working modes is not only very costly but also less reliable; hence, it is better to reduce the number of working modes to a lower level that still supports all the important applications.

For the single carrier option, there are certainly many other modes that are available. However, in a practical way, some modes are more preferable than others, for example, for the same data rate of 10.396 Mbps, 4QAM+0.8 is preferred over 16QAM+0.4 since a 2 level slicer for 4QAM is better than a 4 level slicer for 16QAM in a data directed time domain equalizer. The same is true for the same data rate of 15.593 Mbps that 16QAM+0.6 is preferred over 64QAM+0.4 since a 4 level slicer for 16QAM is better than an 8 level slicer for 64QAM in a data directed time domain equalizer operation [7]. In short, for an all time domain processing approach, a lower constellation with lighter coding is always more preferable than a higher constellation with heaver coding. Additionally, in an all time domain processing approach, the multipath range is not limited by the length of FH, so PN420 is more preferable than PN595, and both PN420 and PN595 are more preferable than PN945. Between PN420 and PN595, since PN595 has no cyclical characteristic, while PN420 has cyclical extensions, PN595 may provide faster convergence process for equalizer in an all time domain processing approach. Further, FH of PN420 and PN945 has twice more power than that of the FB, so even all in C1 mode, the peak to average power ratio of FH of PN420 and PN945 is higher than that of FH of PN595 [7].

For the multi-carrier option, the cyclic characteristic of PN420 and PN945 can transform the linear convolution between the transmitted data and channel impulse response into cyclic convolution, which makes it easier to estimate the channel. Also, the twice average power compared with that of the FB provides higher signal to noise ratio when performing synchronization and channel estimation. Further, in COFDM system, a higher constellation with heavier coding is more suitable because in frequency selective fading channels, some sub-carriers may suffer from lower SNR, but the SNR in other sub-carriers will be even higher. The data in these fading sub-carriers will be recovered by the powerful coding while the higher constellation in the others will provide higher payload data rate.

Based on above considerations as well as extensive and comprehensive test both in the lab and on the field, which covered stationary and mobile, indoor and outdoor reception, as well as at low and high data rates, China's State Administration of Radio, Film and Television (SARFT) has made the final decision that the seven working modes listed in Table 2 are the primary working modes for different applications.

In the seven modes listed above, the payload data rate varies from 9.626 Mbps to 25.989 Mbps which provides the operator with great flexibility. The high data rate modes are used for fixed reception, transmitting 10~12 SDTV programs or 1~2 HDTV programs in one 8 MHz channel; while the other modes are used for mobile reception, transmitting 2~5 SDTV programs in one 8 MHz channel.

Table 2. Primary working modes in CTTB

Mode	Carrier	FEC	Constellation	FH	Pilots	Interleaver	Payload Data Rate (Mbps)
1	C=3780	0.4	16QAM	PN945	OFF	720	9.626
2	C=1	0.8	4QAM	PN595	ON	720	10.396
3	C=3780	0.6	16QAM	PN945	OFF	720	14.438
4	C=1	0.8	16QAM	PN595	ON	720	20.791
5	C=3780	0.8	16QAM	PN420	OFF	720	21.658
6	C=3780	0.6	64QAM	PN420	OFF	720	24.365
7	C=1	0.8	32QAM	PN595	ON	720	25.989

The seven primary working modes listed in Table 2 can be divided into two categories by carrier mode. Compared with multi-carrier modes, the single carrier modes have two unique advantages. One is that it has a lower Peak to Average Power Ratio (PAPR) which means that the transmitter's power amplifier requires a smaller linear range to support a given average power and enables the use of a cheaper power amplifier. This feature is very beneficial especially for the rural area since the power amplifier may be one of the most expansive components in the whole DTV broadcasting system. The dual pilots added in single carrier modes are also very helpful to carrier and timing recovery especially in severe multi-path distortions [7].

1.1.5 Receiver Design

Although the receiver signal processing technology is not defined as part of the standard, the time domain modulated FH and the numbers of the carries (C1, C3780) are the most significant options in the standard which have direct impacts on receiver performance and complexity. Thus, two kinds of channel estimation and compensation methods are developed. One is an all time domain processing approach (TD), which employs a time domain, code enhanced, data directed adaptive decision feedback equalizer with LMS algorithms [12]; the other is a hybrid time and frequency domain processing approach (HTFD), which implements a channel estimator in time domain with the known FH, and compensates the FB in frequency domain [13]. Next, we will present a block diagram and a brief introduction to each approach respectively.

1. Time domain approach

The block diagram for single carrier mode receiver is shown in Fig.7.

In Fig.7, the RF signal is down converted into IF signal by the tuner and then the ADC converts it to a digital signal, which becomes the input of the demodulator. The demodulator can be divided into two parts, one is signal processing called the inner receiver and the other is information processing called the outer receiver. The inner

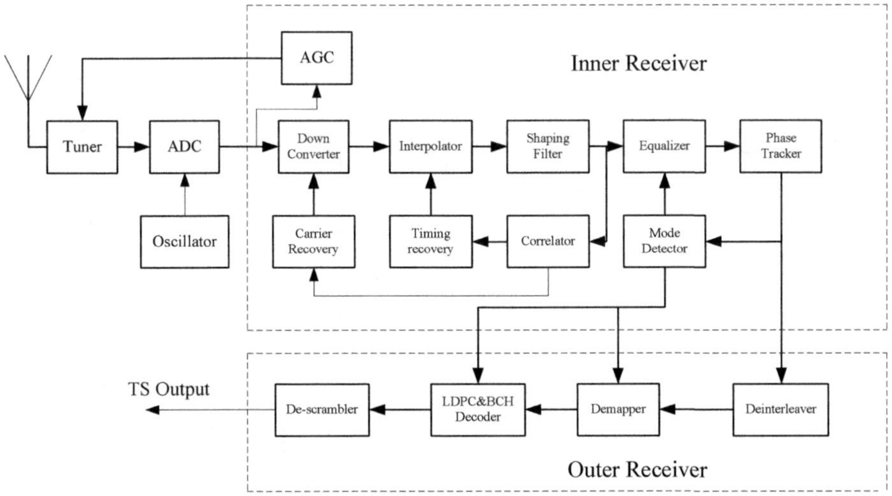

Fig. 7. Basic diagram of demodulator for single carrier mode in CTTB

receiver first adjusts the gain of the amplifier in the tuner automatically by inspecting the amplitude of the input data, then performs carrier and timing recovery by Numerically Controlled Oscillator (NCO) and interpolator respectively and obtains the synchronized baseband symbols by passing through a shaping filter. The baseband data can be used to extract the residual carrier and timing offset by correlating with the known frame header [14].

In addition to synchronization, another significant function of the inner receiver is channel equalization, which usually adjusts its coefficients adaptively by Least Mean Square (LMS) criteria to mitigate ISI [12]. Especially in 4QAM-NR mode, the NR decoder can be combined with the equalizer to provide more accurate input to DFE which will greatly improve equalizer performance in severely distorted multi-path channels and make it possible to support high speed mobile reception [11]. To eliminate the residual carrier frequency offset and phase noise, a phase tracker is often concatenated with the equalizer [15].

There is also a mode detector in the inner receiver, which correlate the equalized symbols with known Walsh sequence and then find the most suited mode information. The mode information will be fed back to the equalizer to make sure that it goes into the correct data directed mode. In the mean time, the deinterleaver, demapper and decoder in the outer receiver will use the mode information to perform corresponding processing and send out the transport stream.

2. Hybrid time and frequency domain processing approach

The basic block diagram of the demodulator for multi-carrier mode in CTTB can be found in Fig.8. Thanks to the PN420 and PN945 inserted in time domain, the synchronized part, including carrier recovery, timing recovery, down convertor, interpolator, shaping filtering and so on shares the same architecture and resources with those

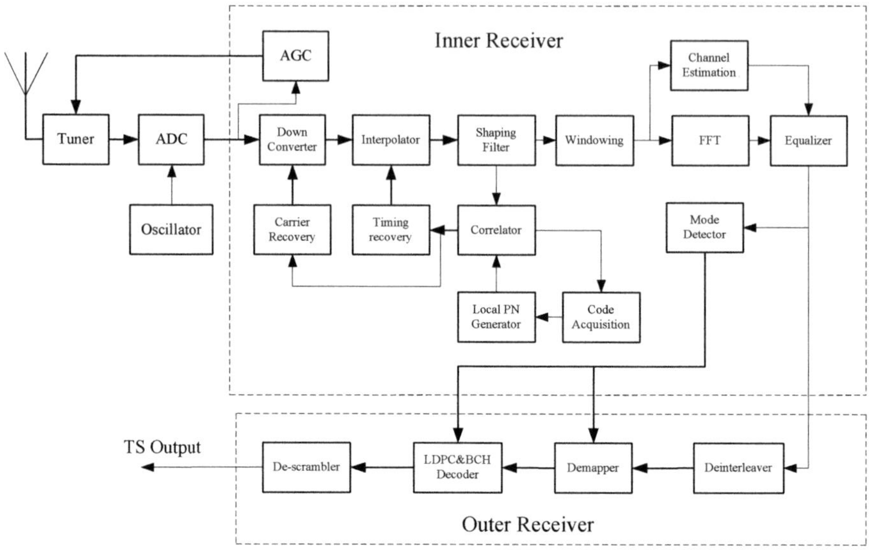

Fig. 8. Basic diagram of demodulator for multi-carrier mode in CTTB

in the single carrier mode with the exception of a PN phase acquisition procedure due to the initial phase variation within a super frame in PN420 and PN945 mode [16].

As for channel equalization, the common method for ODFM system is to perform channel estimation according to the pilots in frequency domain, then convert the receiving signal to frequency domain, and perform frequency domain equalization [17]. However, for CTTB, this common method does not work, for there is no pilot signal in frequency domain at all. The reasonable method is to estimate the channel impulse response using the FH and then do linear convolution with known FH to attain the interference of FH to FB after passing through the channel. Before frequency domain equalization, interference of FH to FB from the received signal needs to be eliminated, and the cyclic convolution of FB with the channel needs to be reconstructed [13].

In addition, a mode detector with the equalized data as its input is also needed to get the proper mode information for the outer receiver, which can be approximately the same with that of singe carrier mode.

1.2 Applications of CTTB

The launch of Digital Television Terrestrial Broadcasting in China has enabled traditional terrestrial services, such as large scale fixed reception (HDTV / multiple SDTV programs) as well as many new services including mobile, portable and high speed mobile applications.

1.2.1 Traditional Terrestrial Broadcasting

Nowadays in China, the main application for terrestrial television is fixed reception, which has an enormous market potential. There are 349 million families in China, of which 67%, or 235 million, watch TV by terrestrial reception [5]. These viewers are scattered in suburban areas and the countryside. Due to the distance these viewers are from the city, it is too expensive for cable TV network to reach them. Most of these viewers can receive only 5 analog TV programs, which are mainly public information. These audiences are eager for more content. With high bandwidth and compression efficiency, China's DTTB standard can transmit 20-26Mbps per 8MHz TV channel. This will greatly improve the effective coverage in the vast central and western regions of China, as well as the rural area of most Chinese cities. The result will be the delivery of 40-50 stable and clear digital TV programs compared to today's unstable and noisy 2-5 analog TV programs [5]. Thanks to the standard's strong receive capabilities in both indoor and outdoor environments, strong resistance to co-channel and adjacent channel interference and large scale Single Frequency Network (SFN) coverage, terrestrial users can now use outdoor or indoor antennas and receivers with digital terrestrial tuners to watch the digital terrestrial TV programs.

China's government has also improved the TV viewing experience by sponsoring and launching a project called "village to village TV coverage." As an example, Chongming Island, a county of Shanghai, participates in one of the digital terrestrial reference projects. Previously, the residences in Chongming could only receive 5 unstable analog TV programs. Now, the digital terrestrial network adopting single-carrier modulation delivers 18 SDTV programs in two 8 MHz channels, dramatically improving TV coverage for these viewers [5]. The low-cost system upgrade and reliable signal coverage provides a reference terrestrial broadcasting solution. More and more cities across China are establishing the digital terrestrial systems now.

China is also encouraging its larger cities to begin delivering free HDTV terrestrial broadcast content, which will be an important trigger for the digital terrestrial market. This will help drive the entire HDTV industry including high definition flat panel displays, chipsets, transmitters, software, content production and so on. China Central Televison (CCTV), the most biggest TV station in China, first launched the terrestrial broadcast of high definition program in January 1st, 2008, using the recommended single carrier mode 4 in Table 2. Since 2008 Beijing Olympics games, besides Beijing, the other big cities including Shanghai, Shenzhen and Guangzhou also have started HDTV live broadcasting.

1.2.2 Mobile Reception

Perhaps the biggest difference between analog and digital terrestrial television is mobile reception which allows TV and other multimedia content to be viewed in buses, taxis and private automobiles. The 20 million vehicles in China comprise a non-neglect part of mobile TV market [5].

Considering the tall and tightly-located building in China's big cities, the shadows from high-rises and terrain shielding can cause receiving difficulties. Using efficient equalization technology, CTTB supports stable mobile reception under either single tower or single frequency network environment. As a typical mobile digital terrestrial network, from the mid of 2004, a 5-tower digital terrestrial SFN was put into

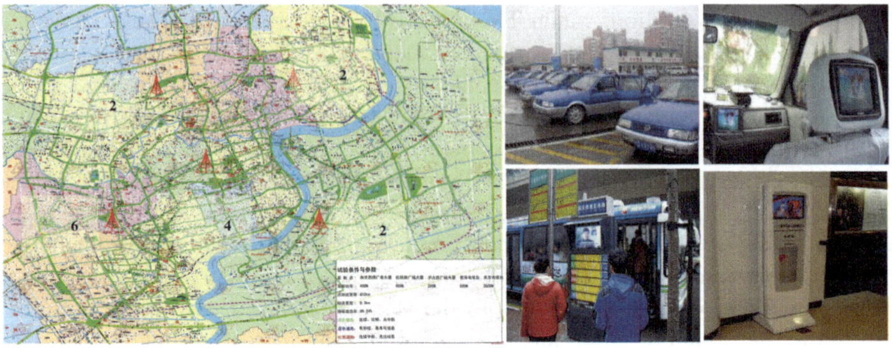

Fig. 9. Mobile Digital Terrestrial Multimedia Services in Shanghai

commercial operation in Shanghai, which provided mobile TV cabs, electronic bus stations and multimedia donation boxes in downtown area [5]. Fig.9 shows the network structure of the SFN and some typical applications.

A potential application is high speed mobile reception in train. Railway transportation is the most popular vehicle in China. It was reported that there is an up to 2.2 billion population which travels on the 90,000 kilometers railway lines in China every year [18]. In the past, the railroad was a blank area for TV broadcasting because of the restriction of technology in analog TV age and real-time TV program broadcasting could not reach the running train. A recent survey showed that the most grateful service on the high speed trains which represents the wish of 57% people was the real-time TV program broadcasting. Therefore, the TV signal coverage of railway lines has become an extremely urgent problem.

The signal coverage scheme for railway lines is quite different from traditional terrestrial broadcasting coverage. Generally, administrative territorial coverage is the common way for TV signal coverage and the average vehicular speed is below 100km/h. On the other hand, the railway lines are obviously linear distribution. The broadcasting signal has to cover a long narrow area with different topography. Moreover, the future speed promotion up to over 300km/h of train speed is also a great challenge for the transmission of terrestrial broadcasting system. On account of the facts that the railway system is a close system, the programs are only broadcasted along the railway lines, and the audiences of railway TV are all on the trains which run on the railway lines [18].

To solve the signal coverage issue for railway lines, a unique SFN system names as CTTB-R (Railway) has been developed. The CTTB-R system is a countrywide SFN network which consists of satellite distribution network, terrestrial coverage network, train transmission network and GPRS surveillant signal feedback network [18]. The framework structure of this system is shown in Fig.10.

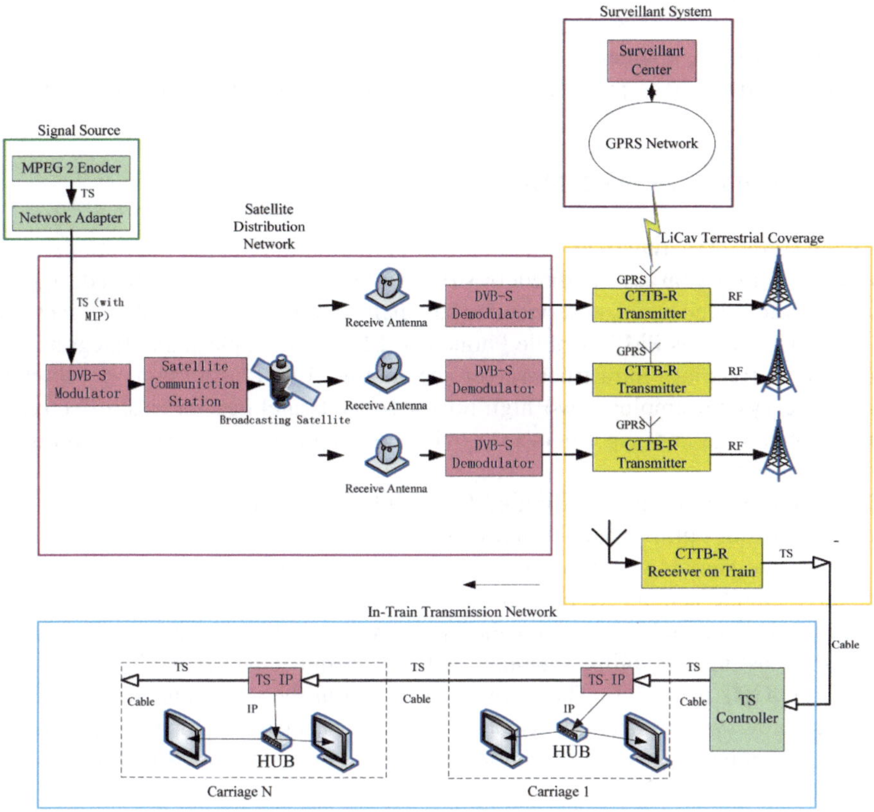

Fig. 10. Framework structure of the SFN distribution network

2 CMMB Standard

Mobile TV services have shown an increasing mass media market and the competition is becoming more and more fierce all over the world. At present, there are already several commercial mobile TV standards in the world, which include Terrestrial Digital Media Broadcasting (T-DMB) in Korea [19], MediaFLO developed by Qualcomm [20], and Digital Video Broadcast for Handhelds (DVB-H) proposed by the DVB organization in Europe [21].

In China, due to the successful network construction and commercial operation, CMMB [22] standard issued by China's broadcast industry regulator SARFT plays a leading role. A CMMB workgroup was also organized to promote this standard led by SARFT. Currently, CMMB workgroup has more than 100 members including broadcasting operators, mobile communication operators, handset manufactories, chip vendors and so on. During the 2008 Beijing Olympic Games, mobile TV services based on CMMB have been available in Beijing and other Olympic cities. Moreover, network construction and signal coverage with five to nine programs had been deployed in 37 cities before August 2008 [24], and this figure will keep increasing in the next few years. Furthermore, amount of trial test results and feedback from equipment manufactories and users have provided precious experiences for the future deployment of CMMB.

2.1 Technical Review of CMMB

2.1.1 System Infrastructure

CMMB system is a mixed satellite/terrestrial wireless broadcasting system designed to provide audio, video and data service for handheld receivers with less than 7 inch wide LCD display, such as PMP, Mobile Phone and PDA [23]. It plans an "Integrated Satellite and Terrestrial" infrastructure for the mobile TV transmission. As shown in Fig.11, the system employs two high-power S-band satellites for cost-effective nationwide coverage and gap fillers to provide complementary terrestrial coverage for dense urban areas.

The information is first transmitted from Uplink Transmission System to satellite via broadcasting uplink channel and distributing uplink channel. Then the satellite broadcasts the signals via broadcasting downlink channel and distributing downlink channel. Complementary terrestrial network receives the signal via distributing downlink channel and retransmits the signal via broadcasting retransmit channel. The broadcasting downlink channel and retransmit channel both work at the 2.635-2.660GHz band [23]. The signal transmitted in these two channels keeps time and frequency synchronization, which ensures the successful signal reception in the SFN. Furthermore, this system supports interactive services by cooperating with the

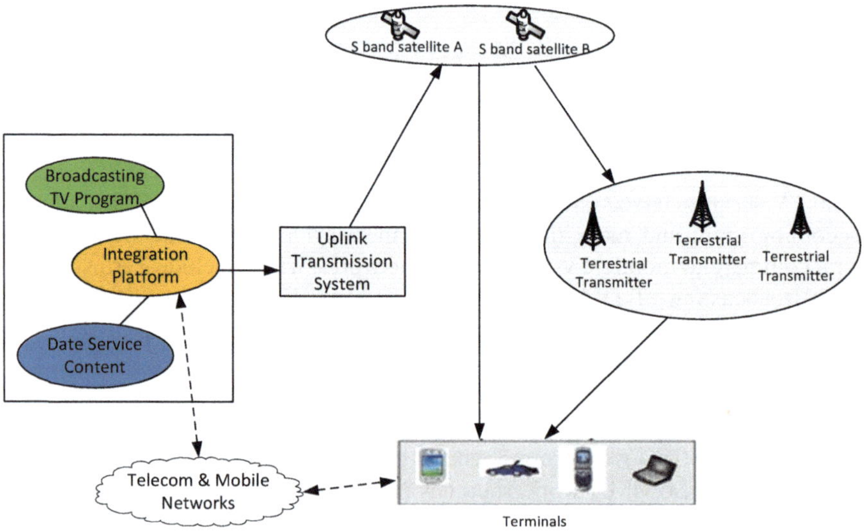

Fig. 11. The designed CMMB system infrastructure

telecom network [23]. The aim of CMMB is to provide seamless mobile multimedia services for all over the country.

2.1.2 Key Techniques of CMMB System

CMMB system is designed for mobile handheld reception, high-sensitivity, mobility and batter supply. As shown in Fig.12, the physical layer structure of CMMB system is analogous to other digital video systems such as DVB-H [21], which also uses OFDM modulation. The input data stream from upper layer is processed by FEC which consists of Reed-Solomon and LDPC as outer code and inner code [24], respectively. After interleaving and constellation mapping, the data is then multiplexed with scatter and continual pilot to help channel estimation and synchronization. Afterwards, the data is processed by OFDM modulation and then the frame headers are inserted to form the frame in the physical layer. The CMMB interface supports frequency bandwidths of 2MHz and 8MHz, depending on spectrum availability and allocation. Accordingly, CMMB utilizes OFDM with 4K mode for 8MHz bandwidth and 1K mode for 2MHz bandwidth [23].

On the other hand, CMMB adopts a frame structure which is coincident with the absolute time. As shown in Fig.13, each frame has a duration equal to 1s and consists of 40 time slots numbered from 0 to 39 [23]. Each time slot has a duration of 25ms and consists of a beacon and 53 OFDM symbols. Based on this frame structure, time slotting technology is applied to optimize power consumption.

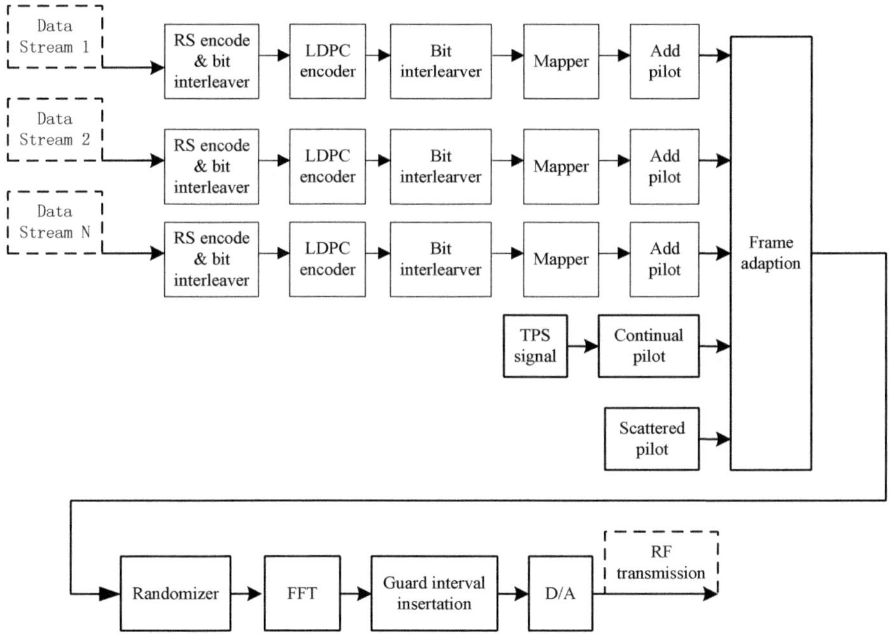

Fig. 12. Physical layer structure of CMMB

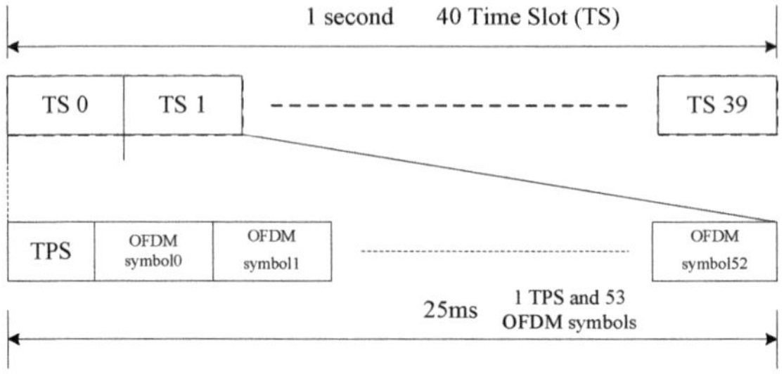

Fig. 13. Frame structure of CMMB system

An overview of the main features of CMMB is given in Table 3.

Table 3. Main features of CMMB

Parameter	Description
OFDM	4K for 8MHz, 1K for 2MHz
Bandwidth	8MHz, 2MHz
Outer Code	Reed-Solomon
Outer Interleaver	Byte Block Interleaver
Inner Code	LDPC(1/2, 3/4)
Inner Interleaver	Bit Block Interleaver
Constellation	BPSK, QPSK, 16QAM
Scramble Code	Pseudo BPSK sequence
Frame Structure	Timeslot-based frame

Compared with DVB-H, CMMB shows its difference mainly in three aspects.

Firstly, it uses LDPC code other than convolutional code. Thus it can obtain higher coding gain and greatly improve the reception performance [24].

Secondly, the guard interval (GI) design in the CMMB system is different from that in DVB-H. As shown in Fig.14, besides the cyclic prefix, it adds cyclic postfix to eliminate the effects of the forward filter of channel impulse response (CIR) [24].

Finally, the pilot design in CMMB is different, too. The pilot pattern consists of continual pilot and scattered pilot, which is the same as that of DVB-H. But the density of pilot in CMMB is bigger than that in DVB-H. As shown in Fig.15, in the timetable line, pilot signal is inserted every two OFDM symbols in the sub-carrier position, while in DVB-H it is inserted every four symbols. Thus we can conclude that it is superior to DVB-H in time-varying channels from the viewpoint of pilot pattern [24].

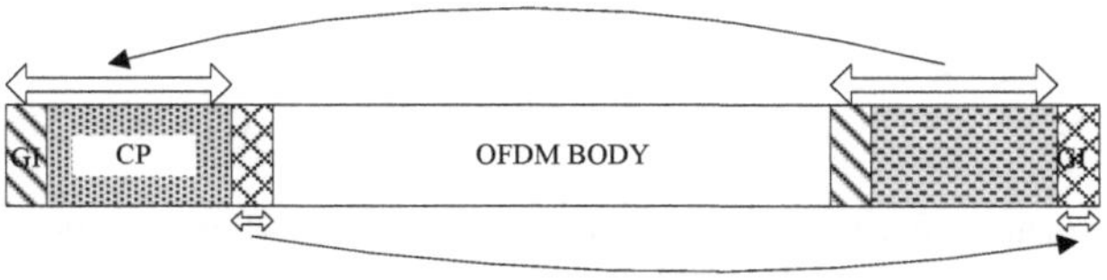

Fig. 14. The guard interval design in OFDM symbol of CMMB

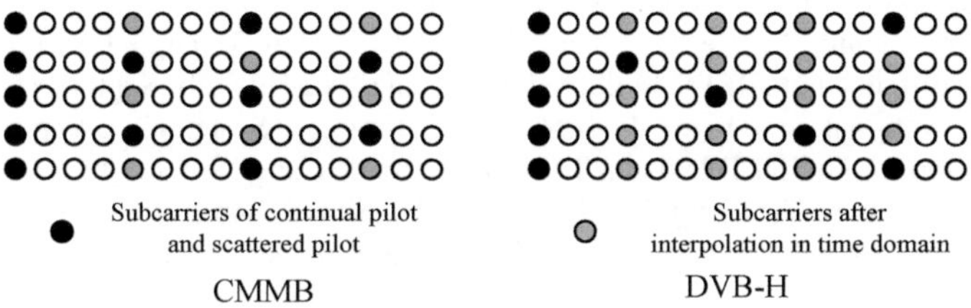

Fig. 15. Pilot pattern of CMMB and DVB-H

2.2 Applications of CMMB

The 2008 Beijing Olympic Games is an important milestone in the development of CMMB. To make more people can receive the live signals of Olympic Games at anytime and anywhere, SARFT organized large-scale experimental network test in 37 cities before the Beijing Olympic Games, which greatly accelerate the spread of CMMB service. Because of the favorable signal coverage deployment and support of the chip and terminal manufactures, CMMB experienced a booming period in the year of 2008, especially during the Beijing Olympic Games. The service of renting mobile phones with CMMB function to audiences and visitors during Olympics has won reputation for the standard [24].

It is the first time in Olympic history to use mobile multimedia broadcasting television and thus has drawn much attention. Through the experiments in Olympics, CMMB shows its validity and establishes good foundation for the future mobile multimedia broadcasting TV industry [24].

In the year of 2009, the terrestrial coverage will be spread to more cities. Until January 12, 2009, there are totally 150 cities where CMMB signal coverage has been accomplished and the next target will be 300 cities [24].

In the meantime, the signal coverage is also improved. For example, in Shanghai, three more transmitters are planned to be deployed in the future. Signal coverage for 13 Metro lines is also on the schedule now [24]. Additionally, three CMMB experimental stores and more than 100 CMMB sales outlets will be set up in this year [24]. The Oriental Pearl (Group) CO., Ltd., which is the unique operator of CMMB in Shanghai, tries to develop more than 80,000 subscribers and set up cooperation with China Mobile in the meantime [24]. In the near future, there will be more and more people enjoying the live TV experience by CMMB service.

References

[1] Advanced Television Systems Committee. ATSC Digital Television Standard, Document A/53 (1995)
[2] ETSI Document 300 744. Framing structure, channel coding and modulation for digital terrestrial television, Ver.1.4.1 (2001)
[3] ARIB. Terrestrial integrated services digital broadcasting (ISDB-T)—Specifications of channel coding, framing structure, and modulation (1998)
[4] Wu, Y., Hirakawa, S., Reimers, U.H., Whitaker, J.: Overview of digitial television development worldwide. Proceeding of IEEE 94(1), 8–21 (2006)
[5] Liang, W., Zhang, W., He, D., Guan, Y., Wang, Y., Sun, J.: Digital terrestrial television broadcasting in China. IEEE Multimedia Maganize, 92–97 (July-September 2007)
[6] Chinese National Standard GB 20600-2006. Framing Structure, Channel Coding and Modulation for Digital Television Terrestrial Broadcasting System (2006)
[7] Zhang, W., Guan, Y., Liang, W., He, D., Ju, F., Sun, J.: An introduction of the Chinese DTTB standard and analysis of the PN595 working modes. IEEE Transactions on Broadcasting 53(1), 8–13 (2007)
[8] Song, J., Yang, Z., Yang, L., Gong, K., Pan, C., Wang, J., Wu, Y.: Technical review on Chinese digital terrestrial television broadcasting standard and measurements on some working modes. IEEE Transactions on Broadcasting 53(1), 1–7 (2007)
[9] Liang, W., Guan, Y., Zhang, W., He, D., Ju, F., Sun, J.: Comments on Chinese DTTB Standard. In: IEEE International Symposium on Broadband Multimedia Systems and Broadcasting (2007)
[10] ATSC. Guide to the use of the ATSC digital television standard, ATSC Doc. A/53 (1995)
[11] He, D., Guan, Y., Liang, W., Ju, F., Zhang, W.: Combined NR decoding With decision feedback equalizer for Chinese DTTB receiver. IEEE Transactions on Broadcasting 54(3), 477–481 (2008)
[12] Haykin, S.: Adaptive Filter Theory, 4th edn. Pubilshing House of Electronics Industry (2003)
[13] Song, B., Gui, L., Guan, Y., Zhang, W.: On channel estimation and equalization in TDS-OFDM based terrestrial HDTV broadcasting system. IEEE Transactions on Consumer Electronics 51(3), 790–797 (2005)
[14] Liang, W., Chai, J., Guan, Y., Zhang, W., He, D.: A robust and adaptive carrier recovery method for Chinese DTTB receiver. IEEE Transcations on Broadcasting 54(1), 146–151 (2008)
[15] He, D., Liang, W., Zhang, W., Huang, G., Guan, Y., Ju, F.: Error rotated decision feedback equalizer for Chinese DTTB Receiver. In: IEEE International Symposium on Broadband Multimedia Systems and Broadcasting (2008)
[16] Wang, J., Yang, Z., Pan, C., Han, M., Yang, L.: A combined code acquisition and symbol timing recovery method for TDS-OFDM. IEEE Transactions on Broadcasting 49(1), 304–308 (2003)
[17] Morelli, M.M., Mengali, U.U.: A comparison of pilot-aided channel estimation methods for OFDM systems. IEEE Transactions on Signal Processing 49, 3065–3073 (2001)
[18] Zhang, W., Gui, L., Ma, W., Liu, B., Xiong, J.: The television broadcasting network of Chinese high speed railway. In: IEEE International Symposium on Broadband Multimedia Systems and Broadcasting (2008)
[19] Lee, G., Cho, S., Yang, K., Hahm, Y., Lee, S.: Development of terrestrial DMB transmission system based on Eureka-147 DAB system. IEEE Transactions on Consumer Electronics 51(1), 63–68 (2005)

[20] Murali, R., Chari, L.F., Ashok, M., Raghuraman, K., Rajiv, V., Kent, G., Rob, C.: FLO Physical Layer: An Overview. IEEE Transactions on Broadcasting 53(1), 145–160 (2007)
[21] ETSI. Digital Video Broadcasting (DVB): DVB-H Implementation Guidelines. TR 102 377 V1.2.1(2005-11)
[22] Mobile Multimedia Broadcasting Part I: Frame Structure, Channel Coding and Modulation for Broadcasting Channel, Chinese Standard of Radio, Film and Television industry. GY220.1-2006 (2006)
[23] Burger, R.: A Survey of Digital TV Standards China. In: IEEE Proceedings of 2nd International Conference on Communications and Networking, pp. 687–696 (2007)
[24] Zhang, W., Gui, L., Liu, B., Xiong, J., Lin, D.: Services and Trial Test of CMMB System. In: IEEE International Symposium on Broadband Multimedia Systems and Broadcasting (2009)

Network Topology Inference for Multimedia Streaming

Xing Jin and S.-H. Gary Chan

[1] Oracle USA, Inc., Redwood Shores, CA 94065, USA
xing.jin@oracle.com

[2] Department of Computer Science and Engineering, The Hong Kong University of Science and Technology, Clear Water Bay, Kowloon, Hong Kong
gchan@cse.ust.hk

Abstract. With the rapid growth of the Internet, multimedia streaming service has been widely studied and deployed. Good streaming service requires high delivery rate and low end-to-end delay. These are not easy to achieve without the knowledge of the underlying network topology. In this chapter, we explore the inference techniques for different network topologies. We discuss the key issues and state-of-the-art approaches for each topology. We also study two typical examples of topology-aware streaming protocols.

1 Introduction

Multimedia streaming has gone through quick growth in the recent years. Typical services include on-demand video streaming that allows users to choose and watch favorite movies anytime, and Internet Protocol television (IPTV) and live streaming that provide live TV service. Recently, peer-to-peer (P2P) streaming has been proposed and deployed to overcome limitations in traditional server-based streaming. In a P2P streaming system, cooperative peers self-organize themselves into an overlay network via unicast connections. They cache and relay data for each other, thereby eliminating the need for powerful servers from the system. One of the pioneering P2P streaming softwares, CoolStreaming, has reported to attract more than $25,000$ concurrent peers for one streaming channel [41]. Another streaming software PPLive reported more than $400,000$ concurrent peers for its over 300 channels [23].

Streaming service is bandwidth-demanding. A single streaming connection may require several hundred to several thousand Kbps delivery rate. And a physical network link may be used by multiple streaming connections (which is often the case in P2P streaming), leading to higher network bandwidth requirement. Hence, a key issue of large-scale streaming is how to schedule streaming connections to achieve high delivery rate on limited network bandwidth. In addition, it is also important to achieve low end-to-end delay.

In order to achieve that, network topology information is important. For example, two streaming connections that are disjoint on the application layer may share common links on the network layer. Without knowledge of the topology, it is difficult to identify or circumvent network bottlenecks.

C.W. Chen et al. (Eds.): Intel. Multimedia Communication: Tech. and Appli., SCI 280, pp. 425–441.
springerlink.com

In this chapter, we explore network topology inference issues for multimedia streaming. We classify four types of network topologies, i.e., router-level topology, autonomous system (AS) level topology, layer-2 topology and virtual topology. For each type, we investigate the key challenges in inference and the state-or-the-art approaches. Then we discuss advanced issues and future research directions in topology inference.

We also study how to use topology information to improve streaming service. We discuss and compare several representative P2P streaming protocols that make use of topology information. These study shows that it is important to infer topology information for constructing an efficient streaming system.

The rest of the chapter is organized as follows. In Chapter 2, we study how to infer network topologies. In Chapter 3, we discuss topology-aware streaming protocols. Finally, we conclude in Chapter 4.

2 Network Topology Inference over the Internet

2.1 An Overview

Network topology information can be used in a large and diverse set of applications besides media streaming. For example, it enables topology-aware protocols and algorithms, which assume the knowledge of network topologies [13, 19]. It also helps explore the Internet properties. As an example of topology study, researchers found that the Internet follows simple power laws, which has guided the modeling of the Internet and the design of network topology generators [36].

In general, the topology of a network is a graph representing network elements and their interconnections. There are several types of network topologies. In RFC 2922, a *physical topology* is defined as the topology model for layer-1 of the Open Systems Interconnection (OSI) stack (i.e., the physical layer). Physical topology consists of the devices in the network and how they are physically interconnected. These devices include communication infrastructure devices, such as hubs, switches, and routers, as well as "leaf" devices such as workstations, printers, and servers. On top of the physical topology, there are two types of topologies on the higher layers of the OSI stack: *layer-2 topology* and *layer-3 topology*. A layer-2 topology consists of network elements visible on the link layer, e.g., hubs, bridges and switches. It also consists of routers and end hosts. A layer-3 topology consists of network elements visible on the network layer, typically routers and end hosts. Edges in layer-2 or layer-3 topologies are determined by routing paths between packet sources and receivers. Hence, these topologies are only stable during a certain period when no routing path changes. Clearly, these topologies display less information than physical topologies, because they do not indicate all connectivity or network elements. However, they provide some important information that cannot be obtained from physical topologies: they identify the paths traversed by packets from sources to receivers and indicate where the paths diverge or merge. This information is useful for applications such as network scheduling, failure diagnosis and loss recovery.

Besides the above topologies, there is a class of *AS-level topologies*. In the current Internet, routers are organized into thousands of ASs. Each AS is represented by a 16-bit

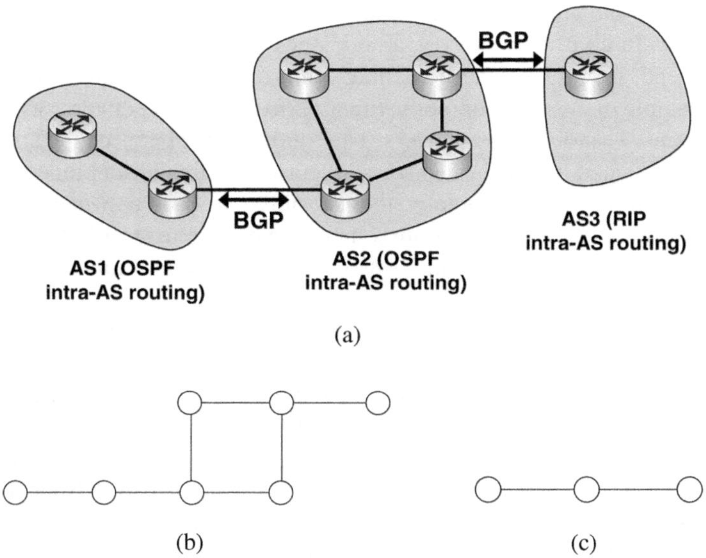

Fig. 1. An example of network topologies. (a) A real network. (b) A router-level topology. (c) An AS-level topology.

AS number, which brings to a total of 65,536 possible ASs. Routers within the same AS all run the same routing algorithm, e.g., Routing Information Protocol (RIP) or Open Shortest Path First protocol (OSPF). One or more of the routers in an AS are responsible for forwarding packets to destinations outside the AS. These routers are called gateway routers. Gateway routers run an inter-AS routing protocol (e.g., Border Gateway Protocol, or BGP in short) to determine routing paths among ASs. Through this hierarchical routing mechanism, any two hosts connected to the Internet can communicate with each other.

Figure 1(a) shows an example of a network, where routers form three ASs. Within an AS, RIP or OSPF is used for routing. Between ASs, BGP routing is used. Figure 1(b) shows a corresponding router-level topology. Each node in the topology is a router, and edges between nodes indicate the connections between routers. Figure 1(c) shows a corresponding AS-level topology. In the topology, each node is an AS, and edges indicate the connections between ASs.

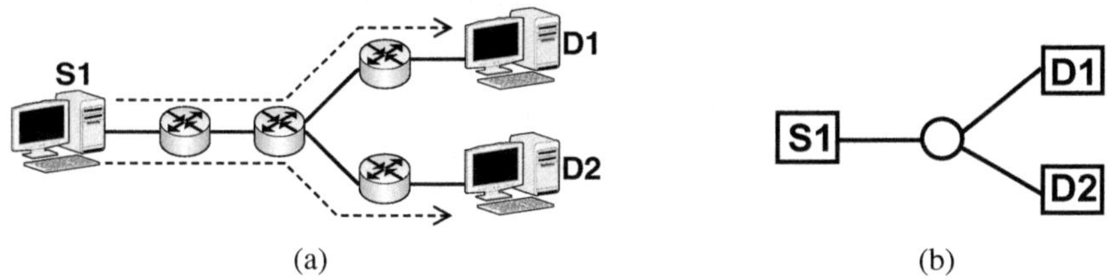

Fig. 2. An example of a virtual topology. (a) A real network. (b) A virtual topology.

Different from the above topologies, there is another class of topologies called *virtual topologies*. In a virtual topology, a node does not represent any physical network element. Instead, it indicates the branching point of different routing paths. Figure 2 shows an example of a virtual topology. Figure 2(a) is the real network, where S_1 is the source host, and D_1 and D_2 are receivers. The routing paths between the source and the receivers are indicated by the dashed lines. A corresponding virtual topology is shown in Figure 2(b). In the virtual topology, there is only one node between the hosts. This node does not represent a router, or any other network element. It indicates that paths $S_1 - D_1$ and $S_1 - D2$ are branching. Clearly, there may be multiple virtual topologies corresponding to a real network.

A network topology may be annotated with additional details such as delay, residual bandwidth or loss rate. In this chapter, we focus on the network elements and their interconnections in the topology. We investigate the key challenges in inferring the different types of topologies as well as the state-of-the-art approaches.

In the following, we discuss how to infer a layer-3 router-level topology and an AS-level topology in Chapter 2.2 and Chapter 2.3, respectively. These two types of topologies are the most commonly used in network applications. In Chapter 2.4, we discuss how to infer a layer-2 topology. In Chapter 2.5, we discuss how to infer a virtual topology. Finally, we give a summary and comparison in Chapter 2.6.

2.2 Layer-3 Router-Level Topology Inference

To obtain the router-level connectivity information between two hosts, traceroute-like tools are often used [1]. Traceroute is implemented with Internet Control Message Protocol (ICMP) messages. Each time, the source sends an IP datagram with a certain time-to-live (TTL) value to the destination. Each router that handles the datagram decrements the TTL value by one. When a router receives an IP datagram whose TTL is 1, it throws away the datagram and returns an ICMP "time exceeded" error message to the source host. The IP datagram containing this ICMP message has the router's name, IP address and round-trip time (RTT) to the source. In another case, if the datagram arrives at the destination with an unused port number (usually larger than 30,000), the destination host generates an ICMP "port unreachable" error message and returns it to the source host. Therefore, in traceroute, the source host sends a series of IP datagrams with increasing TTL to the destination and each datagram can identify one router in the path. The whole router-level path is hence identified.

Traceroute-like tools have been widely used in network topology inference. According to the size of the target network, the inference work can be divided into three categories.

- *Measuring the Internet topology*: The earliest attempt traced paths to 5000 destinations from a single host in 1995 [32]. Later on, the Skitter project deployed around 20 active monitors around the Internet to track global IP level connectivity [2]. Another project, Mercator, started from a single host to infer the Internet topology [22]. The goal of these projects is to infer a complete Internet topology including all routers and inter-router links. The measurements often last for several months or years. They hence do not have critical requirements on measurement time or bandwidth consumption.

- *Measuring an Internet Service Provider (ISP) topology*: Rocketfuel uses information from BGP routing tables and Domain Name System (DNS) to guide traceroutes in order to infer the topology of a single ISP [34]. As compared to the whole Internet, the target of an ISP network is more specific. Hence, the inferred topology is often more complete.
- *Measuring the topology among a group of hosts*: Max-Delta focuses on the topology among a certain group of hosts which are arbitrarily distributed around the Internet [26, 28]. The inferred topology is mainly used for overlay applications. Overlay applications often do not need fully complete and accurate topologies. But they require short measurement time and light measurement traffic. The key issue is hence how to achieve proper tradeoff between topology completeness and measurement cost.

We discuss the key issues in traceroute measurements as follows.

2.2.1 Source Selection

As traceroute can start from any host connecting to the Internet, the selection of traceroute sources is not difficult. Skitter deploys around 20 monitors as the traceroute sources. Mercator can use an arbitrary host as the source. Rocketfuel makes use of 294 public traceroute servers listed at the *www.traceroute.org* webpage. In Max-Delta, each end host in the group is a source and may conduct traceroutes to others.

Barford et al. study the relationship between topology completeness and number of traceroute sources [5]. Given a list of destinations, they find that the marginal utility of adding additional traceroute sources declines rapidly after the second or third one. In other words, the first two or three sources can discover the majority of the complete topology by tracerouting the destinations. Here a complete topology is obtained by combining traceroute results from all sources to the given destinations. These results have partially supported the measurement methodologies of Skitter and Mercator. On the other hand, in Barford's experiments, the number of sources is much less than the number of destinations (e.g., 8 sources and 1277 destinations, or 12 sources and 313,709 destinations). Their conclusion hence may not hold in Max-Delta, where the number of sources is equal to the number of destinations.

2.2.2 Path Selection

The selection of traceroute paths or traceroute destinations is not easy. Skitter aims to infer a complete Internet topology. It hence generates a destination list including as many IP addresses as possible. The list comes from a wide range of sources such as ISP packet traces, data from other projects like NeTraMet and NetGeo, data from CAIDA website, etc. It then filters out addresses that do not want to be probed and that do not respond. The final list contains about 800K destinations. Certainly, this list is not complete. According to their estimation, there are over 16 million potential IPv4 addresses, and about 4 million of them are currently routable.

Mercator does not need any external databases to derive its destination list. It uses a technique called *informed random address probing* to estimate which portions of the entire IP address space contain routable hosts. It first estimates routable IP prefixes. From each routable prefix, it randomly selects an IP address as the target. According to [22], Mercator took about 3 weeks to discover nearly 150K router interfaces and 200K edges. As compared to Skitter, Mercator has a less complete destination list. However, its destination list is automatically generated and does not need manual work. Its deployment is hence much simpler.

Rocketfuel generates its destination list with 120K IP address prefixes from Route-Views BGP tables [3]. However, since their target is a specific ISP network, they do not traceroute to all these IP addresses. They first use a technique called *directed probing* to identify traceroutes that will transit the specific ISP network and skip the reminder. For example, all traceroutes to the prefixes originated by the ISP should transit the ISP. Secondly, they note that traceroute paths contain redundant information. If two traceroutes enter and leave the ISP network at the same points, only one of them needs to be conducted. They hence propose several ways to reduce the redundancy. They have shown that brute-force search from all traceroute sources to all BGP-advertised prefixes would require 90-150 million traceroutes for a specific ISP. With their path selection mechanism, less than 0.1% of these traces are measured. Their results further show that the loss of accuracy introduced by path selection is much smaller than the gain in cost reduction.

Max-Delta has the highest requirement on measurement efficiency. In the scheme, hosts first use a tool to estimate their network coordinates and report them to a central server. The inference process is then divided into multiple iterations. In each iteration, the server selects a target for each host to traceroute. The target is selected based on host coordinates and currently available traceroute results in order to discover as many inter-router connections as possible. Hosts then traceroute their targets and report the results to the server. Such process is repeated until certain measurement accuracy or measurement cost is achieved. It has been shown that Max-Delta can infer a highly accurate topology with a small number of traceroutes.

2.2.3 Reducing Measurement Cost

Careful path selection can reduce measurement cost. Besides that, there are other techniques for cost reduction. In Mercator, not all traceroutes start at TTL 1. Instead, from the results of traceroutes to prefix P, Mercator computes the farthest router R in the path. Subsequent traceroutes to P start at the TTL corresponding to R. If the first response is from R, Mercator continues the path probe. Otherwise it backtracks the path probe to TTL=1. This technique allows Mercator to avoid, where possible, rediscovering router adjacencies. It also reduces the probing overhead in the vicinity of the traceroute source.

This technique was extended by Donnet later [17]. Donnet et al. propose a Doubletree algorithm to reduce repeated router discovery. Given a source and a destination, the traceroute starts at some intermediate point between them. The probing then proceeds towards the destination and backwards towards the source. In either case, the probing stops whenever an already discovered router is met.

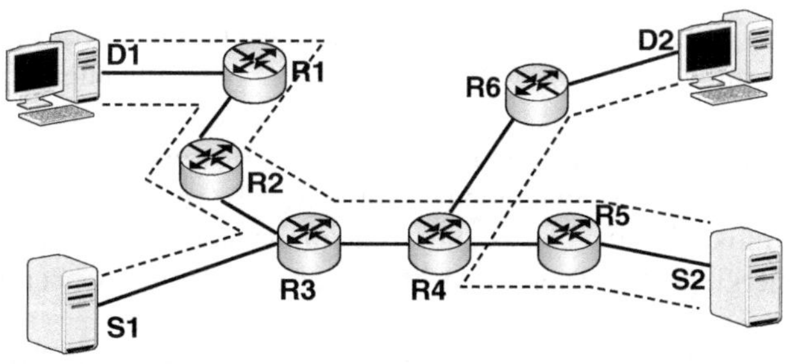

Fig. 3. An example of repeated router discovery. A dashed line indicates the router-level path between a source and a destination.

Figure 3 shows an example of repeated router discovery. In the figure, S_1 and S_2 are two traceroute sources. D_1 and D_2 are two destinations. R_1 to R_6 are routers. A dashed line indicates the router-level path between a source and a destination. In one case, if both two sources traceroute the same destination D_1, they will discover paths $S_1 - R_3 - R_2 - R_1 - D_1$ and $S_2 - R_5 - R_4 - R_3 - R_2 - R_1 - D_1$, respectively. These two paths overlap over $R_3 - R_2 - R_1 - D_1$. In another case, if S_2 traceroutes two destinations D_1 and D_2, the two paths also overlap (over $S_2 - R_5 - R_4$). To reduce such repeated router discovery, Doubletree selects a certain starting TTL other than 1 for traceroute, say, 3 in this example. Consider the case that S_2 traceroutes D_1 and D_2. Suppose that S_2 first traceroutes D_1. The traceroute starts with TTL=3. The first router discovered is hence R_3. The traceroute then continues towards D_1 to discover R_2 and R_1. It also proceeds backwards towards S_2 and discovers R_4 and R_5. The next traceroute from S_2 to D_2 also starts with TTL=3, thereby discovering $R6$ first. The traceroute then proceeds towards D_2 and ends. The traceroute also proceeds backwards towards S_2 and discovers R_4. Since R_4 has been discovered in the path from S_2 to D_1, the traceroute stops. In this way, router R_5 will not be probed twice in the two traceroutes.

According to [17], the simulation results based on Skitter topologies show that Doubletree can reduce the measurement cost by 76% while maintaining the router and connection coverage above 90%. Doubletree can be applied to various traceroute measurements. For example, in Max-Delta, the system can first select traceroute targets and then use Doubletree to supervise the start and stop of traceroutes [27].

2.2.4 Reducing Measurement Noise

- *Anonymous Routers*: In the real Internet, not all routers properly respond to ICMP request messages. Some routers do not return ICMP error messages, and some return ICMP error messages only when their loads are light. These routers do not show valid IP addresses in traceroute results. They are called *anonymous routers* [39]. In traceroute measurement, the presence of anonymous routers is inevitable. Broido et al. report that nearly one third of probed paths contain anonymous, private or invalid routers [9]. They propose two methods (i.e., adding arcs or placeholders) to address it. However, the resultant topology either hides much topology information or suffers high router inflation.

Yao et al. study how to infer a topology consistent with the given traceroute results and containing the minimum number of anonymous routers [39]. They show that producing either an exact or an approximate solution is NP-hard. They then propose a heuristic to merge anonymous routers. However, as shown in [28], their heuristic suffers high computational complexity. To reduce the complexity, some other algorithms relax the consistency constraints and allow a small portion of inconsistent merging [28]. These algorithms trade off topology accuracy with inference speed.

- *Router Alias*: Another type of measurement noise is router alias. Traceroute probes actually discover router interfaces. It is hence possible to discover more than one interface belonging to the same router (i.e., multiple aliases for one router). The basic alias resolution technique sends traceroute-like probes (to a high-numbered UDP port with a TTL of 255) to the potentially aliased IP address. If the router has been configured to send the "UDP port unreachable" response with the address of the outgoing interface as the source address, two aliases will respond with the same source. In practice, this approach is often used with additional refinement. For example, Mercator also uses source-routed alias probes [22], and Rocketfuel uses IP identifiers and ICMP rate limiting techniques [34].

2.3 Autonomous System Level Topology Inference

The Border Gateway Protocol version 4 (BGP4, or simply BGP), specified in RFC 1771, is the de facto standard inter-domain routing protocol in today's Internet. BGP is a distributed path vector protocol, where gateway routers exchange detailed path information with their directly connected gateway routers. Global information about routes to distant destinations hence propagates in an AS-by-AS manner between pairs of directly connected gateway routers.

There is no publicly available information about inter-AS connectivities. ISPs do not register their relationships to the Internet registries such as American Registry for Internet Numbers (ARIN). Contractual agreements between ISPs are proprietary and companies are unwilling to reveal even the names of their ISPs. Therefore, there are limited data to infer AS topologies. Currently, the most common way to infer AS-level topologies is to use BGP routing tables or BGP update messages. Table 1 shows a snippet of a BGP routing table from RouteViews [3]. The table has two entries for two destination prefixes 4.17.88.0/21 and 12.0.19.0/24. The AS has three candidate routes to the first prefix: AS path (293 10886 6059) via next_hop 134.55.20.229, AS path (1224 11537

Table 1. A Snippet of a BGP Routing Table

Network	Next Hop	Path
4.17.88.0/21	134.55.20.229	293 10886 6059 i
	141.142.12.1	1224 11537 10886 6059 i
	198.32.8.196	11537 10886 6059 i
12.0.19.0/24	209.123.12.51	8001 7018 27487 i
	216.191.65.118	15290 7018 27487 i
	... ...	... ...

10886 6059) via next_hop 141.142.12.1, and AS path (11537 10886 6059) via next_hop 198.32.8.196. Each AS path consists of a list of ASs that will be traversed to reach the prefix. ASs closer to the destination are to the right of the path. Similarly, the AS has two candidate routes to the second prefix 12.0.19.0/24. In practice, BGP routing tables are usually not public. Fortunately, some organizations have provided public access to their BGP routing tables, e.g., University of Oregon's RouteViews project [3] or RIPE's Routing Information Service (see *http://www.ris.ripe.net*). AS connectivity information can also be obtained from BGP update messages. A BGP update message contains a set of address prefixes and is exchanged between BGP gateway routers. For each prefix, it either indicates the sender's inability to reach destination in the prefix, or shows an AS path to the destination in the prefix.

2.3.1 Topology Completeness

As mentioned, AS paths in BGP tables or BGP update messages have implicitly indicated AS connectivities. An AS topology can be accordingly inferred. In an early work Govindan et al. derived an AS topology from routing updates collected over 21 days [21]. However, the inferred topology may yield a very incomplete picture of the whole Internet. As discussed, ISPs would not like to publish their BGP tables. Even if all BGP tables are available, it is impossible to build a complete AS topology, because BGP routing tables do not show backup links connecting multi-homed ASs. Therefore, it is important to explore the completeness of an inferred AS topology.

Chang et al. have tried to quantify the completeness of AS topologies inferred from RouteViews [11]. Besides RouteViews BGP tables, they also obtain BGP tables from Swiss Network Operators Group and some commercial route servers. They further access a set of Internet Looking Glass sites and the Internet Routing Registry. They finally collect around 40 BGP views from different ASs, and accordingly infer their AS topologies. The results show that their topologies have typically 25%-50% more edges than the topologies based on only RouteViews data. Their findings suggest that the Internet maintains much richer connectivity at the AS level than has been previously reported.

Zhang et al. further refine Chang's work by accumulating topological information from routing tables over time [40]. They note that a snapshot of routing tables only captures the best paths to destinations at the time of the snapshot. As the best paths may change over time (due to various reasons, e.g., link failure), they collect BGP routing tables over time and strive to discover nodes and edges that occur in any best path seen during the observation period. Their results show that by integrating two months of routing tables, they can improve the snapshot topology by 16% of edges and 2% of nodes.

2.3.2 Inferring Reachability Instead of Connectivity

The commercial agreements between pairs of administrative domains can be classified into customer-provider, peering, mutual-transit, and mutual-backup agreements. For example, a customer pays its provider for connectivity to the rest of the Internet. Hence, a provider transits traffic for its customers, but a customer does not transit traffic for its providers. These contractual commercial agreements between administrative domains play a crucial role in shaping the structure of the Internet. For example, suppose two

ISPs A and B are connected to their customer, ISP C, respectively. Although A and B are connected through C, A cannot reach B via C. Because C as a customer does not provide transit services between its providers. Therefore, connectivity does not imply reachability. Since routing between ASs is controlled by policy-based BGP, a global picture of AS relationships besides connectivity becomes important.

Gao first studies the reachability issue between ASs in [20]. She classifies the relationship between a pair of interconnected ASs into customer-provider, peering, and sibling. Two connected ASs are peering (or sibling) if they do not (or do) transit traffic for each other. AS u is a provider of AS v iff u transits traffic for v and v does not transit traffic for u. Gao then proposes some heuristics to infer the AS relationships from BGP tables. The heuristics are based on the intuition that a provider typically has a larger size than its customer and the size of an AS is typically proportional to its degree in the AS topology (the latter has been verified in [21]). The heuristic then goes through the BGP tables, and finds the AS with the highest degree as the top provider. Then it starts from the top provider to infer the relationships between other ASs.

Subramanian et al. consider similar AS relationships in [35]. They propose a technique for combining data from multiple vantage points in the Internet to construct a complete topology with AS relationships. Each vantage point offers a partial view of the Internet topology. They generate a directed AS-level topology from each vantage point and assign a rank to each AS based on its position. Then, each AS is represented by the vector that contains its rank. They infer the relationship between two ASs by comparing their vectors. Based on these relationships, they construct a new directed AS topology and examine the AS-level hierarchy of the Internet. They finally present a five-level classification of ASs, where the top-most level consists of a rich set of peering relationships between 20 so-called tier-1 providers.

Clearly, a topology with AS relationships provide much more information than the one with only connectivities. However, the inference of AS relationships is even more difficult than connectivity inference. Lacking direct measurement data support, current inference mechanisms are mostly heuristics. There are no standards or methods to verify or compare these heuristics. There is much improvement space on this work.

2.3.3 Inferring AS Topologies from Traceroutes

Chang et al. study how to discover AS-level topologies from router-level topologies in [12]. Given a router-level path, they determine the AS of each router hop and extract AS adjacency information from the path. To achieve this, they first construct a mapping table that shows the ASs and the corresponding address prefixes. Based on the table, the AS of a router interface can be determined by identifying the longest address prefix that matches the interface address. Similar approach is adopted in [24]. The method is further refined in [31] by using a large collection of BGP tables and resolving anonymous routers as well as unmapped IP-level hops.

Clearly, this method requires an accurate enough router-level topology. This will increase the inference load. As compared to BGP-based inference, it infers an AS topology at a finer granularity (e.g., multiple connections between a pair of ASs). It also infers a more complete topology, as BGP routing tables do not capture all existing AS paths.

2.4 Layer-2 Topology Inference

While traceroute can discover a layer-3 router-level topology, it cannot capture the interconnections between layer-2 elements (e.g., switches, bridges, or hubs). Nowadays, many switches are deployed to provide bandwidth through subnet microsegmentation. Such portions of the network infrastructure that are invisible to a layer-3 topology will continue to grow.

Some vendors have developed proprietary tools and protocols for layer-2 topology inference. Examples include Cisco's Discovery Protocol and Bay Networks' Optivity Enterprise. These tools are based on vendor-specific developments and are not useful in a heterogeneous network comprising elements from multiple vendors. Other approaches such as Peregrine's Infratools and Riversoft's NMOS use proprietary technology and many details are undiscovered.

Current layer-2 topology inference is mainly based on management information base (MIB) in Simple Network Management Protocol (SNMP) [10]. We briefly review SNMP and MIB as follows. A piece of network equipment including its software is called a managed device. It might be a host, router, bridge, hub, printer or modem. Within a managed device, there may be several so-called managed objects, which are the actual pieces of hardware in the managed device as well as the configuration parameters. These managed objects have pieces of information associated with them that are collected into a local MIB. A managed device follows a certain management protocol to communicate with a centralized network management station, and takes local actions under the command of the station. Nowadays, the standard management protocol is SNMP. In summary, MIB stores information of network equipments, which can be queried or set by SNMP messages.

Breitbart et al. propose to discover routers and switches based on standard address forwarding tables (AFT) in MIB [8]. Assuming that at least one router's IP address is known, the algorithm repeatedly discovers neighboring routers of the currently known routers (through MIB information) until no new routers are discovered. It then discovers switches as follows. For each interface of a router R, it computes a set D by enumerating the set of IP addresses in the subnet corresponding to the IP addresses of the interface. Once D is computed, for each IP address in D, it checks for the presence of the Bridge MIB to determine whether the address corresponds to a switch. After all routers and switches are discovered, it then identifies the connections between routers and switches through AFT information.

The above algorithm assumes that AFT information is available from every element in the network and, thus, cannot cope with hubs or uncooperative switches. Lowekamp et al. then explore how to handle dumb/uncooperative elements given incomplete AFT information [30]. However, their algorithm works only in a single subnet network and may easily fail when multiple subnets are present. A further extension for large multi-subset networks are studied in [6].

Another piece of work by Black et al. does not rely on MIB for inference [7]. They note that Ethernet has some interesting properties. For example, in Ethernet, an equipment can freely fake the source address of its packets, and equipments are often connected to a share media and can overhear each other's transmissions. Based on these properties, they inject probing packets into the network to actively infer the network topology. Although it does not need MIB information, the method has its own limitations. It fails in the presence of 802.1X port-based access control, as it relies on sending packets with arbitrary addresses. It also requires switches to implement IEEE 802.1D spanning tree protocol, which is not available in many switches. Finally, it requires equipments to run a certain detecting process and to inject packets when necessary. The deployment of the process may be a problem.

Up to today, study on layer-2 topology inference is limited. Layer-2 equipments are completely transparent to layer-3 routers. There are only few ways to detect these equipments. Among them, most methods rely on MIB information, but a topology inferred from MIB is often incomplete and of limited interest. This is because many devices do not implement or turn on SNMP support, and MIB only contains information on recently active equipments. In addition, some secured network equipments need passwords for access. Finally, layer-2 equipments are mostly deployed in LANs, but many applications are more interested in a broader network such as WAN or the Internet.

2.5 Inferring a Virtual Topology by Network Tomography

A virtual topology is often inferred by so-called network tomography. Network tomography sends probing traffic into the network and exploits the performance in correlation to infer network properties. This technique can estimate link-level parameters (such as topology, loss rate or link delay) based on end-to-end path-level traffic measurement.

Tomography-based topology inference often identifies a tree structure connecting a single sender to multiple receivers, or a combination of multiple trees. The key idea is to collect measurements at pairs of receivers that behave (in an average sense) as a monotonic, increasing function of the number of shared links or queues in the paths to the two receivers [16]. For example, consider the paths to a pair of receivers from the same source. The covariance between end-to-end delays of the paths is highly correlated to the number of shared links in the two paths (assuming that delays are not correlated on unshared links). The more shared links, the larger covariance. We can then measure the covariance to estimate the shared links. In tomography, probing packets may be sent by unicast (packet pairs or stripes sent to multiple receivers at the same time) [15, 33] or multicast [18]. A multicast approach is more efficient. In multicast measurement, a probing packet is transmitted only once on each link, and by sending one packet, the correlation between any two hosts in the group can be measured. However, IP multicast is not always available in the Internet.

Tomography may use loss, delay correlation, delay differences or other utilization as measurement metrics. Given a set of measurement statistics from different pairs of receivers, topology inference can be cast as a maximum likelihood estimation problem. That is, from the forest of all possible tree topologies connecting the sender to the receivers, the one with the highest probability performing as observed will be identified. A difficult part of such likelihood optimization is the computational complexity for optimal solutions. Up to now, there is no method for computing the global maximum except by a brute-force examination of each tree in the forest. But a loose lower bound on the size of the forest is $N!/2$, where N is the number of receivers. Much research has turned to suboptimal algorithms instead, for example, the deterministic binary tree classification algorithm [18], or markov chain monte carlo based inference [15].

Network tomography is based on end-to-end measurements and does not need any assistance from routers or other network elements. However, the inferred topology is far from realistic use. Firstly, the metrics used (such as loss rate or delay) are often unstable and inaccurate. The resultant topology is hence not accurate. Secondly, the topology can only be a single tree or a combination of multiple trees, and the target network size cannot be large. Thirdly, it is difficult to verify the accuracy of the inferred topology. Therefore, it will take significant effort to refine the current tomography methodology in order to infer practically useful topologies.

2.6 Comparison and Discussion

In general, the inference of router-level and AS-level topologies has developed practical and efficient methodologies. Router-level topology inference can use traceroute. AS-level topology inference can use BGP tables/updates, or be deduced from traceroute results. These methods can infer topologies of different sizes, up to the whole Internet. The most important, these two types of topologies are inferred from routing information. The inferred topologies can hence be used to improve the performance of network applications or analyze the Internet properties. Much research has been conducted to refine the inference methodologies, e.g., how to collect the most complete measurement data, or how to reduce measurement cost such as measurement time or traffic.

Layer-2 topology inference has not attracted much interest. There are several reasons for it. Firstly, layer-2 topology inference takes much effort to identify link-layer elements such as switches and hubs, which are mostly deployed in LAN. However, network applications often span much larger networks such as WAN or the Internet. People are hence more interested in topologies of larger networks, where router-level or AS-level topologies have contained enough information. Secondly, the inference method for link-layer elements is less developed. A typical method relies on SNMP and MIB information. But not all network elements support SNMP and MIB. The inferred topology is hence often incomplete. A major usage of layer-2 topologies is to provide information to network managers for troubleshooting or assessing the network. Therefore, large companies such as Cisco and Intel have taken effort to develop their own, often proprietary, technologies to infer layer-2 topologies.

Virtual topology inference by network tomography has been quickly developed recently. Network tomography is purely end-to-end and does not need network assistance. As a comparison, traceroute requires routers to support ICMP and AS topology inference needs BGP tables/updates from gateway routers. However, with no router assistance, its inferred topology is often inaccurate, small and can only be in the form of trees. Such topologies are far from practical use. Tomography methodology needs significant refinement in the future.

3 Topology-Aware Multimedia Streaming

In this section, we discuss how to use topology information to build efficient streaming systems. We select P2P streaming as example environment. We investigate two representative P2P streaming protocols that make use of router-level topology information.

3.1 Reducing Delivery Delay

Topology-Aware Grouping (TAG) is a topology-aware P2P protocol which aims at constructing a low-delay overlay tree [29]. In TAG, a new host uses traceroute to measure the router-level path from the source to itself before joining the system, which is called the *spath* of the host. Each host then selects as its parent the host whose spath has the maximal overlap with its own spath. The selection process works as follows. Each time, the new host N selects a host C for examination. The first host being examined is the source. N compares the spaths of C's children with its own spath. If C has a child whose spath has larger overlap with N than C, N moves to this child of C and continues examining it. Otherwise, if C has a child whose spath contains the whole spath of N (in terms of routers), N joins the tree between C and its child, i.e., this child becomes N's child and N becomes C's child. If none of the above two cases happens, N joins as C's child. In this way, TAG reduces the numbers of underlay hops and hence the delay over unicast paths.

From simulation studies, TAG shows low end-to-end delay. A promising property of TAG is that each host only needs to conduct one traceroute during tree construction. As a comparison, in many other P2P protocols, a host needs to ping dozens of peers in order to select a close neighbor (e.g., [4, 14]). Hence, through traceroute-based topology inference, TAG achieves significantly low measurement overhead. This study indicates that knowing underlay topology is good to construct efficient overlay trees, and lots of work continues in this area [37, 38, 42]. However, a limitation of these approaches is that many of them are empirical study without rigorous mathematical support. Much more work in this area, especially theoretical study, is desired.

3.2 Improving Delivery Rate

As discussed, a key issue in multimedia streaming is how to schedule streaming connections to achieve high delivery rate. Jin et al. study this issue given router-level topology information. The topology is measured from traceroutes with link connectivity and delay information. An additional option is residual bandwidth along links, which can be obtained through bandwidth measurement tools such as Pathload [25].

Suppose a set of hosts are given. The target is to build an overlay tree spanning all hosts with the maximum tree bandwidth. Here tree bandwidth is defined as the minimum path-bandwidth of an overlay tree. The study considers that hosts are in position before tree construction. For example, there are a group of Internet participants waiting for a video conference, or a content distribution network (CDN) needs to distribute data to some pre-deployed proxies. They investigate two problems as follows:

- In the absence of link bandwidth
  Define *link stress* as the number of copies of a packet transmitted over a certain physical link [14]. They consider minimizing the maximal link stress in a tree. They formulate the problem as building a *Minimum Stress Multicast Tree (MSMT)* on a given topology. Their study shows that it is NP-hard and is not approximable within a certain factor, unless $P = NP$. They then propose a centralized approximation algorithm to address it.

- In the presence of link bandwidth
  When link bandwidth is available, they build a *Maximum Bandwidth Multicast Tree (MBMT)* on the topology. This problem is also NP-hard since it is equivalent to MSMT if all the links have the same bandwidth. They extend the approximation algorithm for MSMT to address the MBMT problem.

Their study gives the theoretical performance bound of topology-aware applications. The algorithm performance can serve as a benchmark. Jin et al. conducted simulations on Internet-like topologies as well as Internet measurements on PlanetLab to evaluate the algorithms. The results show that they can achieve higher tree-bandwidth and lower link stress than traditional tree-based protocols with no topology information. The study shows that underlay information can help build a highly efficient overlay tree. Currently, many P2P streaming softwares do not take topology information into consideration. It is possible to achieve significant performance improvement if they measure topologies and accordingly build streaming overlays.

However, many practical issues are still open following Jin's study. For example, they assume that all hosts are in position before tree construction. But in practice, hosts may dynamically join or leave the system. A practical protocol should be adaptive to host dynamics. In addition, the study assumes a central server for tree construction. For a large system with hundreds of thousands of peers, a distributed tree construction method is more useful.

4 Conclusion

We discuss in this chapter network topology inference issues for multimedia streaming. Network topology information can improve streaming quality such as delivery rate or end-to-end delay. We classify four types of network topologies. For each of them, we investigate the challenging issues and typical approaches. We then discuss some advanced issues and future research directions. We also explore two example streaming protocols which make use of underlying topology information. The study indicates that topology information is indeed important to form efficient streaming overlay for data distribution.

References

1. Traceroute, `http://www.traceroute.org/`
2. Skitter, `http://www.caida.org/tools/measurement/skitter/`
3. Route Views, `http://www.routeviews.org/`
4. Banerjee, S., Bhattacharjee, B., Kommareddy, C.: Scalable application layer multicast. In: Proc. ACM SIGCOMM 2002, pp. 205–217 (2002)
5. Barford, P., Bestavros, A., Byers, J., Crovella, M.: On the marginal utility of network topology measurements. In: Proc. ACM SIGCOMM Internet Measurement Workshop (IMW 2001), pp. 5–17 (2001)
6. Bejerano, Y., Breitbart, Y., Minos Garofalakis, R.R.: Physical topology discovery for large multi-subnet networks. In: Proc. IEEE INFOCOM 2003, pp. 342–352 (2003)

7. Black, R., Donnelly, A., Fournet, C.: Ethernet topology discovery without network assistance. In: Proc. IEEE Int'l Conf. Network Protocols (ICNP 2004), pp. 328–339 (2004)
8. Breitbart, Y., Garofalakis, M., Jai, B., Martin, C., Rastogi, R., Silberschatz, A.: Topology discovery in heterogeneous IP networks: the NetInventory system. IEEE/ACM Trans. Networking 12(3), 401–414 (2004)
9. Broido, A., claffy, k.: Internet topology: Connectivity of IP graphs. In: Proc. SPIE ITCom 2001 (2001)
10. Case, J., Fedor, M., Schoffstall, M., Davin, J.: A simple network management protocol (SNMP). RFC 1157 (1990)
11. Chang, H., Govindan, R., Jamin, S., Shenker, S.J., Willinger, W.: Towards capturing representative AS-level Internet topologies. Computer Networks 44(6), 737–755 (2004)
12. Chang, H., Jamin, S., Willinger, W.: Inferring AS-level Internet topology from router-level path traces. In: Proc. SPIE ITCom 2001 (2001)
13. Chen, Y., Bindel, D., Song, H., Katz, R.: An algebraic approach to practical and scalable overlay network monitoring. In: Proc. ACM SIGCOMM 2004, pp. 55–66 (2004)
14. Chu, Y.H., Rao, S., Seshan, S., Zhang, H.: A case for end system multicast. IEEE J. Sel. Areas Comm. 20(8), 1456–1471 (2002)
15. Coates, M., Castro, R., Nowak, R., Gadhiok, M., King, R., Tsang, Y.: Maximum likelihood network topology identification from edge-based unicast measurements. In: Proc. ACM SIGMETRICS 2002, pp. 11–20 (2002)
16. Coates, M., Hero, A., Nowak, R., Yu, B.: Internet tomography. IEEE Signal Processing Magazine 19(3), 47–65 (2002)
17. Donnet, B., Raoult, P., Friedman, T., Crovella, M.: Deployment of an algorithm for large-scale topology discovery. IEEE J. Sel. Areas Comm. 24(12), 2210–2220 (2006)
18. Duffield, N., Horowitz, J., Lo Presti, F., Towsley, D.: Multicast topology inference from end-to-end measurements. In: Proc. ITC Seminar on IP Traffic, Measurement and Modelling (2000)
19. Fei, T., Tao, S., Gao, L., Guerin, R.: How to select a good alternate path in large peer-to-peer systems? In: Proc. IEEE INFOCOM 2006 (2006)
20. Gao, L.: On inferring autonomous system relationships in the Internet. IEEE/ACM Trans. Networking 9(6), 733–745 (2001)
21. Govindan, R., Reddy, A.: An analysis of Internet inter-domain topology and route stability. In: Proc. IEEE INFOCOM 1997, pp. 850–857 (1997)
22. Govindan, R., Tangmunarunkit, H.: Heuristics for Internet map discovery. In: Proc. IEEE INFOCOM 2000, pp. 1371–1380 (2000)
23. Hei, X., Liang, C., Liang, J., Liu, Y., Ross, K.W.: A measurement study of a large-scale P2P IPTV system. IEEE Trans. Multimedia 9(8), 1672–1687 (2007)
24. Hyun, Y., Broido, A., claffy, k.: Traceroute and BGP AS path incongruities. Tech. Report, CAIDA (2003)
25. Jain, M., Dovrolis, C.: End-to-end available bandwidth: Measurement methodology, dynamics, and relation with TCP throughput. In: Proc. ACM SIGCOMM 2002, pp. 295–308 (2002)
26. Jin, X., Wang, Y., Chan, S.H.G.: Fast overlay tree based on efficient end-to-end measurements. In: Proc. IEEE Int'l Conf. Comm. (ICC 2005), pp. 1319–1323 (2005)
27. Jin, X., Yiu, W.P.K., Chan, S.H.G.: Improving the efficiency of end-to-end network topology inference. In: Proc. IEEE Int'l Conf. Comm. (ICC 2007), pp. 6454–6459 (2007)
28. Jin, X., Yiu, W.P.K., Chan, S.H.G., Wang, Y.: Network topology inference based on end-to-end measurements. IEEE J. Sel. Areas Comm. 24(12), 2182–2195 (2006)
29. Kwon, M., Fahmy, S.: Topology-aware overlay networks for group communication. In: Proc. ACM Int'l Workshop Network and Operating Systems Support for Digital Audio and Video (NOSSDAV 2002), pp. 127–136 (2002)

30. Lowekamp, B., O'Hallaron, D., Gross, T.: Topology discovery for large Ethernet networks. In: ACM SIGCOMM 2001, pp. 237–248 (2001)
31. Mao, Z.M., Rexford, J., Wang, J., Katz, R.H.: Towards an accurate AS-level traceroute tool. In: ACM SIGCOMM 2003, pp. 365–378 (2003)
32. Pansiot, J.J., Grad, D.: On routes and multicast trees in the Internet. ACM SIGCOMM Computer Comm. Review 28(1), 41–50 (1998)
33. Rabbat, M.G., Coates, M.J., Nowak, R.D.: Multiple-source Internet tomography. IEEE J. Sel. Areas Comm. 24(12), 2221–2234 (2006)
34. Spring, N., Mahajan, R., Wetherall, D.: Measuring ISP topologies with Rocketfuel. In: Proc. ACM SIGCOMM 2002, pp. 133–145 (2002)
35. Subramanian, L., Agarwal, S., Rexford, J., Katz, R.H.: Characterizing the Internet hierarchy from multiple vantage points. In: Proc. IEEE INFOCOM 2002, pp. 618–627 (2002)
36. Tangmunarunkit, H., Govindan, R., Jamin, S., Shenker, S., Willinger, W.: Network topology generators: degree-based vs. structural. In: Proc. ACM SIGCOMM 2002, pp. 147–159 (2002)
37. Waldvogel, M., Rinaldi, R.: Efficient topology-aware overlay network. ACM SIGCOMM Computer Comm. Review 33(1), 101–106 (2003)
38. Winter, R., Zahn, T., Schiller, J.: Topology-aware overlay construction in dynamic networks. In: Proc. IEEE Int'l Conf. Networking, ICN 2004 (2004)
39. Yao, B., Viswanathan, R., Chang, F., Waddington, D.G.: Topology inference in the presence of anonymous routers. In: Proc. IEEE INFOCOM 2003, pp. 353–363 (2003)
40. Zhang, B., Liu, R., Massey, D., Zhang, L.: Collecting the Internet AS-level topology. ACM SIGCOMM Computer Comm. Review 35(1), 53–61 (2005)
41. Zhang, X., Liu, J., Li, B.: On large scale peer-to-peer video streaming: Experiments and empirical studies. In: Proc. IEEE Int'l Workshop on Multimedia Signal Processing, MMSP 2005 (2005)
42. Zhang, X.Y., Zhang, Q., Zhang, Z., Song, G., Zhu, W.: A construction of locality-aware overlay network: mOverlay and its performance. IEEE J. Sel. Areas Comm. 22(1), 18–28 (2004)

Resolution-Improvement Scheme for Wireless Video Transmission

Liang Zhou[1], Athanasios Vasilakos[2], Yan Zhang[3], and Gabiel-Miro Muntean[4]

[1] UEI, ENSTA-ParisTech, Paris, France
liang.zhou@ieee.org
[2] Department of Computer and Telecommunications Engineering, University of Western Macedonia, Greece
vasilako@ath.forthnet.gr
[3] Simula Research Laboratory, Norway
yan.zhang@ieee.org
[4] School of Electronic Engineering, Dublin City University, Ireland
munteangateeng.dcu.ie

Abstract. In recent years, wireless video transmission has emerged as one of the high growth applications of wireless communication technology. However, this error-prone network is packet based where many potential reasons may result in packet loss which has a devastating effect on the visual quality of images at the receiver. In this work, we study a coordinated application of Error-Resilient (ER) and Super-Resolution (SR) to enhance the resolution of image transmitted over wireless networks. Compressed video bitstreams require protection from channel errors in a wireless channel. The 3-D set partitioning in hierarchical trees (SPIHT) coder has proved its efficiency and its real-time capability in compression of video. A forward-error-correcting (FEC) channel (RCPC) code combined with a single automatic-repeat request (ARQ) proved to be an effective means for protecting the bitstream. Furthermore, a robust SR algorithm is proposed in the presence of different kinds of packet loss rate to enhance the image resolution. Experimental results indicate that the proposed robust resolution-enhancement scheme outperforms the competing methods from the aspects of PSNR (Peak-Signal-to-Noise Ratio) and visual quality under different packet loss rates.

1 Introduction

In recent years, wireless video transmission has emerged as one of the high growth applications of wireless communication technology. However, this error-prone network is packet based where many potential reasons may result in packet loss which has a devastating effect on the visual quality of images at the receiver. Furthermore, in most electronic imaging applications, image with high resolution (HR) is desired and often required because HR image can offer more details that may be critical in various applications. Unfortunately, it is challenging to provide HR image transmitted over wireless networks where no Quality of Service (QoS)

C.W. Chen et al. (Eds.): Intel. Multimedia Communication: Tech. and Appli., SCI 280, pp. 443–464.
springerlink.com

is guaranteed at the network level. In order for HR, it would be essential to solve the following primary technical challenges [3]:

• Trade-off between coding efficiency and error resilience. Most current standardized video codecs, including MPEG-2/4 and H.263/4 are designed to achieve high compression efficiency at the expense of error resilience. The coding efficiency in these codecs is achieved by using motion-compensated prediction to reduce the temporal and statistical redundancy between the video frames. This brings a severe problem, namely error propagation, where errors due to packet loss in a reference frame propagate to all of the dependent frames leading to visual artifacts that can be long lasting and annoying [1].

Therefore, Error resilience is needed to achieve robust video transmission. One strategy to achieve resilience is to insert redundant information systematically into compressed video signals so that the decoder can compensate transmission errors. The redundant information can be error correction codes [2],[4] or multiple descriptions [5],[6]. The former one combined with layered coding can provide good performance in prioritized networks while the latter is suitable for delivery over multiple channels to enhance reliability. However, error resilience is achieved at the expense of coding efficiency in both methods. Another way can be achieved with feedback mechanism to request retransmission or adjust encoding modes according to conditions. The methods proposed in [7], [8] rely on feedback from the decoder and are, therefore, application-limited.

• Enhance resolution under the scenario of packet loss. In order for HR image, one promising approach, which is called super resolution (SR), uses signal processing techniques to obtain a HR image from observed multiple low resolution (LR) images [9]. In the past few decades, a variety of SR methods have been proposed for estimating the HR image from a set of LR images without taking into account the packet loss during the transmission. The critical requirement for traditional SR approach is that the observations contain different but related views of the scene [9], however, it can not be guaranteed under the framework of error-prone networks where the packet loss destroys the correlation between the related views of scene. Therefore, how to apply the SR approach to the packet loss scenario is still an open problem. Here, it is necessary to differentiate error concealment (EC) with SR. EC hides or recovers the errors by using correctly received image information without modifying source or channel coding schemes [10], which can only produce a visually acceptable (rather than exact) image from the available data, and can not enhance the physical resolution of the image. While SR extracts the exact detail information hidden among the different but related video frames to enhance the image resolution.

To meet these challenges, in this chapter, we propose an entire scheme to get HR video transmitted over wireless networks by integrating efficient ER strategy with robust SR algorithm, which not only provides relatively efficient compression and transport performance but also provides robust resolution-enhancement performance in the presence of various packet loss rates.

2 Preliminaries

In this section, we first overview system framework of the video transmission and processing, and then present some related technical preliminaries used in this work, such as shifted 3-D SPIHT algorithm and multiple description coding.

2.1 System Overview

The total architecture of video transmission and processing illustrated in Fig.1 is composed of three processes, such as image degradation, image transmission over error-prone networks and the image SR reconstruction process.

Generally, all of the video sequences we observe are LR images comparing to the real-world scenes which are viewed as the original HR images. That is because the degradation process affects the quality of images acquired by digital video camera which results from the lens' physical limits, such as motion warping, optical blur and additive noise. In addition, these LR images are usually down-sampled convenient for transmission or storage. Next, the observed LR images are encoded and packetized preparing for transmission over error-prone networks. In this chapter, the method of encoding and packetizing is based on the shifted 3-D SPIHT algorithm to generate variable descriptions (substreams) at the sender,

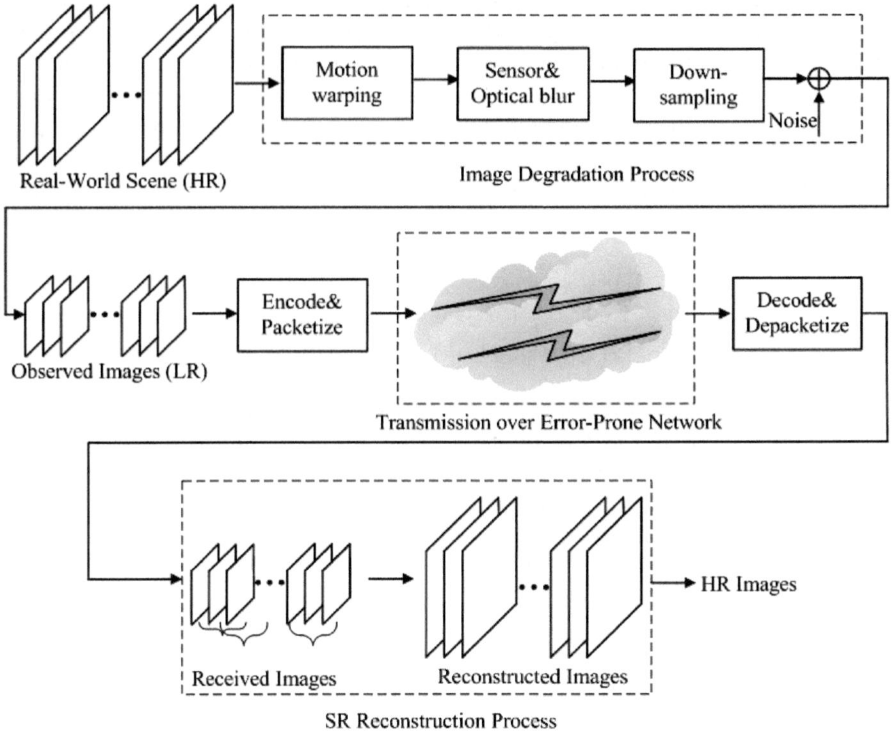

Fig. 1. The total architecture of video transmission and processing

and different descriptions employ different error protection strategies according to its priority. To specify the wireless networks, we employ a classic wireless sensor networks (WSN) here. As to the WSN system, suppose that there are n nodes in the system and n' $(1 \leq n' < n)$senders (sources) stream complementary substreams to a single receiver (destination) over different paths. In this system, sender-n'' streams substreams-n'' to the receiver over path-$n''(n'' = 1, 2..., n')$.At the receiver, the received images can be reconstructed after depacketizing and decoding the received data.

2.2 Shifted 3-D SPIHT Algorithm

Wavelet zero-tree image-coding technique was first addressed by Shapiro [11], and further developed by Said and Pearlman [12], and have provided unprecedented high performance in image compression with low complexity. Later, extended 2-D zero-tree wavelet coding (EZW) by [12] had been introduced to three dimensions (3-D EZW) by Chen and Pearlman [13], and had shown promise of an effective and computationally simple video-coding system without motion compensation, obtaining excellent visual results. And then, Kim and Pearlman developed the 3-D SPIHT [14] coding algorithm based on the 3-D EZW mentioned in [13].

3-D SPIHT algorithm provides excellent rate-distortion performance along with a low encoding complexity, and spatiotemporal trees are defined as groups of wavelet transform coefficients organized into trees rooted in the lowest frequency subband and the spatially and temporally related coefficients in the higher frequency subbands, which is helpful to reduce the error propagation [14]. Fig.2(a) shows how coefficients in a 3-D transform are related according to their spatial and temporal domains. The parent-children linkage except at the highest and lowest pyramid levels (which do not have offspring) is:

$$\begin{aligned} O(i,j,k) = \quad & \{(2i,2j,2k),(2i,2j+1,2k),(2i,2j,2k+1),\\ & (2i+1,2j,2k),(2i+1,2j+1,2k),(2i+1,2j,2k+1),\\ & (2i,2j+1,2k+1),(2i+1,2j+1,2k+1)\} \end{aligned} \tag{1}$$

where $O(i, j, k)$ represents a 3-D of coordinates of all the offspring at node (i, j, k).

Although the SPIHT-coded bitstream has many advantages, it is also sensitive to data losses because of the dependence among wavelet coefficients in constructing a significance map. Motivated by the packetized 2-D SPIHT coding algorithm [15] partitions the wavelet coefficients into independent packets to combat error propagation by shifting 2-D wavelet tree structure, here we apply it to 3-D case by shifting the 3-D wavelet tree structure. The essential aim is the wavelet coefficients from different sub-bands are interleaved to form independent packets that can be decoded independently. The formation of the shifted wavelet trees is shown in Fig.2(b). It should be noted that although a lot of multi-interlaced partitioning and packetizing methods have been presented in recent years [24], they could enhance the encoding burden greatly. Therefore, we don't employ them here.

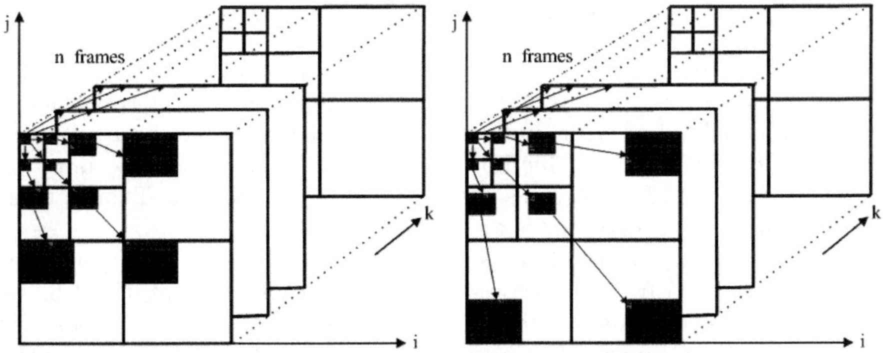

(a) Traditional 3-D SPIHT (b) Shifted 3-D SPIHT

Fig. 2. Structure of the spatiotemporal relation of 3-D SPIHT

2.3 Multiple Description Coding

As to the way to protect data from packet losses induced by the error-prone channels, one of the most common ways is to add the redundant information at the bitstream so that the original video can be recovered in presence of the packet loss, and the code should be devised in such a way that the decoder is able to recover the lost information or conceal the irrecoverable errors.

One such popular approach is multiple description coding (MDC), which is similar in spirit to the multiple-substream approach. The fundamental principle of MDC is to generate multiple correlated descriptions of the source such that each description approximates the source with a certain level of fidelity [16]. MDC does not impose any dependency among its descriptions so that each extra successfully received description improves the quality further regardless of what has been received so far.

The benefits of using MDC in video streaming can be further amplified when MDC is combined with path diversity (PD) [17]. In this approach, each substream (or description) is explicitly transmitted over an independent path to receiver. PD exploits the fact that the probability of having all the paths simultaneously congested is relatively low. As a result, the use of PD in video streaming can achieve higher throughput and increase tolerance to packet loss.

3 Adaptive Error-Resilient Strategy

In this section, a novel error-resilient strategy is proposed based on partitioning the GOF (group of frames) into variable substreams with different priority levels adapting to the current network condition.

3.1 Unequal Error Protection

Progressive bitstreams provide a natural basis for unequal error protection (UEP), by which perceptually more important parts of the bitstream are assigned a greater level of error protection [18]. In this work, we propose a novel UEP based on the expected lifetime of the path to guarantee the important part (high priority) of substreams are received at the receiver. That is to say, important substream corresponds to the "good" path.

Here, we set the priority of the substreams according to the importance of the subband derived from the wavelet decomposition.In general, the LL_k provides more visual information than HH_k,HL_k and LH_k(where k denotes the decomposition level),so we set priority level of LL_k as 2^{2k},while the other three subbands are $2^{2(k-1)}$. To the priority of path, we use the ratio of remaining energy (E) to transmit power (P_T) to denote the expected lifetime of the node. So the higher the ratio is the more lifetimes of the node. In order to maximize the lifetime of the whole network system, to an arbitrary node i, the criterion of its next hop should satisfy that

$$R(i) = \max(E_i/P_{T_i}) \tag{2}$$

where $R(i)$ denotes the maximum expected lifetime for node i corresponding to the optimal energy transmission to next hop which locates on the desired path. Assuming that the discovered path j consists of nodes such as $i, i+1, ...i+n$ $(n \in N^+)$, R_j, the energy consumer function of path j, here we define

$$R_j = \min\{R(i), R(i+1), ...R(i+n)\} \tag{3}$$

This means that the minimum expected lifetime of each node determines the whole lifetime of the path. Because once one of the nodes in the path dies, the whole path dies as well. The more R_j the higher priority of the path, and correspond to higher level of encoded substream. Note that, in order to realize the proposed UEP, we modify the traditional DSR (Dynamic Source Routing) Protocol as follows: besides adding the node's ID to the request packet, each node also adds the information of transmit power and remaining energy to the request packet, if the node receives packet, the information of the received power is also added to the packet. So when the sender node receives the request packet, the packet should consist of the route nodes' ID, the remaining energy of each route node and each node's transmit power and received power.

3.2 Flexible MDC

We focus on how to design the MDC according to the networking condition: one is how many descriptions are needed to guarantee the reconstructed video quality as well as keep the total bitstream as little as possible, while the other one is how to distribute these encoded data to the determinate substreams.

Obviously, the more substream the more data received at the destination, but it is infeasible for the practical wireless network. Here, we give an oversimplified method to compute the minimum needed substream number according to the packet loss rate (P_L) of the obtained channels.

With regard to the channel model, we use a two-state Markov model (i.e. Gilbert model) to simulate the bursty packet loss behavior [19]. The two states of this model are denoted as G (good) and B (bad). In state G, packets are received correctly and timely, whereas, in state B, packets are assumed to be lost. This model can be described by the transition probabilities p from state G to B and q from state B to G. Then the average P_L is given by

$$P_L = \frac{p}{p+q} \tag{4}$$

And the average length of burst errors L_B is given by

$$L_B = \frac{1}{p} \tag{5}$$

So the average channel P_L of path j is $p_j/p_j + q_j$. Suppose that all the transmit probability of potential paths are independent, the minimum number of substreams L is

$$\min_{L \in N+} \prod_{j=1}^{L} \left(\frac{p_j}{p_j + q_j}\right) \leq Thr \tag{6}$$

where Thr is the tolerance P_L threshold which depends on the practical application requirements. Usually, its value varies from $10^{-1} \sim 10^{-4}$ [23], in this paper, we set $Thr = 10^{-2}$.

In the case of data distribution, it also contains two aspects: one is the decision of the wavelet decomposition level, and the other one is the data distribution among these determinate paths. As to this point, we employ three basic principles:

- High priority level of the source data corresponding to high priority of the path, which is called equity principle.
- The whole transmission system only guarantees the most important source data, which has the highest priority level.
- As to other parts of the data, we use the best-effort strategy to transmit.

If a K-level dyadic wavelet decomposition is used, the number of wavelet coefficients in the k ($0 < k \leq K$) level spatial-frequency subband $C \in \{LL, LH, HL, HH\}$ is given by

$$C_k = \left(\frac{X}{2^k}\right) \times \left(\frac{Y}{2^k}\right) \tag{7}$$

where X and Y represent the frame width and height respectively. The level of the wavelet decomposition is decided by the expected lifetime of the highest priority path R_H. The essential requirement is that the GOF of the most important data should be guaranteed to transmit from source to destination over the highest priority path, so

$$\min_{K\geq 1} \lceil \frac{(LL_K/XY)\times r_b \times N_{GOF}}{F\times S} \rceil \leq R_H \tag{8}$$

where LL_K/XY denotes the ratio of the most important part in the total data streaming; N_{GOF} is the frame number of one GOF; r_b is the total source coding rate in bytes/s; F is the frame rate in frames/s; S is the packet size in bytes.

The fundamental distribution rule of data distribution is that the highest priority level data should be distributed at the each path while the others only distributed once at the path. Assuming $C' \in \{LH, HL, HH\}$, and expected life of path j is R_j, so each path should satisfy that

$$\lceil \frac{((LL_K + \sum_{k=K}^{1} C'_k)/XY)\times r_b \times N_{GOF}}{F\times S} \rceil \leq R_j \tag{9}$$

If there is the $C'_k (1 \leq k \leq K-1)$ which not transmitted by this path, it is transmitted by the other potential paths with lower priority level.

4 Spatial-Domain MAP Estimator

In this section we use maximum a posteriori probability (MAP) estimator to construct an estimated original HR video frame based on spatial-domain. Let us consider the LR sensor plane with $m \times n$ sensors elements. Suppose that the down-sampling parameter is q in both the horizontal and vertical direction. Then the HR frame is of size $qm \times qn$. The HR frame has intensity values $f_{i,j}$, for $i = 0, ..., qm-1, j = 0, ..., qn-1$. To simplify the computation, let f_i be a vector of size $q^2mn \times 1$ containing the intensity of the HR frame in lexicographically order, g_i be the $mn \times 1$ lexicographically ordered vector containing the intensity value of the blurred, decimated and noisy LR frame. Then the matrix form can be written as: where D is a decimation matrix of size $mn \times q^2mn$, H_i is a blurring matrix of size $q^2mn \times q^2mn$ and n_i is a $mn \times 1$ noise vector. Considering the estimated HR frame as a Markov random field (MRF), the probability distribution of the estimation object f_i and the degraded frame g_i (LR) is formulated by using Bayesian rule

$$p(f_i/g_i) = \frac{p(g_i/f_i)p(f_i)}{p(g_i)} \tag{10}$$

In order to maximize $p(f_i/g_i)$, it is equivalent to the constrained optimization problem

$$\sup\{\ln(p(g_i/f_i)) + \ln p(f_i)\}, subject\ to\ U(f_i) \leq E \tag{11}$$

where $U(f_i)$ is the energy function, subjecting to the Gibbs distribution and E is an unknown scalar. By using Lagrange multiplier technique, we obtained the non-constraint optimal problem as follows:

$$\arg\sup\{\ln(p(g_i/f_i)) + \ln p(f_i) - \alpha U(f_i)\} \tag{12}$$

where $\alpha \in [0, 1]$ is the Lagrange multiplier. According to the MAP estimation technique [10], we find that (12) can be re-written as

$$\arg\min_{f_i}\left\{\sum_{i=1}^{p}\|g_i - DH_i f_i\|_2^2 + \beta\|Lf_i\|_2^2\right\} \tag{13}$$

where p is the number of observed LR frames, β is a regularization parameter, L is the first-order finite-difference matrix and $L^T L$ is the discrete Laplacian matrix. In the above formulation, the noise variance term is absorbed in the regularization parameter β, and energy function $U(f_i)$ is included in the first-order finite-difference matrix. The minimization of the above equation is equivalent to solving the following linear system:

$$(\sum_{i=1}^{p} H_i^T D^T D H_i + \beta L^T L) f_i = \sum_{i=1}^{p} H_i^T D^T g_i \tag{14}$$

Generally, the blur process is the aperiodic boundary condition. We denote that T_i be the block-Toeplitz-Toeplitz-block (BTTB) matrix, and $L_e^T L_e$ to be the discrete Laplacian matrix with the zero boundary condition. Then the system (14) becomes

$$(\sum_{i=1}^{p} T_i^T D^T D T_i + \beta L_e^T L_e) f_i = \sum_{i=1}^{p} T_i^T D^T g_i \tag{15}$$

In this case, we employ a circulant matrix C_i to approximate the Toeplitz matrix T_i, similarly, we use $L_c^T L_c$ to to be the discrete Laplacian matrix with the periodic boundary condition to approximate $L_e^T L_e$.Then (15) becomes

$$(\sum_{i=1}^{p} C_i^T D^T D C_i + \beta L_c^T L_c) f_i = \sum_{i=1}^{p} C_i^T D^T g_i \tag{16}$$

where C_i is a block-circulant-circulant-block (BCCB) blurring matrix and $L_c^T L_c$ is a Laplacian matrix in BCCB structure. Notice that $C_i^T D^T D C_i$ is singular for all i, and $L_c^T L_c$ is positive semi-definite but it is has only one zero eigenvalue. This shows that the coefficient matrix $(\sum_{i=1}^{p} C_i^T D^T D C_i + \beta L_c^T L_c)$ is nonsingular. Therefore, the (8) can be solved and the HR frame can be reconstructed in theory.

However, it is difficult to get the solution of the HR frame f_i directly. Motivated by the PCG method mentioned in [16], to get an approximate solve instead of optimizing the MAP objective function, an iterative algorithm can be used:

$$\widetilde{f}_i^{t+1} = \widetilde{f}_i^{t} - \varepsilon G_i(g_i) \tag{17}$$

where $\widetilde{f}$ is the estimation of f, index t denotes the iterative times, the gradient G_i is

$$G_i = \frac{\sum_{i=1}^{p} C_i^T D^T D C_i + \beta L_c^T L_c}{\sum_{i=1}^{p} C_i^T D^T} \tag{18}$$

A good choice for the regularization parameter is:

$$\beta = \frac{\sum_{i=1}^{p} C_i^T D^T}{\sum_{i=1}^{p} C_i^T D^T D C_i} \tag{19}$$

and

$$\varepsilon = \frac{2}{p}\left(\frac{C_i^T D^T}{(C_i^T D^T)\phi_{max}(L_c^T L_c) + 1}\right) \tag{20}$$

where $\phi_{max}(\cdot)$ is the maximum eigenvalue of matrix.

5 Robust Super-Resolution Algorithm

Although adaptive error-resilient video transmission can reduce the transmission distortion, it can not exterminate the packet loss which has a devastating effect on visual quality. In this section, we propose a robust SR algorithm taking into consideration the various packet loss scenarios to enhance the resolution of received image. At first, we propose a simplified estimator to estimate the lost wavelet coefficients. And then, a series of convex sets which extract the exact detail information hidden among the adjacent images are constructed by taking advantage of the correlation of the wavelet coefficients.

5.1 Coefficient Estimator

Motivated by the model of [20], here we propose a simplified estimator to estimate the lost coefficients. Since LL subband provide basic information (low frequency) for original image, missing samples in approximation subband LL are estimated prior to processing the high frequency subbands in LH, HL, and HH. Therefore, different strategies are employed to deal with the different kinds of packet loss.

In the case of the LL subband packet loss, as the correlation of the wavelet coefficients is much less than the high-frequency subbands, it is difficult to use a common interpolation method to estimate the lost coefficients precisely. As a result, we propose a low-complexity solution which can replace the commonly used single interpolation mask by a set of masks. In this case, each of them is adapted to a specific direction of the predominant spatial correlation. Here we define the correlation from the aspects of horizontal and vertical direction respectively and use general 5×5 interpolation mask presented in Fig.3(a). The

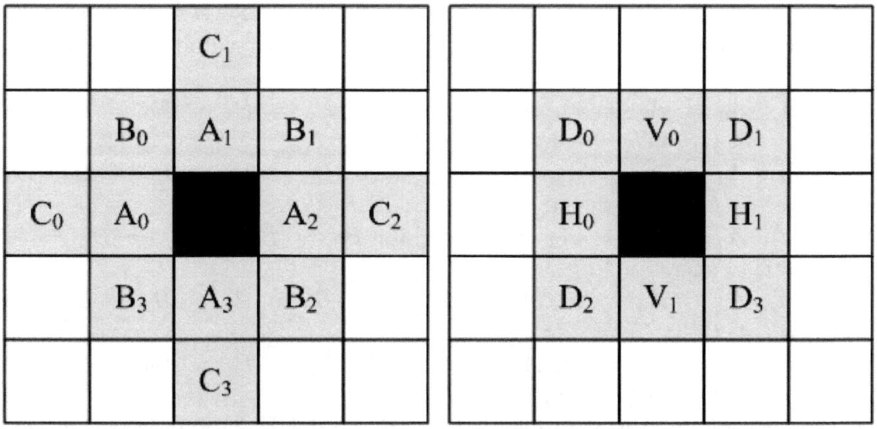

(a) Mask used in low-frequency subband (b) Mask used in high-frequency subband

Fig. 3. Labeling of the weights used in calculation at the missing sample.

parameters $A_0 \sim A_3$, $B_0 \sim B_3$ and $C_0 \sim C_3$ are weights for the corresponding neighboring coefficients. The weights in each mask are chosen according to the predominant correlation direction (horizontal or vertical) and according to the degree of correlation (strong or weak). In order to measure the predominant correlation direction and the degree of correlation we propose the following algorithm:

After the wavelet decomposition, the sender bi-linearly interpolates each scaling coefficients as if it is lost. This process is done twice: a first approximation is obtained by using a horizontal interpolation; a second approximation by using a vertical interpolation. The horizontal interpolation pass only calculates the mean value of the left and right neighbors and the vertical interpolation pass simply calculates the mean value of the upper and lower neighbors. In this way, two approximations of the subband LL_k are created: LL_{k_h} and LL_{k_v}. The sender then calculates the sum of the absolute differences (SAD) values for these two subbands compared to the original LL_k subband: SAD_v and SAD_h,

$$SAD = \sum_{i \in W} |I(i) - O(i)| \tag{21}$$

where I and O denote the interpolated and original coefficients respectively, and W is the comparison region, here is a 5×5 mask. We define a directional correlation measure as follows:

$$G_{h-v} = SAD_h - SAD_v \tag{22}$$

The value of G_{h-v} tells us how much one direction is better for interpolation than the other one. We define five classes: (a) strong horizontal correlation ($G_{h-v} > A$), (b) weak horizontal correlation ($A \geq G_{h-v} > B$), (c) isotropic

$(B \geq G_{h-v} \geq -B)$, (d) weak vertical correlation $(-B > G_{h-v} \geq -A)$ and (e) strong vertical correlation $(-A > G_{h-v})$. The resulting values of A and B are $A = 15$ and $B = 5$.

• Mask (a): $A_0 = A_2 = 30$, $A_1 = A_3 = 20$; $B_0 = B_1 = B_2 = B_3 = -8$; $C_0 = C_3 = -5$, $C_1 = C_2 = -4$;
• Mask (b): $A_0 = A_2 = 35$, $A_1 = A_3 = 25$; $B_0 = B_1 = B_2 = B_3 = -7$; $C_0 = C_3 = -5$, $C_1 = C_2 = -4$;
• Mask (c): $A_0 = A_1 = A_2 = A_3 = 35$; $B_0 = B_1 = B_2 = B_3 = -10$; $C_0 = C_1 = C_2 = C_3 = -4$;
• Mask (d): $A_0 = A_2 = 25$, $A_1 = A_3 = 35$; $B_0 = B_1 = B_2 = B_3 = -7$; $C_0 = C_3 = -4$, $C_1 = C_2 = -5$;
• Mask (e): $A_0 = A_2 = 20$, $A_1 = A_3 = 30$; $B_0 = B_1 = B_2 = B_3 = -8$; $C_0 = C_3 = -4$, $C_1 = C_2 = -5$;

In the case of missing samples in LH, HL, and HH subbands, we label the weighting factors as shown in Fig.3(b). That is, weights connecting horizontal neighbors are labeled H_0 and H_1, those connecting vertical neighbors V_0 and V_1, diagonal neighbors D_0, D_1, D_2, and D_3. Note that the direction of low-pass filtering is the direction in which high correlation of samples may be expected. To exploit the correlation between the wavelet coefficients in high-frequency subbands, a linear estimated model is adopted to estimate the missing samples. So the weighting factors can be set that

• Mask (LH): In LH, $H_0 = H_1 = 6$, $V_0 = V_1 = D_0 = D_1 = D_2 = D_3 = 1$;
• Mask (HL): In HL, $V_0 = V_1 = 6$, $H_0 = H_1 = D_0 = D_1 = D_2 = D_3 = 1$;
• Mask (HH): In HH, $D_0 = D_1 = D_2 = D_3 = 4$, $H_0 = H_1 = V_0 = V_1 = 1$;

The value of G_{h-v} can be sent as three bits (which represent the five classes) in each packet. This causes nearly no additional transmission overhead and no computational overhead for the receiver. All calculations are done by the sender and for the scaling coefficients only. According to a number of experiments, for a wavelet decomposition with depth is 3, the scaling coefficients only 1.6% of the total number of the coefficients. It should be emphasized that these factors are obtained empirically on the basis of considerable experimentation.

5.2 Projection onto Convex Sets

In this subsection, a projection procedure is utilized to extract information hidden in a group of video frames to update the wavelet coefficients. Since these coefficients correspond to the high frequency information in the spatial domain, the exacted fine features from other frames augment the individual LR frame to a HR frame. The constructed convex set should satisfy the following two points: 1) enhance the resolution of the received images, 2) reduce the artifacts generated during the projection process. We account for them in the following and present the relevant notation descriptions in Table 1.

Table 1. Notation

Symbol	Description
$f(n,t)$	observed low-resolution frame
$f_c(n,t)$	the current constructed low-resolution frame
$f_r(n,t)$	the reference low-resolution frame
$f'(n,t)$	the original high-resolution frame
$\widetilde{f}'(n,t)$	the estimation of the original high-resolution frame
C_{inter}	the convex set using the inter-frame projection
C_{intra}	the convex set using the intra-frame projection

Let $f'(n,t) \in L^2(R^2)$ denotes the original HR image[1], which can be expanded as a sum of approximation component in the LL band and three detail components in the LH, HL and HH bands.

$$\begin{aligned} f'(n,t) \equiv & \underbrace{\sum_{j,k\in Z} \alpha_{j,k}\psi_{j,k}(n,t)}_{LL} + \underbrace{\sum_{j,k\in Z} \beta^h_{j,k}\varphi^h_{j,k}(n,t)}_{LH} \\ & + \underbrace{\sum_{j,k\in Z} \beta^v_{j,k}\varphi^v_{j,k}(n,t)}_{HL} + \underbrace{\sum_{j,k\in Z} \beta^d_{j,k}\varphi^d_{j,k}(n,t)}_{HH} \end{aligned} \quad (23)$$

where $\varphi^h_{j,k}(n,t)$, $\varphi^v_{j,k}(n,t)$ and $\varphi^d_{j,k}(n,t)$ are the translated wavelets at the next coarse scale level that capture detail information in the horizontal, vertical and diagonal directions respectively, and $\psi_{j,k}(n,t)$is the translated coarse scaling function. The approximation and detail wavelet coefficients are given by

$$\alpha_{j,k} = \iint f'(n,t)\psi_{j,k}(n,t)dndt \quad (24)$$

$$\beta^h_{j,k} = \iint f'(n,t)\varphi^h_{j,k}(n,t)dndt \quad (25)$$

$$\beta^v_{j,k} = \iint f'(n,t)\varphi^v_{j,k}(n,t)dndt \quad (26)$$

$$\beta^d_{j,k} = \iint f'(n,t)\varphi^d_{j,k}(n,t)dndt \quad (27)$$

Let $f_c(n,t)$ the current constructed LR frame and $f_r(n,t)$ be the reference LR image. The convex set can be defined from the above four aspects:

$$C_{inter} = \{\widetilde{f}'(n,t)|C^{\alpha}_{inter}, C^h_{inter}, C^v_{inter}, C^d_{inter}\} \quad (28)$$

where

[1] The original value of $f'(n,t)$ can be achieved by expanding the LR image $f(n,t)$ using bi-linear method.

$$C^{\alpha}_{inter} = \{\iint \widetilde{f}'\psi_{j,k} dndt = f_c\} \tag{29}$$

$$C^{h}_{inter} = \{\iint \widetilde{f}'\varphi^{h}_{j,k} dndt = \widetilde{\beta}^{h}_{j,k} = \iint f_r\varphi^{h}_{j,k} dndt\} \tag{30}$$

$$C^{v}_{inter} = \{\iint \widetilde{f}'\varphi^{v}_{j,k} dndt = \widetilde{\beta}^{v}_{j,k} = \iint f_r\varphi^{v}_{j,k} dndt\} \tag{31}$$

$$C^{d}_{inter} = \{\iint \widetilde{f}'\varphi^{d}_{j,k} dndt = \widetilde{\beta}^{d}_{j,k} = \iint f_r\varphi^{d}_{j,k} dndt\} \tag{32}$$

$\widetilde{f}'(n,t)$ denotes the estimated value of the $f'(n,t)$. So the projection of $\widetilde{f}'(n,t)$ to C_{inter} (denoted by P_{inter}) is defined as

$$\begin{aligned} f'(n,t) &\equiv P_{inter}[\widetilde{f}'(n,t)] \\ &\equiv \sum_{j,k\in Z} f_c\psi_{j,k}(n,t) + \sum_{j,k\in Z} \widetilde{\beta}^{h}_{j,k}\varphi^{h}_{j,k}(n,t) \\ &+ \sum_{j,k\in Z} \widetilde{\beta}^{v}_{j,k}\varphi^{v}_{j,k}(n,t) + \sum_{j,k\in Z} \widetilde{\beta}^{d}_{j,k}\varphi^{d}_{j,k}(n,t) \end{aligned} \tag{33}$$

Next, two steps of wavelet transform operation are implemented on the image $f'(n,t)$ which is produced after the inter-frame projection. Firstly, we don't employ the down-sampling, and get the three high frequency bands such as LH_1, HL_1 and HH_1. Secondly, we employ the down-sampling, and get the low frequency band LL and other three high frequency bands LH_2, HL_2 and HH_2. In addition, wavelet transform operation is implemented on the LL band and non-sampling, so we can get three bands of LH_3, HL_3 and HH_3. At last, the LS (Least Square) estimation of prediction is utilized to get the three bands of $\widetilde{LH}$, $\widetilde{HL}$ and $\widetilde{HH}$. The convex can be constructed as follow:

$$C_{intra} = \{\widetilde{f}''(n,t)|C^{LH}_{intra}, C^{HL}_{intra}, C^{HH}_{intra}\} \tag{34}$$

where

$$C^{LH}_{intra} = \{[r_{LH} = \widetilde{LH} - \sum_{j,k\in Z} \widetilde{\beta}^{h}_{j,k}\varphi^{h}_{j,k}] \in [-\delta,\delta]\} \tag{35}$$

$$C^{HL}_{intra} = \{[r_{HL} = \widetilde{HL} - \sum_{j,k\in Z} \widetilde{\beta}^{v}_{j,k}\varphi^{v}_{j,k}] \in [-\delta,\delta]\} \tag{36}$$

$$C^{HH}_{intra} = \{[r_{HH} = \widetilde{HH} - \sum_{j,k\in Z} \widetilde{\beta}^{d}_{j,k}\varphi^{d}_{j,k}] \in [-\delta,\delta]\} \tag{37}$$

where r_{LH}, r_{HL} and r_{HH} denote the residue between the estimated HR image through degradation and the original LR image in LH, HL and HH subband, respectively. $[-\delta,\delta]$ shows the boundary of the noise magnitude. After projecting by P_{intra}, we get (here we only take the LH part for example, the other two bands are similar to this)

$$(LH)' = P_{intra}[\widetilde{LH}] = \widetilde{LH} + \begin{cases} \widetilde{LH} - \sum\limits_{j,k\in Z} \widetilde{\beta}^h_{j,k}\varphi^h_{j,k}, & r_{LH} > \delta \\ \widetilde{LH} + \sum\limits_{j,k\in Z} \widetilde{\beta}^h_{j,k}\varphi^h_{j,k}, & r_{LH} \le -\delta \\ 0, & else \end{cases} \tag{38}$$

From the $(LH)'$, $(HL)'$, $(HH)'$ and the known LL, the HR image can be get easily.

6 Simulation Results and Discussions

In this section, we conduct simulation experiments to study the performance of the proposed robust resolution-enhancement scheme in a distributed video streaming framework. First of all, we describe the simulation environment. Secondly, we present the main simulation results where we show the objective and subjective results of the performance of the proposed system under different scenarios. Finally, we conclude this section by summarizing the conclusions to be drawn based on the selected simulation results described.

In order to provide a meaningful comparison between our proposed approach and other alternative approaches, we consider use of a recent unbalanced MDC with UEP mentioned in [16] as a comparison system. This paper presents a distributed video streaming framework using unbalanced MDC and UEP under the video streaming framework that two senders simultaneously stream complementary descriptions to a single receiver over different paths. To minimize the overall distortion and exploit the benefits of multi-path transport when the characteristics of each path are different, this paper also propose an unbalanced MDC method for wavelet-based coders combined with a TCP-friendly rate allocation algorithm. In addition, three-level wavelet decomposition is applied to a group of 16 frames and the 3-D wavelet coefficients are divided into two unequal-sized groups according to the priority level. Each substream is independently protected using the Forward Error Correction (FEC) based UEP algorithm proposed in [21]. Because this method can not enhance the resolution of image, we utilize the proposed SR method to give a fair comparison. For simplicity, we note this method as unbalanced method. In addition, a fixed wavelet decomposition $K = 2$ (the remaining is the same as our proposed approach) is compared with the flexible MDC; and traditional bilinear method (the remaining is the same as our proposed approach) is employed to compare with our SR algorithm. Similarly, the above two methods are noted as fixed method and bilinear method, respectively.

6.1 Simulation Environment

For these experiments, the two standard video sequences, Foreman and Weather forecast, are encoded with shifted 3-D SPIHT algorithm. These video sequences are 352×288 pixels per frame; frame rate $F = 30 frames/s$; down-sampling parameter $q = 2$; the blur is Gaussian blur where the support of the blurring kernel is 29; frame number of one GOF $N_{GOF} = 16$. At the receiver, the desired HR frame is reconstructed from several LR 176×144 degraded frames transmitted by MANETs. In order for objective comparison, PSNR at the receiver relative to the original HR video sequence is used and its definition is

$$PSNR(dB) = 10 \log_{10}(\frac{255^2}{MSE}) \tag{39}$$

where MSE is the mean-square error between the original the reconstructed luminance frame. To the network part, 30 nodes that move randomly at maximum speed 2m/s are dispersed in the area 200×200 square meters, and the routing algorithm is based on MRDSR (Multiple Route Dynamic Source Routing)[22]. In addition, the packet size $S = 512$bytes and initial energy of nodes is 1 Joule with uniform distribution. It should be noted that all the simulation results in this section have been obtained using 30 runs in order to obtain statistically meaningful average values.

6.2 Selected Simulation Results and Discussions

In Fig.4, we illustrate a plot of PSNR versus the packet loss rate P_L with burst length $L_B = 4$ for the Foreman sequence at $r_b = 96$Kbps, which is the range of low-bit services. The proposed method can be seen to achieve a much higher performance in terms of end-to-end PSNR compared to the representative fixed method which apply fixed wavelet decomposition level regardless of the condition

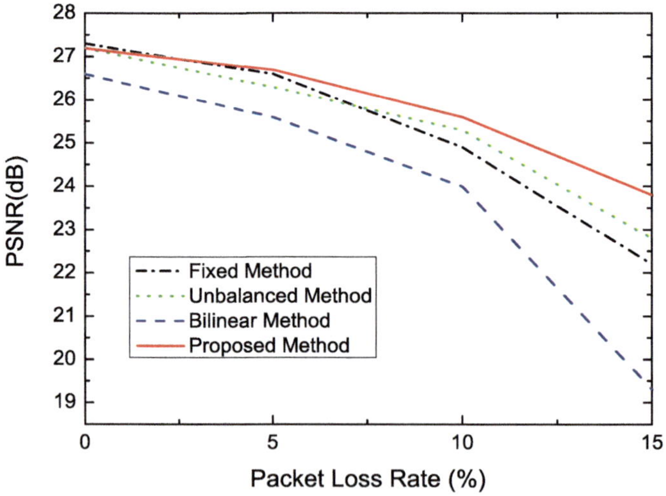

Fig. 4. Performance achieved by proposed method for the Foreman sequence, at the $r_b = 96$Kbps and $L_B = 4$.

of the network and unbalanced method which compulsively divided stream into two unbalanced substreams. For the above two methods, there is a considerable performance disadvantage due to the improper use of the decomposition level and data distribution. Obviously, for little loss transmission ($P_L \leq 5\%$) the fixed method achieves almost the same performance compared to the proposed scheme, and proposed method only gets marginal performance gain compared to unbalanced method when ($8\% \leq P_L \leq 10\%$). However, as P_L increases, the performance gap is increased dramatically since when the packet loss rate is high, more packets are unsuccessfully delivered obviously, non-adaptive scheme, including fixed and unbalanced method, can not adapt transmission strategy to this case while proposed method employ more decomposition to guarantee the most important data successfully transmitted. For example, when $P_L = 5\%$, the 2-level decomposition is adopted by the proposed method, so the gap between the proposed method with the fixed method is only about 0.1 dB, and when $P_L = 10\%$, the decomposition is adaptive to 3-level, the gap between the proposed method with the unbalanced method is about 0.3 dB, while the gap become 1.0dB when the $P_L = 15\%$ and 4-level decomposition is employed.

It should also be noted from Fig.4 that the performance achieved by the proposed method is also super to the traditional bilinear method to get higher resolution. Though proposed SR method can only achieve little gain compared to the traditional bilinear method in the case of lossless transmission ($P_L = 0$), the performance gap also increases sharply as the P_L increases. In particular, as can be seen from Fig.4, the performance gap between proposed SR methods with bilinear comparison method is more than 1.5dB when $P_L \geq 10\%$. The reason is that, when $P_L \leq 5\%$, few packets are lost and correlation of the adjacent

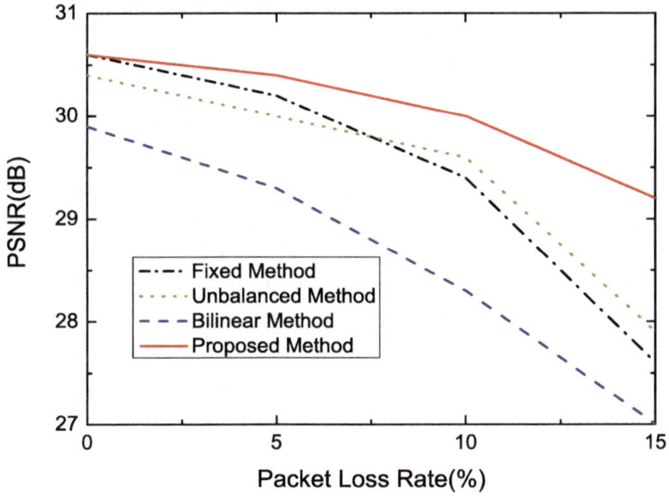

Fig. 5. Performance achieved by proposed method for the Foreman sequence, at the $r_b = 256$Kbps and $L_B = 4$.

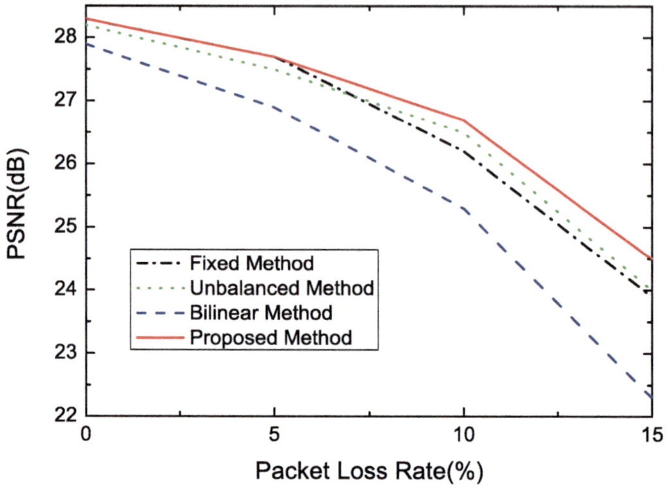

Fig. 6. Performance achieved by proposed method for the Weather forecast sequence, at the $r_b = 96$Kbps and $L_B = 4$.

packets is strong, bilinear as a simply interpolation method can also provide a relatively accurate estimation, while when $P_L > 5\%$ (especially $P_L \geq 10\%$), the packet correlation is so weak that bilinear can't performs well in this case. To our proposed SR method, adjacent and current frames information is utilized to enhance the resolution of the received image, therefore, it can achieve a more satisfying performance whether packet loss rate high or not.

In Fig.5, we repeat the results for Foreman sequence with $r_b = 256$Kbps and $L_B = 4$. It can be seen that the performance gap between the fixed or unbalanced method with the proposed method is much larger than Fig.4, which is the case for low-bit rate services, while Fig.5 is for high-rate services. This follows the fact that, when the available bit rate is high, the reconstructed video quality benefits considerably from the use of flexible MDC substreams and SR reconstruction technique which uses adaptive strategy to adapt current network condition and correlation information hidden in adjacent frames to enhance the received image resolution. Still, it can be seen that when $P_L = 10\%$, the proposed method can achieve a performance gain of approximately 0.4dB compared to the unbalanced method, and 1.7dB compared to the bilinear interpolation method, and when $P_L = 15\%$, the gap increases to 1.3dB and 2.2dB, respectively.

As discussed previously, the Foreman sequence is a high-motion sequence, while the Weather forecast sequence is a low-motion sequence. In order to provide a more comprehensive evaluation of the proposed adaptive scheme, we repeat the results for Weather forecast sequence in Fig.6 and Fig.7, for $r_b = 96$Kbps and $r_b = 256$Kbps, respectively, and similar behaviors as in Fig.4 and Fig.5 are observed. It should be note that the performance gap between the proposed adaptive scheme and the compared methods are less obvious than that in Fig.4 and Fig.5 for the high-motion Foreman sequence, because the correlation between

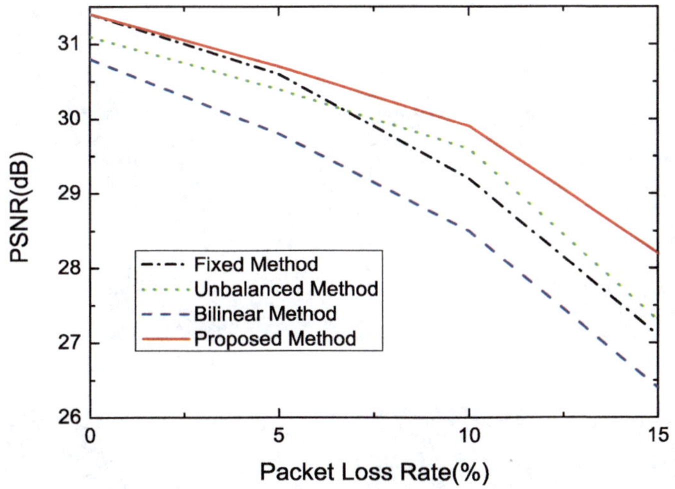

Fig. 7. Performance achieved by proposed method for the Weather forecast sequence, at the $r_b = 256$Kbps and $L_B = 4$.

the adjacent frames increases as the low-motion sequence implies. However, the proposed adaptive scheme also has the advantage of other alternative methods no matter what the P_L is.

Some subjective results to demonstrate the relative performance achieved by the proposed method are illustrated in Fig.8 corresponding to the objective results in Fig.8 for $P_L = 15\%$. Clearly, in Fig.8, the subjective results are consistent with the objective results in Fig.7. In particular, the proposed method can maintain more visual information comparing to the other alternative methods.

6.3 Observations

Based on the selected objective and subjective simulation results described above, there are several main observations:

- The adaptive error-resilient strategy has played an important role in the whole video transmission system. Comparing to the fixed method or the unbalanced method, the performance of the adaptive strategy is greater than those of them, especially when the packet loss rate is high.
- The proposed SR algorithm actually can enhance the resolution of the received image, and much detail information is maintained by taking advantage of the projection onto convex sets. Comparing to the traditional bilinear method, the proposed SR algorithm at least has the 0.5dB PSNR gain, furthermore, the advantage gets more evident as the packet loss rate increases.
- The proposed method has great robust. No matter the video sequence is high-motion or low-motion, the packet loss rate is high or low, the proposed method can perform well all the time.

(a) PSNR=26.4 dB

(b) PSNR=27.1 dB

(c) PSNR=27.3 dB

(d) PSNR=28.2 dB

Fig. 8. Subjective results achieved by proposed method and other comparison schemes, for the Weather forecast at $r_b = 256$Kbps, $L_B = 4$ and $P_L = 15\%$. (a) bilinear method; (b) fixed method; (c) unbalanced method; (d) proposed method.

7 Conclusions and Future Work

In this work, we propose a robust resolution-enhancement scheme for video stream transmission over wireless networks. As detailed in the paper, the scheme can adaptively respond to the dynamic network condition by adjusting the coding fashion at the encoder and error protection strategy during the transmission process. Furthermore, the SR algorithm used in the proposed scheme is so robust that it performs well in presence of different kinds of packet loss rates. Experiment results demonstrate that the proposed scheme outperforms the competing methods under the basis of the same simulation condition.

For future work, there are some potential research topics on resolution-improvement for wireless video transmission: (1) How to reduce the computation complexity of super-resolution algorithm to satisfy real-time video service. (2) Recently, there are some new emerging wireless systems (i.e., LTE, relay network, multi-radio network etc.), how to adapt the proposed resolution-improvement scheme to these emerging networks is imperative. (3) How to extend this work to pervasive video-related media service (i.e., IPTV, Mobile TV etc.) is also an interesting topic.

References

1. Kung, W., Kim, C., Kuo, C.: Spatial and temporal error concealment techniques for video transmission over noisy channels. IEEE Transactions on Circuits and Systems for Video Technology 167, 789–802 (2006)
2. Hwang, J., Chen, J., Huang, Y., Tsai, T.: Layered video coding based on displaced frame difference prediction and multi-resolution block matching. IEEE Transactions on Communications 529, 1504–1513 (2004)
3. Zhou, L., Zheng, B., Wei, A., Geller, B., Cui, J.: A Robust Resolution-Enhancement Scheme for Video Transmission over Mobile Ad-Hoc Networks. IEEE Transactions on Broadcasting 542, 312–321 (2008)
4. Mathew, R., Arnold, J.: Efficient layered video coding using data partitioning. Signal Process: Image Communication 14, 761–782 (1999)
5. Kim, C., Lee, S.: Multiple description coding of motion fields for robust video transmission. IEEE Transactions on Circuits System Video Technology 119, 999–1010 (2001)
6. Wang, Y., Lin, S.: Error-resilient video coding using multiple description motion compensation. IEEE Transactions on Circuit System Video Technology 122, 438–452 (2002)
7. Zhang, R., Regunathan, S., Rose, K.: Video coding with optimal inter/intra-mode switching for packet loss resilience. IEEE Journal of Selected Areas Communication 186, 966–976 (2000)
8. Steinbach, E., Farber, N., Girod, B.: Standard compatible extension of H. 263 for robust video transmission in mobile environment. IEEE Transactions on Circuit System Video Technology 76, 872–881 (1997)
9. Park, S., Kang, M.: Super-Resolution Image Reconstruction: A Technical Overview. IEEE Signal Processing Magazine 203, 21–36 (2003)
10. Lie, W., Gao, Z.: Video Error Concealment by Integrating Greedy suboptimization and Kalman Filtering Techniques. IEEE Transactions on Circuits and Systems Video Technology 168, 982–992 (2006)
11. Shapiro, J.: Embedded image coding using zerotrees of wavelet coefficient. IEEE Transactions on Signal Processing 416, 3445–3462 (1993)
12. Said, A., Pearlman, W.: A New, fast and efficient image codec based on set partitioning in hierarchical trees. IEEE Transactions on Circuits System, Video Technology 64, 243–250 (1996)
13. Chen, Y., Pearlman, W.: Three-dimensional subband coding of video using the zero-tree method. In: SPIE Visual Communication Image Processing, pp. 1302–1309 (1996)
14. Kim, B., Pearlman, W.: An embedded wavelet video coder using three-dimensional set partitioning in hierarchical trees. In: Proceeding Data Compression Conference, pp. 251–260 (1997)
15. Sriraja, Y., Karp, T.: A packetized SPIHT algorithm with over complete wavelet coefficients for increased robustness. Eurasip journal on applied signal processing, special issue on frames and over complete representations in signal processing, communications, and information theory 13, 112–120 (2005)

16. Kim, J., Mersereau, R., Altunbasak, Y.: Distributed Video Streaming Using Multiple Description Coding and Unequal Error Protection. IEEE Transactions on Image Processing 147, 849–861 (2005)
17. Gogate, N., Chung, D., Panwar, S., Wang: Supporting image and video applications in a multihop radio environment using path diversity and multiple description coding. IEEE Transactions on Circuits System Video Technology 126, 777–792 (2002)
18. Mohr, A., Riskin, E., Ladner, R.: Unequal loss protection: Graceful degradation of image quality over packet erasure channel through forward error correction. IEEE Journal Selected Areas Communication 186, 818–828 (2000)
19. Horn, U., Stuhlmuller, K., Girod, B.: Robust internet video transmission based on scalable coding and unequal error protection. Signal Processing: Image Communication 152, 77–94 (1999)
20. Bajic, I.: Adaptive MAP Error Concealment for Dispersively Packetized Wavelet-Coded Images. IEEE Transactions on Image Processing 155, 1226–1235 (2006)
21. Kim, J., Mersereau, R., Altunbasak, Y.: Error-resilient image and video transmission over the internet using unequal error protection. IEEE Trans. on Image Processing 122, 121–131 (2003)
22. Nasipuri, A., Castaneda, R., Das, S.: Performance of multipath routing for on-demand protocols in mobile ad-hoc networks. ACM/Kluwer mobile networks and applications 62, 339–349 (2001)
23. Kang, M., Alouini, M.: Transmission of Multiple Description Codes Over Wireless Channels Using Channel Balancing. IEEE Transactions on Wireless Communications 45, 2070–2075 (2005)
24. Bajic, I., Woods, J.: Domain-based multiple description coding of images and video. IEEE Transactions on Image Processing 1210, 1211–1225 (2003)

Human-Centered Face Computing in Multimedia Interaction and Communication

Yun Fu[1], Hao Tang[2], Jilin Tu[3], Hai Tao[4], and Thomas S. Huang[3]

[1] Department of Computer Science and Engineering, University at Buffalo, 201 Bell Hall, NY 14260, USA
yunfu@buffalo.edu
[2] Beckman Institute, University of Illinois at Urbana-Champaign, IL 61801, USA
{haotang2,huang}@ifp.uiuc.edu
[3] Visualization and Computer Vision Lab, General Electric, Niskayuna, NY 12308, USA
tujilin@ge.com
[4] Department of Computer Engineering, Baskin School of Engineering, University of California, Santa Cruz, CA 95064
tao@soe.ucsc.edu

1 Introduction

Facial image computing has been an extensively studied topic since its wide applications in human-centered computing, human-avatar interaction, virtual reality, and multimedia communication. Successful systems have been equipped with realistic face models, efficient compression algorithms, reliable animation techniques and user friendly interaction schemes. In this chapter, we will mainly focus on techniques, algorithms, models, applications and real-world systems. Comprehensive summarization of the state-of-the-art works will be presented as well as our experiences and contributions in the field, especially several real prototype systems developed in our group, such as the online interactive gaming system hMouse, humanoid emotive audio-visual avatar, and 3D face/head tracking based video compression. Performances of these three systems are also illustrated based on standard evaluations.

2 Online Interactive Gaming System: hMouse [25]

In recent years, new technologies and dedicated systems in Human-Computer Interaction (HCI) and Computer Vision (CV) have brought a great opportunity to the improvement of human life and independent living of the disabled and elder people. Those multidisciplinary researches from engineering sciences and human sciences provide a significant technical support for designing "*touch-free*" Perceptual User Interfaces (PUI), e.g. vision-based PUI. The basic function of such PUIs and human interface devices require moderately reliable and fast

C.W. Chen et al. (Eds.): Intel. Multimedia Communication: Tech. and Appli., SCI 280, pp. 465–505.
springerlink.com

system performance to handle a variety of complicated user interactions. Much attention of both academia and industry has been focused on the scenario of tracking user's movements with a video camera and translating the motion parameters into semantic indicatory symbols to manipulate machine operations. Camera-based perceptual user system can provide an alternative solution for convenient device control, which encourages the application of interactive computer/internet games, very low bit-rate human-to-human communication, intelligent avatar-based kiosks, and film-making, etc. It also affords a significant help to severely physically handicapped or elder people to avoid the communication gap between them and normal people, and work with voluntary actions [8, 25, 62].

As mentioned above, for significant number of people, natural access and independent control over a computer system is highly desirable. The camera mouse system [25, 62] is such a dedicated touch-free PUI that can track human body movement in video to manipulate virtual computer pointing devices to respond user intentions. In particular, since human face and head embody the focus of attention, human instincts naturally suggest using face and head as tools to communicate with machines. Therefore, head or face tracking based hand-free computer mouse is a comparatively intuitive and convenient PUI solution. Great progresses in the field of face detection or tracking and recent advances in hardware have made it possible of building real-time tracking system on common PCs and laptops.

In view of the foregoing background, our research goal is to develop a computer vision system that makes computers perceptive to the user's natural communicative cues relative to head pose and motion. Specifically, we develop the '*hMouse*' prototype system, which is a general head controlled camera mouse framework via real-time visual tracking techniques.

Related Work

Over the last decade, alternative PUIs have been developed for users who cannot control a computer through a standard mouse and keyboard. The Infrared Head-Operated Joystick [21] uses infrared LED and photo-detectors to determine head position that is subsequently converted into signals to emulate a computer mouse. This system, as a commercial product in U.K., is mainly developed for people with disabilities. The implementation of Tilt Sensors [12] in headset is invented for economical head-operated computer mouse. The tilt sensors detect the lateral and vertical head-motion to drive the left/right and up/down displacements with respect to the mouse displacement. A touch switch device contacts gently with user's cheek to trigger the device to perform single click, double clicks, and drag commands. Without ignoring the innovative and pioneering ideas of such works, the inconvenient aspect that users need to wear accessory instruments is also visible.

Active researches have proposed to navigate cursor and trigger mouse click with the movement of eyes, nose, and face [26, 28, 8, 32] in a touch-free manner. The basic idea of such system is to detect and track user's face in real-time

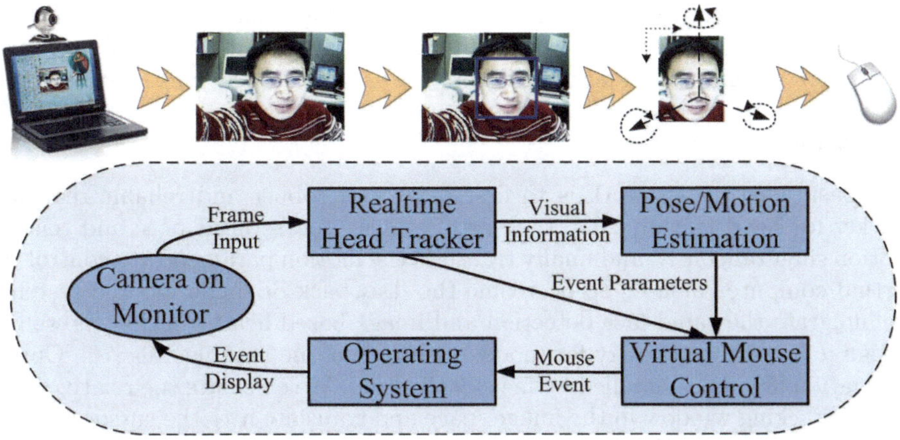

Fig. 1. Framework and system setup of hMouse.

and analyze the eye blinks, nose-tip movements, face pose and facial gesture to manipulate a virtual computer mouse. The advantage of such vision-based technique is that people can dynamically configure and control it by face or head motions without wearing any professional instruments. Some commercially available vision-based products were already developed in industry [11, 44, 39].

One of the most related work to hMouse is the 3-D tracking model based camera mouse [61, 62]. The system is basically a functional extension of the PBVD 3-D tracker [60], which is not fully automatic and requires manually tracker initialization. The PBVD tracker provides real-time performance for detailed facial motion tracking, but it will sometimes lose global tracking when users make large head movements. The definition of the virtual mouse events in the 3-D camera mouse system mainly depends on the motion of mouth, lip, nose and face, which actually restricts the user's focus of attention and other activities to some extent.

In general, the above mentioned vision systems require reliable and fast face tracking strategy, which allows users the comfortable and sufficient motion control. Some existing detection and tracking solutions still can not achieve such desired robustness and precision. Rapid face detection, such as boosted cascade of Haar-like features [66, 42], achieves high speed for frontal and near frontal faces in large size variation cases. However it also has high missing rates for large head rotations. Face / head tracking provides much higher speed performance and lower missing rate and false alarm rate than face detection [49, 76], but it can neither automatically initialize itself, nor handle large size variation cases. In particular, model-based tracking defines 2-D features or 3-D geometric mesh topology and precisely tracks action unit or subtle expression [60, 13]. However it has high computational cost for mesh localization and fails from large head motion. Feature-based tracking can achieve pixel accuracy at the cost of requiring high-resolution cameras and low robustness for head rotation and scaling. Image-based tracking, considering global facial cues, is robust to head rotation and scaling or low quality frames [4], but it is typically a coarse estimation scheme which lacks precision.

Our Work

The basic idea of our work is to first develop a robust and reliable real-time tracker for head tracking. We then estimate the coarse head pose and relative motion simultaneously, and finally translate the motion parameters to control the virtual computer mouse. To overcome the drawback of single module systems, we integrate the rapid face detection and image-based head tracking, as well as design a transition strategy for mode switch when one module fails [76]. Cursor navigation [56, 48] is handled by face localization. We calculate the relative position of tracking window in the image space and translate it to the cursor position in the screen space. Head tilting and yawing rotations are further used to fine tune the result. After mouse cursor is navigated to the desired location, virtual mouse button clicks are triggered by retrieving the head roll rotation. We also define the hold and release mouse events by inferring the threshold of tracking window size. The result is a prototype camera mouse system—'hMouse'—driven by virtual head tracking. Inheriting the good PUI concepts of the 3-D camera mouse, hMouse provides more flexibility and reliability in both visual tracking and virtual mouse controlling at the cost of some speed slowdown for mouse operations.

2.1 'hMouse' System Framework

Our hMouse system, implemented on a typical PC or laptop, provides a specific virtual human interface for hand-free computer mouse controlled by the user's head movements. The system setup and the framework of hMouse are illustrated in Fig. 1. As the same as the general camera mouse, hMouse consists of a virtual tracking module and a mouse control module. The system uses a common consumer camera with un-calibrated lens, and it can process each frame of the captured video in real-time. The user's face or head is first automatically detected and tracked by a robust and reliable head tracker. The head pose and motion parameters are then estimated by the visual tracking module from analyzing visual cues, such as motion, location, orientation, and object size. With basic synchronization control and spatio-temporal smoothing, mouse control module navigates cursor and controls virtual mouse buttons according to the received motion parameters. This PUI then passes all events to the underlying operating system to response. The current version of hMouse is not embedded into the operating system. Users need to first run an executable file to start the hMouse.

2.2 Robust Real-Time Head Tracking in a Loop

To implement a practical real-time system, such as hMouse, we have to consider two issues. One concern is the system resource usage (CPU and memory). Since hMouse is an accessorial system for OS, we need to guarantee the

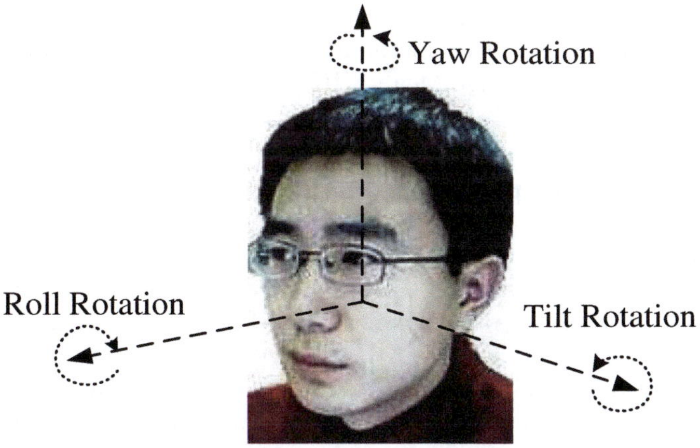

Fig. 2. 3-D head pose and parameters.

low resource cost in designing the hMouse framework. The majority of system resources should be released for other usage when the hMouse runs. Another concern is the reliability and robustness of the hMouse. The hMouse could be part of OS which needs to be sufficiently controllable for usage in any common case currently fulfilled by the hand-controlled mouse. Indeed, more system resource should be expended when necessary for complicated algorithm design to ensure robustness. To better balance the tradeoff, we'd better take into account the multi-model system design. In view of this balance, the hMouse tracker is designed to enable detection-tracking complementary switching, which exhibits very reliable head tracking performance.

Some existing pure frontal or profile face detection systems show robustness against occlusion, illumination variation, and small head yawing and tilting. However, they often lose detection in the case of users head rolling, side view, extreme motions, and part users' faces partially out of the camera viewport. Some existing pure skin-color based head tracking systems proceed very fast and survive from the cases when detection fails. However, they are too sensitive to handle illumination variation and skin-color occlusion (e.g. hand occlusion on the face). The detection and tracking module obviously can be complementary to each other for handling fairly difficult cases, e.g. users jumping, coming in and out, large degree rotations, users' faces partially out of the camera viewport, turning around, and multi-user occlusion.

We developed a robust real-time head tracker, which has an interactive detection-tracking switching loop, to drive the hMouse. In each tracking cycle, for a captured video frame, hMouse first processes automatic real-time face detection. If there are detected faces, the system will choose the most likely user (a single user) according to experiential rules and previous tracking positions. The first detection window determined when the hMouse is turned on is then selected as the initialization of the real-time tracking module. A pose or motion estimation module is further activated accompanied with head tracking. Our fundamental detection-tracking algorithm is a hybrid approach inheriting the advantages of two popular vision methods, boosted cascade of Haar-like features [66] and Continuously Adaptive Mean SHIFT (CAMSHIFT) [4, 5], proposed by Viola & Jones and Bradski, respectively.

2.3 Head Pose and Motion Estimation

There are typically six head pose or motion parameters, which record the estimation of rolling, tilting, yawing, scaling, horizontal, and vertical head motion respectively. Fig. 2 illustrates the definition of 3-D head pose with tilting, yawing and rolling rotations.

In hMouse system, the head rolling angle is estimated by the CAMSHIFT [4] algorithm. A neat code of this algorithm is available in OpenCV [50], which runs very fast (about 190 $\sim$ 200 fps on our computer system) and is reliable for head rolling angle estimation. In order to save as much the system computational resource as possible, we did not design a 3-D model to estimate the tilting and yawing angles. With adequately satisfying performance for real-time operation, the following straightforward motion-based algorithm is adopted to approximately estimate the coarse tilting and yawing angles.

- Calculate and threshold differences between frames to get the silhouette mask.
- Update the Motion History Image (MHI) [50].
- Calculate the derivatives of MHI and compute the gradient orientation.
- In the face tracking window, build the orientation histogram and find the basic orientation.
- Calculate the shift relative to the basic orientation and estimate the general motion direction.
- Accumulate the motion intensity (time shifts) in general motion directions of several adjacent frames, relative to the neutral position ($\theta_{\text{tilt}} = \theta_{\text{yaw}} = \theta_{\text{roll}} = 0$), to determine the tilting and yawing angles.

The scaling parameter is relative to the size of tracking window. Since hMouse has a detection-tracking mode-selective scheme, the tracking window yields severe jitters at some mode-switching points. In order to overcome the unexpected jitters by spatio-temporal smoothing, we restrict the tracking bounding boxes to be all square shapes. Based on each estimated object size $(length, width)$ by raw tracking in which the estimated tracking window can be rectangular, we re-scale the window size to be $(length + width)/2$. Also, the final tracking window size determined in each frame is smoothed by the average of three adjacent frames including previous two frames and the current frame. The square window is easily determined as long as the centroid coordinates of the object are calculated. The effect of our proposed simple smoothing method with much lower system resource cost is comparable to that of Kalman filter [36].

The horizontal and vertical motion are calculated differently in the two modes of the head tracker. In the detection mode, the reference motion point is defined

Table 1. Virtual Mouse Event Description. (Table reproduced from [25] with permission.)

Virtual Mouse Event	Description
Hold and Release	Use/activate/control and disuse/stop/quit
Move Cursor	Vertical and horizontal navigation, coarse localization
Point	Up and down fine tune, precise localization
Click Button	Left and right virtual mouse button down and up

as $R_m = \left[\frac{w_{\text{left}}+w_{\text{right}}}{2}, \frac{w_{\text{top}}+w_{\text{bottom}}}{2}\right]$, where w denotes the detection window coordinate in the frame space. In the tracking mode, the reference motion point R_m is represented by the object centroid. In the same way as scaling, the final reference motion point coordinates determined in each frame are smoothed by the average of three adjacent frames.

2.4 Virtual Mouse Control

Once the visual tracking module generates the six head pose or motion parameters, the virtual mouse control module first translates the parameters into virtual mouse events, and then navigates the cursor or triggers the button clicks with respect to the virtual mouse event description.

The descriptions of four basic virtual mouse events are summarized in Table 1. These events emulate general functional aspects of the hand-held computer mouse manipulation. We define the descriptions based on the sufficient concerning about users' convenient and flexible usage. The user is allowed to look at anywhere on the screen since the cursor is uncorrelated with user's focus of attention. When user tentatively wants to stop the hMouse, he or she can release it by moving his or her head backwards enough. The user can also exit the camera viewport for a while. When he or she comes in again, the system will automatically continue working without any further user interventions.

The forward and backward head movement against monitor screen is defined to emulate moving a hand-held mouse across the surface of the physical desktop and activating switches. If the tracking window width is larger than 80 for the 320×240 video size, the system triggers the 'hold' event. Otherwise, the system triggers the 'release' event. These two events are important for virtual mouse control and often triggered for two cases:

- Like a common hand-held computer mouse, hMouse can also retrieve the cursor back to the center of the monitor screen when it is navigated to the screen boundary.
- Since the virtual button click is controlled by users' head rolling, this small movement can sometimes influence the cursor navigation controlled by head

motion. It is obviously undesirable to have cursor position changed during a virtual mouse button click event. In hMouse, we define the hold and release virtual mouse event to be only operating on cursor navigation, but not on left and right button click event.

The vertical and horizontal object centroid movements are used for coarse mouse pointer relocation. We calculate the relative object centroid motion with its temporal coordinates difference and define a large scope cursor movement function to transform the object movement into cursor movement [25].

This event is dependent on the hold and release virtual mouse events. Releasing the virtual mouse can suspend the cursor moving temporarily. Empirically, we set parameters $\varepsilon = 6$, $\alpha = 0.9$ and $\beta = 18$ as default for common hMouse applications. The cursor control scheme is the integration of Joystick mode and Differential mode [61], which overcomes the drawbacks of any single modes. In small motion ranges, the hMouse controls cursor to move accurately and frequently; while in large motion range, the cursor moves fast and leapingly. This is prone to match the most with human habit when using hand-controlled mouse.

When the cursor is navigated close by the target location, small head movements can manipulate the cursor to point. We specify the up and down precise mouse localization with tilting rotation, and left and right with yaw rotation, to fine tune the cursor location. Since the tilting and yawing angles are calculated only in the tracking area relative to neutral position, the actual movement is the relative motion difference between local area and global area of the tracking window. This event is dependent on the hold and release virtual mouse events. Only in the state of holding mouse can the cursor navigation be enabled.

When the cursor is moved to the target location, user's head rolling is defined as the virtual button click event. Different from the common hand-held computer mouse, hMouse treats virtual mouse button down and up as a single event so that users do not need to move their head too frequently. We set the threshold of button click trigger with ± 0.6 radian rolling angles. Note that the left and right button click events are independent of the hold and release virtual mouse events, which means user can first release the virtual mouse to disable the cursor navigation function, and then trigger the virtual button click.

In the system implementations, we find that some users may somehow rotate their heads slightly slowly and tremblingly, especially the elderly and the people with disabilities. To overcome multiple mouse button clicks owing to the above situation, we improve the button click operation with temporal smoothing. Moreover, some constraints are also applied to make the click operation more naturally. If the rolling angle continues showing the value larger than ± 0.6 radian, the system only processes a single click event. Double button click can only be triggered when the rolling angle shows two times of changes from a smaller value to larger value than the threshold in a default short time interval.

2.5 System Performance and Experiments

Our hMouse system is implemented in C++ on a standard PC, using VisGenie [70] as the video based information visualization platform and a typical web camera mounted on top of the monitor as video input. The system interface is composed of a main window and four sub-windows. The four sub-windows render original video data, intermediate tracking results, intermediate histogram analysis, and final tracking results. Two circular indicators are drawn in real-time at the top left corner of the final tracking result window to point out the head motion direction and roll angle. It processes $11 \sim 13$ fps on a desktop with 2.80GHz Pentium 4 CPU and 1.0G RAM. The pure detection part processes $13 \sim 14$ fps without any system optimization. Our current system can work on Windows XP operating system with 320×240 video frame size in RGB channels. The average CPU usage is approximately $35\% \sim 43\%$.

On-Screen Typing and Interactive Games

Driven by the robust head tracking, hMouse can control the Microsoft Windows OS in a hands-free fashion. Users can activate the hMouse by moving close to the computer monitor and moving or rotating their face up/down and left/right to navigate cursor. Users can also change head roll orientation to click left or right mouse button at desired cursor position. Fig. 3-(a) shows an on-screen typing example for Windows Paint text input. The text "hMouse" on the right Paint canvas is written by hMouse typing on the Windows on-screen keyboard shown in the lower-left part of the figure. Fig. 3-(b) shows the user playing the Windows game Minesweeper with the hMouse. The left figure shows the user navigating cursor with head motion while right figure shows the user rolling his head to click the left mouse button down and up. The left black and white image shows the scaled MHI under detection only mode while the right one shows the back projection of the hue histogram under tracking only mode. Fig. 3-(c) and (d) show the user playing popular internet games, YetiSports 2 - Orca Slap [31] and YetiSports 1 - Pingu Throw [30], with the hMouse. These games do not need any cursor navigation. Only left button click event needs to be triggered during the game. Our hMouse can efficiently manipulate the game and bring much fun for the users.

System Performance Comparison

We test the controllability of hMouse with an experimental setup similar to the one in [62]. The 4×4 arrays of square buttons are generated on the screen with the button width specified among $0.3, 0.5, 1.0, 1.5, 2.0$ inches. Fig. 4 shows the first 3 arrays of square buttons with the button width $0.3, 0.5$, and 1.0, respectively. The blue and red '+' denote the centers of the previous and current activated buttons, respectively. The green '+' denotes the position where subject clicks the button. During the experiments, all the buttons will be randomly activated and highlighted one by one after the subject clicks the previous highlighted button. We carries out the experiment 5 times for each array of different button size. The average time of carrying out button clicking indicates how convenient for the subject to navigate the mouse cursor to the specified area and carry

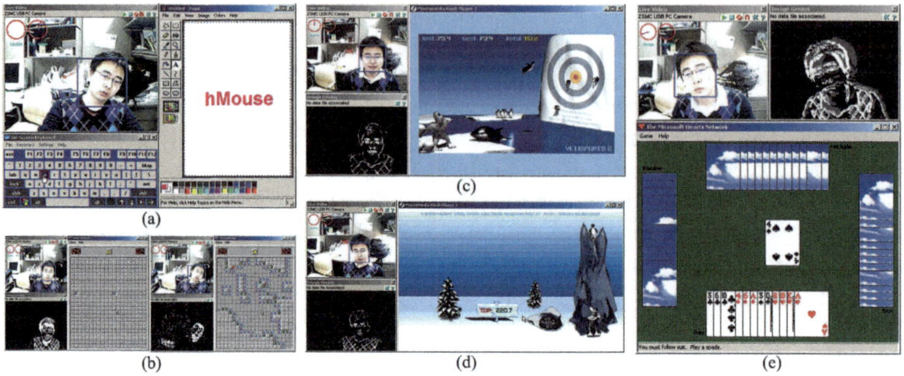

Fig. 3. On-screen typing and interactive games. (a) On-screen typing for Windows Paint text input. The text "hMouse" on the right Paint canvas is written by the hMouse typing on Windows on-screen keyboard shown in the lower-left part of the figure. (b) Playing Windows Minesweeper game with the hMouse. The left figure shows the cursor navigation while the right figure shows the left mouse button down and up click event. (c) Playing internet game, YetiSports 2 - Orca Slap [31], with the hMouse. (d) Playing internet game, YetiSports 1 - Pingu Throw [30], with the hMouse. (e) Playing Windows Hearts game with the hMouse.

out click operation. We compare the average button clicking time of hMouse with that of hand-controlled Microsoft Windows MouseKeys (left ALT + left SHIFT + NUMLOCK) and the 3 modes of 3-D camera mouse [61] in Fig. 5. The performance of hMouse with entirely hand-free control is comparable to the hand-controlled Microsoft Windows MouseKeys.

The 3-D face tracking driven camera mouse developed in [61] is the state-of-the-art system. It works fast and easily for manipulation, yet it needs manual tracker initialization. The system sometimes suffers from the conflict of global and local constraints and will lose tracking if the user makes large head movement. Unlike the 3-D camera mouse, our hMouse developed with 2-D head tracking techniques is totally automatic, and has low system costs. The description of mouse control events are based on the entire head movements rather than the motion units of facial features. Our hMouse has more reliable tracking performance with hand-free initialization but less precise cursor control than the 3-D camera mouse as shown in Fig. 5. Hence, hMouse provides more flexibility and reliability for user-friendly usage at the cost of slightly slower speed for mouse operations. In reality, this is still tolerable for severely physically handicapped or elderly people who can not even move their hands.

2.6 Summary

We have proposed the hMouse system, a novel head controlled camera mouse via real-time visual tracking. Comprehensive experiments and simulations have demonstrated the robustness and effectiveness of hMouse as a new hand-free

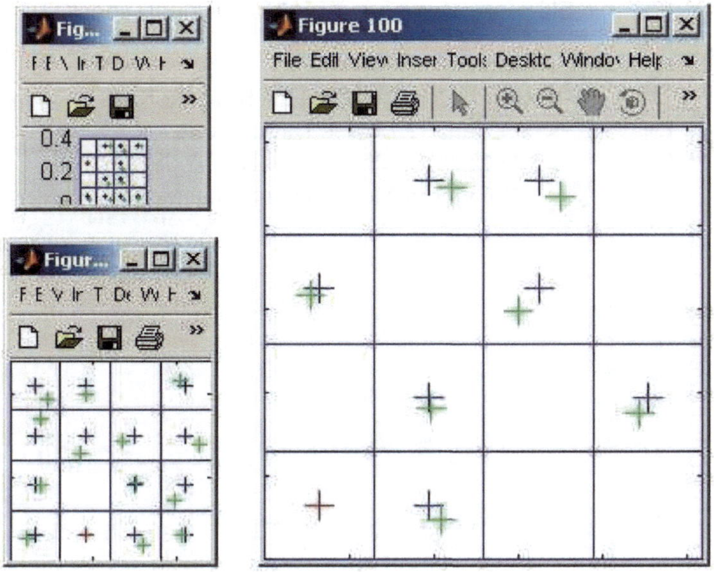

Fig. 4. Performance comparison experiments of the first 3 arrays of square buttons with the button width 0.3, 0.5, and 1.0, respectively. The blue '+' denotes the center of the previous activated button. The green '+' denotes the position where subject clicks the button. The red '+' denotes the center of the current activated button.

PUI. Several successful use cases of hMouse in popular interactive games reveal potential industry value and promising game marketing development. The function of this new mouse is consistent with the conventional mechanical mouse, and exhibits some new features in addition.

The current PC-based hMouse may inspire the applications of interactive computer/internet games, very low bit-rate human-to-human communication, intelligent avatar-based kiosks, and film-making. It is also a potential alternative OS solution for future portable devices, such as cell phone, mini PC, PDA, and GPS, etc. Moreover, a large potential market for the hMouse is the computer/machine access for severely physically handicapped and elderly people with limited motion ability. It affords a significant help to those people to avoid the communication gap between them and normal people in the same community.

Although the latest version of hMouse is effective enough for several listed realistic applications, it is still limited by its cursor control mode and head pose estimation mode to some extent. In our future work, we need to further improve the smoothness of cursor navigation algorithm according to the knowledge of kinesics, so that users may feel more at easy getting used to manipulating the cursor with their head. Furthermore, since the background may contains undesirable motion noises, the head pose estimation module may be severely affected since it is partially based on the motion features of users' head. The future

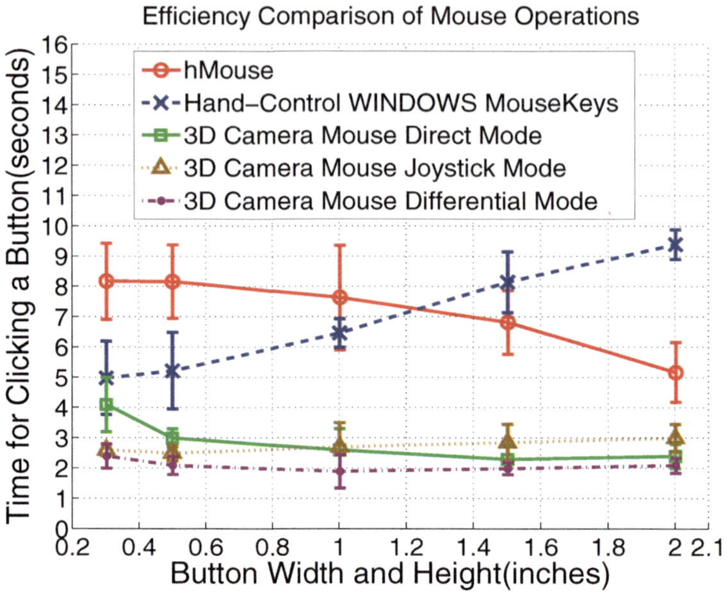

Fig. 5. Performance comparison of different mouse operations. (Figure reproduced from [25] with permission.)

tracker design will focus on tracking with local and global combined facial features [69], which are independent of head motion and only consider the facial features, such as eye and nose positions in the image.

3 Humanoid Emotive Audiovisual Avatar [58]

We have been conducting research in 3D face modeling, analysis, and synthesis [73], with applications to cellphone communication, broadcasting, entertainment industry, and desktop human-computer interfaces. Specifically, we have been carrying out basic research leading to the development of methodologies and algorithms for constructing an emotional, synthetic 3D talking avatar driven by text in real-time [58, 59, 57].

Avatars, a.k.a. synthetic talking faces, are virtual agents for intelligent human computer interaction. They introduce the presence of an individual to capture attention, mediate conversational cues, and communicate emotional affect, personality and identity in many scenarios. They provide realistic facial expressions and natural sounding speech to enhance the understanding of the content and intent of a message. If computers are embodied and represented by avatars the quality of human computer interaction can be significantly improved. Customized avatars, as personal clones or virtual actors, may also serve as assistive communication devices that enable individuals with hearing and speech/voice problems to participate in spoken communication interactions.

The use of avatars is emerging in the marketplace of many areas, but the understanding of human communication has not advanced to the point where it is possible to make realistic avatars that demonstrate interaction of speech and emotional expressions. Accordingly, avatars do not communicate subtle or complex emotions. Absent this, avatars can not communicate personality, culture, or communication specific to a social context. The key research issues to enable avatars to communicate subtle or complex emotions are how to synthesize natural sounding emotive speech, how to animate realistic emotional facial expressions, and how to combine speech gestures (i.e. lip movements) and facial expressions naturally and realistically. This requires the integration of engineering, psychological and speech and hearing science perspectives, establishment of a platform of basic research, and preparation of the prerequisite data and tools to realize naturalistic and compelling avatars.

It is desirable to make avatars mimic human communication in its entirety as human communication consists of more than just words [45]. To this end, ideally, avatars should be emotive, capable of producing emotive speech and emotional facial expressions. However, research into this particular topic is sparse. The difficulty lies in the fact that identification of the most important cues for realistic synthesis of emotive speech remains unclear and little is known about how speech gestures co-articulate with facial expressions. In addition, there is a great and increasing demand that audio-visual avatars should be in three dimensional (3D) forms and should resemble the actual faces of the persons whom they virtually represent.

We have been conducting basic research leading to the development of methodologies and algorithms for constructing a text-driven 3D humanoid emotive audio-visual avatar. This is a multi-disciplinary research area overlapping with speech, computer graphics and computer vision. The success of this research is believed to advance significantly the state of the art of human computer interaction. Specifically, it makes the interaction between humans and avatars more like human-to-human interaction. Equally exciting is the fact that this technology is a powerful tool for cognitive scientists to carry out research in the perception of emotional expression combined with speech. Results from these cognitive studies can in turn help technologists to design better avatars and human computer interfaces. Potentially, there will be considerable impact on various applications, such as 3D model-based very low bit rate video coding, gaming, and video teleconferencing.

The immediate applications of this research involve the production of emotionally-expressive avatars that demonstrate naturalistic movements for speech and expressions combined. These avatars could also demonstrate personality types through eye-gaze and expressions. Although the first use of avatars might be to facilitate person-to-person communication, another important application is as a product interface. The use of realistic avatars with personalities could be a compelling feature of seamless mobility: an avatar that remains with the consumer, no matter what particular device the consumer is using.

3.1 Technological Framework

In our research toward building a text-driven 3D humanoid emotive audio-visual avatar, we basically follow a technological framework as illustrated in Fig. 6. In this technological framework, text is converted into emotive speech by emotive speech synthesis. Simultaneously, a phoneme sequence with timing is produced. The phoneme sequence is mapped into a viseme sequence that defines the speech gestures. The emotional state decides the facial expressions, which will be combined with speech gestures synchronously and naturally. The key frame technique is used to animate the sequence.

Our primary research is focused on 3D face modeling (including 3D facial geometry modeling, 3D facial deformation modeling, and 3D facial animation), facial expression synthesis, viseme synthesis, and emotional text-to-speech synthesis.

3.2 3D Face Modeling

3D face modeling has been an active research topic for computer graphics and computer vision. The task of 3D face modeling can be divided into three components, namely, 3D facial geometry modeling, 3D facial deformation modeling, and 3D facial animation. All these three components are equally important for 3D face modeling.

3D Facial Geometry Modeling

In the back old days, people (artists) have used interactive tools to design geometric 3D face models under the guidance of prior knowledge. As laser-based

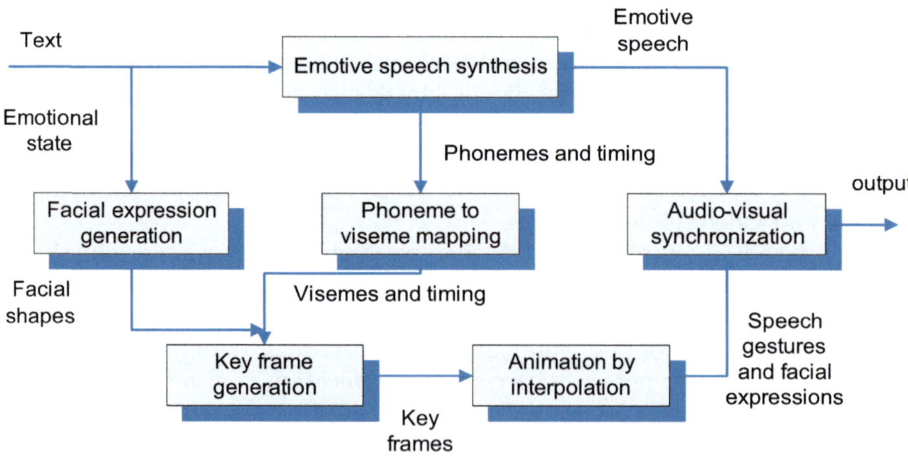

Fig. 6. Technological framework for constructing a text-driven humanoid emotive audio-visual avatar. (Figure reproduced from [58] with permission.)

range scanners such as the Cyberware^{TM} scanner have become commercially available, people have been able to measure the 3D geometry of human faces. The Cyberware^{TM} scanner shines a low-intensity laser beam onto the face of a subject and rotates around his or her head for 360 degrees. The 3D shape of the face can be recovered from the lighted profiles of every angle. With the measurement of 3D face geometry available, geometric 3D face models can be constructed. Alternatively, a number of researchers have proposed to build 3D face models from 2D images using computer vision techniques [54]. Usually, in order to reconstruct 3D face geometry, multiple 2D images from different views are required. 3D range scanners, though very expensive, are able to provide high-quality measurement of 3D face geometry and thus enable us to construct realistic 3D face models. Reconstruction from 2D images, despite its noisy results, provides an alternative lost-cost approach to build 3D face models.

A 3D face model describes the 3D geometry of the facial surface as well as the skin textures. The 3D facial geometry in the 3D space is normally approximated by a 3D triangular or polygonal mesh with tens of thousands of vertices and triangles or polygons. As long as the number of triangles or polygons is sufficient enough, the approximated facial surface will be very smooth. Fig. 7 shows a 3D face model approximated by a dense polygonal mesh.

3D Facial Deformation Modeling

A geometrical 3D face model determines the 3D geometry of a static facial surface. By spatially and temporally deforming the geometrical 3D face model we can obtain different facial expressions. Such facial deformation is described by a 3D facial deformation model. Although there exist several facial deformation models, including physics-based facial deformation models, muscle-based facial

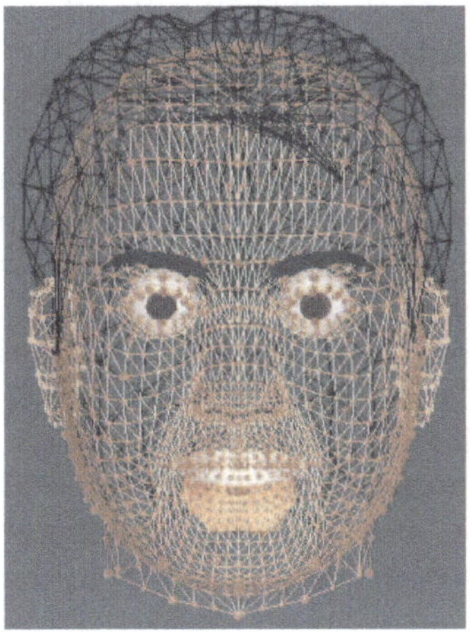

Fig. 7. A 3D polygonal mesh based 3D face model. (Figure reproduced from [58] with permission.)

deformation models and so forth, the free-form facial deformation model is one of the most popular facial deformation models in the literature. In a free-form facial deformation model, as its name suggests, the coordinates of the 3D mesh vertices can be deformed in a free-form manner by changing the positions of some control points while these control points can either belong to the mesh vertices or not. One example is the Piecewise Bezier Volume Deformation proposed by Tao and Huang (1999) [60]. Specifically, a set of control points are defined over the 3D geometry of the face model and facial deformation is achieved through proper displacements of these control points. The displacements of the facial geometry are obtained by a certain interpolation scheme.

To approximate the facial deformation space, linear subspace methods have been proposed. One example is the Facial Action Coding System (FACS) which describes arbitrary facial deformation as a linear combination of Action Units (AUs) [22]. AUs are qualitatively defined based on anatomical studies of facial muscular activities that cause facial movements. Often, extensive manual work has to be done to create AUs. As more and more facial motion capture data becomes available, it is possible to learn the subspace bases from the data which are able to capture the characteristics of real facial deformations. Hong et al. applied Principal Component Analysis (PCA) to real facial motion capture data and derived a few bases called Motion Units (MUs) [33]. Any facial deformation can then be approximated by a linear combination of the MUs. Compared with AUs, MUs are automatically derived from data. Therefore, labor-intensive manual work can be avoided. In addition, MUs have proven to yield smaller reconstruction error than AUs.

3D Facial Animation

3D facial animation is the process of generating continuous facial deformations over time [51]. A popular way of doing so is to use the key-frame interpolation technique. The basic idea is that we place particular facial deformations at particular time instants (key-frames), and automatically obtain facial deformations in between the key-frames by a certain interpolation scheme (e.g., cosine or a sophisticated interpolation model learned from motion capture data). As long as the number of facial deformations (i.e. frames) is sufficiently large (> 24 per second), the facial animation will be continuous and smooth.

Although basically we can consider to place any arbitrary facial deformations at the key-frames. However, those arbitrary facial deformations which are not anatomically correct, or those which are not semantically meaningful, are not usually considered to be placed at the key-frames. Two basic types of key-frames which are directly related to facial animation and which are frequently used in producing facial animation are facial expressions and visemes. We will cover them in the subsequent subsections.

3.3 The UIUC iFace System

We at the Image Formation and Processing (IFP) Group at the University of Illinois at Urbana-Champaign have developed a 3D face modeling tool which lays a foundation of many of the subsequent research. In this subsection, we will give an overview of this 3D face modeling tool, the UIUC iFace system [34]. The UIUC iFace system takes as input the Cyberware^{TM} scanner data of a person's face and fits the data with a generic head model. The output is a customized geometric 3D face model ready for animation.

The generic head model, as shown in Fig. 8, consists of nearly all the components of the head (i.e., face, eyes, ears, teeth, tongue, etc.). The surfaces of these components are approximated by densely connected triangular meshes. There are a total of 2240 vertices and 2946 triangles to ensure the closeness of the approximation. In order to customize the generic head model for a particular person, we need to obtain the range map (Fig. 9, left) and texture map (Fig. 9, right) of that person by scanning his or her face using a Cyberware^{TM} scanner.

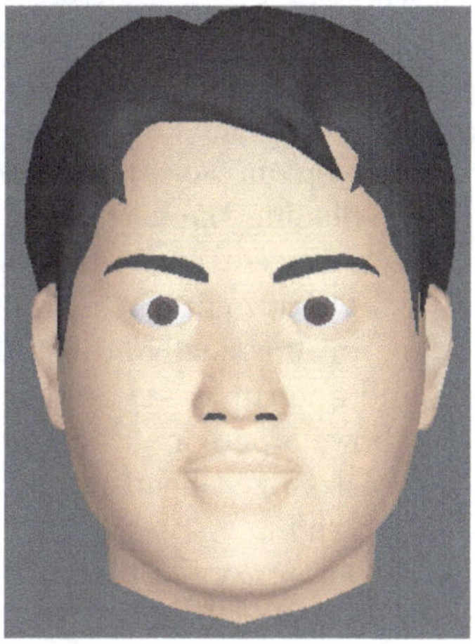

Fig. 8. The generic 3D face model in the UIUC iFace system. (Figure reproduced from [58] with permission.)

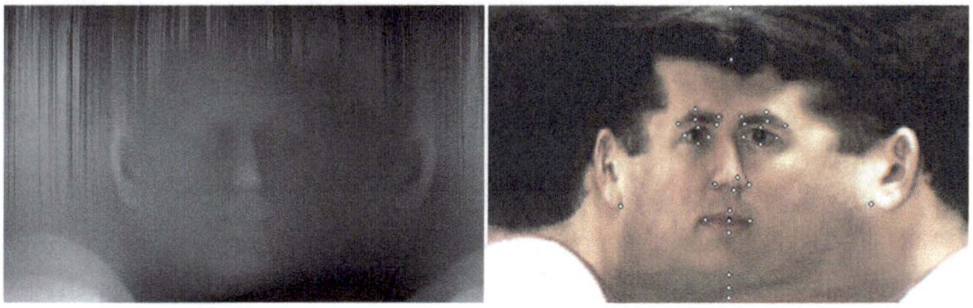

Fig. 9. Example of Cyberware^{TM} scanner data. (Left): range map. (Right): texture map with 35 feature points defined on it. (Figure reproduced from [58] with permission.)

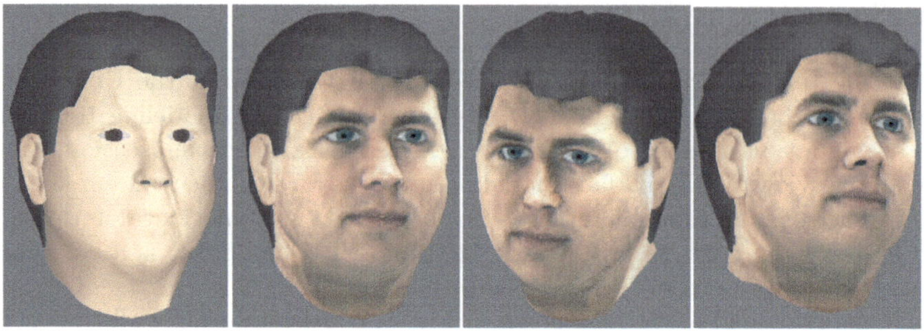

Fig. 10. Customized face model. (Left): Shape model. (Right): Texture is mapped onto the shape model which is rotated to several different poses. (Figure reproduced from [58] with permission.)

On the face component of the generic head model, 35 feature points are explicitly defined. If we were to unfold the face component onto a 2D plane, those feature points would triangulate the whole face into multiple local patches. By manually selecting 35 corresponding feature points on the texture map (and thus on the range map), as shown in Fig. 9, right, we can compute the 2D positions of all the vertices of the face meshes in the range and texture maps. As the range map contains 3D face geometry of the person, we can then deform the face component of the generic head model by displacing the vertices of the triangular meshes by certain amounts as determined by the corresponding range information collected at the corresponding positions. The remaining components of the generic head model (hair, eyebrows, etc.) are automatically adjusted by shifting, rotating and scaling. Manual adjustments and fine tunes are needed where the scanner data is missing or noisy. Fig. 10, left, shows a personalized head model. In addition, texture can be mapped onto the customized head model to achieve photo-realistic appearance, as shown in Fig. 10, right.

Based on the this 3D face modeling tool, a text-driven humanoid 3D emotive audio-visual avatar can be built. We will detail the various technical aspects of research leading to a text-driven humanoid 3D emotive audio-visual avatar in the following subsections. Our primary work is focused on facial expression synthesis, viseme synthesis, and emotional text-to-speech synthesis.

3.4 Facial Expression Synthesis

Facial expressions can be synthesized using templates. We use the geometrical 3D face model described earlier to animate the various facial expressions. On the 3D face model, a control model is defined, as shown in Fig. 11. The control model consists of 101 vertices and 164 triangles which cover the whole face region and divide it into local patches. By dragging the vertices of the control model, one can deform the 3D face model to arbitrary shapes. A set of basic facial shapes corresponding to different fullblown emotional facial expressions (Fig. 12) are manually designed and parameterized as the displacements of all the vertices of

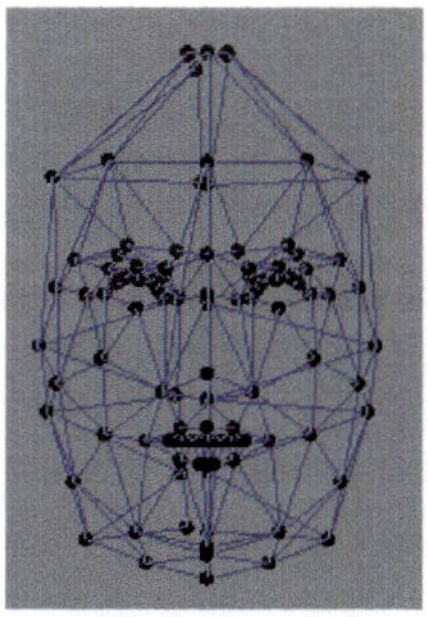

Fig. 11. A control model defined over the 3D face model. (Figure reproduced from [58] with permission.)

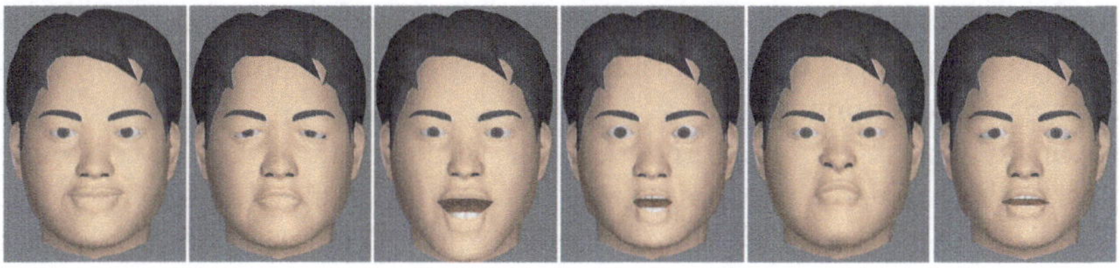

Fig. 12. A set of basic facial shapes corresponding to different fullblown emotional facial expressions (neutral, sad, happy, angry, disgusted, and afraid). (Figure reproduced from [58] with permission.)

the face model from their initial values corresponding to neutral state. Let those displacements be $\Delta\mathbf{V}_i, i = 1, 2, ..., N$ where N is the number of vertices of the face model. The vertices of the deformed face model are given by

$$\mathbf{V}_{0i} + \alpha \times \Delta\mathbf{V}_i, \quad i = 1, 2, ..., N \tag{1}$$

where $\mathbf{V}_{0i}$ are the vertices of the neutral face and $0 \leq \alpha \leq 1$ is the magnitude coefficient representing the intensity of the facial expression. $\alpha = 1$ corresponds to the fullblown emotional facial expression and $\alpha = 0$ corresponds to the neutral expression.

3.5 Viseme Synthesis

The word viseme was introduced by C. Fisher at the University of Iowa in 1968 [24]. A compound word of "visual" and "phoneme", a viseme is deemed as the visual counterpart of a phoneme. It describes the facial shapes of a person when he or she is uttering a particular sound. Fig. 13 illustrates this concept. Therefore, a set of visemes can be defined in the visual domain that basically covers the facial movements that occur over a complete set of phonemes in a particular language.

The mapping from phonemes to visemes is however not one-to-one. Several phonemes may share the same viseme if they produce a similar look on the face.

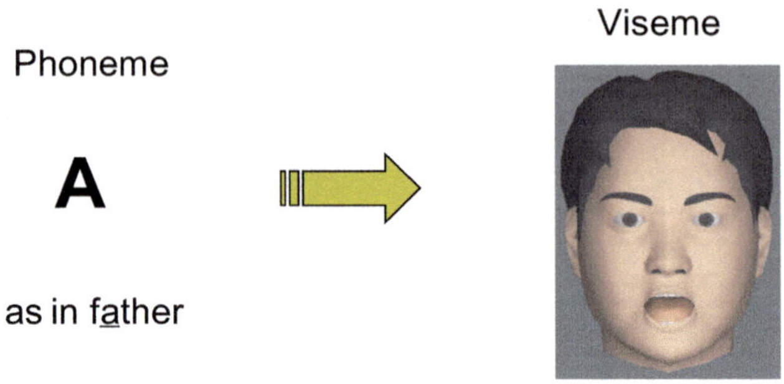

Fig. 13. Viseme: visual counterpart of phoneme.

For example, the sound /k/, /g/, /ng/, and /tS/, /S/, /dZ/, /Z/. This reduces the number of visemes that need to be implemented yet does not introduce distinguishable differences in perception. In American English, as few as 14 visemes may cover the 40 phonemes.

Viseme synthesis are implemented via table lookup. We have defined 17 visemes, each of which corresponds to one or more of the 40 phonemes in American English. Table 2 lists all the phonemes and the corresponding visemes that they are mapped to.

Table 2. Phonemes, examples, and corresponding visemes. Viseme synthesis is done by simple table lookup. (Table reproduced from [58] with permission.)

phoneme	example	viseme	phoneme	example	viseme
i	beet	IY	I	bit	AX
E	bet	AY	{	bat	AE
r=	above	AX	u	boot	W
U	book	AX	V	above	AX
O	caught	AO	A	father	AA
@	butter	AX	EI	bay	NE
AI	bye	AY	OI	boy	T
aU	about	AY	@U	boat	OW
p	pan	M	t	tan	T
k	can	T	b	ban	M
d	dan	T	g	gander	T
m	me	M	n	knee	T
N	sing	AY	f	fine	F
T	thigh	TH	s	sign	T
S	assure	SH	h	hope	AY
v	vine	F	D	thy	TH
z	resign	T	Z	azure	SH
tS	church	SH	dZ	judge	SH
l	lent	LL	r	rent	R
j	yes	T	w	went	W
_	(silent)	NE			

3.6 Emotional Text-to-Speech Synthesis

The process of automatic generation of speech from arbitrary text by machine, referred to as text-to-speech (TTS) synthesis, is a highly complex problem [17]. The quality of the synthetic speech generated by a TTS synthesis system is usually described by two measures: intelligibility and naturalness. Major developments in these two aspects have taken place since the early rule-based or formant synthesizers in the 1970's and 1980's. Further improvements have been achieved by the use of diphone synthesizers in the 1990's. The latest corpus-based or unit-selection synthesizers have demonstrated to be able to produce human-quality synthetic speech, especially in limited domains such as weather forecast or stock analysis. Nonetheless, the quality of TTS synthesis is by no means considered completely satisfactory, mainly in the aspect of naturalness. It seems that the human ears are far more critical than the human eyes in that the human eyes are easily cheated by an illusion. This probably can explain in part that why computer graphics has been widely accepted by the movie industry while TTS synthesis still has a very long way to go until the synthetic speech can replace the real speech of an actor [53].

The unsatisfactory naturalness is largely due to a lack of emotional affect in the speech output of current state-of-the-art TTS synthesis systems. The emotional affect is apparently a very important part of naturalness that can help people enhance their understanding of the content and intent of a message. Researchers have realized that in order to improve naturalness, work must be performed on the simulation of emotions in synthetic speech. This involves investigation into the perception of emotions and how to effectively reproduce emotions in the spoken language. Attempts to add emotional affect to synthetic speech have existed for over a decade now. However, emotive TTS synthesis is by far still an open research issue.

We concern the process of building an emotive TTS synthesis system using a diphone synthesizer. Diphone synthesis [17] has a small footprint which is ideal for low-cost implementation in resource-sensitive mobile or embedded devices such as cellphones and Personal Digital Assistants (PDA's), yet is able to produce higher-quality synthetic speech than formant synthesis. Aside from the emotive audiovisual avatar, it is also desirable to build an emotive TTS synthesis system using a diphone synthesizer to meet the increased demands of speech-enabled mobile applications by today's society.

In this subsection, we present a general framework of emotive TTS synthesis using a diphone synthesizer that makes possible the R&D of emotive TTS synthesis systems in a unified manner. The validity of the framework is demonstrated by the implementation of a rule-based emotive TTS synthesis system based on the Festival-MBROLA architecture.

Framework

In order to produce high-quality synthetic speech, it is necessary for a TTS synthesis system to take a number of essential processing steps. In general, a TTS

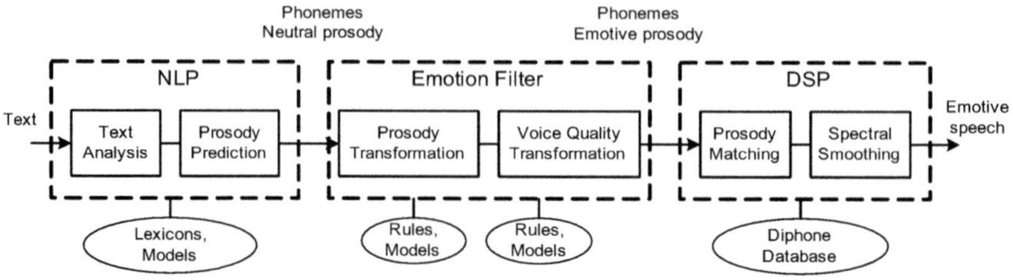

Fig. 14. A general framework of emotive TTS synthesis using a diphone synthesizer. (Figure reproduced from [58] with permission.)

synthesis system consists of two main components. The first component is a Natural Language Processing (NLP) module. The NLP module performs necessary text analysis and prosody prediction procedures to convert orthographical text into proper phonetic transcription (i.e., phonemes) together with the desired intonation and rhythm (i.e., prosody). The second component is a Digital Signal Processing (DSP) module. The DSP module takes as input the phonetic transcription and prosodic description which are the output of the NLP module, and transforms them into a speech signal.

Fig. 14 illustrates a general framework of emotive TTS synthesis using a diphone synthesizer. In this framework, an emotion filter is inserted into the functional diagram of a general TTS synthesis system and plays a role that bridges the NLP and DSP modules. The NLP module generates phonemes and prosody corresponding to the neutral state (i.e., neutral prosody). The emotion filter is aimed at transforming the neutral prosody into the desired emotive prosody as well as transforming the voice quality of the synthetic speech. The phonemes and emotive prosody are then passed on to the DSP module to synthesize emotive speech using a diphone synthesizer.

Prosody Transformation through a Differential Approach

One major mission of the emotion filter is to transform neutral prosody into emotive prosody. Such prosody transformation is achieved through a differential approach, which aims at finding the differences of emotional states with respect to the neutral state [64]. The emotion filter either maintains a set of prosody manipulation rules defining the variations of prosodic parameters between the neutral prosody and the emotive one, or trains a set of statistical prosody prediction models that predict these parameter variations given the features representing a particular textual context. At synthesis time, the rules or the prediction outputs of the models will be applied to the neutral prosody, obtained via the NLP module. The emotive prosody can thus be obtained by summing up the neutral prosody and the predicted variations. There are obvious advantages of using a differential approach over the use of a full approach that aims at finding the emotive prosody directly. One advantage is that the differences of the prosodic parameters between the neutral prosody and the emotive one have far smaller dynamic ranges than the prosodic parameters themselves and therefore the differential approach requires far less data to train the models than the full approach (e.g., 15 minutes versus several hours). Another advantage is that the differential approach makes possible the derivation of the prosody manipulation rules which are impossible to obtained in the full approach.

One way for prosody transformation is to apply a set of prosody manipulation rules. The prosody of speech can be manipulated at various levels. High-level prosody modification refers to the change of symbolic prosodic description of a sentence (i.e., human readable intonation and rhythm) which by itself does not contribute much to emotive speech synthesis unless there is a cooperative prosody synthesis model (i.e., a model that maps high-level symbolic prosodic description to low-level prosodic parameters). Low-level prosody modification refers to the direct manipulation of prosodic parameters, such as f_0, duration, and energy, which can lead to immediate perception of emotional affect. Macro-prosody (suprasegmental prosody) modification refers to the adjustment of global prosody settings such as f_0 mean, range and variability as well as the speaking rate, while micro-prosody (segmental prosody) modification refers to the modification to local segmental prosodic parameters which are probably related to the underlying linguistic structure of a sentence. It should be noted that not all levels of prosody manipulations can be described by rules. For instance, due to the highly dynamic nature of prosody variation, micro-prosody modification cannot be summarized by a set of rules.

Macro-prosody modification can be effectively described by a set of prosody manipulation rules. These rules can be either derived from the literature or learned through corpus analysis, i.e., analysis of small databases of neutral and emotive speech recorded using the same script. Fig. 15-(a) shows that the prosody manipulation rules can be obtained by a corpus analysis process.

One drawback of the above prosody manipulation rules is that they are applied at the macro-prosody level and thus cannot capture the subtlety and dynamic nature of the local variation of the emotive prosody. On the contrary, micro-prosody embodies all the subtle but complex variation of emotive prosody that is related to the linguistic structure of a sentence. However, no manual rules are able to describe prosody modification at the micro-prosody level. Thus, we need to seek a data-driven approach to build statistical prosody prediction models based on recorded human emotive speech [64]. This process is illustrated in Fig. 15-(b) and is detailed as follows.

1. Statistical models can be trained and used to predict emotive prosody from text. Instead of predicting the emotive prosodic parameters directly we adopt a differential approach. That is, we construct statistical prosody prediction models to predict the deviations of the prosodic parameters from their values corresponding to the neutral state to new values corresponding to a specific emotional state given a set of features representing a texture context.

2. We record a small but both phonetically and prosodically rich speech database for the neutral state as well as for every emotional state (happy, sad, angry, etc.). Each database will contain approximately 15 minutes of speech

recorded by an actor who speaks a script emotively. The waveforms in the database will be forced aligned with the text in the script at both the word and phoneme levels. A context window of five syllables (the current syllable, two to the left and two to the right) will be used to extract a set of text-based features. These text-based features, denoted as a vector $\mathbf{F}_i$, along with the prosodic deviations (deviations of the f_0 contour, duration and energy) computed from the waveforms, denoted as $\mathbf{P}_i(f_0) = \{\Delta \mathbf{f}_{0i}\}$, $\mathbf{P}_i(\text{duration}) = \{\Delta \text{duration}_i\}$, and $\mathbf{P}_i(\text{energy}) = \{\Delta \text{energy}_i\}$, for the current syllable i, comprise an observation O_i.

3. After the observations $O_i, i = 1, 2, ..., N$, where N is the number of syllables in the database, are collected, a set of statistical classification and regression tree (CART) [7] models (i.e., f_0 model, duration model and energy model) can be trained based on these observations. CART is a binary decision tree which at each interior node asks a question regarding the text-based features and at each leaf node stores a distribution of the observations (assuming Gaussian).

4. Once the CART-based prosody prediction models are trained, we can then use it for prosody prediction. Based on the text-based features, the neutral prosody is predicted by a full prosody prediction model which is already existing and well-trained in many (commercial) TTS synthesis systems, while the prosodic deviations are predicted by the differential model of our approach. The emotional prosody is then the sum of the predicted neutral prosody and prosody deviation.

It should be worth mentioning that this approach has one more advantage. It has the potential to model a continuum of gradually changing emotional states. If we can find a function (linear or nonlinear) $f(D(p), k)$ which is continuous in k, where p is a vector of the neutral prosodic parameters, $D(p)$ is the deviation of the prosodic parameters from p, and k, $0 \leq k \leq 1$, is the magnitude coefficient for the emotion, the corresponding emotional prosody can be given by $p + f(D(p), k)$.

Voice Quality Transformation

Whether prosody transformation alone is enough for the synthesis of emotive speech has been a fundamental question for every attempt to use a diphone synthesizer to synthesize emotive speech. Interestingly, very different answers exist

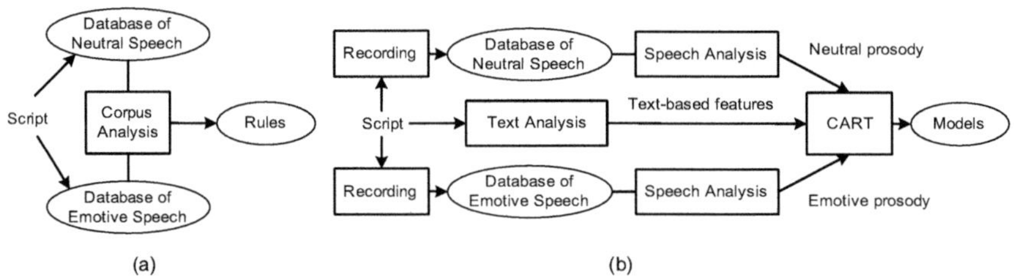

Fig. 15. Prosody prediction for emotive speech. (a) prosody manipulation rules found by corpus analysis. (b) statistical prosody prediction model based on classification and regression tree (CART).

in the literature. Some studies indicate that in addition to prosody transformation, voice quality transformation is important for synthesizing emotive speech and may be indispensable for some emotion categories (emotion-dependent). Other studies find that there seem to be speaker strategies replying mostly on prosody for expressing some emotions (speaker-dependent). However, whether this is true for all types of emotions is not yet clear.

In diphone synthesis, it is particularly not easy to control voice quality, as it is very difficult to modify the voice quality of a diphone database. However, one partial remedy of this is to record separate diphone databases with different vocal efforts. During synthesis, the system switches among different voice-quality diphone databases and select the diphone units from the appropriate database. Another low-cost partial remedy is to use jitters to simulate voice quality transformation. Jitters are fast fluctuations of the f_0 contour. Thus, adding jitters is essentially equivalent to adding noise to the f_0 contour. By jitter simulation we can observe voice quality change in synthetic speech to a certain degree.

3.7 Experimental Results

The previously described frameworks and approaches have resulted in a fully functional system of text-driven 3D humanoid emotive audio-visual avatar. We have obtained preliminary but promising experiment results for rendering neutral as well as several basic emotions: happy, joyful, sad, angry, and fearful. Demos are available at http://www.ifp.uiuc.edu/~haotang2. Fig. 16 shows two examples of the results: The animation sequence of emotional facial expressions along with the associated waveforms of synthetic emotive speech generated for happy and sad emotional states. That is, the avatar says "This is happy voice" and "This is sad voice" respectively.

Informal evaluations within our group show that for emotive speech, negative emotions (e.g., sad, angry, fear) are more successfully synthesized than positive emotions (happy, joyful). For some emotions such as joyful and fear, artifacts are

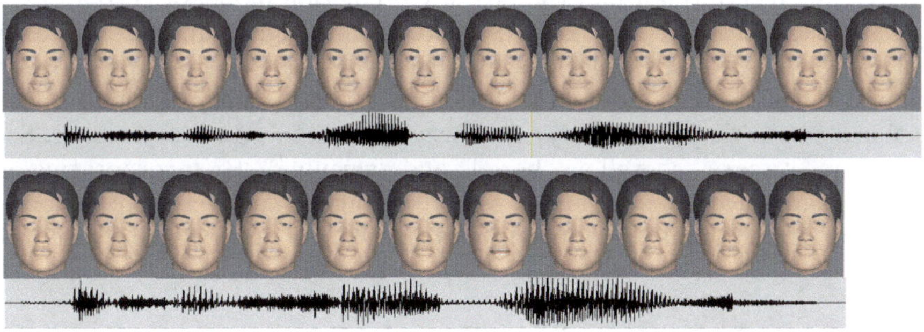

Fig. 16. Examples of animation of emotive audio-visual avatar with associated synthetic emotive speech waveform. (Top): The avatar says "This is happy voice." (Bottom): The avatar says "This is sad voice." (Figure reproduced from [58] with permission.)

easily observed in the synthesized speech due to extreme prosody modification by signal processing. While the synthetic emotive speech alone cannot be always recognized by a listener, it can be easily distinguished when combined with compatible emotional facial expressions.

We have conducted a series of subjective listening experiments to evaluate the expressiveness of our system. We have derived the parameters (rate factors, gradients, etc.) for the manipulation rules that control the various aspects of the global prosody settings and voice quality from the literature and manually adjusted these parameters by listening tests. The system is capable of synthesizing eight basic emotions: neutral, happy, joyful, sad, angry, afraid, bored, and yawning. We designed and conducted three subjective listening experiments on the results generated by the system. The experiment 1 uses eight speech files corresponding to the eight distinct emotions, synthesized by the system using the same semantically neutral sentence (i.e., a number). The experiment 2 also uses eight speech files, but each of the files was synthesized using a semantically meaningful sentence appropriate for the emotion that it carries. The experiment 3 incorporates the visual channel based on experiment 1. That is, each speech file was provided with an emotive talking head showing emotional facial expressions and lip movements consistent and synchronized with the speech. The experiment 4 incorporates the visual channel based on experiment 2. That is, each speech file consists of semantically neutral sentence, semantically meaningful verbal content appropriate for the emotion, and synchronized emotional facial expressions. 20 subjects (10 males and 10 females) were asked to listen to the speech files (with the help of other channels such as verbal content and facial expressions if possible), determine the emotion for each speech file (by forced-choice), and rate his or her decision with a confidence score (1: not sure, 2: likely, 3: very likely, 4: sure, 5: very sure). The results are shown in Table 3. In the table, recognition rate is the ratio of the correct choices, and average score is the average confidence score computed for the correct choices.

It is obviously shown in the results of the subjective listening experiments that our system has achieved a certain degree of expressiveness despite of the relatively poor performance of the prosody prediction model of Festival. As can be clearly seen, negative emotions (e.g, sad, afraid, angry) are more successfully synthesized than positive emotions (e.g, happy). This observation is consistent with what has been found in the literature. In addition, we found that the emotional states "happy" and "joyful" are often mistaken for each other. So are "afraid" and "angry" as well as "bored" and "yawning". By incorporating other channels that convey emotions such as verbal content and facial expressions, the perception of emotional affect in synthetic speech can be significantly improved.

4 3D Face/Head Tracking Based Video Compression [63]

For people's daily communication, facial expressions convey very important information. To make the communication effective among people who are teleconferencing at remote distances, large volume of video data from each site has to be

Table 3. Results of subjective listening experiments. $R \longleftrightarrow$Recognition Rate. $S \longleftrightarrow$Average Confidence Score. Experiment 1$\longleftrightarrow$Speech only. Experiment 2 $\longleftrightarrow$Speech+verbal content. Experiment 3$\longleftrightarrow$Speech+facial expression. Experiment 4$\longleftrightarrow$Speech+verbal content+facial expression. (Table reproduced from [58] with permission.)

Emotion	Experiment 1		Experiment 2		Experiment 3		Experiment 4	
	R	S	R	S	R	S	R	S
neutral	54.54%	2.83	100%	4.36	90.90%	4.1	100%	4.81
happy	18.18%	2.5	63.63%	4.57	63.63%	4.43	54.54%	4.5
joyful	45.45%	2.8	63.63%	4.86	63.63%	4.29	45.45%	5
sad	36.36%	2.75	81.81%	4.67	100%	4.18	100%	4.81
angry	18.18%	2	90.90%	4.3	90.90%	4	100%	4.63
afraid	45.45%	2.4	81.81%	4.89	72.72%	4	81.81%	5
bored	45.45%	3	81.81%	4.44	-	-	-	-
yawning	18.18%	3.5	72.72%	3.75	-	-	-	-

transmitted to the other sites, and causes traffic congestion problems in communication channels. One of the solution is to code the videos at very low bit rate. As the major foreground object is the face of the person who is talking and the background are more or less static in most of the videos, it is possible to model the facial motions and the static background individually so that the video can be encoded at very low bit rate. We proposed an efficient and robust very low bit rate face coding method via visual face tracking. After the face is located in the video frame, the generic facial geometric model is adapted to the face, and facial texture for the model is extracted from the first frame of the video. The facial motion is then tracked and synthesized. The facial motion parameters, synthesized residual error and video background are transmitted at very low bit rate. Experiments showed that this method can achieve better PSNR around facial area than H.26L at about the same low bit rate and can achieve better subjective visual effects.

4.1 Introduction

We rely on our faces to convey subtle emotions in our daily communication and we read people's face when we try to interpret the subtleties in people's speech. Therefore, it is always preferable to have video phones, or some teleconferencing applications for people communicating at remote distance. The difficulty is how to transmit the video data across the communication channels at low bit rate so that there is no delay in the video streaming. For the teleconferencing scenario, very low bit rate transmission is possible because the foreground of the video is usually human face and shoulder, and the background is usually simple and static.

For the purpose of transmitting large volume of video data over limited channel, video coding protocols, such as MPEG1, MPEG2, MPEG4 and H.26x [52] [72] [29] have been developed. These protocols are characterized by coding schemes that consist of block-based motion-compensated prediction, and quantization of prediction error with discrete cosine transform. As these approaches take advantages of the spatial-temporal redundancy statistics in the video frames without a prior knowledge of the semantic content of the video, they accommodate the need of general purpose video data compression.

For the teleconferencing applications where the major foreground object is the human face, people have proposed model-based coding techniques for further enhancing the coding efficiency [3][47][74]. These approaches characterize the human face geometry and texture by 3D mesh models, and the facial motions can be formulated in terms of the rotation and translation of the head (the rigid head motion), and the action unit coefficiencts of the face (the non-rigid facial motions). Very low bit rate video transmission can be achieved by transmitting these parameters together with the video background. While this idea seems simple and intuitive, it is difficult to extract these facial geometry and motion parameters from video automatically and robustly.

In [37], the facial features, such as the eyes, the mouth, the nose, the chin, etc, are localized frame by frame through image segmentation and deformable template matching, and are fitted to a face model (Candide) so that the 3D face model parameters (the size, the face orientation and the local facial feature geometry) can be estimated based on the perspective projection and human facial symmetry assumptions. In [14] [68] [41], DeMenthon's successive scaled orthographic approximation [18] was applied to iteratively estimate the head pose and facial expressions according to the estimation of the key facial feature locations. By computing the optical flows from the facial motions, the head poses can be estimated either as angular changes by solving linearized differential equations [23], [60], [19], or as rotation matrix with orthonormality constraints by factorization approaches [6]. The nonrigid facial deformation is typically modeled as a linear combination of some key facial deformations (action units) and is uniquely defined by the action unit coefficients. Similar to the head pose estimation, the action unit coefficients can be retrieved by solving differential equations [23], [60], [19], or by factorization with rank deficiency constraints [6].

In this chapter, we introduce an efficient and robust algorithm that achieves very low bit rate face coding at real-time via visual face tracking. We first localize the facial features of the face in the first frame of the video. We then fit a generic facial geometric model to the face, and acquire the facial texture from the fitted face. The facial motions can then be tracked and synthesized thereafter. The facial motion parameters, synthesized residual error and the video background are transmitted at very low bit rates. Experiments show that our approach can achieve a better PSNR around the facial area than that of H.26L at about the same low bit rate. The subjective visual effects also look much better.

4.2 The System Framework

Fig. 17 shows the framework of our system. A face tracker is first utilized to track the human face in the video and extract the facial motion parameters. Based on the motion parameters and the facial texture acquired at the first frame, the facial appearances caused by the facial motions can be synthesized frame by frame. Based on the synthesized facial appearances, the original video frame can be split into the foreground (facial area) and the background regions. The residuals of the synthesized facial appearances can then be evaluated. The frame background and synthesized foreground residuals are then sent to a H.264 coder, respectively. The frame background can be coded as ordinary video frames by H.264 codec, and the foreground residuals can be coded by $Intra\_16X16$ mode as there is no significant temporal correlation in the synthesis residuals. The model parameters and the encoded video frames are then transmitted through the communication channel. At the receiving side, the decoder synthesizes the facial appearances frame by frame according to the received face motion parameters. The foreground region can be reconstructed by summing up the synthesized face and the decoded foreground synthesis residuals. Combining the foreground region and the background region decoded by a H.264 decoder, the decoded video frame can be obtained. As most of the facial motion details are carried in the facial motion parameters, the foreground residual tends to have small amplitude, we can choose to code the video with very lot bit rates without losing much information of the foreground.

In the next section, we will introduce how the face tracker is implemented in details.

4.3 3D Model Based Face Tracking[55]

Perspective camera model

The geometry of a human face can be modeled by a set of vertices in 3D space. The actual 2D face image in video frames is considered as a rendering of the projection of the 3D face model to the image plane in a camera model.

Consider a vertex on the face model with world coordinate $X_w = (x_w, y_w, z_w)$; we first transform the coordinate to camera coordinate system through rotation R and translation T.

$$\begin{pmatrix} x \\ y \\ z \end{pmatrix} = R \begin{pmatrix} x_w \\ y_w \\ z_w \end{pmatrix} + T \tag{2}$$

Then the perspective projection M is

$$\begin{pmatrix} X_u \\ Y_u \end{pmatrix} = \begin{pmatrix} f/z & 0 & 0 \\ 0 & f/z & 0 \end{pmatrix} \begin{pmatrix} x \\ y \\ z \end{pmatrix} \tag{3}$$

where f is the focus length of the camera lens.

Sometimes a weak perspective camera model is favored if the variations in depth of the object are small compared to the average depth z_{avg}. The depth z is replaced by constant z_{avg}, and we simply have $\begin{pmatrix} X_u \\ Y_u \end{pmatrix} = C \begin{pmatrix} x \\ y \end{pmatrix}$ where $C = \frac{f}{z_{\text{avg}}}$ is a constant.

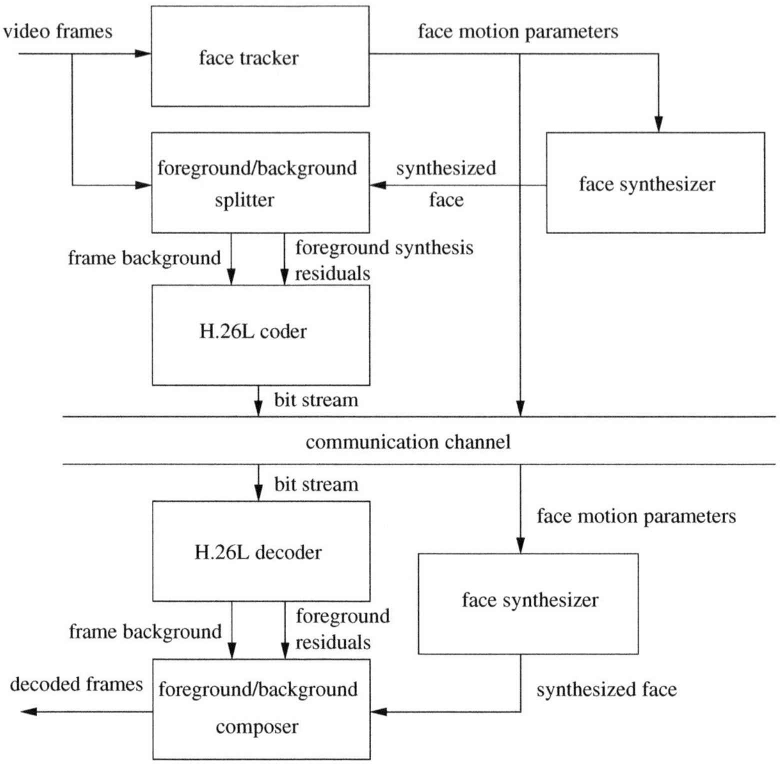

Fig. 17. System framework. (Figure reproduced from [63] with permission.)

In real-world applications, camera lens radial distortion correction needs to be carried out, and the perspective projection is transformed into 2D pixel coordinate afterwards. The details are not introduced here as they are not relevant to 3D face modeling.

3D face models

The 3D geometry of human facial surface can be represented by a set of vertices $\{(x_i, y_i, z_i)|i = 1, ..., N\}$ in space, where N is the total number of vertices. Some 3D key facial deformations are defined as the displacements of the vertices from neutral facial surface, i.e., $\{(\Delta x_i, \Delta y_i, \Delta z_i)|i = 1, ..., N\}$. An arbitrary face articulation can be formulated as a linear combination of the neutral facial surface and the key facial deformations, i.e., $V = \bar{V} + Lp$, where $\bar{V}$ is the neutral facial surface, $L_{3N \times K}$ is a matrix that contains the key facial deformations, and p is the linear combination coefficients.

Taking head rotation $R(\theta, \phi, \psi)$ and translation $T = [T_x T_y T_z]'$ into account, the projection of each vertex $V_i, i = 1, ..., N$ on face surface to image plane with weak perspective camera model M can be formulated as

$$V_i^{\text{Image}} = M(R(\theta, \phi, \psi)(\bar{V}_i + L_i p) + T) \tag{4}$$

where V_i^{Image} is the projection of the i-th vertex node on face surface. Therefore the facial motion is characterized by rotation $W = [\theta, \phi, \psi]$/translation $T = [T_x T_y T_z]$ (rigid motion parameters), and the key facial deformation coefficient vector p (nonrigid facial motion parameter).

For the purpose of efficient MPEG-4 video coding, people developed MPEG-4 standard for face modeling and animation, which includes Facial Definition Parameters (FDP) and Facial Animation Parameter Units (FAPUs). With FDP, generic 3D face model can be customized to describe faces in video, and with FAPUs, the facial deformation is decomposed into facial expressions, visemes (key facial deformations corresponding to phonemes), and key facial component movements (eyeballs, pupils, eyelids, etc.). The FDP parameters can be transmitted over the channel and the 3D faces can be rendered at remote sites efficiently [75], [20], [46].

For the purpose of facial expression analysis, Ekman and W. Friesen proposed Facial Action Coding System (FACS) [22]. According to analysis of the contraction of each facial muscle, FACS decomposes observed human facial expressions into a set of key facial deformations, so called action units (AUs). The scores for a facial expression consist of the list of AUs possibly with duration, intensity, and asymmetry, etc. FACS provides only descriptive analysis. In [60], a Piecewise Bezier Volume Deformation (PBVD) model was developed so that 12 AUs can be manually crafted for facial expression analysis. The crafted 12 action units are listed in Table 4.

By varying the AU coefficients p, we can synthesize arbitrary facial deformations. An illustration of such synthesis is shown in Fig. 18.

Tracking by Kalman filtering

As the rendering of a 2D face image from a 3D face model can be characterized by Equation (4), the face model parameters (the rotation $W = [\theta, \phi, \psi]$ and

Table 4. Action units. (Table reproduced from [61] with permission.)

AU	Description
1	Vertical movement of the center of upper lip
2	Vertical movement of the center of lower lip
3	Horizontal movement of left mouth corner
4	Vertical movement of left mouth corner
5	Horizontal movement of right mouth corner
6	Vertical movement of right mouth corner
7	Vertical movement of right eyebrow
8	Vertical movement of left eyebrow
9	Lifting of right cheek
10	Lifting of left cheek
11	Blinking of right eye
12	Blinking of left eye

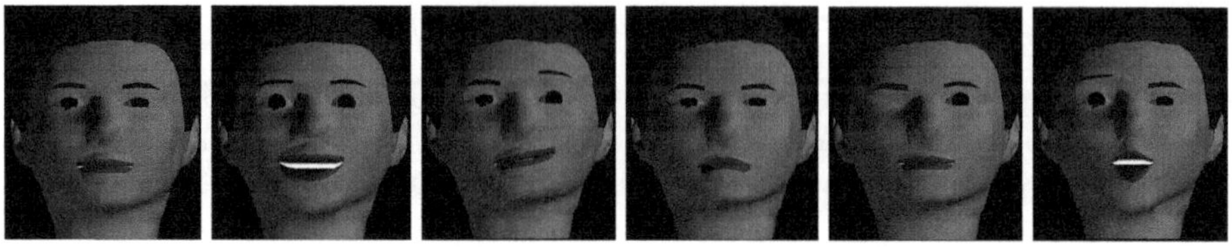

Fig. 18. Some facial articulations synthesized by changing the weights of the key facial deformations.

translation $T = [T_x T_y T_z]$, and the AU coefficient vector p) can be estimated in an extended Kalman filtering framework.

Denoting $x(t) = (W(t), T(t), p(t))$, and $y(t) = \{V_i^{\text{Image}}\}$, the facial motions can be formulated as a dynamic process as follows [2]:

$$x(t + \Delta t) = \Psi(\Delta t)x(t) + \xi(t) \tag{5}$$
$$y(t) = h(x(t)) + \nu(t) \tag{6}$$

where $\xi(t)$ and $\nu(t)$ are process noise and measurement noise respectively, $\Psi(\Delta t)$ is a dynamic state transition matrix, and the nonlinear observation function $h(\cdot)$ is defined by Equation (4).

Denote $\hat{x}(t_1|t_2)$ the linear least square estimate of $x(t_1)$ given observation $y(\tau)$, $\tau < t_2$, and $P(t_1|t_2) = E\{(x(t_1) - \hat{x}(t_1|t_2))(x(t_1) - \hat{x}(t_1|t_2))^T\}$ as the error covariance.

The state prediction is defined by

$$\hat{x}(t + \Delta t|t) = \Psi(\Delta t)\hat{x}(t|t) \tag{7}$$
$$P(t + \Delta t|t) = \Psi(\Delta t)P(t|t)\Psi(\Delta t)^T + Q(t) \tag{8}$$

where $Q(t) = E\nu(t)\nu(t)^T$.

And the EKF update is defined by

$$\hat{x}(t|t) = \Psi(\Delta t)\hat{x}(t|t - \Delta t) + K(t)(y(t) - h(\hat{x}(t|t - 1)) \tag{9}$$
$$P(t|t) = (I - K(t)C(t))P(t|t - \Delta t) \tag{10}$$

where

$$K(t) = P(t|t - \Delta t)C^T(t)(C(t)P(t|t - \Delta t)C^T(t) + R(t))^{-1}$$

is the Kalman gain, $R(t) = E\xi(t)\xi(t)^T$ and $C(t) = \frac{\partial h}{\partial x}|_{x=\hat{x}(t|t-\Delta t)}$.

The implementation

As the extended Kalman filtering is an iterative process, an initial state is usually provided by manual labeling or by automatic detection of the face to be tracked in the first frame of the video.

We initialize the tracking by automatically localizing the facial features of the face in the first still frame of the video using an automatic face localization procedure based on Adaboost algorithm [67] and ASM techniques [15], [65]. The generic 3D face model is then adjusted to fit to the detected 2D facial features.

In order to achieve real-time tracking performance, we simplify the Kalman filtering process by ignoring the process and measurement noises and assuming $\Psi(\Delta t) = I$. The state prediction and update becomes

$$\hat{x}(t|t) = \hat{x}(t - \Delta t|t - \Delta t) + \widehat{\Delta x}(t)$$

where $y(t) - h(\hat{x}(t-1|t-1)) = C(t)\widehat{\Delta x}(t)$

As $y(t) - h(\hat{x}(t-1|t-1))$ is the displacement of $V_i^{\text{Image}}, i = 1, ..., N$ which can be obtained by optical flow $\Delta V_i^{\text{Image}} = [\Delta X_i, \Delta Y_i]^T, i = 1, ..., N$, the derivative of the state variable $\Delta x = [dW, dT, dp](dW = [d\theta, d\phi, d\psi])$ can be computed by solving the following linearized differential equations [60]:

$$\begin{aligned}\Delta V_i^{\text{Image}} &= \frac{\partial[M(R(\bar{V}_i + Lp))]}{\partial[T\,W\,p]}\begin{bmatrix} dT \\ dW \\ dp \end{bmatrix} \\ &= \begin{bmatrix} \frac{fs}{z_i} & 0 \\ 0 & \frac{fs}{z_i} \end{bmatrix} \times \\ &\quad \left[\begin{bmatrix} 1 & 0 & \frac{-x_i}{z_i} \\ 0 & 1 & \frac{-y_i}{z_i} \end{bmatrix} \begin{bmatrix} [G]_0 - \frac{x_i}{z_i}[G]_2 \\ [G]_1 - \frac{y_i}{z_i}[G]_2 \end{bmatrix} \begin{bmatrix} [RL]_0 - \frac{x_i}{z_i}[RL]_2 \\ [RL]_0 - \frac{x_i}{z_i}[RL]_2 \end{bmatrix}\right] \begin{bmatrix} dT \\ dW \\ dp \end{bmatrix}\end{aligned}$$

for $i = 1, ...N$, where

$$G = \begin{bmatrix} 0 & \hat{z}_i & -\hat{y}_i \\ -\hat{z}_i & 0 & \hat{x}_i \\ \hat{y}_i & -\hat{x}_i & 0 \end{bmatrix}, \begin{pmatrix} \hat{x}_i \\ \hat{y}_i \\ \hat{z}_i \end{pmatrix} = \bar{V} + Lp, \begin{pmatrix} x_i \\ y_i \\ z_i \end{pmatrix} = R \begin{pmatrix} \hat{x}_i \\ \hat{y}_i \\ \hat{z}_i \end{pmatrix} + T$$

and $[G]_i$ and $[RL]_i$ denote the i-th row of the matrix G and RL, respectively.

As there are K+6 parameters (K =12 in our case), and the face geometry is defined with N (N=500) facial vertex nodes, a LAPACK [38] linear least square (LLS) robust solver is utilized to find the solution of this overdetermined linear system of equations.

The whole tracking procedure is illustrated by Fig. 19. At the first frame of the video, the model parameter is initialized. Since the second frame, the optical flow at each vertex on the face surface is computed and the displacement of the model parameters is estimated. Then the model parameters can be updated at real time. This process iterates in a coarse-to-fine fashion in a multi-resolution framework frame by frame.

4.4 Embedding Synthesized Face in H.264 Codec

After the synthesized face is obtained, the synthesis residual error can be calculated. And the video frame is divided into foreground residual region and

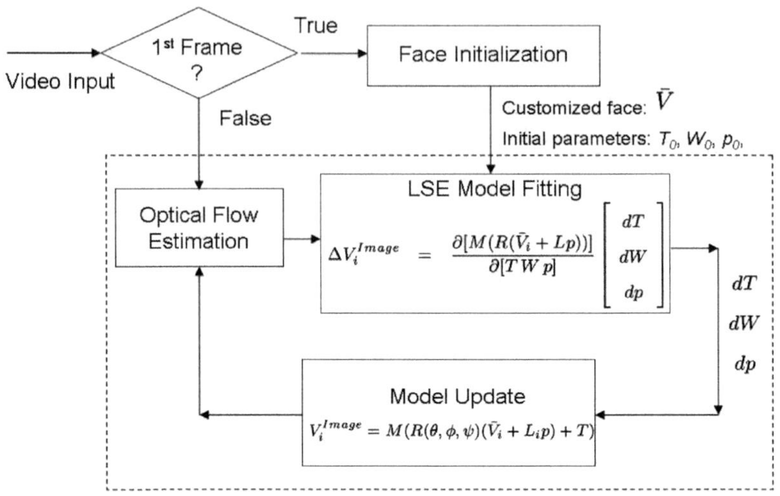

Fig. 19. Flowchart of our tracking procedure. (Figure reproduced from [61] with permission.)

background region. For the background, we can still employ H.26L coding based on spatial-temporal redundancy analysis. For the foreground residual, it contains mostly high frequency signal components, therefore we specify $Intra\_16X16$ mode as the coding mode for macroblocks in the foreground residual region. The advantage of this is that the facial motion details mostly captured by the motion parameters will not be lost when the video is coded at the lowest bit rate.

4.5 Experiments

The tracking system runs on a Dell workstation with dual CPU of 2 GHz and SCSI RAID hard drive. The face tracker runs at 25 frame per second (fps) in rigid tracking mode and 14 fps in non-rigid tracking mode. In Fig. 20, we show the tracking result of the head pose parameters and the AU coefficients related to mouth and eyes in a video of about 10 seconds. The figure indicates that, in the first 40 frames, the user shifted his head horizontally (in x coordinate) and back-forth (in z coordinate), rotated his head about y axis (yaw movement); he then opened his mouth wide at about the 80-th frame, and blinked his eyes at about the 20-th, 40-th, and 85-th frame.

Based on the face tracking module and we can encode and decode some teleconferencing video sequences using the framework described in section 2. A video sequence of about 147 frames is encoded and decoded by the H.26L codec with and without the face tracking component. Some reconstructed video frames are shown in Fig. 21. The video frame is of size 352 by 240. The video is coded at about bit rate of 18kb/s. The first column shows the key frames of the input video. The second column shows the synthesized facial appearances based on the model parameters computed by the visual face tracking component. The third

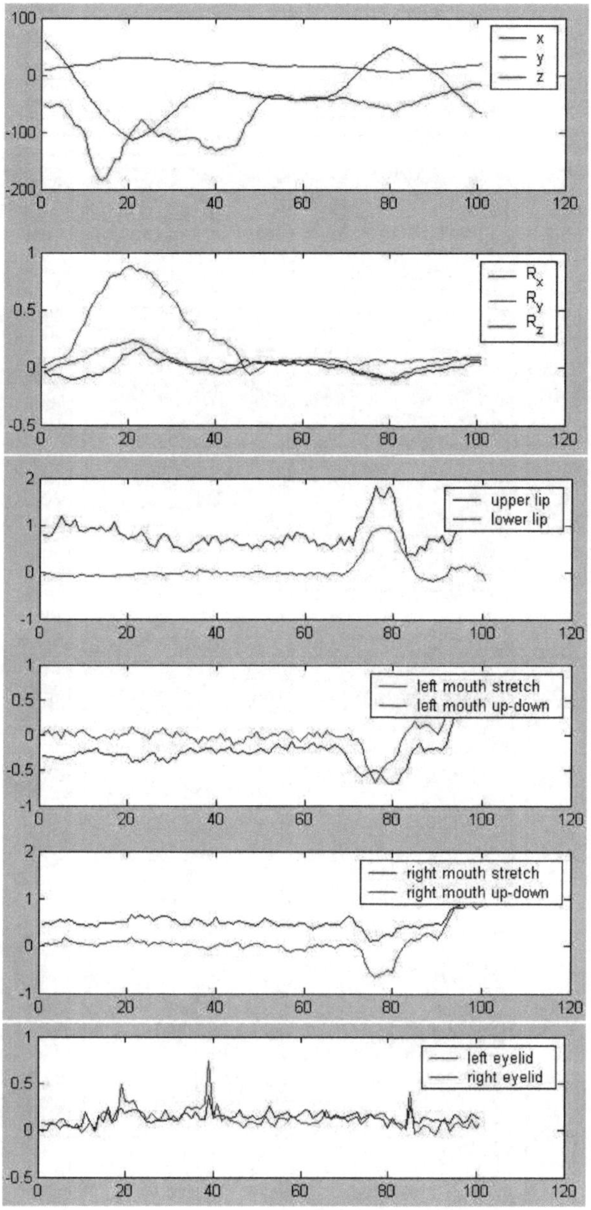

Fig. 20. Motion parameters indicated that the user moved his head horizontally (in x coordinate) and back-forth (in z coordinate), rotated his head about y axis (yaw movement) in the first 40 frames, opened his mouth wide at about the 80-th frame, and blinked his eyes at about the 20-th, 40-th, and 85-th frame. (Figure reproduced from [61] with permission.)

(a) (b) (c) (d)

Fig. 21. Comparison of some video key frames. (a) The original frames; (b) The synthesized frames; (c) The H.26L decoded frames without face tracking; (d) The H.26L decoded frames with face tracking.

column shows the decoded frames solely by the H.26L codec. And the fourth column shows the decoded frames by our proposed approach. We then compute the Pixel Signal to Noise Ratio (PSNR) in the facial area for both coding schemes. The PSNR of facial area is 27.35 if we code the video solely by H.26L. And the PSNR is 29.28 if we code the video using the model based approach. By visual inspection, it is shown that the facial appearance details are preserved better by the H.26L codec with model than that without model.

4.6 Summary

We presented an efficient and robust very low bit rate face coding method via visual face tracking in this chapter. For the teleconferencing videos where the face of the conference attendee is the major foreground object and the background clutter of the video is mostly static, we can extract the facial motion parameters by visual face tracking, and transmit the facial motion parameters, the encoded video background and the facial motion synthesis residuals through the communication channel. The video can then be reconstructed by putting together the decoded background region, the synthesis residuals and the synthesized facial motion appearances based on the facial motion parameters in reverse order. We discussed the details of the 3D face modeling techniques, and described the mechanism of the model based face tracking component. We then demonstrated that very low bit rate video coding can be achieved with better PSNR at the facial region using our proposed approach.

5 Future Work

Case studies of human-centered face computing in multimedia interaction and communication are conducted in this chapter by introducing three real-world systems developed in our group: the online interactive gaming system hMouse, humanoid emotive audio-visual avatar, and 3D face/head tracker based video compression. There are some potential research topics and open issues remaining. For example, how to associate face contents to the whole multimedia contents? How to benefit the multimedia communication from face computing? How to reduce the semantic gap in the multimedia analysis and video content understanding by means of human-centered interactions and services.

References

1. Ahlberg, J.: CANDIDE-3 – An updated parameterized face. Dept. of Electrical Engineering, Linkping University, Sweden, Tech. Rep. LiTH-ISY-R-2326 (2001)
2. Azarbayejani, A., Horowitz, B., Pentland, A.: Recursive estimation of structure and motion using relative orientation constraints. In: Proc. of the IEEE Conference on Computer Vision and Pattern Recognition, pp. 294–299 (1993)

3. Aizawa, K., Huang, T.S.: Model based image coding: Advanced video coding techniques for very low bit-rate applications. Proc. IEEE 83, 259–271 (1995)
4. Bradski, G.R.: Real time face and object tracking as a component of a perceptual user interface. IEEE WACV, 214–219 (1998)
5. Bradski, G.R.: Computer vision face tracking for use in a perceptual user interface. Intel Technology Journal (1998)
6. Brand, M., Bhotika, R.: Flexible flow for 3D nonrigid tracking and shape recovery. In: Proc. of the IEEE Conference on Computer Vision and Pattern Recognition, vol. 1, pp. 315–322 (2001)
7. Breiman, L., Friedman, J., Olshen, R., Stone, C.: Classification and regression trees. Wadsworth, Belmont (1984)
8. Betke, M., Gips, J., Fleming, P.: The camera mouse: visual tracking of body features to provide computer access for people with severe disabilities. IEEE Trans. on NSRE 10(1), 1–10 (2002)
9. Bergasa, L.M., Mazo, M., Gardel, A., Barea, R., Boquete, L.: Commands generation by face movements applied to the guidance of a wheelchair for handicapped people. In: ICPR, pp. 660–663 (2000)
10. Blanz, V., Vetter, T.: A morphable model for the synthesis of 3D faces. In: SIGGRAPH, 187–194 (1999)
11. CameraMouse, Inc., `http://www.cameramouse.com/`
12. Chen, Y.L.: Application of tilt sensors in human-computer mouse interface for people with disabilities. IEEE Trans. on NSRE 9(3), 289–294 (2001)
13. Cascia, M.L., Sclaroff, S., Athitsos, V.: Fast, reliable head tracking under varying illumination: an approach based on registration of texture-mapped 3D models. IEEE Trans. on PAMI 22(4), 322–336 (2000)
14. Chang, C.C., Tsai, W.H.: Determination of head pose and facial expression from a single perspective view by successive scaled orthographic approximations. Intl. Journal of Computer Vision 46(3), 179–199 (2002)
15. Cootes, T.F., Taylor, C.J.: Active shape model search using local grey-level models: A quantitative evaluation. In: 4th British Machine Vision Conference, pp. 639–648 (1993)
16. DAZ3D, `http://www.daz3d.com/`
17. Dutoit, T.: An introduction to text-to-speech synthesis. Kluwer Academic Publishers, Dordrecht (1997)
18. DeMenthon, D., Davis, L.: Model-based object psoe in 25 lines of codes. In: European Conference on Computer Vision, pp. 335–343 (1992)
19. DeCarlo, D., Metaxas, D.: Optical flow constraints on deformable models with applications to face tracking. International Journal of Computer Vision 38(2), 99–127 (2000)
20. Dalong, J., Zhiguo, L., Zhaoqi, W., Wen, G.: Animating 3d facial models with MPEG-4 FaceDefTables. In: Proc. of the 35th Annual Simulation Symposium, p. 395 (2002)
21. Evans, D.G., Drew, R., Blenkhorn, P.: Controlling mouse pointer position using an infrared head-operated joystick. IEEE Trans. on NSRE 8(1), 107–117 (2000)
22. Ekman, P., Friesen, W.V.: The facial action coding system. Consult. Psychological Press, Palo Alto (1978)
23. Eisert, P., Wiegand, T., Girod, B.: Model-aided coding: A new approach to incorporate facial animation into motion-compensated video coding. IEEE Transactions on Circuits and Systems for Video Technology 10(3), 344–358 (2000)
24. Fisher, C.G.: Confusions among visually perceived consonants. Journal of Speech and Hearing Research 11(4), 796–804 (1968)

25. Fu, Y., Huang, T.S.: hMouse: head tracking driven virtual Computer Mouse. IEEE WACV (2007)
26. Grauman, K., Betke, M., Gips, J., Bradski, G.R.: Communication via eye blinks—detection and duration analysis in real time. IEEE CVPR 1, 1010–1017 (2001)
27. Guo, H., Jiang, J., Zhang, L.: Building a 3D morphable face model by using thin plate splines for face reconstruction. In: Advances in Biometric Person Authentication, pp. 258–267. Springer, Heidelberg (2004)
28. Gorodnichy, D.O., Malik, S., Roth, G.: Nouse 'use your nose as a mouse' - a new technology for hands-free games and interfaces, vol. VI, pp. 354–360 (2002)
29. H.264: Advanced video coding for generic audiovisual services. International Telecommunication Union, Tech. Rep. (2007)
30. Hilgert, C.: Pingu Throw, YetiSports 1, `http://www.yetisports.org/`
31. Hilgert, C.: Orca Slap, YetiSports 2, `http://www.yetisports.org/`
32. Hong, P., Huang, T.S.: Natural mouse-a novel human computer interface. IEEE ICIP 1, 653–656 (1999)
33. Hong, P., Wen, Z., Huang, T.S.: iFACE: a 3D synthetic talking face. International Journal of Image and Graphics 1(1), 1–8 (2001)
34. Hong, P., Wen, Z., Huang, T.S.: Real-time speech-driven expressive synthetic talking faces using neural networks. IEEE Transaction on Neural Networks 13(4), 916–927 (2002)
35. Java mobile 3D, `http://www.jcp.org/en/jsr/detail?id=184`
36. Kalman, R.E.: A new approach to linear filtering and prediction problems. Trans. of the ASME–Journal of Basic Engineering 82(D), 35–45 (1960)
37. Kampmann, M.: Automatic 3-d face model adaptation for model-based coding of videophone sequences. IEEE Trans. on Circuits and Systems for Video Technology 12(3), 172–182 (2002)
38. Lapack, LAPACK – Linear Algebra Package, `http://www.netlib.org/lapack/`
39. Logitech, Inc., `http://www.logitech.com/`
40. Li, R.X.: A technical report on Motorola avatar framework, TR-2006, Applications Research Center, Motorola Labs, Schaumburg, IL (2006)
41. Lu, L., Dai, X., Hager, G.: A particle filter without dynamics for robust 3D face tracking. In: Proc. of the IEEE Conference on Computer Vision and Pattern Recognition Workshop, vol. 5, p. 70 (2004)
42. Lienhart, R., Maydt, J.: An extended set of Haar-like features for rapid object detection. IEEE ICIP 1, 900–903 (2002)
43. Lee, Y.C., Terzopoulos, D., Waters, K.: Realistic modeling for facial animation. In: SIGGRAPH, pp. 55–62 (1995)
44. MouseVision, Inc., `http://www.mousevision.com/`
45. Mehrabian, A.: Communication without words. Psychology Today 2(4), 53–56 (1968)
46. Massaro, D.W.: Perceiving talking faces: from speech perception to a behavioral principle. MIT Press, Cambridge (1998)
47. Musmann, H.: A layered coding system for very low bit rate video coding. Signal Processing: Image Communication 7, 267–278 (1995)
48. Morris, T., Chauhan, V.: Facial feature tracking for cursor control. Journal of Network and Computer Applications 29(1), 62–80 (2006)
49. Nanda, H., Fujimura, K.: A robust elliptical head tracker. In: IEEE FGR, pp. 469–474 (2004)
50. Open source computer vision library, Intel Corporation, `http://www.intel.com/technology/computing/opencv/index.htm`

51. Parke, F.I., Waters, K.: Computer facial animation. AK Peters, Ltd., Wellesley (1996)
52. Roy, B., Bouyssou, D.: Video coding for low bitrate communication. ITU-T Recommendation H.263 (1996)
53. Schroder, M.: Speech and emotion research: an overview of research frameworks and a dimensional approach to emotional speech synthesis (Ph.D thesis). Phonus, Research Report of the Institute of Phonetics, 7, Saarland University (2004)
54. Salas, S.L., Hille, E.: Computational imaging and vision. In: A survey on 3D modeling of human faces for face recognition, ch. 2. Springer, Netherlands (2007)
55. Tu, J.: Visual face tracking and its applications (Ph.D thesis), University of Illinois at Urbana-Champaign (2007)
56. Toyama, K.: 'Look, ma - no hands!' hands-free cursor control with real-time 3D face tracking. In: Proc. Workshop on Perceptual User Interfaces, San Fransisco (1998)
57. Tang, H., Fu, Y., Tu, J., Huang, T.S., Hasegawa-Johnson, M.: EAVA: a 3D emotive audio-visual avatar. In: 2008 IEEE Workshop on Applications of Computer Vision, Copper Mountain, CO (2008)
58. Tang, H., Fu, Y., Tu, J., Hasegawa-Johnson, M., Huang, T.S.: Humanoid audio-visual avatar with emotive text-to-speech synthesis. IEEE Transactions on Multimedia 10(6), 969–981 (2008)
59. Tang, H., Hu, Y., Fu, Y., Hasegawa-Johnson, M., Huang, T.S.: Real-time conversion from a single 2D face image to a 3D text-driven emotive audio-visual avatar. In: 2008 IEEE International Conference on Multimedia & Expo., Hannover, Germany (2008)
60. Tao, H., Huang, T.S.: Explanation-based facial motion tracking using a piecewise Bézier volume deformation model. IEEE CVPR 1, 611–617 (1999)
61. Tu, J., Huang, T.S., Tao, H.: Face as mouse through visual face tracking. In: IEEE CRV, pp. 339–346 (2005)
62. Tu, J., Tao, H., Huang, T.S.: Face as mouse through visual face tracking. CVIU Special Issue on Vision for Human-Computer Interaction 108(1-2), 35–40 (2007)
63. Tu, J., Wen, Z., Tao, H., Huang, T.S.: Coding face at very low bit rate via visual face tracking. In: Proc. of Picture Coding Symposium, Saint Malo, France, pp. 301–304 (2003)
64. Tang, H., Zhou, X., Odisio, M., Hasegawa-Johnson, M., Huang, T.S.: Two-stage prosody prediction for emotional text-to-speech synthesis. In: INTERSPEECH 2008, Brisbane, Australia (2008)
65. Tu, J., Zhang, Z., Zeng, Z., Huang, T.S.: Face Localization via Hierarchical CONDENSATION with Fisher Boosting Feature Selection. In: Proc. of the IEEE Conference on Computer Vision and Pattern Recognition, vol. 2, pp. 719–724 (2004)
66. Viola, P., Jones, M.: Rapid object detection using a boosted cascade of simple features. IEEE CVPR, 511–518 (2001)
67. Viola, P., Jones, M.: Robust real-time object detection. International Journal of Computer Vision 747 (2002)
68. Vaccetti, L., Lepetit, V., Fua, P.: Fusing online and o2ine information for stable 3D tracking in real-time. In: Proc. of the IEEE Conference on Computer Vision and Pattern Recognition, pp. II: 241–248 (2003)
69. Wenzel, M.T., Schiffmann, W.H.: Head pose estimation of partially occluded faces. IEEE CRV, 353–360 (2005)
70. Wang, Y.: VisGenie, http://www.ee.columbia.edu/~ywang/Research/VisGenie/
71. Waters, K.: A muscle model for animating three-dimensional facial expressions. Computer Graphics 21(4), 17–24 (1987)

72. Watkinson, J.: The MPEG handbook: MPEG1, MPEG2, MPEG4. Focal Press (2001)
73. Wen, Z., Huang, T.S.: 3D face processing: modeling, analysis, and synthesis. Kluwer Academic Publishers, Boston (2004)
74. Welsh, W.J., Searby, S., Waite, J.B.: Model-based image coding. British Telecom Technology Journal 8(3), 94–106 (1990)
75. Xiao, B., Zhang, Q., Wei, X.: A NURBS facial model based on MPEG-4. In: 16th International Conference on Artificial Reality and Telexistence–Workshops, pp. 491–495 (2006)
76. Yang, T., Li, S.Z., Pan, Q., Li, J., Zhao, C.H.: Reliable and fast tracking of faces under varying pose. IEEE FGR, 421–428 (2006)

Author Index

GPSR Compliance

The European Union's (EU) General Product Safety Regulation (GPSR) is a set of rules that requires consumer products to be safe and our obligations to ensure this.

If you have any concerns about our products, you can contact us on ProductSafety@springernature.com

In case Publisher is established outside the EU, the EU authorized representative is:

Springer Nature Customer Service Center GmbH
Europaplatz 3
69115 Heidelberg, Germany

Batch number: 10614733

Printed by Printforce, the Netherlands